Pattern Dynamics of Marine Plankton Behavior

Shu Tang Liu · Li Zhang

Pattern Dynamics of Marine Plankton Behavior

Nonlinear Dynamics Applications

Springer

Shu Tang Liu
College of Control Science and Engineering
Shandong University
Jinan, Shandong, China

Li Zhang
Business School
College of Control Science
and Engineering, Shandong University
of Political Science and Law
Jinan, Shandong, China

ISBN 978-981-97-5368-0 ISBN 978-981-97-5369-7 (eBook)
https://doi.org/10.1007/978-981-97-5369-7

This Springer imprint is published by the registered company Springer Nature Singapore Pte Ltd.
The registered company address is: 152 Beach Road, #21-01/04 Gateway East, Singapore 189721, Singapore

Preface

Heterogeneous patterns in the spatial distributions of individuals are almost universal across living organisms, mainly marine organisms. The aggregation usually presents specific features that are visually striking to human eyes. Extensive studies have been conducted to uncover the mechanism behind this phenomenon, but the fundamental questions about how and why this aggression emerges still need to be unified theory. Pattern formation in the behavior of marine organisms is of great research value since spatial pattern features may affect the population's evolutionary direction. Moreover, understanding the discipline of pattern dynamics in an aquatic ecosystem will promote the design and regulation of ecological control strategy.

Previous studies on spatial patterns of marine organisms mainly depend on the statistical analysis of data, which lacked the quantitative characterization of pattern changes and therefore is difficult to form a standard theoretical framework. Mathematical modeling is a promising avenue to address these issues. Through developing ecological models by considering different factors, the similarities among pattern dynamics of marine organisms can be surveyed, and we can grasp the commonness of aggregation. In this book, we study the pattern dynamics along the progressive clues: firstly, we mainly focus on the permanence and extinction of marine organisms since population survival is the prerequisite of aggression and spatial pattern formation; secondly, we concentrate on the impact of different issues such as water temperature, organism size, delay, etc. on the stability of marine ecosystems; finally, we discuss the feature of pattern dynamics of marine organism behavior and the cause of change and provide the theoretical analytic results and the corresponding simulation results. The specific structure of this work is shown in the following figure.

This book systematically investigates the behavior of the marine organisms from the perspective of mathematical analysis and tests the relevant conjectures about the emergence of the pattern. The results in this book provide a new understanding of the ability of marine organisms to resist external interference, the conditions for holding the stability of the marine ecosystem and the mechanism behind the pattern transformation of marine populations. Hence, this book is an attractive grumous treatise.

It is noteworthy that the overall contents of this book are innovative achievements of our research team and most of them have been published in the International Journal of Bifurcation and Chaos, Nonlinear Dynamics, Applied Mathematics and Computation, Mathematical Methods in the Applied Science, and so on. The contents focus on the current hot topics of marine ecology and contain the latest and influential discovery. Therefore, it will promote not only the solution of practical ecological problems in marine but also the development of nonlinear dynamics in mathematics. The results can be generalized to study other population problems.

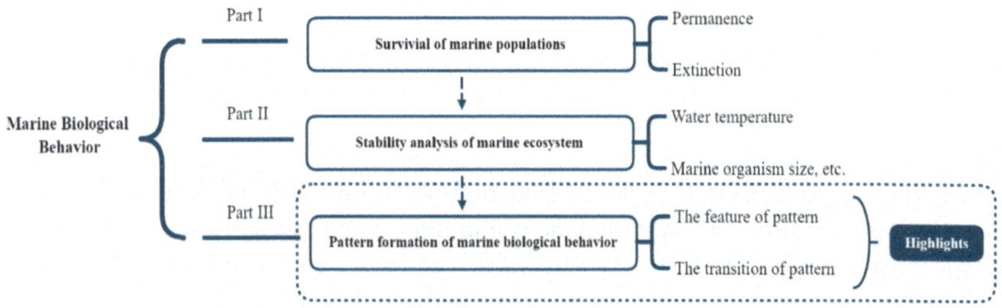

We would like to express our appreciation for the great efforts of Drs. Qiuyue Zhao, Zhibin Liu, Zhimin Bi, Yupin Wang, Da Wang, and Dadong Tian. Finally, we are grateful for the sequential support of the National Natural Science Foundation of China-Shandong joint fund (Number U1806203) and the Key Program of the National Natural Science Foundation of China (Number 61533011), the Foundation for the Author of National Excellent Doctoral Dissertation of China (Number 200444), the National Natural Science Foundation of China (Numbers 60472112, 60372028, 600874009, 10971120, 61273088), the Shandong Province Natural Science Foundation (Numbers ZR2010FM010, Y98-A02005), the China Postdoctoral Science Foundation (Number 2004035036), Project supported by the Scientific Research Staring Foundation for the Returned Overseas Chinese Scholars, Ministry of Education of China, and the K. C. Wong Education Foundation, Hong Kong, China, which ensure the smooth developments of this research project.

Jinan, China

Shu Tang Liu
Li Zhang

Contents

Abbreviations

HAB	Harmful Algal Blooms
inf	Infimum
lim	Limit
max	Maximum
min	Minimum
sup	Supremum

Symbols

$\mathbb{R}, \mathbb{R}^n, \mathbb{R}^{n\times m}$	Set of real numbers, set of n-dimensional real vectors, and set of $n \times m$ real matrices
$\mathbb{C}, \mathbb{C}^n, \mathbb{C}^{n\times m}$	Set of complex numbers, set of n-dimensional complex vectors, and set of $n \times m$ complex matrices
$\mathbb{R}^+$	Set of non-negative real numbers
$\mathbb{Z}^+$	Set of non-negative integers
$Re(s)$	Real part of $s \in \mathbb{C}$
$Im(s)$	Imaginary part of $s \in \mathbb{C}$
A^T	Transpose of matrix A
A^{-1}	Inverse of matrix A
A^{-T}	Shorthand for $(A^{-1})^T$
I_n	$n \times n$ Identify matrix (the subscript is omitted if no confusion will occur)
$diag\{A_1, \ldots, A_n\}$	Diagonal matrix with A_i as its ith diagonal element
$\begin{bmatrix} X & Y \\ * & Z \end{bmatrix}$	Symmetric matrix $\begin{bmatrix} X & Y \\ Y^T & Z \end{bmatrix}$
$A > 0\ (< 0)$	Symmetric positive (negative) definite matrix
$A \geq 0 (\leq 0)$	Symmetric positive (negative) semi-definite matrix

$det(A)$	Determinant of matrix A
$Tr\{A\}$	Trace of matrix A
$\lambda(A)$	Eigenvalue of matrix A
$\lambda_{max}(A)$	Largest eigenvalue of matrix A
$\lambda_{min}(A)$	Smallest eigenvalue of matrix A
$\sigma_{max}(A)$	Largest singular value of matrix A
$\lvert\cdot\rvert$	Absolute value (or modulus)
$\lVert\cdot\rVert$	Euclidean norm of a vector or spectral norm of a matrix
$\lVert\cdot\rVert_\infty$	Included l_∞−norm
$\mathcal{C}([a,b],\mathbb{R}^n)$	Family of continuous functions ϕ from $[a,b]$ to $\mathbb{R}^n$
$\lVert\phi\rVert_c$	Continuous norm $\sup_{a\leq t\leq b}\lVert\phi(t)\rVert$ for $\phi\in\mathcal{C}([a,b],\mathbb{R}^n)$
$\mathcal{E}$	Mathematical expectation
$\left[\begin{array}{c\|c} A & B \\ \hline C & D \end{array}\right]$	Shorthand for state space realization $C(sI-A)^{-1}B+D$ for a continuous-time system or $C(zI-A)^{-1}B+D$ for a discrete-time system
$\forall$	For all
$\in$	Belongs to
$\exists$	There exists
$\subseteq$	Is a subset of
$\cup$	Union
$\cap$	Intersection
$\rightarrow$	Tends toward or is mapped into (case sensitive)
$\Rightarrow$	Implies
$\frac{\partial V}{\partial t}$	Partial derivative of function V
$\mathrm{grad}V$	Gradient of function V
$\prod_{i=1}^{n}\lambda_i$	Product of $\lambda_1,\lambda_2,\ldots,\lambda_n$
ΔA	Amount of change in matrix A
$\square$	End of proof

Chapter 1
Introduction

Oceans cover 71% of the Earth's surface and provide us with oxygen, food, and jobs. But they are probably the least understood, most biologically diverse, and most undervalued of all ecosystems. Marine organisms have not been studied as extensively as their counterparts living on land. Focus issues throughout the ecological oceanography today are: What are the spatial distributions of marine organisms? How do these distributions vary with time and region? What are the relationships between distribution and function (i.e., population characteristics, environmental factors, and chemical substances)? What are the mechanisms that drive these processes?

Plankton is the most abundant organisms in the ocean and can be divided into two main categories: phytoplankton and zooplankton. Phytoplankton, the microscopic plants that float with water, drives all marine ecological communities and the life within them. Phytoplankton can absorb half of the world's carbon dioxide and generate half of the oxygen that humanity needs, which is said to be one of the main factors controlling the further development of the world's climate. In addition, the vital role of phytoplankton also reflects in its recycling of phosphorus, nitrogen, and sulfur compounds, which helps to reduce the eutrophication of water [1]. Zooplankton plays the role of predator in the plankton. They feed on phytoplankton but also cannibalism. Plankton is an integral part of the ocean, forming the food web's basis. In this way, understanding the evolution of planktonic populations will become increasingly important. However, plankton has some adverse ecological problems, and the most typical one is harmful algal blooms shown in Fig. 1.1. The leading causes of harmful algal blooms are over the aggregation of organisms; hence, the need to determine how plankton's spatial and temporal distributions are more and more prominent. On this basis, researchers attempt to control the development trend of blooms, which is one of the hottest topics today. This work addresses these critical issues and contains the latest research from the perspective of nonlinear science. The mathematical model is consistent with the actual ecological background, and theoretical results can be applied to deal with the marine ecological problems.

S. T. Liu and L. Zhang, *Pattern Dynamics of Marine Plankton Behavior*,
https://doi.org/10.1007/978-981-97-5369-7_1

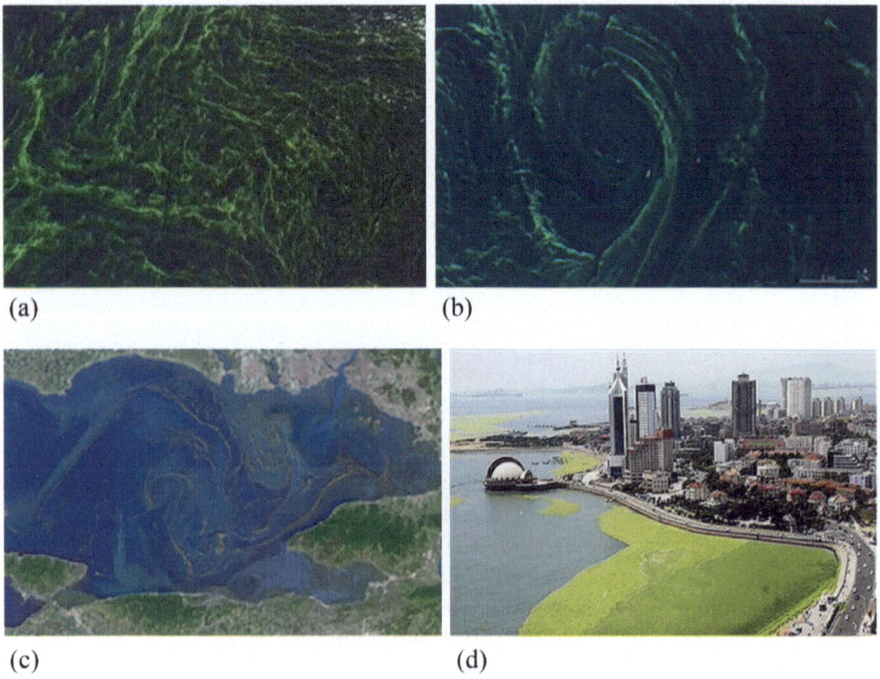

Fig. 1.1 The cases of harmful algal blooms

Marine biologists have led the study of spatial patterns of plankton for several decades, but the nonlinear scientific theory has not been widely used in this field. In general, ecologists mostly proceed with the specific phenomenon and summarize common law contained in them by ecological experiment. However, this research method is challenging to uncover the unified mechanism behind the aggregation phenomenon. A nonlinear scientific theory can make up this default and allow us to proceed beyond the specific characteristic to discern the mechanisms that govern the distribution of marine organisms.

Pattern dynamics is a new branch of modern mathematics and nonlinear science. Because of its excellent performance in analyzing the self-organization phenomenon, it has been widely concerned in physics, chemistry, ecology, and other fields. The development of pattern dynamics provides a powerful tool for studying the common law of spatial structure of complex system. It can not only answer how the non-uniform spatial structure is formed but also clarify the choice behavior of spatial structure under different parameters. The pattern theory has been confirmed by experiments in arid and semi-arid vegetation ecosystems. The analysis shows that the pattern characteristics are essential to the ecological function of the population system [2]. Please see Fig. 1.2 for some examples of pattern formation.

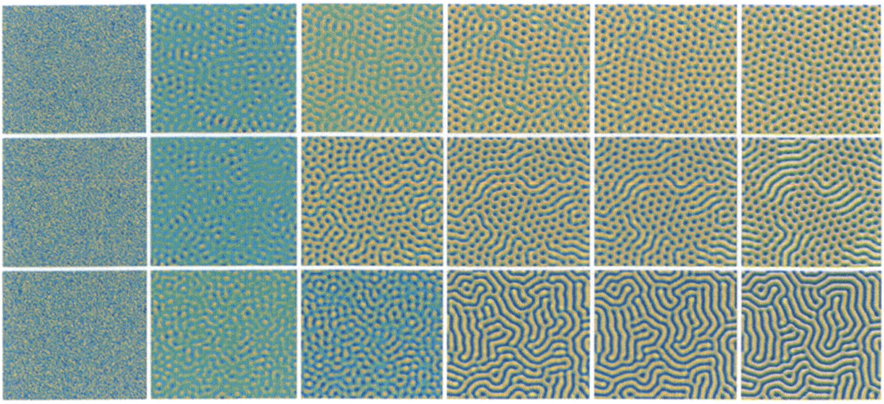

Fig. 1.2 Examples of pattern formation

In marine ecosystem, the phenomenon of uneven spatial density of the population can be reflected as the "patchy" stationary solution of pattern theory. In this book, pattern theory is applied to study the aggregation phenomenon of marine organisms. According to the theoretical results, the prediction of the marine ecosystem's future development and the design of ecological control strategy can be realized. The results of this book not only analyze the causes of pattern formation in marine biological behavior but also give the dynamic mechanism of pattern changes. In addition, we study the influence of internal and external factors on the stability of the ecosystem and put forward theoretical suggestions for maintaining a stable operation. For the persistence and extinction of marine life, this book also gives relevant conclusions that positively affect ecological protection. To sum up, this book provides new ideas and means to study marine organism's behavior and will promote the development of interdisciplinary theory and application.

Chapter 2
Preliminaries

In this chapter, we introduce some preliminaries about pattern formation which are needed for the following part.

2.1 Stability of Perturbed Linear Systems

This section gives the characterizations of stability for linear systems and the results applied to the stability of perturbed linear systems.

Definition 1 [3] We call the solution $x = 0$ of the following system: $\dot{x} = f(t, x),\ f(t, 0) \equiv 0,\ t \in [0, \infty)$ is called
(1) Lyapunov stable if for any $\epsilon > 0$ and any $t_0 \geq 0$, there is a $\delta = \delta(\epsilon, t_o)$ such that $|x_0| < \delta$ implies $|x(t, t_0, x_0)| < \epsilon$ for $t \in [t_0, \infty)$, where $f : [0, \infty) \times R^n \to R^n$.
(2) Uniformly stable if it is stable and δ can be chosen independent of $t_0 \geq 0$.
(3) Asymptotically stable if it is stable and there exists a $b = b(t_0)$ such that $|x_0| < b$ implies $|x(t, t_0, x_0)| \to 0$ as $t \to \infty$.
(4) Uniformly asymptotically stable if it is uniformly stable, b in the definition of asymptotic stability can be chosen independent of $t_0 \geq 0$, and for every $|\eta| > 0$ there is a $T(\eta) > 0$ such that $|x_0| < b$ implies $|x(t, t_0, x_0)| < \eta$ if $t \geq t_0 + T(\eta)$.
(5) Unstable if it is not stable.

Consider the linear system of n first-order equations:

$$\dot{x_j} = \sum_{k=1}^{n} a_{jk}(t)x_k + h_j(t),\ j = 1, 2, \ldots, n, \tag{2.1}$$

where a_{jk} and h_j are continuous real- or complex-valued functions on the interval $(-\infty, +\infty)$, $j,\ k = 1, 2, \ldots, n$. In matrix notation, Eq. (2.1) can be rewritten as the following more compact form:

S. T. Liu and L. Zhang, *Pattern Dynamics of Marine Plankton Behavior*,
https://doi.org/10.1007/978-981-97-5369-7_2

$$\dot{x} = A(t)x + h(t), \tag{2.2}$$

where $A = (a_{jk}),\ j, k = 1, 2, \ldots, n;\ h = \mathrm{col}(h_1, \ldots, h_n)$. The principle of superposition implies that $x - y$ is a solution of the homogeneous equation

$$\dot{x} = A(t)x, \tag{2.3}$$

where x and y are solutions of (2.2) corresponding to the forcing function h.

Theorem 1 [3] *Every solution of (2.2) is deduced by the following formula:*

$$x(t) = X(t)[X^{-1}(\tau)x(\tau) + \int_r^t X^{-1}(s)h(s)ds],$$

as X is a fundamental matrix solution of (2.3) and the real number τ is in $(-\infty, +\infty)$.

Theorem 2 [3] *If $X(t)$ is a fundamental matrix solution of (2.3) and β is any number in $(-\infty, +\infty)$, then system (2.3) is*
(i) stable for any t_0 in $(-\infty, +\infty)$ if and only if there is a $K = K(t_0) > 0$ such that

$$|X(t)| \leq K,\quad t_0 \leq t < \infty;$$

(ii) uniformly stable for $t_0 \geq \beta$ if and only if there is a $K = K(\beta) > 0$ such that

$$|X(t)X^{-1}(s)| \leq K,\ t_0 \leq s \leq t < \infty;$$

(iii) asymptotically stable for any t_0 in $(-\infty, +\infty)$ if and only if

$$|X(t)| \to 0,\ ast \to \infty;$$

(iv) uniformly asymptotically stable for $t_0 \geq \beta$ if and only if it is exponentially asymptotically stable; that is, there are $K = K(\beta) > 0,\ \alpha = \alpha(\beta) > 0$ such that

$$|X(t)X^{-1}(s)| \leq Ke^{-a(t-s)},\ \beta \leq s \leq t \leq \infty.$$

Theorem 3 [3] *Let β be given in $(-\infty, \infty)$ and the system (2.3) be uniformly stable for $t_0 \geq \beta$. If the $n \times n$ continuous matrix function $B(t)$ satisfies $\int_\beta^\infty |B(t)|dt < \infty$, then the system*

$$\dot{x} = [A(t) + B(t)]x \tag{2.4}$$

is uniformly stable.

Theorem 4 [3] *System (2.4) is uniformly asymptotically stable if $\gamma < r$ when*

$$\int_{t_0}^t |B(s)|ds \leq \gamma(t - t_0) + \tau,\ t \geq t_0 \geq \beta,$$

for some constants $\tau = \tau(\beta)$, $\gamma = \gamma(\beta)$, where β is given in $(-\infty, \infty)$ and system (2.3) is uniformly asymptotically stable for $t_0 \geq \beta$, $B(t)$ is the $n \times n$ continuous matrix function. Then there is an $r > 0$ such that

Theorem 5 [3] *The solution $x = 0$ of*

$$x = A(t)x + f(t, x)$$

is uniformly asymptotically stable for $t_0 \geq \beta$, as there is a $\sigma > 0$ for any $\epsilon > 0$ satisfying the listed below:

$$|f(t, x)| \leq \epsilon |x| \text{ for } |x| < \sigma, \ t \text{ in } R,$$

where β is given in $(-\infty, \infty)$, system (2.3) is uniformly asymptotically stable for $t_0 \geq \beta$, and $f(t, x)$ is continuous for (t, x) in $\mathbb{R} \times \mathbb{R}^n$.

2.2 Center Manifold

The corresponding linear system is derived by considering the local behavior of nonlinear term. However, analyzing a linear system has drawbacks in understanding the new states appearing with the nonlinear phase change. So the nonlinear system should also be studied.

Assume O is a structurally unstable equilibrium in an n-dimensional space. Consequently, we focus on the situation that certain characteristic exponents of O are located on the imaginary axis, and some other characteristic exponents have negative genuine parts:

$$\operatorname{Re}\lambda_1 = \cdots = \operatorname{Re}\lambda_m = 0, \ \operatorname{Re}\lambda_{m+1} < 0, \cdots, \operatorname{Re}\lambda_n < 0.$$

Therefore, the system is described in terms of the following form:

$$\begin{cases} \dot{u} = Au + f(u, v), \\ \dot{v} = Bv + g(u, v), \end{cases} \tag{2.5}$$

where $m \in \mathbb{R}^m$, $y \in \mathbb{R}^{m-n}$, $A = \{\lambda_{m+1}, \cdots, \lambda_n\}$ is the spectrum, f and g mean $\mathbb{C}^r$-smooth functions and their first derivatives are zero at origin. We consider the following system:

$$\begin{cases} \dot{u} = Au + f(u, v, \mu), \\ \dot{v} = Bv + g(u, v, \mu), \end{cases} \tag{2.6}$$

where $\mu = (\mu_1, \cdots, \mu_p)$.

Theorem 6 [5] *Let f and g depend continuously on μ along with all derivatives with respect to (u, v), and $f(0,0,0) = 0,\ g(0,0,0) = 0,\ (f, g)'_{(u,v)}(0,0,0) = 0$, and system (2.6) has, for each small μ, an m-dimensional $\mathbb{C}^r$-smooth invariant local center manifold $W^C_{loc} : u = \psi(u, v)$ (here the function depends continuously on μ along with all its derivatives with respect to v) which is tangent at O to the $v-space$ at $\mu = 0$ ($\psi(0,0) = 0,\ \frac{\partial\psi}{\partial v}(0,0) = 0$). Then for each μ, the center manifold contains all trajectories that stay in a small neighborhood of O for all times.*

The following is reduced from the basis of computational normalization.

Theorem 7 [5] *Suppose $F \in \mathbb{C}^{r-1}$, $G \in \mathbb{C}^r$*

$$F(0,0) = 0,\ G(0,0) = 0,\ G'(0) = 0.$$

Then there exists a $\mathbb{C}^{r-1}$-smooth transformation of coordinates ($\mathbb{C}^1$-close to the identity near the origin) which brings system (2.6) to the form

$$\begin{cases} \dot{u} = (A + F(u, v))u, \\ \dot{v} = Bv + g(v). \end{cases} \tag{2.7}$$

The center manifold theory is mainly based on the following theorem.

Theorem 8 [5] *In a small neighborhood of O there exists an $(m+k)$-dimensional invariant center stable manifold $W^{sC}_{loc} : x = \psi^{sC}(v, u)$ of class $\mathbb{C}^r$, which contains O and which is tangent to the subspace $x = 0$ at O. The manifold W^{sC}_{loc} contains all trajectories which stay in a small neighborhood of O for all positive times. Though the center stable manifold is not defined uniquely, for any two manifolds W^{sC}_l and W^{sC}_2 the functions ψ^{sC}_1 and ψ^{sC}_2 have the same Taylor expansion at O (and at each point whose trajectory stays in a small neighborhood of O for all $t \geq 0$):*

Next, applying the theorem on the center stable manifold to the system obtained from

$$\begin{cases} \dot{u} = Au + f(u, v, x), \\ \dot{x} = Cx + h(u, v, x), \\ \dot{v} = Bv + g(u, v, x), \end{cases} \tag{2.8}$$

by a reversion of time, we can obtain the following theorem on a center unstable manifold:

Theorem 9 [5] *In a small neighborhood of O there exists an $(n-k)$-dimensional $\mathbb{C}^r$-smooth invariant manifold $W^{uC}_{loc} : u = \psi^{uC}(v, x)$ which contains O and which is tangent to the subspace $u = 0$ at O. The center unstable manifold contains all the trajectories which stay in a small neighborhood of O for all negative times. For any two manifolds W^{uC}_1 and W^{uC}_2 the functions ψ^{uC}_1 and ψ^{uC}_2 have the same Taylor expansion at O (and at each point whose trajectory stays in a small neighborhood of O for all $t \leq 0$). In the case where the system is $\mathbb{C}^\infty$-smooth, the center unstable* *manifold has, in general, only finite smoothness, but if every trajectory of* W^{uC}_{loc} *tends to* O *as* $t \to -\infty$, *then* W^{uC}_{loc} *is* $\mathbb{C}^{\infty}$*-smooth.*

The straightening of the center unstable and the center stable manifolds together with the straightening of the strong unstable and strong stable invariant foliations on these manifolds leads to the following result.

Theorem 10 [5] *By a* $\mathbb{C}^{r-1}$*-smooth transformation system (2.8) can be locally reduced to the form*

$$\begin{cases} \dot{u} = (A + F(u, v, x))u, \\ \dot{x} = (C + H(u, v, x))x, \\ \dot{v} = Bv + G_0(x) + G_1(u, v, x)u + G_2(u, v, x)x, \end{cases}$$

where G_0 *is a* $\mathbb{C}^r$*-smooth function vanishing at* $x = 0$ *along with its first derivative,* F *and* H *are* $\mathbb{C}^r$*-functions vanishing at the origin,* $G_{1,2} \in \mathbb{C}^{r-1}$ *and* G_1 *vanishes identically at* $z = 0$ *and* G_2 *vanishes identically at* $y = 0$.

2.3 Conservative and Dissipative Systems

Consider a causal nonlinear system $(\Sigma) : u(t) \mapsto y(t);\ u(t) \in \mathfrak{L}_{pe},\ y(t) \in \mathfrak{L}_{pe}$, represented by the following state-space representation affine in the input:

$$(\Sigma) \begin{cases} \dot{x} = f(x(t)) + g(x(t))u(t), \\ y(t) = h(x(t) + j(x(t))u(t)), \\ x(0) = x_0, \end{cases} \tag{2.9}$$

where $x(t) \in \mathbb{R}^n, u(t), y(t) \in \mathbb{R}^m,\ f(\cdot), g(\cdot), h(\cdot)$, and $j(\cdot)$ possess sufficient regularity so that the system with inputs in $\mathfrak{L}_{2,e}$ is well-posed, and $f(0) = h(0) = 0$.

Definition 2 [6] The system (Σ) is said to be dissipative if there exists a so-called storage function $V(x) \geq 0$ such that the following dissipation inequality holds:

$$V(x(t)) < V(x(0)) + \int_0^t \omega(u(s), y(s))ds$$

along all possible trajectories of (Σ) starting at $x(0)$, for all $x(0), t \geq 0$ (said differently: for all admissible controllers $u(\cdot)$ that drive the state from $x(0)$ to $x(t)$ on the interval $[0, t]$).

Definition 3 [6] The system (Σ) is said quasi-dissipative with respect to the supply rate $\omega(u, y)$ if there exists a constant $\alpha \geq 0$ such that it is dissipative with respect to the supply rate $\omega(u, y) + \alpha$.

Theorem 11 [6] *For some $X_1 \subseteq X$, $G(X_1)$ is weakly dissipative if and only if there exists a function $V : X_1 \to \mathbb{R}$, with $V(x) \geq 0$ for all $x \in X$, such that*

$$V(x_1) + \int_{t_0}^{t} \omega(u(s), y(s))ds \geq V(x_2)$$

for all $x_1 \in X_1$, all admissible $u(\cdot) \in U$, all $t \geq t_0$, with $y(t) = G_{x_1}(u(t))$ and $x(t) = x_2$ is the state starting at x_1 at t_0.

Definition 4 [6] Let us consider a dissipative system, with supply rate

$$\omega(u, y) = y^T Q y + u^T R u + 2y^T S u \tag{2.10}$$

with $Q = Q^T$, $R = R^T$. If $Q = 0$, $R = -\epsilon Im$, $\epsilon > 0$, $S = \frac{1}{2}Im$, the system is said to be input strictly passive (ISP), i.e.,

$$\int_0^t y^T(s)u(s)ds \geq \beta + \epsilon \int_0^t u^T(s)u(s)ds.$$

If $R = 0$, $Q = -\delta Im$, $\delta > 0$, $S = \frac{1}{2}Im$, the system is said to be output strictly passive (OSP), i.e.,

$$\int_0^T y^T(s)u(s) \geq \beta + \delta \int_0^t y^T(s)y(s)ds.$$

If $Q = -\delta Im$, $\delta > 0$, $R = -\epsilon Im$, $\epsilon > 0$, $S = \frac{1}{2}Im$, the system is said to be very-strictly passive (VSP), i.e.,

$$\int_0^t y^T(s)u(s)ds + \beta \geq \delta \int_0^t y^T(s)y(s)ds + \epsilon \int_0^t u^T(s)u(s)ds.$$

Theorem 12 [6] *The system is dissipative with respect to the general supply rate in (2.10) with zero bias ($\beta = 0$) and with $Q < 0$, if and only if it is finite-gain stable.*

In mathematics, a conservative system is a dynamic system which stands in contrast to a dissipative system. Roughly speaking, such systems have no friction or other mechanism to dissipate the dynamics; thus, their phase space does not shrink over time.

2.4 Multiscale Analysis

In this section, we give some theorems about using what are known as asymptotic expansions to find approximate solutions of differential equations.

Theorem 13 [7] *Given a function $f(\epsilon)$, suppose its $(n+1)$st derivative $f^{(n+1)}$ is continuous for $\epsilon_a < \epsilon < \epsilon_b$. In this case, if ϵ_0 and ϵ are points in the interval (ϵ_a, ϵ_b), then*

$$f(\epsilon) = f(\epsilon_0) + (\epsilon - \epsilon_0)f'(\epsilon_0) + \cdots + \frac{1}{n!}(\epsilon - \epsilon_0)f^{(n)}(\epsilon_0) + R_{n+1},$$

where

$$R_{n+1} = \frac{1}{(n+1)!}(\epsilon - \epsilon_0)^{n+1} f^{(n+1)}(\xi)$$

and ξ is a point between ϵ_0 and ϵ.

When the first $n+1$ terms from Taylor series are used as an approximation of $f(\epsilon)$, it is possible to estimate the error.

Theorem 14 [7] *Suppose $f(\epsilon)$ and $\phi(\epsilon)$ are differentiable on the interval (ϵ_0, ϵ_b) and $\phi'(\epsilon) \neq 0$ in this interval. Also suppose*

$$\lim_{\epsilon \to \epsilon_0} \frac{f'(\epsilon)}{\phi'(\epsilon)} = A,$$

where $-\infty \leq A \leq \infty$. In this case,

$$\lim_{\epsilon \to \epsilon_0} \frac{f(\epsilon)}{\phi(\epsilon)} = A,$$

if either one of the following conditions holds:

1. *$f \to 0$ and $\phi \to 0$ as $\epsilon \to \epsilon_0$, or*
2. *$\phi \to \infty$ as $\epsilon \to \epsilon_0$.*

To define an asymptotic approximation, we need to introduce order.

Definition 5 [7]

1. $f = O(\phi)$ as $\epsilon \to \epsilon_0$ means that there are constants k_0 and ϵ_1 (independent of ϵ) so that
$$|f(\epsilon)| \leq k_0|\phi(\epsilon)| \ for \ \epsilon_0 < \epsilon < \epsilon_1.$$
We say that f is big Oh of ϕ as $\epsilon \to \epsilon_0$.
2. $f = o(\phi)$ as $\epsilon \to \epsilon_0$ means that for every positive δ there is an ϵ_2 (independent of ϵ) so that
$$|f(\epsilon)| \leq \delta|\phi(\epsilon)| \ for \ \epsilon_0 < \epsilon < \epsilon_2.$$
We say that f is little oh of ϕ' as $\epsilon \to \epsilon_0$.

Definition 6 [7]

1. If
$$\lim_{\epsilon \to \epsilon_0} \frac{f(\epsilon)}{\phi(\epsilon)} = L,$$
where $-\infty < L < \infty$, then $f = O(\phi)$ as $\epsilon \to \epsilon_0$.
2. If
$$\lim_{\epsilon \to \epsilon_0} \frac{f(\epsilon)}{\phi(\epsilon)} = 0,$$
then $f = o(\phi)$ as $\epsilon \to \epsilon_0$.

Definition 7 [7]

1. The functions $\phi_1(\epsilon), \phi_2(\epsilon), \ldots$ form an asymptotic sequence, or are well ordered, as $\epsilon \to \epsilon_0$ if and only if $\phi_{m+1} = o(\phi_m)$ as $\epsilon \to \epsilon_0$ for all m.
2. If $\phi_1(\epsilon), \phi_2(\epsilon), \ldots$ is an asymptotic sequence, then $f(\epsilon)$ has an asymptotic expansion to n terms, with respect to this sequence, if and only if
$$f = \sum_{k=1}^{m} a_k \phi_k + o(\phi_m) \ for \ m = 1, 2, \ldots, n \ as \ \epsilon \to \epsilon_0,$$
where the a_k are independent of ϵ. In this case, we write
$$f \sim a_1\phi_1(\epsilon) + a_2\phi_2(\epsilon) + \ldots + a_n\phi_n(\epsilon) \ as \ \epsilon \to \epsilon_0.$$
The ϕ_k are called the scale or gauge or basis functions.

2.5 Normal Form and Resonance Mode

Consider a C^{r+1} differential equation, $r \geq 2$:

$$\dot{x} = Ax + h(x), \ x \in \mathbb{C}^n, \tag{2.11}$$

where $A \in \mathbb{C}^{n\times n}$, the $n \times n$ matrices with complex entries, and $h(x) = O(|x|^2)$ as $|x| \to 0$. We write $h(x)$ as a formal power series using superscripts to denote the order of the homogeneous terms

$$h(x) = h^2(x) + h^3(x) + \cdots,$$

where for each $k \geq 2$, $h^k \in H_n^k$, the vector space of homogeneous polynomials of order k in n variables with values in $\mathbb{C}^n$. Let $\mathcal{R}_k$ be the range of L_A^k in H_n^k and $\prec^k$ be any complementary subspace to $\mathcal{R}_k$ in H_n^k. Therefore,

$$H_k^n = \mathcal{R}_k \bigoplus \prec^k, \; k \geq 2. \tag{2.12}$$

The following theorem gives the desired simplification of (2.11).

Theorem 15 [8] *Let $X : \mathbb{C}^n \to \mathbb{C}^n$ be a C^{r+1} vector field with $X(0) = 0$ and $DX(0) = A$. Let the decomposition (2.12) of H_n^k be given for $k = 2, \ldots, r$. Then there exists a sequence of near-identity transformations $x = y + \xi^k(y)$, $y \in \Omega_k$, where $\xi^k \in H_n^k$ and Ω_k is a neighborhood of the origin, $\Omega_{k+1} \subseteq \Omega_k$, $k = 2, \ldots, r$, such that Eq. (2.11) is transformed into*

$$\dot{y} = Ay + g^2(y) + \cdots + g^r(y) + O(|y|^{r+1}), \; y \in \Omega_r,$$

where $g_k \in \phi^k$ for $k = 2, \ldots, r$.

Definition 8 [8] If $\sigma(A) = \{\lambda_1, \ldots, \lambda_n\}$ is the spectrum of A, then the following relations are called resonant conditions:

$$\lambda \cdot \alpha - \lambda_j = 0, \tag{2.13}$$

where $\lambda = (\lambda_1, \ldots, \lambda_n)^T$, $\alpha = (\alpha_1, \ldots, \alpha_n)^T$, $|\alpha| \geq 2$, $1 \leq j \leq n$, and $\lambda \cdot \alpha = \lambda_1 \alpha_1 + \cdots + \lambda_n \alpha_n$. Let $(x_1, \ldots, x_n)$ be coordinates with respect to the standard basis $\{e_1, \ldots, e_n)$ of $\mathbb{C}^n$ in which the matrix A is in Jordan normal form with diagonal elements $(\lambda_1, \ldots, \lambda_n)$. Then a monomial $x^\alpha e_j (|\alpha| = k \geq 2$ and $1 \leq j \leq n)$ is called a resonant monomial of order k if and only if (2.13) holds for α and j.

Theorem 16 [8] *Let $A = diag(\lambda_1, \ldots, \lambda_n)$. Then an A-normal form up to order $r \geq 2$ can be chosen so that its nonlinear part consists of all resonant monomials up to order r.*

Theorem 17 [8] *Operator $L_{A^*}^k$ is the adjoint operator of L_A^k with respect to the inner product $< \cdot, \cdot >$ in H_n^k for each $k \geq 2$, where A^* is the adjoint operator of A with respect to the usual product $< \cdot, \cdot >$ in $\mathbb{C}^n$.*

Corollary 1 *[8] $Ker(L_{A^*}^k)$ is the orthogonal complementary subspace to $\mathcal{R}^k$ with respect to the inner product $< \cdot, \cdot >$ in H_n^k for each $k \geq 2$.*

2.6 Reaction–Diffusion Equation

In this section, we give some theorems to study the initial boundary value problem for nonlinear parabolic equations of the form

$$\begin{aligned} u_t(x,t) - \Delta u(x,t) &= F(x,t,\mu) \text{ for } x \in \Omega, t < 0, \\ u(x,t) &= g(x,t) \text{ for } x \in \partial\Omega, t > 0, \\ u(x,0) &= f(x) \text{ for } x \in \Omega, \end{aligned} \tag{2.14}$$

with given (continuous and smooth) functions g, f and a Lipschitz continuous function F.

Theorem 18 [9] *Let* $\Omega \subset \mathbb{R}^d$ *be a bounded domain of class* C^2, *and let*

$$g \in C^0(\partial\Omega \times [0, t_0]), \quad f \in C^0(\bar{\Omega}),$$

$$with\ g(x, 0) = f(x)\ for\ x \in \partial\Omega$$

and let

$$F \in C^0(\bar{\Omega} \times [0, t_0] \times \mathbb{R})$$

be locally bounded, i.e., given $\eta > 0$ *and* $f \in C^0(\bar{\Omega})$, *there exists* $M = M(\eta)$ *with*

$$|F(x, t, v(x))| \leq M \ \ for\ x \in \bar{\Omega}, t \in [0, t_0], \ |v(x) - f(x)| \leq \eta,$$

and locally Lipschitz continuous w.r.t. u, i.e., there exists a constant $L = L(\eta)$ *with*

$$\begin{aligned}&|F(x, t, u_1(x)) - F(x, t, u_2(x))| \leq L|u_1(x) - u_2(x)|,\\ &for\ x \in \bar{\Omega},\ t \in [0, t_0], ||u_1 - f||_{C^0(\bar{\Omega})}, ||u_2 - f||_{C^0(\bar{\Omega})} < \eta.\end{aligned}$$

Then there exists some $t_1 \leq t_0$ *for which the initial boundary value problem*

$$\begin{aligned} u_t(x, t) - \Delta u(x, t) &= F(x, t, \mu)\ \ for\ \ x \in \Omega, 0 < t \leq t_1,\\ u(x, t) &= g(x, t)\ \ for\ \ x \in \partial\Omega, 0 < t \leq t_1,\\ u(x, 0) &= f(x)\ \ for\ \ x \in \Omega \end{aligned} \tag{2.15}$$

admits a unique solution that is continuous on $\bar{\Omega} \times [0, t_1]$.

Lemma 1 *[9] Let* $u_1(x, t)$, $u_2(x, t) \in C^0(\bar{\Omega} \times [0, T])$ *be solutions of (2.15) with* $u_i(x, t) = g(x, t)$ *for* $x \in \partial\Omega$, $0 \leq t \leq T$, $|u_i(x, 0) - f(x)| \leq \frac{\eta}{2}$ *for* $x \in \bar{\Omega}$, $i = 1, 2$. *Then there exists a constant* $K = K(\eta)$ *with*

$$\sup_{x \in \bar{\Omega}} |u_1(x, t) - u_2(x, t)| \leq e^{Kt} \sup_{x \in \bar{\Omega}} |u_1(x, 0) - u_2(x, 0)|\ for\ 0 \leq t \leq T.$$

Lemma 2 *[9] Let the integrable function* $\phi : [0, T] \to \mathbb{R}^+$ *satisfy*

$$\phi(t) \leq \phi(0) + c \int_0^t \phi(\tau) d\tau$$

for all $0 \leq t \leq T$ *and some constant c. Then for* $0 \leq t \leq T$

$$\phi(t) \leq e^{ct} \phi(0).$$

Corollary 2 *[9] Under the assumptions of Theorem 18, suppose that the solution* $u(x, t)$ *of (2.15) satisfies the a priori bound*

$$\sup_{x\in\bar{\Omega},0\leq\tau\leq t} |u(x,\tau)| \leq K$$

for all times t for which it exists, with some fixed constant K*. Then the solution* $u(x, t)$ *exists for all times* $0 \leq t \leq \infty$.

Lemma 3 *[9] Let* u, v *be of class* C^2 *w.r.t.* $x \in \Omega$, *of class* C^1 *w.r.t.* $t \in [0, T]$, *and satisfy*

$$u_t(x,t) - \Delta u(x,t) - F(x,t,\mu) \geq v_t(x,t) - \Delta v(x,t) - F(x,t,v),$$

for $x \in \Omega$, $0 < t \leq T$,

$$\frac{\partial u(x,t)}{\partial v} \geq \frac{\partial v(x,t)}{\partial v}$$

for $x \in \partial\Omega$, $0 < t \leq T$,

$$u(x,0) \geq v(x,0),$$

for $x \in \Omega$, *with the assumptions on F which in addition is assumed to be continuously differentiable w.r.t.* u *with* $\frac{\partial F}{\partial u} \leq 0$. *Then*

$$u(x,t) \geq v(x,t) \;\; \textit{for } x \in \bar{\Omega}, 0 \leq t \leq T.$$

Consider the following nonlinear parabolic systems

$$u_t^\alpha(x,t) - d_\alpha \Delta u^\alpha(x,t) = F^\alpha(x,t,u) \;\; \text{for } x \in \Omega, t > 0, \alpha = 1, \ldots, n, \tag{2.16}$$

for suitable initial and boundary conditions.

Corollary 3 *[9] Let the diffusion constants* d_α *all be positive. Under the assumptions of Theorem 18 for the right-hand side components* F^α, *and with the same type of boundary conditions for the components* u^α, *suppose that the solution* $u(x,t) = (u^1(x,t), \ldots, u^n(x,t))$ *of Eq. (2.16) satisfies the a priori bound*

$$\sup_{x\in\bar{\Omega},0\leq\tau\leq t} |u(x,\tau)| \leq K$$

for all times t for which it exists, with some fixed constant K*. Then the solution* $u(x,t)$ *exists for all times* $0 \leq t < \infty$.

2.7 Comparison Principle

Theorem 19 [10] *Consider the scalar differential equation*

$$\dot{u} = f(t, u), \ u(t_0) = u_0,$$

where $f(t, u)$ is continuous in t and locally Lipschitz in u, for all $t \geq 0$ and all $u \in J \subset \mathbb{R}$. Let $[t_0, T)$ (T could be infinity) be the maximal interval of existence of the solution $u(t)$, and suppose $u(t) \in J$ for all $t \in [t_0, T)$. Let $v(t)$ be a continuous function whose upper right-hand derivative $D^+v(t)$ satisfies the differential inequality

$$D^+v(t) \leq f(t, v(t)), \ v(t_0) \leq u_0$$

with $v(t) \in J$ for all $t \in [t_0, T)$. Then, $v(t) \leq u(t)$ for all $t \in [t_0, T)$.

2.8 Hopf Bifurcation

Hopf bifurcation is the most important and popular category leading to limit cycles. Consider the following general nonlinear system:

$$\dot{\boldsymbol{x}} = \boldsymbol{f}(\boldsymbol{x}, \mu), \ \boldsymbol{x} \in \mathbb{R}^n, \mu \in \mathbb{R}, \boldsymbol{f} : \mathbb{R}^{n+1} \to \mathbb{R}^n, \tag{2.17}$$

where $\boldsymbol{x}$ is an n-dimensional state vector, while μ is a scalar parameter called a bifurcation parameter.

Theorem 20 [11] *Suppose that the system (2.17) has an equilibrium (x_0, μ_0) at which the following properties are satisfied.*

(H1) *$D_x\mathbf{f}(\mathbf{x}_0, \mu_0)$ has a simple pair of purely imaginary eigenvalues and no other parts.*

*Then **(H1)** implies that there is a smooth curve of equilibria $(\mathbf{x}(\mu), \mu)$ with $\mathbf{x}(\mu_0) = \mathbf{x}_0$. The eigenvalues $\lambda(\mu)$, $\bar{\lambda}(\mu)$ of $D_x\mathbf{f}(\mathbf{x}(\mu), \mu_0)$, which are imaginary at $\mu = \mu_0$, vary smoothly with μ. If, moreover,*

(H2) $\frac{d}{d\mu}(Re\lambda(\mu))|_{\mu=\mu_0} = d \neq 0$,

then there is a unique three-dimensional center manifold passing through $(\mathbf{x}_0, \mu_0)$ in $\mathbb{R}^n \times \mathbb{R}$ and a smooth system of coordinates (preserving the planes $\mu = const$) for which the Taylor expansion of degree 3 on the center manifold. If $a \neq 0$, there is a surface of periodic solutions in the center manifold which has quadratic tangency with the eigenspace of $\lambda(\mu_0)$, $\bar{\lambda}(\mu_0)$ agreeing to second order with the paraboloid

$$\mu = -\left(\frac{a}{d}\right)(x^2 + y^2).$$

If $a < 0$, then these periodic solutions are stable limit cycles, while if $a > 0$, the periodic solutions are repelling.

Consider a C^∞ system defined in a neighborhood of the origin in $\mathbb{R}^2$

$$\begin{cases} \dot{x} = f(x, y, \mu), \\ \dot{y} = g(x, y, \mu), \end{cases} \tag{2.18}$$

where $x, y, \mu \in \mathbb{R}^1$.

Suppose that the origin is an equilibrium of (2.18), and, for μ near 0, the linear part of (2.18) around the origin has eigenvalues $\alpha(\mu) \pm i\beta(\mu)$. Then, derive a normal form for Eq. (2.18) for $\mu = 0$. Making a suitable linear change of coordinates and $z = x + iy$, Eq. (2.18) for $\mu = 0$ becomes

$$\begin{aligned} \dot{z} &= i\beta_0 z + F(z, \bar{z}), \\ \dot{\bar{z}} &= -i\beta_0 \bar{z} + \bar{F}(z, \bar{z}). \end{aligned} \tag{2.19}$$

Since the eigenvalues of the linear part of (2.18) for $\mu = 0$ are $\lambda_{1,2} = \pm i\beta_0$, we have the resonances $\lambda_j = k(\lambda_1 + \lambda_2) + \lambda_j$, $j = 1, 2$ and $k = 1, 2, \ldots$. By a polynomial change of variables

$$z = w + \sum_{z \le k+1 \le m} b_{kl} w^k \bar{w}^l,$$

Equation (2.19) takes the form

$$\dot{w} = i\beta_0 w + C_1 w^2 \bar{w} + C_2 w^3 \bar{w}^2 + \cdots + O(|w|^{2k+3}).$$

Theorem 21 [8] *Suppose that*

(H1) $\alpha(0) = 0$, $\beta(0) = \beta_0 \neq 0$.
(H2) $\alpha'(0) \neq 0$.
(H3) $Re(C_1) \neq 0$.

hold. Then there are $\sigma > 0$ and a neighborhood U of $(x, y) = (0, 0)$ such that

(i) *If $|\mu| < \sigma$ and $Re(C_1)\alpha'(0)\mu = 0$, the system (2.18) has exactly one limit cycle inside U.*
(ii) *If $|\mu| < \sigma$ and $Re(C_1)\alpha'(0)\mu \geq 0$, the system (2.18) has no periodic orbits inside U.*

Moreover, the limit cycle is stable (unstable) if $Re(C_1) < 0$ ($Re(C_1) > 0$), and it tends to the equilibrium $(0, 0)$ as $\mu \to 0$.

2.9 Limit Cycle

Let $P(x, y)$, $Q(x, y)$ be the nonlinear functions, for the following nonlinear dynamical system:

$$\begin{cases} \frac{dx}{dt} = P(x, y), \\ \frac{dy}{dt} = Q(x, y), \end{cases} \tag{2.20}$$

we have

Definition 9 [12] Suppose Eq. (2.20) has a closed track C. If in the sufficiently small neighborhood of C, none of the tracks except C is closed, and these non-closed tracks tend to C when $t \to +\infty$, then C is said to be isolated, and this closed track is said to be a limit cycle of (2.20). Therefore, C divides the phase plane (phase space, if higher dimensional) into two regions, the inner domain and the outer domain.

Definition 10 [12]

(1) Inside C, when $t \to +\infty(-\infty)$, the trajectories spiral closer C. Then C is said to be internally stable (internally unstable).
(2) Outside C, when $t \to +\infty(-\infty)$, the trajectories spiral closer C. Then C is said to be externally stable (externally unstable).
(3) When $t \to +\infty(-\infty)$, the trajectories inside and outside C all spiral closer C. Then C is said to be stable (unstable).
(4) When $t \to +\infty(-\infty)$, the stabilities inside and outside C are different. Then C is said to be semi-stable.

We obtain the following criteria for semi-stable limit cycles (Fig. 2.1).

Theorem 22 [12] *For system (2.20), the following are necessary conditions to identify the existence of the semi-stable limit cycle:*

(1) *Suppose P, Q has second partial derivatives. If Eq. (2.20) has a semi-stable limit cycle C that is monotonically close, then the coordinates (x, y) of any point L in C must satisfy the equation*

$$F(x, y) = P^2\frac{\partial Q}{\partial y} - PQ(\frac{\partial P}{\partial y} + \frac{\partial Q}{\partial x}) + Q^2\frac{\partial Q}{\partial x} = 0.$$

(2) *Suppose P, Q has second partial derivatives. If Eq. (2.20) has a semi-stable limit cycle C that is monotonically close, then for coordinates (x, y) of any point L in C, the curvature of $\Gamma(l)$ is 0.*

Theorem 23 [12] *Let P and Q be polynomials of x and y. Then all monotone semi-stable limit cycles of Eq. (2.20) are algebraic curves and they are limited in quantity.*

Theorem 24 [12] *If there is no single closed curve in the curves of $F = 0$, the differential Eq. (2.20) does not have a semi-stable limit cycle that is monotonically close.*

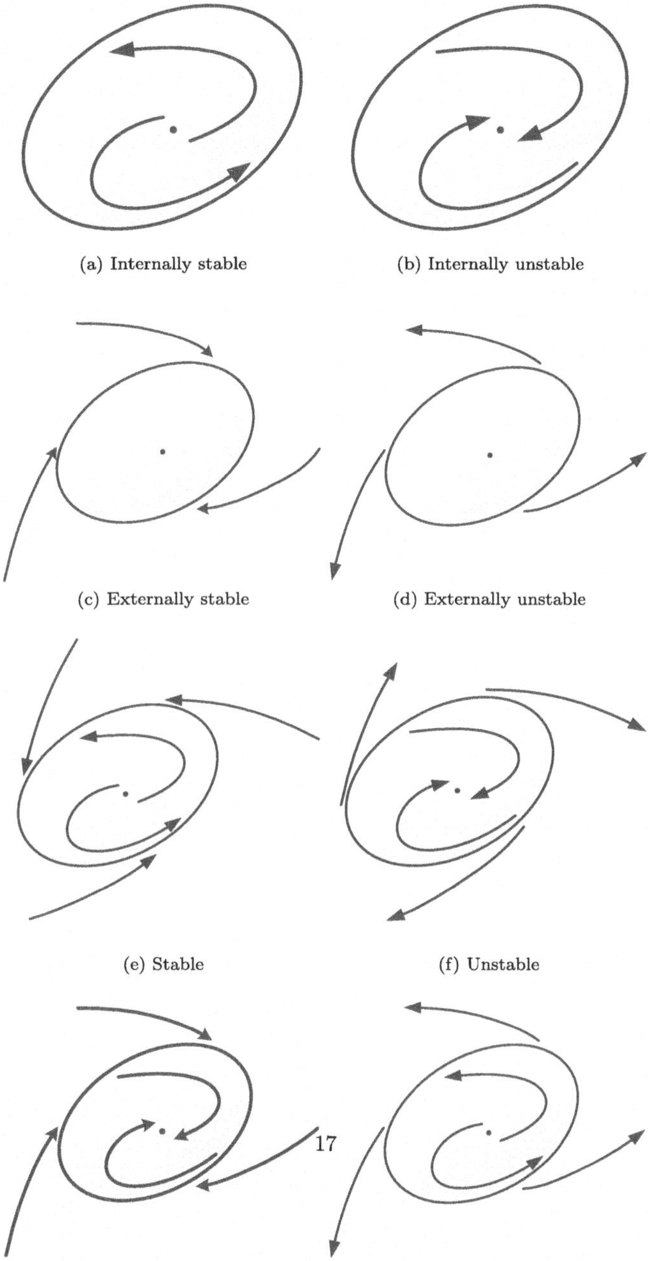

(a) Internally stable
(b) Internally unstable
(c) Externally stable
(d) Externally unstable
(e) Stable
(f) Unstable
(g) External stable but internal unstable
(h) External unstable but internal stable

Fig. 2.1 The stability of limit cycles

Theorem 25 [12] *Let C be a periodic solution of Eq. (2.20). If there exists an open set S satisfying the following three conditions:*

(H1) *There is no odd point in S.*
(H2) *$S \supset C$.*
(H3) *Take any two points K_1 and K_2 in $S - C$. If there always satisfies*

$$(\sin \alpha(K_1))(\sin \alpha(K_2)) > 0,$$

then C is one of the monotonously close semi-stable limit cycles of Eq. (2.20).

Theorem 26 [12] *Suppose that C is a periodic solution of Eq. (2.20). If for any small positive d, there exists two points K_1 and K_2 in $\bigcup_{|\delta|\le d} C_\delta$ satisfying that $(\sin \alpha(K_1))(\sin \alpha(K_2)) < 0$, then C is a non-monotonic approximation semi-stable limit cycle.*

Theorem 27 [12] *If*

$$\psi'(0) < 0\ (> 0),$$

then Γ is stable (unstable) limit cycle. If

$$\psi'(0) = 0,\ \psi''(0) \neq 0,$$

then Γ is semi-stable limit cycle.

Theorem 28 [12] *For closed orbit Γ, if there exists*

$$\psi'(0) = \psi''(0) = \cdots = \psi^{k-1}(0) = 0,\ \psi^k(0) < 0(> 0).$$

Here, k is odd number. Γ is stable (unstable) limit cycle. If

$$\psi'(0) = \psi''(0) = \cdots = \psi^{k-1}(0) = 0,\ \psi^k(0) \neq 0,$$

and k is even number, then Γ is semi-stable limit cycle.

Theorem 29 [12] *For closed orbit Γ of Eq. (2.20), if there exists $\int_0^T (\frac{\partial P}{\partial x} + \frac{\partial Q}{\partial y})dt < 0(> 0)$, then Γ is stable (unstable) limit cycle.*

2.10 Delay Differential Equations

Assume Ω is a subset of $\mathbb{R} \times C$, $f : \Omega \to \mathbb{R}^n$ is a given function, and "$\cdot$" represents the right-hand derivative; then, we call

$$\dot{x}(t) = f(t, x_t) \tag{2.21}$$

a retarded delay differential equation (RDDE(f)) or a retarded functional differential equation (RFDE(f)) on Ω.

Definition 11 [13] Suppose $\Omega \subseteq \mathbb{R} \times C$ is open, $f : \Omega \to \mathbb{R}^n$, $D : \Omega \to \mathbb{R}^n$ are given continuous functions with D atomic at zero. The equation

$$\frac{d}{dt} D(t, X_t) = f(t, x_t) \tag{2.22}$$

is called the neutral delay differential equation NDDE(D, f). The function D is called the difference operator for NDDE(D, f). For a given NDDE(D, f), a function x is said to be a solution of the NDDE(D,f) if there are $\sigma \in R$, $A > 0$, such that

$$x \in C([\sigma - r, \sigma + A], \mathbb{R}^n), \qquad (t, x_t) \in \Omega, \qquad t \in [\sigma, \sigma + A],$$

$D(t, x_t)$ is continuously differentiable and satisfies (2.22) on $[\sigma, \sigma + A]$. For a given $\sigma \in \mathbb{R}, \phi \in C$, and $(\sigma, \phi) \in \Omega$, we say $x(\sigma, \phi)$ is a solution of (2.22) with initial value ϕ at σ, or simply a solution through (σ, ϕ), if there is an $A > 0$ such that $x(\sigma, \phi)$ is a solution of (2.22) on $[\sigma - r, \sigma + A]$ and $x_\sigma(\sigma, \phi) = \phi$.

If $D(t, \phi) = D_0(t, \phi) - g(t)$, $f(t, \phi) = L(t, \phi) + h(t)$, where $D_0(t, \phi)$ and $L(t, \phi)$ are linear in ϕ, then NDDE(D, f) is called linear. It is called linear homogeneous if $g \equiv 0$, $h \equiv 0$, and linear nonhomogeneous if otherwise. If both $D(t, \phi)$ and $f(t, \phi)$ do not depend upon t, we call NDDE(D, f) autonomous; otherwise, we call it nonautonomous.

Theorem 30 [13] *In (2.21), suppose Ω is an open subset in $\mathbb{R} \times C$ and f is continuous on Ω. If $(\sigma, \phi) \in \Omega$, then there is a solution of (2.21) passing through (σ, ϕ).*

We say $f(t, \phi)$ is Lipschitz in ϕ in a compact set K of $R \times C$ if there is a constant $k > 0$ such that, for any $(t, \phi_i) \in K$, $i = 1, 2$,

$$|f(t, \phi_1) - f(t, \phi_2)| \leq k|\phi_1 - \phi_2|.$$

Theorem 31 [13] *Suppose Ω is an open set in $\mathbb{R} \times C$, $f : \Omega \to \mathbb{R}^n$ is continuous, and $f(t, \phi)$ is Lipschitz in ϕ in each compact set in Ω. If $(\sigma, \phi) \in \Omega$, then there is a unique solution of (2.21) through (σ, ϕ).*

Theorem 32 [13] *Suppose $\Omega \subseteq R \times C$ is open, $(\sigma, \phi) \in \Omega$, $f \in C(\Omega, R^n)$, and x is a solution of (2.21) through (σ, ϕ), which exists and is unique on $[\sigma - r, b]$, $b > \sigma - r$. Let $W \subseteq \Omega$ be the compact set defined by*

$$W = \{(t, x_1) : t \in [\sigma, b]\},$$

and let V be a neighborhood of W on which f is bounded. If (σ^k, ϕ^k, f^k), $k = 1, 2, \ldots$, satisfies $\sigma^k \to \sigma$, $\phi^k \to \phi$, and $|f^k - f|_V \to 0$ as $k \to \infty$, then there is a K such that, for $k \geq K$, each solution $x^k = x^k(\sigma^k, \phi^k, f^k)$ through (σ^k, ϕ^k) of

$$\dot{x}(t) = f^k(t, x_1)$$

exists on $[\sigma^k - r, b]$ *and* $x^k \to x$ *uniformly on* $[\sigma - r, b]$.

Theorem 33 [13] *Suppose* Ω *is an open set in* $R \times C$, $f : \Omega \to R^n$ *is completely continuous, and* x *is a noncontinuable solution of (2.21) on* $[\sigma - r, b]$. *Then for any closed bounded set* U *in* $\mathbb{R} \times C$, $U \subset \Omega$, *there is a* t_U *such that* $(t, x_t) \notin U$ *for* $t_U \leq t < b$.

Theorem 34 [13] *If* Ω *is an open set in* $\mathbb{R} \times C$ *and* $(\sigma, \phi) \in \Omega$, *then there exists a solution of the NDDE*(D, f) *through* (σ, ϕ).

Theorem 35 [13] *If* $\Omega \subseteq \mathbb{R} \times C$ *is open and* $f : \Omega \to \mathbb{R}^n$ *is Lipschitz in* ϕ *on compact sets of* Ω, *then, for and* $(\sigma, \phi) \in \Omega$, *there exists a unique solution of the NDDE*(D, f) *through* (σ, ϕ).

Theorem 36 [13] *Suppose* $\Omega \subseteq \mathbb{R} \times C$ *is open,* Λ *is a subset of a Banach space, and* $D : \Omega \times \Lambda \to \mathbb{R}^n$, $f : \Omega \times \Lambda \to R^n$ *satisfy the following conditions:*

(i) $D(t, \phi, \lambda)$ *is atomic at zero for each* $(t, \phi) \in \Omega$ *uniformly with respect to* λ.

(ii) $D(t, \phi, \lambda)$ *and* $f(t, \phi, \lambda)$ *are continuous in* $(t, \phi) \in \Omega$ *for each* $\lambda \in \Lambda$ *and continuous at* (t, ϕ, λ_0) *for* $(t, \phi) \in \Omega$.

(iii) *The NDDE(*$D(\cdot, \lambda_0)$, $f(\cdot, \lambda_0)$*) has a unique solution through* $(\sigma, \phi) \in \Omega$ *that exists on an internal* $[\sigma - r, b]$.

Then, for any $(\sigma', \phi', \lambda') \in N(\sigma, \phi, \lambda_0)$, *there is a neighborhood* $N(\sigma, \phi, \lambda_0)$ *of* $(\sigma, \phi, \lambda_0)$ *such that all solutions* $x(\sigma', \phi', \lambda')$ *of the NDDE(*$D(\cdot, \lambda')$, $f(\cdot, \lambda')$*) through* (σ', ϕ') *exist on* $[\sigma' - r, b]$, *and* $x_t(\sigma', \phi', \lambda')$ *is continuous at* (t, σ', λ_0) *for* $t \in [\sigma, b]$, $(\sigma', \phi', \lambda') \in N(\sigma, \phi, \lambda_0)$.

Consider the following linear autonomous NFDE(D, L) (neutral functional differential equation):

$$\frac{d}{dt} D x_t = L x_t, \tag{2.23}$$

where $D, L : C \to R^n$ are continuous linear functionals on C.

Theorem 37 [13] *In Eq. (2.23), let*

$$D_\phi = \phi(0) - \int_{-r}^{0} \phi(\theta) d\mu(\theta),$$

$$\lim_{s \to 0} \int_{-s}^{0} \times |d\mu(\sigma)| = 0.$$

If

$$\sup\{\mathrm{Re}\lambda : \det\Delta(\lambda) = 0\} < 0,$$

where $\Delta(\lambda) = \lambda D(e^{\lambda} \cdot I) - L(e^{\lambda} \cdot I)$, *then the zero solution of Eq. (2.23) is uniformly asymptotically stable. If* $\mathrm{Re}\lambda > 0$ *for some* λ *satisfying* $\det\Delta(\lambda) = 0$, *then (2.23) is unstable. Moreover, if* $\det\Delta(\lambda) = 0$ *has a nonsimple pure root, then Eq. (2.23) is also unstable.*

2.11 Stochastic Differential Equations

Let us consider the following stochastic differential equation

$$\mathrm{d}X_i = f(t, X_i)dt + G(t, X_t)dW_t, \; X_{t_0} = c, \; t_0 \leq t \leq T < \infty, \tag{2.24}$$

or, in integral form

$$X_t = c + \int_{t_0}^{t} f(s, X_t)ds + \int_{t_0}^{t} G(s, X_s)dW_s, \; t_0 \leq t \leq T < \infty, \tag{2.25}$$

where X_t is an $\mathbb{R}^d$-valued stochastic process (for the time being, assumed known) defined on $[t_0, T]$ and W_t is an $m-$dimensional Wiener process. The $\mathbb{R}^d$-valued function f and the ($d \times m$ matrix)-valued function G are assumed to be defined and measurable on $[t_o, T] \times \mathbb{R}^d$. For fixed (t, x), suppose that $f(t, x)$ and $G(t, x)$ are independent of $\omega \in Q$, i.e., that the random parameter w appears only indirectly in the coefficients in Eq. (2.24) in the form $f(t, X_t(w))$ and $G(t, X_t(w))$.

Definition 12 [14] An equation of the form (2.24) is called an (Ito's) stochastic differential equation. The random variable c is called the initial value at the instant t_o. Here, (2.24), together with the initial value, is only a symbolic way of writing the stochastic integral Eq. (2.25). A stochastic process X_t is called a solution of Eqs. (2.24) or (2.25) on the interval $[t_o, T]$ if it has the following properties:

(a) X_t is $\mathfrak{F}$-measurable, that is, nonanticipating for $t \in [t_0, T]$.
(b) The functions $\bar{f}(t, w) = f(t, X_t(w))$ and $\bar{G}(t, w) = G(t, X_t(w))$ (nonanticipating in accordance with a) are such that, with probability I,

$$\int_{t_0}^{T} |\bar{f}(s, w)|ds < \infty$$

and

$$\int_{t_0}^{T} |\bar{G}(s, w)|^2 ds < \infty.$$

Then, the right-hand member of (2.24) is meaningful.
(c) Equation (2.25) holds for every $t \in [t_0, T]$ with probability I.

Theorem 38 [14] *Suppose that we have a stochastic differential equation*

$$dX_t = f(t, X_t)dt + G(t, X_t)dW_t, \;\; X_{t_0} = c, \;\; t_0 \leq t \leq T < \infty, \tag{2.26}$$

where W_t is an R_m-valued Wiener process and c is a random variable independent of $W_t - W_{t_0}$ for $t \geq t_o$. Suppose that the R^d-valued function $f(t, x)$ and the ($d \times mmatrix$)-valued function $G(t, x)$ are defined and measurable on $[t_o, T] \times R^d$ and have the following properties: there exists a constant $K > 0$ such that

(a) *(Lipschitz condition) For all* $t \in [t_0, T]$, $x \in R^d$, $y \in R^d$,

$$|f(t,x) - f(t,y)| + |G(t,x) - G(t,y)| \le K|x-y| \; (|G|^2 = trGG').$$

(b) *(Restriction on growth) For all* $t \in [t_o, T]$ *and* $x \in R^d$,

$$|f(t,x)|^2 + |G(t,x)|^2 \le K^2(1+|x|^2).$$

Then, Eq. (2.26) has on $[t_0, T]$ *a unique* R^d*-valued solution* X_t*, continuous with probability 1, that satisfies the initial condition* $X_{t_0} = c$*; that is, if* X_t *and* Y_t *are continuous solutions of (2.26) with the same initial value c, then*

$$P|\sup_{t_0 \le t \le T} |X_t - Y_t| > 0| = 0.$$

Next, we assume that the conditions of the existence-and-uniqueness Theorem (38) are satisfied and we shall investigate the moments $E|X_t|^k$ of the solution of

$$dX_t = f(t, X_t)dt + G(t, X_t)dW_t, \; X_{t_0} = c, \tag{2.27}$$

on $[t_0, T]$.

Theorem 39 [14] *Suppose that the assumptions of the existence-and-uniqueness Theorem (38) are satisfied and that*

$$E|c|^{2n} < \infty$$

where n is a positive integer. Then, for the solution X_t *of the stochastic differential Eq. (2.27) on* $[t_o, T]$*, where* $T < \infty$,

$$E|X_t|^{2n} \le (1 + E|c|^{2n})e^{C(t-t_0)}$$

and

$$E|X_t - c|^{2n} \le D(1 + E|c|^{2n})(t - t_0)^n e^{C(t-t_0)}$$

where $C = 2n(2n+1)K^2$ *and D are constants (dependent only on* n, K*, and* $T - t_o$*).*

Again, we assume that the conditions of the existence-and-uniqueness Theorem 38 are satisfied. Let us investigate the sample functions of the solution process X, of

$$X_t = c + \int_{t_0}^{t} f(s, X_s)ds + \int_{t_0}^{t} G(s, X_s)dW_s \tag{2.28}$$

as R^d-valued functions of t on the interval $[t_0, T]$, where $T < \infty$.

Theorem 40 [14] *If X_t is the solution of Eq. (2.28), $t_0 < t_1 < \ldots < t_n = T$ is a partition of the interval $[t_0, T]$, and $\delta_n = \max(t_k - t_{k_1})$, then*

$$\text{st-}\lim_{b_n \to 0} \sum_{k=1}^{n} (X_{t_k} - X_{t_{k-1}})(X_{t_k} - X_{t_{k-1}})' = \int_{t_0}^{T} G(s, X_s)G(s, X_s)' ds,$$

and, in particular,

$$\text{st-}\lim_{b_n \to 0} \sum_{k=1}^{n} |X_{t_k} - X_{t_{k-1}}|^2 = \int_{t_0}^{T} |G(s, X_s)|^2 ds.$$

Theorem 41 [14] *If X_t is the solution of Eq. (2.28) and if f and G are continuous with respect to t for $t \in [t_0, T]$, then, for every fixed $t \in [t_0, T]$,*

$$\limsup_{h \to 0} \frac{|X_{t+h} - X_t|}{\sqrt{2h \log\log 1/h}} = \sqrt{\lambda(t, X_t)},$$

with probability 1, where $\lambda(t, x)$ is the greatest eigenvalue of the (necessarily nonnegative-definite) matrix $G(t, x)G(t, x)'$.

Theorem 42 [14] *If X_t is the solution of Eq. (2.28), if f and G are continuous for $t \in [t_0, T]$, and if the initial value c is almost certainly constant, then, in the case $G(t_0, c) = 0$,*

$$\lim_{t \to t_0} \frac{X_t - c}{t} = f(t_0, c)$$

with probability 1. Thus, differentiability of the solution of a stochastic differential equation is the exception and nondifferentiability is the rule. The formal differential equation

$$\dot{X}_t = f(t, X_t) + G(t, X_t)\xi_t,$$

where ξ_t is an m-dimensional white noise, cannot therefore as a rule be interpreted as an ordinary differential equation for the function X_t.

2.12 Pattern Formation of Nonlinear System

The self-organization phenomenon in the reaction–diffusion system results from the coupling between the chemical dynamic process and the diffusion process. The stable uniform steady states lose stability when they are far away from the thermodynamic equilibrium state. After the non-equilibrium phase transition, a series of spatiotemporal or dynamic patterns are spontaneously produced. From a mathematical point of view, the mechanism of this non-equilibrium phase transition can be divided into two categories. The first type originates from the instability of a uniform steady state in the phase space. This non-equilibrium phase transition can be studied through perturbation analysis to obtain different spatiotemporal symmetry breaking types. The core of perturbation analysis is linear stability analysis. Since perturbation analysis is only applicable to a small phase space, this instability mechanism is local. The following takes a general two-variable reaction–diffusion system as an example to introduce the application of linear stability analysis in reaction–diffusion equations.

Consider the following general reaction–diffusion two-variable equation:

$$\begin{cases} \dfrac{\partial U}{\partial t} = f(U, V, \mu) + D_u \nabla^2 U, \\ \dfrac{\partial V}{\partial t} = g(U, V, \mu) + D_v \nabla^2 V. \end{cases} \tag{2.29}$$

Here, U and V are variables; μ represents the collection of parameters; and D is the diffusion coefficient of reactant. Suppose that there exists only one uniform steady state (U_0, V_0) of the system which satisfies

$$\begin{cases} f(U_0, V_0) = 0, \\ g(U_0, V_0) = 0. \end{cases}$$

Imposing a small perturbation on this uniform steady state, i.e., $U = U_0 + u,\ V = V_0 + v$, and then substituting this form into Eq. (2.29), we can obtain the following linear perturbation equation by applying Taylor expansion and removing the high-order terms:

$$\begin{cases} \dfrac{\partial u}{\partial t} = a_{11}u + a_{12}v + D_u \nabla^2 U, \\ \dfrac{\partial v}{\partial t} = b_{11}u + b_{12}v + D_v \nabla^2 V. \end{cases} \tag{2.30}$$

Here,

$$a_{11} = \frac{\partial f}{\partial U}|_{U_0, V_0}, \qquad a_{12} = \frac{\partial f}{\partial V}|_{U_0, V_0},$$
$$a_{21} = \frac{\partial g}{\partial U}|_{U_0, V_0}, \qquad a_{22} = \frac{\partial g}{\partial V}|_{U_0, V_0}.$$

Expanding the perturbation variable in Fourier space, we have

$$\begin{pmatrix} u \\ v \end{pmatrix} = \sum \begin{pmatrix} c_k^1 \\ c_k^2 \end{pmatrix} e^{\lambda_k t + ikr}.$$

Substituting this solution into Eq. (2.30), the dispersion relation can be written as follows:

$$\lambda_k^2 - Tr_k \lambda_k + \Delta_k = 0,$$

where

$$Tr_k = a_{11} + a_{22} - k^2(D_u + D_v) = Tr_0 - k^2(D_u + D_v),$$

$$\begin{aligned}\Delta_k &= a_{11}a_{22} - a_{21}a_{12} - k^2(a_{11}D_v + a_{22}D_u) + k^4 D_u D_v \\ &= \Delta_0 - k^2(a_{11}D_v + a_{22}D_u) + k^4 D_u D_v.\end{aligned}$$

So we can get

$$\lambda_k = \frac{Tt_k \pm \sqrt{Tr_k^2 - 4\Delta_k}}{2}.$$

Unlike the stability analysis of ordinary differential equations, the uniform steady states of partial differential equations can be stable only when the real parts of all eigenvalues are less than zero. Otherwise, the perturbations will increase with time, leading to the loss of stability of uniform steady states (U_0, V_0).

Part I
Permanence and Extinction of Plankton

Chapter 3
The Effect of Stochastic Marine Environment

Marine ecology includes two primary planktons: one is called phytoplankton at the first trophic level, and the other is zooplankton at the second trophic level. Phytoplankton, commonly unicellular and microscopic, is the leading energy supplier and controls the global carbon cycle. Zooplankton plays an essential regulatory role in the structure and function of the marine ecosystem, which lives on phytoplankton. Variation of plankton density is the main factor affecting the marine environment and the development of marine economy. Due to the difficulty in measuring plankton biomass, many important mathematical models have been proposed to study the complex dynamics of plankton [15, 16, 18–21]. As we all know, the ecological environment constantly changes, such as the seasonal effect, food supply, and harvest [22–24]. For plankton, many factors like the unpredictability of pH, dissolved oxygen, water temperature, and turbulence influence the growth and reproduction of phytoplankton and zooplankton. The environmental variations cause complex relationships among the birth rate, nutrient quota, mortality rate, and consumption rate of species, which can change the population size of organisms significantly in a short time. Therefore, stochastic differential equations (SDEs) could be a more appropriate way to model plankton growth in many circumstances [25–28].

To describe the interaction of plankton more realistic, the formulation of plankton models considers the intrinsic biological disciplines and external relationships, such as additional food [29], stage structure [31], and prey refuge [30]. Considering the critical intrinsic factor of biology, it is obvious that an organism's body size reflects its adaptive modification in response to the prevailing abiotic and biotic environment [32]. The trophic organization of planktonic communities and their ability to export biogenic materials toward the ocean's interior are controlled by the size structure of phytoplankton controls [33], and the metabolism and life history are affected by the body size of zooplankton. Hence, the fluctuation of the dynamic behavior of the entire plankton, the system varies with the cell size of the plankton [34, 35].

In the following, a stochastic nutrient-phytoplankton–zooplankton model with cell size is generated to represent the interaction between internal mechanism of species

S. T. Liu and L. Zhang, *Pattern Dynamics of Marine Plankton Behavior*,
https://doi.org/10.1007/978-981-97-5369-7_3

and the external environment. For this, the existence and uniqueness of the global positive solution with positive initial values are first investigated. With this, sufficient conditions for an ergodic stationary distribution of positive solution are constructed. Once more, we also find out that large noise intensities cause the extinctions of phytoplankton and zooplankton. Finally, numerical simulations are given to verify the effectiveness of theoretical results.

3.1 Preliminaries and Model Description

3.1.1 Preliminaries

Assume that $(\Omega, \mathcal{F}, \{\mathcal{F}_t\}_{t\geq 0}, \mathbb{P})$ is a complete probability space with a filtration $\{\mathcal{F}_t\}_{t\geq 0}$ satisfying the usual conditions. Let $a \wedge b$ denote $\min\{a, b\}$ and $\langle f \langle_t = \frac{1}{t}\int_0^t f(s)ds$, where f is an integrable function on $[0, \infty)$. Almost surely is replaced by a.s. Denote $\mathbb{R}^n_+ = \{X \in \mathbb{R}^n : X_i > 0 \ for \ all \ 1 \leq i \leq n\}$.

The following n-dimensional stochastic differential equation is considered:

$$dX(t) = f(X(t), t)dt + g(X(t), t)dB(t) \quad on \ t \geq t_0, \tag{3.1}$$

where $X(t)$ means the state of (3.1) with initial value $X(t_0) = X_0 \in \mathbb{R}^n$; $B(t)$ stands for a n-dimensional standard Brownian motion on complete probability space. The diffusion matrix is defined as $A(X) = (a_{ij}(X))$, $a_{ij}(X) = g^i(X)g^j(X)$. The differential operator $\mathcal{L}$ associated with 3.1 is constructed by

$$\mathcal{L} = \frac{\partial}{\partial t} + \sum_{i=1}^{n} f_i(X,t)\frac{\partial}{\partial X_i} + \frac{1}{2}\sum_{i=1}^{n}[g^T(X,t)g(X,t)]_{ij}\frac{\partial^2}{\partial X_i \partial X_j}.$$

If $\mathcal{L}$ acts on a function $V \in C^{2,1}(\mathbb{R}_n \times [t_0, +\infty); \mathbb{R}_+)$, then

$$\mathcal{L}V(X,t) = V_t(X,t) + V_X(X,t)f(X,t) + \frac{1}{2}\textbf{trace}[g^T(X,t)V_{XX}(X,t)g(X,t)].$$

According to Itô's formula,

$$dV(X(t),t) = \mathcal{L}V(X(t),t)dt + V_X(X(t),t)g(X(t),t)dB(t).$$

Presenting the definitions of stationary distribution and extinction, a lemma can give a criterion for the existence of an ergodic stationary distribution of (3.1).

Definition 13 (Stationary distribution [37]) Assume $\mathbb{P}(t, X, \cdot)$ is the probability measure induced by $X(t)$ with initial value X_0. Then this yields

$$\mathbb{P}(t, X, B) = \mathbb{P}(X(t) \in B | X_0), \ \ for \ any \ Borel \ set \ B \in \mathbb{R}^n_+.$$

Let there exist a probability measure $\pi(\cdot)$ such that $\lim_{t\to\infty}\mathbb{P}(t, X, B) = \pi(B)$ for all $X \in \mathbb{R}^n_+$. Then (3.1) has a stationary distribution $\pi(\cdot)$.

Definition 14 (Stochastic extinction [38]) A population, $X(t)$, is prescribed as a stochastic extinct if

$$\mathbb{P}\{\lim_{t\to+\infty} X(t) = 0\} = 1.$$

Lemma 4 [37] *Let there is a bounded open set $U \subset \mathbb{R}^n$ with a smooth boundary and it satisfy the conditions as given below:*
(B.1) *There exists a positive number β_0 such that $\sum_{i,j=1} a_{ij}(X)\xi_i\xi_j \geq \beta_0|\xi|^2$, $X \in U$, $\xi \in \mathbb{R}^n$.*
(B.2) *There is a nonnegative C^2-function $V(X)$ and a positive constant C such that $\mathcal{L}V(X) \leq -C$ for $\forall X \in \mathbb{R}^n \setminus U$. For any integrable function $g(\cdot)$ in regard to the measure π, it yields*

$$\mathbb{P}(\lim_{t\to\infty}\frac{1}{t}\int_0^t g(X(t))dt = \int_{\mathbb{R}^n} g(X)\pi(dX)) = 1.$$

Then it can be obtained that the Markov process $X(t)$ has a unique ergodic stationary distribution $\pi(\cdot)$.

3.1.2 Model Description

Based on the experimental analysis of death rates of phytoplankton and zooplankton varying exponentially with water temperature constantly [35], the perturbation can have a random effect on its parameters most influentially. The stochastic perturbations are assumed as the white noise type by considering the stochastic disturbances and cell sizes. Assume that the loss rate of nutrient a, a mortality rate of zooplankton δ, and the mortality rate of phytoplankton m are disturbed as the following form:

$$-a \to -a + \sigma_1\dot{B}_1(t), \qquad -m \to -m + \sigma_2\dot{B}_2(t), \qquad -\delta \to -\delta + \sigma_3\dot{B}_3(t),$$

where $B_i(t)$ $(i = 1, 2, 3)$ means the independent standard Brownian motion with $B_i(0) = 0$, σ_i stands for the intensity of the white noise. Hence, the model related with a single limiting nutrient, phytoplankton and zooplankton are deduced as given below:

$$\begin{cases} dN(t) = (I - aN(t) - \mu(x)\frac{N(t)}{N(t)+K}P(t)Q(x))dt + \sigma_1 N(t)dB_1(t), \\ dP(t) = P(t)(\mu(x)\frac{N(t)}{N(t)+K} - m - s(x) - C(x, y)Z(t))dt + \sigma_2 P(t)dB_2(t), \\ dZ(t) = Z(t)(\phi\frac{Q(x)}{q(y)}C(x, y)P(t) - \delta)dt + \sigma_3 Z(t)dB_3(t), \end{cases} \tag{3.2}$$

where $N(t)$, $Z(t)$, and $P(t)$ represent the nutrient concentration, zooplankton at time t, and the densities of phytoplankton, respectively. x and y denote the cell sizes of phytoplankton and zooplankton, respectively, which are all positive in the biological sense. The nutrient concentration includes the following parts: a constant nutrient inflow I as the environmental productivity, aN as the nutrient loss of water outflow, and the nutrient uptake by phytoplankton $\mu(x)Q(x)\frac{NP}{N+K}$. The following factors influence the dynamics of phytoplankton: nutrient conversion $\mu(x)\frac{NP}{N+K}$, per capita mortality of phytoplankton mP, per capita sinking of phytoplankton $s(x)P$, and consumption of zooplankton $C(x, y)PZ$. The following factors influence the dynamics of zooplankton: per capita mortality of zooplankton δZ and the assimilation of phytoplankton biomass $\frac{Q(x)}{q(y)}C(x, y)PZ$. Table 3.1 yields the parameters and their biological interpretations. In some cases, model 3.2 can be deduced to fit with other biologic systems. For example, as $\sigma_i = 0$ and $x = y = 0$, model 3.2 degenerates to the simply NPZ model (e.g., [16]). Let $\sigma_i = 0$ and $x \neq 0$, $y \neq 0$, model 3.2 degenerates to the model of [39]. Moreover, the consumption of zooplankton PZ

Table 3.1 Parameters and their biological meanings

Parameters	Biological meanings
x	Phytoplankton cell size
y	Zooplankton cell size
I	Inflow of nutrient
a	Outflow rate of nutrient
$\mu(x) = \frac{x}{c_1x^2+c_2x+c_3}$ [35]	Maximum phytoplankton growth rate
(c_1, c_2, c_3)	Coefficients of $\mu(x)$
$\frac{N}{N+K}$	Nutrient limitation for phytoplankton growth
K	Half-saturation constant
$Q(x) = \beta x^{b_1}$	Nutrient quota of phytoplankton
β	Coefficient of nutrient quota
b_1	Exponent of nutrient quota
m	Mortality rate of phytoplankton
$s(x) = \alpha x^2$ [36]	Sinking rate of phytoplankton
α	Coefficient of sinking rate
$C(x, y) = C_m e^{-\frac{1}{\lambda}(x-\kappa y)^2}$ [35]	Consumption rate of zooplankton
λ	Coefficient of consumption rate
κ	Optimal predator–prey ratio
C_m	Maximum consumption rate
ϕ	Conversion efficiency of phytoplankton
$q(y) = \rho y^{b_2}$	Nutrient quota of zooplankton
ρ	Coefficient of nutrient quota
b_2	Exponent of nutrient quota
δ	Mortality rate of zooplankton
$B_i(t)$ $(i = 1, 2, 3)$	Mutually independent standard Brownian motion with initial value $B_i(0) = 0$
$\sigma_i > 0$ $(i = 1, 2, 3)$	Intensity of white noise

can be the bilinear assimilation rate, while other nonlinear conversion functions can be proposed by researchers (e.g., $\frac{PZ}{1+P}$, $\frac{P^2Z}{1+P^2}$, $\frac{PZ}{1+P+Z}$). $\mu(x)$ can represent a unimodal relationship between growth rates and cell size. And $C(x, y) = C_m e^{-\frac{1}{\lambda}(x-\kappa y)^2}$ can be maximum as zooplankton with an optimal size ratio κ of phytoplankton.

3.2 Stationary Distribution and Ergodicity

For the stochastic NPZ model, consider the existence of stationary distribution which implies the persistence of the population. In the following, the stationary distribution which is ergodic of model (3.2) is proved as the white noises are limited in some range. Next, the existence of a unique global positive solution is indicated.

Theorem 43 *There exists a unique global positive solution $(N(t), P(t), Z(t))$ of (3.2) for all $t \geq 0$ almost surely, for any initial value $(N_0, P_0, Z_0) \in \mathbb{R}^3_+$.*

Proof Please see Appendix A. □

Theorem 44 *Let*

$$4(\mu(x)I - \alpha(aK + I)) > \sigma_1 a^2(\alpha - \mu(x))^2, 2a > \sigma_1^2, \quad 2(m + s(x)) > \sigma_2^2, \quad 2\delta > \sigma_3^2, \qquad (3.3)$$

where $\alpha = m + s(x) + \frac{1}{2}\sigma_2^2$. Then the solution of model (3.2) is ergodic and satisfies a unique stationary distribution for any initial value $(N_0, P_0, Z_0) \in \mathbb{R}^3_+$.

Proof In order to prove the ergodic stationary distribution, conditions (B1) and (B2) of Lemma 4 should be verified.

In the following the condition (B1) is verified firstly. The diffusion matrix of model 3.2 is defined as $A = diag\{\sigma_1^2 N^2, \sigma_2^2 P^2, \sigma_3^2 Z^2\}$. It is obvious that there exists $\beta_0 = \min_{(N,P,Z)\in\bar{U}} \{\sigma_1^2 N^2, \sigma_2^2 P^2, \sigma_3^2 Z^2\} > 0$ such that

$$\sum_{i,j=1}^{3} a_{ij}(N, P, Z)\xi_i\xi_j = \sigma_1^2 N^2 \xi_1^2 + \sigma_2^2 P^2 \xi_2^2 + \sigma_3^2 Z^2 \xi_3^2 \geq \beta_0 \parallel \xi \parallel^2,$$

where $(N, P, Z) \in \bar{U}$ and $\xi = (\xi_1, \xi_2, \xi_3) \in \mathbb{R}^3_+$. Hence condition (B.1) of Lemma 4 holds.

Next a C^2-function $W(N, P, Z) : \mathbb{R}^3_+ \to \mathbb{R}_+$ is constructed as follows:

$$\begin{aligned} W(N, P, Z) &= M(N - \tfrac{I}{a} - \tfrac{I}{a}\log\tfrac{Na}{I} - \sigma_1 \log P) \\ &- \log N + Z - \log Z + \tfrac{1}{2}(N + PQ(x) + \tfrac{q(y)}{\phi}Z)^2. \end{aligned}$$

After selecting a suitable constant $M > 0$ such that $-\lambda M + F \leq -2$, it yields that

$$
\begin{aligned}
G(N) = &-(N-\frac{I}{a})^2+\sigma_1 a(m+s(x)+\frac{1}{2}\sigma_2^2-\mu(x))(N-\frac{I}{a})\\
&+\sigma_1 a(m+s(x)+\frac{1}{2}\sigma_2^2)K+\sigma_1 I(m+s(x)+\frac{1}{2}\sigma_2^2-\mu(x)),\\
F = &\sup_{(N,P,Z)\in\mathbb{R}_+^3}\left\{M\frac{\mu(x)PQ(x)I}{aK}+\sigma_1 MC(x,y)Z+a+\mu(x)\frac{PQ(x)}{K}+\frac{1}{2}\sigma_1^2\right.\\
&-\frac{1}{4}(2a-\sigma_1^2)N^2-\frac{1}{8}(2m+2s(x)-\sigma_2^2)(Q(x)P)^2\\
&\left.-\frac{1}{8}(2\delta-\sigma_3^2)(\frac{q(y)}{\phi}Z)^2+E\right\}<\infty,\\
E = &\sup_{(N,P,Z)\in\mathbb{R}_+^3}\left\{I(N+Q(x)P+\frac{q(y)}{\phi}Z)-\frac{1}{4}(2a-\sigma_1^2)N^2\right.\\
&\left.-\frac{1}{4}(2m+2s(x)-\sigma_2^2)(Q(x)P)^2-\frac{1}{4}(2\delta-\sigma_3^2)(\frac{q(y)}{\phi}Z)^2\right\}<\infty,
\end{aligned}
$$

where $\lambda=-\sup_{N\in(0,\infty)}\frac{G(N)}{a(N+K)}-\frac{I}{2a}\sigma_1^2>0$. By differentiating $W(N,P,Z)$, the following system is derived:

$$
\begin{cases}
\frac{\partial W}{\partial N}=M(1-\frac{I}{Na})-\frac{1}{N}+(N+PQ(x)+\frac{q(y)}{\phi}Z),\\
\frac{\partial W}{\partial P}=-M\sigma_1\frac{1}{P}+(N+PQ(x)+\frac{q(y)}{\phi}Z)Q(x),\\
\frac{\partial W}{\partial Z}=1-\frac{1}{Z}+(N+PQ(x)+\frac{q(y)}{\phi}Z)\frac{q(y)}{\phi}.
\end{cases}
$$

After simple calculation, it is obvious that $W(N,P,Z)$ has a unique positive stationary point (N^*,P^*,Z^*), where $P^*=\frac{MN^*a\sigma_1}{Q(x)(a+M(I-N^*a))}$, $Z^*=\frac{N^*a\phi}{(a+M(I-N^*a))q(y)+N^*a\phi}$ and N^* satisfies

$$
M(Na-I)-a+N^2a+\frac{N^2aM\sigma_1}{a+M(I-Na)}+\frac{N^2a^2q(y)}{(a+M(I-Na))q(y)+Na\phi}=0.
$$

Based on the virtue of the monotonicity of $W(N,P,Z)$, (N^*,P^*,Z^*) as the minimum point of $W(N,P,Z)$ can be obtained easily.

Next the following nonnegative Lyapunov function $V(N,P,Z)$ is defined as

$$
\begin{aligned}
&V(N,P,Z)=W(N,P,Z)-W(N^*,P^*,Z^*)+\log Z\\
&=M(N-\frac{I}{a}-\frac{I}{a}\log\frac{Na}{I}-\sigma_1\log P)-\log N+Z\\
&+\frac{1}{2}(N+PQ(x)+\frac{q(y)}{\phi}Z)^2-W(N^*,P^*,Z^*)\\
&=MV_1(N,P,Z)+V_2(N,P,Z)+V_3(N,P,Z)-W(N^*,P^*,Z^*).
\end{aligned}
$$

By taking into account Itô's formula, the following is obtained:

$$
\begin{aligned}
\mathcal{L}V_1 &= \mathcal{L}(N - \tfrac{I}{a} - \tfrac{I}{a}\log\tfrac{Na}{I} - \sigma_1 \log P)\\
&= \tfrac{aN-I}{aN}(I - aN - \mu(x)\tfrac{N}{N+K}PQ(x)) + \tfrac{I}{2a}\sigma_1^2\\
&+\sigma_1(m + s(x) + C(x,y)Z - \mu(x)\tfrac{N}{N+K} + \tfrac{1}{2}\sigma_2^2)\\
&\le -\tfrac{(aN-I)^2}{a^3N} + \tfrac{I\mu(x)PQ(x)}{a(N+K)} + \tfrac{I}{2a}\sigma_1^2 + \sigma_1(m+s(x)+\tfrac{1}{2}\sigma_2^2 - \tfrac{\mu(x)N}{N+K} + C(x,y)Z)\\
&\le -\tfrac{(aN-I)^2}{a^3(N+K)} + \tfrac{I\mu(x)PQ(x)}{aK} + \tfrac{I}{2a}\sigma_1^2 + \tfrac{\sigma_1(m+s(x)+\frac{1}{2}\sigma_2^2-\mu(x))(aN-I)}{a(N+K)}\\
&+\tfrac{\sigma_1(m+s(x)+\frac{1}{2}\sigma_2^2)K}{N+K} + \tfrac{\sigma_1(m+s(x)+\frac{1}{2}\sigma_2^2-\mu(x))I}{a(N+K)} + \sigma_1 C(x,y)Z\\
&= \tfrac{-(aN-I)^2+\sigma_1 a^2(m+s(x)+\frac{1}{2}\sigma_2^2-\mu(x))t(aN-I)+\sigma_1 a^3(m+s(x)+\frac{1}{2}\sigma_2^2)K}{a^3(N+K)}\\
&+\tfrac{\sigma_1 I(m+s(x)+\frac{1}{2}\sigma_2^2-\mu(x))}{a(N+K)} + \tfrac{I\mu(x)PQ(x)}{aK} + \tfrac{I}{2a}\sigma_1^2 + \sigma_1 C(x,y)Z\\
&= \tfrac{G(N)}{a(N+K)} + \tfrac{\mu(x)PQ(x)I}{aK} + \tfrac{I}{2a}\sigma_1^2 + \sigma_1 C(x,y)Z.
\end{aligned}
$$

With (3.3), the negativeness of the discriminant of $G(N)$ is obtained, implying $G(N) < 0$ with $N > 0$ and

$$
\mathcal{L}V_1 \le -\lambda + \frac{\mu(x)PQ(x)I}{aK} + \sigma_1 C(x,y)Z.
$$

Additionally, it yields

$$
\begin{aligned}
\mathcal{L}V_2 &= \mathcal{L}(-\log N + Z)\\
&= -\frac{I}{N} + a + \mu(x)\frac{PQ(x)}{N+K} + \frac{1}{2}\sigma_1^2 + \phi\frac{Q(x)}{q(y)}C(x,y)PZ - \delta Z\\
&\le -\frac{I}{N} + a + \mu(x)\frac{PQ(x)}{K} + \frac{1}{2}\sigma_1^2 + \phi\frac{Q(x)}{q(y)}C(x,y)PZ,
\end{aligned}
$$

and

$$
\begin{aligned}
\mathcal{L}V_3 &= \mathcal{L}(\tfrac{1}{2}(N + PQ(x) + \tfrac{q(y)}{\phi}Z)^2)\\
&= (N + Q(x)P + \tfrac{q(y)}{\phi}Zt)(I - aN - (m+s(x))Q(x)P - \tfrac{q(y)}{\phi}\delta Z)\\
&+\tfrac{1}{2}(\sigma_1^2N^2 + \sigma_2^2(Q(x)P)^2 + \sigma_3^2(\tfrac{q(y)}{\phi}Zt)^2),\\
&\le I(N + Q(x)P + \tfrac{q(y)}{\phi}Z) - (m+s(x))(Q(x)P)^2 - aN^2 - \delta(\tfrac{q(y)}{\phi}Z)^2\\
&+\tfrac{1}{2}(\sigma_1^2N^2 + \sigma_2^2(Q(x)P)^2 + \sigma_3^2(\tfrac{q(y)}{\phi}Z)^2),\\
&\le -\tfrac{1}{4}(2a - \sigma_1^2)N^2 - \tfrac{1}{4}(2m + 2s(x) - \sigma_2^2)(Q(x)P)^2\\
&-\tfrac{1}{4}(2\delta - \sigma_3^2)(\tfrac{q(y)}{\phi}Z)^2 + E.
\end{aligned}
$$

Therefore,

$$
\begin{aligned}
\mathcal{L}V &\le M(-\lambda + \tfrac{I\mu(x)PQ(x)}{aK} + \sigma_1 C(x,y)Z) - \tfrac{I}{N} + a + \mu(x)\tfrac{PQ(x)}{K}\\
&+\tfrac{1}{2}\sigma_1^2 + \phi\tfrac{Q(x)}{q(y)}C(x,y)PZ - \tfrac{1}{4}(2a - \sigma_1^2)N^2 - \tfrac{1}{4}(2m + 2s(x) - \sigma_2^2)(Q(x)P)^2\\
&-\tfrac{1}{4}(2\delta - \sigma_3^2)(\tfrac{q(y)}{\phi}Z)^2 + E.
\end{aligned}
$$

Next $\mathcal{L}V(N, P, Z) \leq -1$ is shown on some U^c. At last, a bounded open subset U is constructed by

$$U = \{(N, P, Z) \in \mathbb{R}_+^3 : \varepsilon < N < \frac{1}{\varepsilon}, \varepsilon < P < \frac{1}{\varepsilon}, \varepsilon < Z < \frac{1}{\varepsilon}\},$$

where $0 < \varepsilon < \frac{q(y)}{4\phi Q(x)C(x,y)}((2m + 2s(x) - \sigma_2^2)Q(x)^2 \wedge (2\delta - \sigma_3^2)(\frac{q(y)}{\phi})^2)$ is a sufficiently small number and follows the following conditions:

$$\begin{cases} -\lambda M + \phi\frac{Q(x)}{q(y)}C(x, y)\varepsilon + F \leq -1, \\ -\frac{I}{\varepsilon} + H \leq -1, \\ -\frac{1}{8\varepsilon^2}(2a - \sigma_1^2) + H \leq -1, \\ -\frac{1}{8\varepsilon^2}(2m + 2s(x) - \sigma_2^2)Q(x)^2 + H \leq -1, \\ -\frac{1}{8\varepsilon^2}(2\delta - \sigma_3^2)(\frac{q(y)}{\phi})^2 + H \leq -1, \end{cases} \tag{3.4}$$

where

$$\begin{aligned} H = \sup_{(N,P,Z)\in\mathbb{R}_+^3} \Big\{ & M\frac{\mu(x)PQ(x)I}{aK} + \sigma_1 MC(x, y)Z + \phi\frac{Q(x)}{q(y)}C(x, y)PZ + a \\ & +\mu(x)\frac{PQ(x)}{K} + \frac{1}{2}\sigma_1^2 - \frac{1}{4}(2a - \sigma_1^2)N^2 - \frac{1}{4}(2m + 2s(x) - \sigma_2^2)(Q(x)P)^2 \\ & -\frac{1}{4}(2\delta - \sigma_3^2)(\frac{q(y)}{\phi}Z)^2 + E \Big\} < \infty. \end{aligned}$$

Further the following relationship is introduced

$$U_1 = \{(N, P, Z) \in \mathbb{R}_+^3, 0 < Z \leq \varepsilon\}, \qquad U_2 = \{(N, P, Z) \in \mathbb{R}_+^3, 0 < P \leq \varepsilon\},$$

$$U_3 = \{(N, P, Z) \in \mathbb{R}_+^3, 0 < N \leq \varepsilon\}, \qquad U_4 = \{(N, P, Z) \in \mathbb{R}_+^3, N \geq \frac{1}{\varepsilon}\},$$

$$U_5 = \{(N, P, Z) \in \mathbb{R}_+^3, P \geq \frac{1}{\varepsilon}\}, \qquad U_6 = \{(N, P, Z) \in \mathbb{R}_+^3, Z \geq \frac{1}{\varepsilon}\}.$$

Clearly, $U^c = U_1 \cup U_2 \cup U_3 \cup U_4 \cup U_5 \cup U_6$. The proof $\mathcal{L}V \leq -1$ on U^c is given in appendix. Based on Lemma 4, it is clear that model (3.2) is ergodic and follows a unique stationary distribution. □

Remark (a) In the proof of Theorem 44, it yields that the selection of ε has effect of the existing region on a stationary distribution and plays important role in the dynamics structure of phytoplankton and zooplankton. In order to guarantee the positiveness of the Lyapunov function $V(N, P, Z)$, function $W(N, P, Z)$ is analyzed and λ is obtained as term reflecting the impact of environmental white noise.

(b) As the disturbances of environment are small enough, the ergodic stationary distribution is unique, representing that the biomass of phytoplankton and zooplankton can be persistent and coexistent in the long term. Moreover, the limit of the adaptability of plankton to environmental disturbances is also shown.

3.3 Extinction

In this section, the extinction of the population with probability one would be shown as the intensities of noises are sufficiently large.

Theorem 45 *If* $2I < a[(m + s(x) + \frac{1}{2}\sigma_2^2) \wedge (\delta + \frac{1}{2}\sigma_3^2)]$ *holds, then the phytoplankton and zooplankton populations of model (3.2) tend to be extinct for any initial value* $(N_0, P_0, Z_0) \in \mathbb{R}_+^3$.

Proof As the first equation of (3.2), it yields

$$\frac{N(t) - N(0)}{t} = I - a < N >_t - \mu(x)Q(x) < \frac{P}{N + K} >_t + \frac{\sigma_1}{t} \int_0^t N(s)dB_1(s),$$

which means

$$a < N >_t = -\frac{N(t) - N(0)}{t} + I - \mu(x)Q(x) < \frac{P}{N + K} >_t + \frac{\sigma_1}{t} \int_0^t N(s)dB_1(s).$$

Based on Theorem 43 [40], it yields

$$\lim_{t\to\infty} (-\frac{N(t) - N(0)}{t} + \frac{\sigma_1}{t} \int_0^t N(s)dB_1(s)) = 0, \quad a.s.$$

Then

$$\limsup_{t\to\infty} < N >_t \le \frac{I}{a} \quad a.s.$$

In order to prove the convergence to zero of P and Z, let $S(t) = Q(x)P(t) + \frac{q(y)}{\phi}Z(t)$. After applying Itô's formula to $\log S(t)$, it yields

$$\begin{aligned} d\log S(t) &= \{\frac{Q(x)\mu(x)\frac{NP}{N+K} - Q(x)(m+s(x))P - \frac{q(y)}{\phi}\delta Z}{Q(x)P + \frac{q(y)}{\phi}Z} - \frac{\sigma_2^2(Q(x)P)^2 + \sigma_3^2(\frac{q(y)}{\phi}Z)^2}{2(Q(x)P + \frac{q(y)}{\phi}Z)^2}\}dt \\ &+ \frac{\sigma_2 Q(x)P}{Q(x)P + \frac{q(y)}{\phi}Z}dB_2(t) + \frac{\sigma_3 q(y)Z}{Q(x)P\phi + q(y)Z}dB_3(t) \\ &\le \mu(x)Ndt - \frac{Q(x)^2P^2(m+s(x)+\frac{1}{2}\sigma_2^2) + (\frac{q(y)}{\phi})^2Z^2(\delta + \frac{1}{2}\sigma_3^2)}{(Q(x)P + \frac{q(y)}{\phi}Z)^2}dt \\ &+ \frac{\sigma_2 Q(x)P}{Q(x)P + \frac{q(y)}{\phi}Z}dB_2(t) + \frac{\sigma_3 q(y)Z}{Q(x)P\phi + q(y)Z}dB_3(t) \\ &\le \mu(x)Ndt - \frac{1}{2}[(m + s(x) + \frac{1}{2}\sigma_2^2) \wedge (\delta + \frac{1}{2}\sigma_3^2)]dt \\ &+ \frac{\sigma_2 Q(x)P}{Q(x)P + \frac{q(y)}{\phi}Z}dB_2(t) + \frac{\sigma_3 q(y)Z}{Q(x)P\phi + q(y)Z}dB_3(t). \end{aligned} \tag{3.5}$$

After integrating (3.5) from 0 to t and dividing by t on both sides, it yields

$$\begin{aligned}\frac{\log S(t)}{t} - \frac{\log S(0)}{t} &\le \mu(x) < N >_t - \tfrac{1}{2}[(m + s(x) + \tfrac{1}{2}\sigma_2^2) \wedge (\delta + \tfrac{1}{2}\sigma_3^2)] \\ &+ \tfrac{1}{t}\int_0^t \frac{\sigma_2 Q(x)P(s)}{Q(x)P(s) + \frac{q(y)}{\phi}Z(s)} dB_2(s) + \tfrac{1}{t}\int_0^t \frac{\sigma_3 q(y)Z(s)}{Q(x)P(s)\phi + q(y)Z(s)} dB_3(s) \\ &\le \mu(x)\tfrac{I}{a} - \tfrac{1}{2}[(m + s(x) + \tfrac{1}{2}\sigma_2^2) \wedge (\delta + \tfrac{1}{2}\sigma_3^2)] + \tfrac{M_3(t)}{t} + \tfrac{M_4(t)}{t},\end{aligned}$$

where

$$M_3(t) = \int_0^t \frac{\sigma_2 Q(x)P(s)}{Q(x)P(s) + \frac{q(y)}{\phi}Z(s)} dB_2(s)$$

and

$$M_4(t) = \int_0^t \frac{\sigma_3 q(y)Z(s)}{Q(x)P(s)\phi + q(y)Z(s)} dB_3(s)$$

stands for local martingales, in which the quadratic variations are defined as

$$< M_3, M_3 >_t = \sigma_2^2 \int_0^t \left(\frac{Q(x)P(s)}{Q(x)P(s) + \frac{q(y)}{\phi}Z(s)}\right)^2 ds,$$

$$< M_4, M_4 >_t = \sigma_3^2 \int_0^t \left(\frac{\sigma_3 q(y)Z(s)}{Q(x)P(s)\phi + q(y)Z(s)}\right)^2 ds.$$

Using the strong law of large numbers leads to $\lim_{t\to\infty} \frac{M_i(t)}{t} = 0, \quad a.s., \quad i = 3, 4.$ Taking the limit supremum of (3.6) as $t \to \infty$ and (3.3), the following is obtained:

$$\lim_{t\to\infty} sup \frac{\log S(t)}{t} \le \mu(x)\frac{I}{a} - \frac{1}{2}[(m + s(x) + \frac{1}{2}\sigma_2^2) \wedge (\delta + \frac{1}{2}\sigma_3^2)] < 0, \quad a.s.,$$

which implies

$$\lim_{t\to\infty} P(t) = 0, \quad \lim_{t\to\infty} Z(t) = 0 \;\; a.s.$$

This indicates that the phytoplankton and zooplankton die out eventually. □

Remark Obviously, the phytoplankton and zooplankton will be extinct exponentially with probability one when the noise intensities are large enough. At the same time, the extinctions of phytoplankton and zooplankton modeled by (3.2) depend on stochastic noises σ_2 and σ_3, while σ_1 has no impact on the extinction of the species.

3.4 Numerical Simulations

In this section, some numerical simulations are given to verify the feasibility of our theoretical results by choosing appropriate parameters. Consider the model as given below:

$$\begin{cases} dN = (1.4 - 0.5N - \frac{x}{0.1x^2+0.2x+0.3} \times 0.8x^2 \times \frac{NP}{N+0.5})dt + \sigma_1 N dB_1(t), \\ dP = P(\frac{x}{0.1x^2+0.2x+0.3} \times \frac{N}{N+0.5} - 0.35 - x^2 - 0.3\exp(-\frac{1}{2}(x-0.2y)^2)Z)dt \\ +\sigma_2 P dB_2(t), \\ dZ = Z(0.9 \times \frac{0.8x^2}{0.1y^{1.5}} \times 0.3\exp(-\frac{1}{2}(x-0.2y)^2)P - 0.4)dt + \sigma_3 Z dB_3(t). \end{cases} \tag{3.6}$$

Case I. Let the intensities of white noises be $\sigma_1 = \sigma_2 = \sigma_3 = 0.5$, the initial condition $(N(0), P(0), Z(0)) = (1.5, 1.2, 0.5)$ and $x = 0.8$, $y = 1.5$. Then it yields $0.05 = \sigma_1 < \frac{4(\mu(x)\frac{I}{a} - (m+s(x)+\frac{1}{2}\sigma_2^2)(K+\frac{I}{a}))}{a(m+s(x)+\frac{1}{2}\sigma_2^2-\mu(x))^2} = 302.2305$, $1 = 2a > \sigma_1^2 = 0.0025$, $1.98 = 2(m+s(x)) > \sigma_2^2 = 0.0025$, $0.8 = 2\delta > \sigma_3^2 = 0.0025$, meaning that the conditions of Theorem 44 are all satisfied. Then model (3.6) is concluded as stationary distribution and ergodic (see Fig. 3.1). The left subgraph of Fig. 3.1 stands for the densities of nutrient, phytoplankton and zooplankton, meanwhile the right ones represent the frequency histograms for $N(t)$, $P(t)$ and $Z(t)$ at time $t = 100s$. Moreover, the increasing of the intensities of white noises to $\sigma_1 = \sigma_2 = \sigma_3 = 0.1$ in Fig. 3.2 also offers the stationary distribution. Comparing Fig. 3.1 with 3.2, it is easy to see that the increasing of white noises makes the larger fluctuation regions of the solutions.

Case II. In order to verify the effects of cell sizes x and y on the stochastic NPZ model, assume other coefficients as the same in Case I. Assume $x = 1.2$, $y = 1.5$ and $x = 1.2$, $y = 2$. Then left subgraph and right subgraph of Fig. 3.3 are obtained, showing that the reason of the extinction of species is the large cells of phytoplankton and zooplankton. In Fig. 3.4, let $x = 0.8$, $y = 0.9$ and $x = 0.5$, $y = 0.9$, respectively. Then the populations of phytoplankton and zooplankton can keep coexistence. Based on those figures, it is obvious that the persistence and extinction of species are affected by different cell sizes of phytoplankton and zooplankton and the plankton population density will be extinct as the increasing cell sizes of phytoplankton and zooplankton.

Case III. In order to obtain the effects of white noises on the extinction of phytoplankton and zooplankton, assume $\sigma_1 = 0.05$, $\sigma_2 = 3.05$, $\sigma_3 = 3.25$, and other parameters as the same in Case I. After computing, it yields $2.8206 = \frac{1}{2}(m + s(x) + \frac{1}{2}\sigma_2^2) > \frac{I}{a} = 2.8$ and $2.8406 = \frac{1}{2}(\delta + \frac{1}{2}\sigma_3^2) > \frac{I}{a} = 2.8$, satisfying the conditions of Theorem 45 and meaning that both the phytoplankton and zooplankton are extinct (see Fig. 3.5).

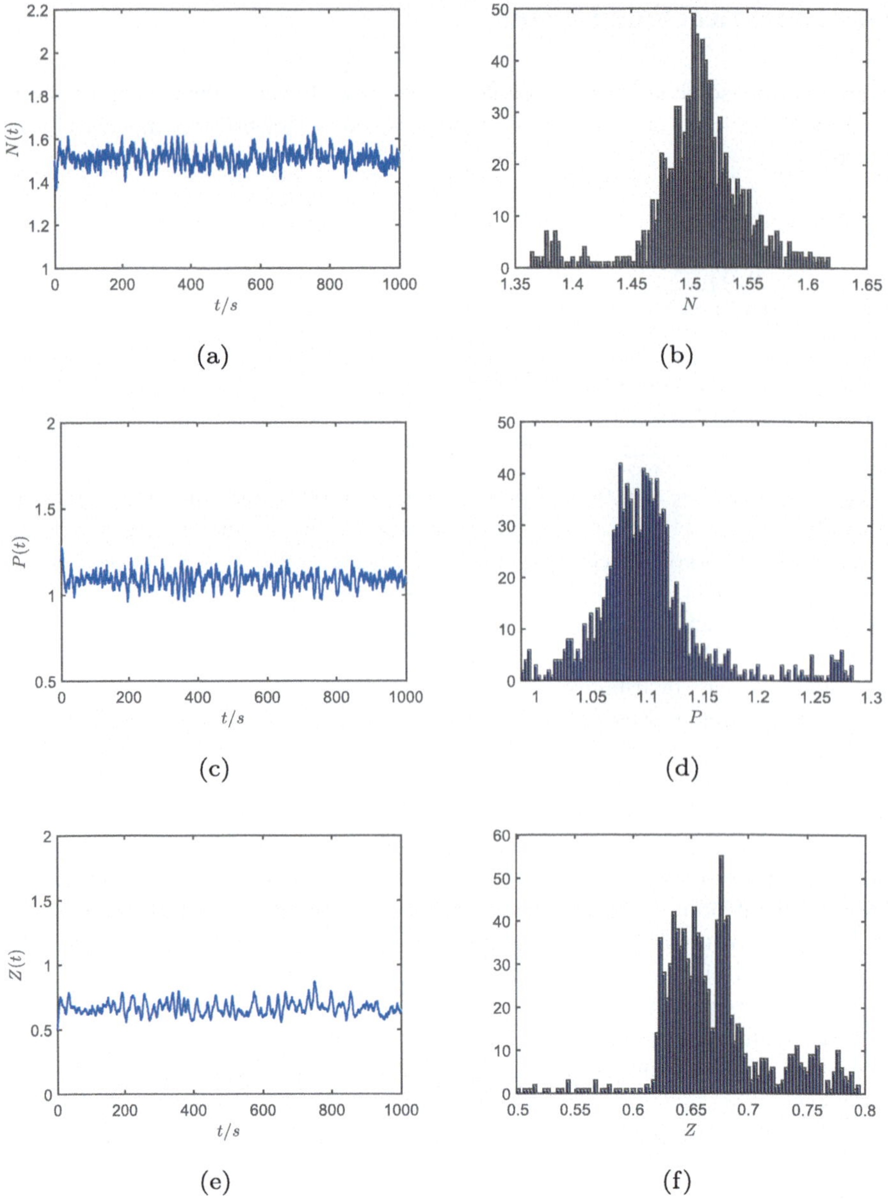

Fig. 3.1 (Left) The solution and (right) the density function of the model (3.6), where $\sigma_1 = \sigma_2 = \sigma_3 = 0.05$, $x = 0.8$, $y = 1.5$ and initial condition (1.5, 1.2, 0.5)

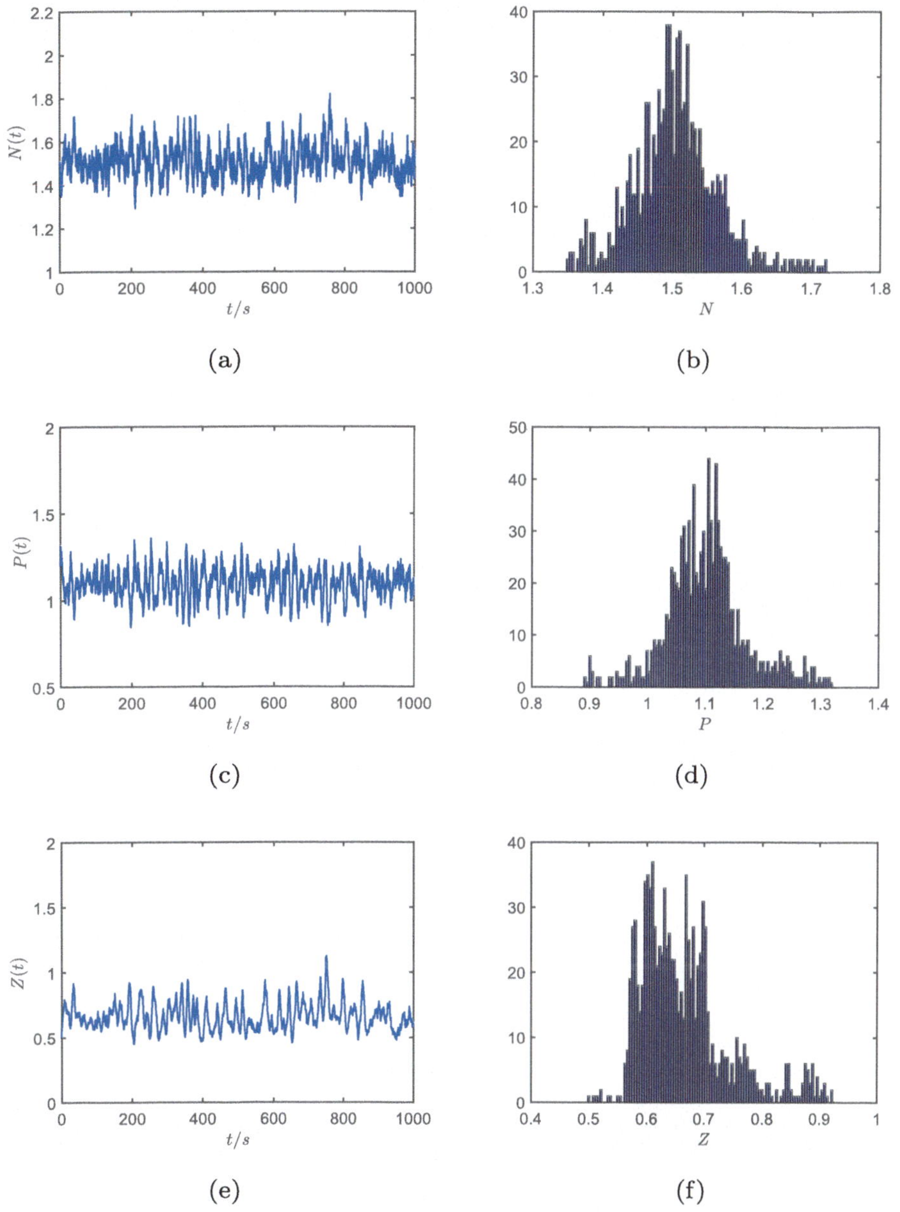

Fig. 3.2 (Left) The solution of model (3.6) and (right) the density function of the solution, where $\sigma_1 = \sigma_2 = \sigma_3 = 0.1$, $x = 0.8$, $y = 1.5$ and initial condition (1.5, 1.2, 0.5)

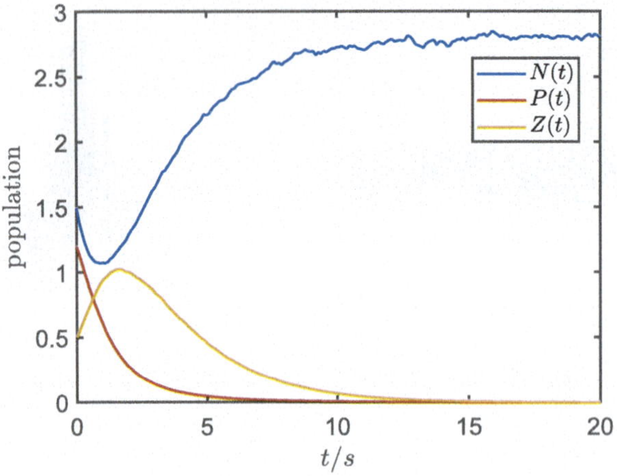

(a)

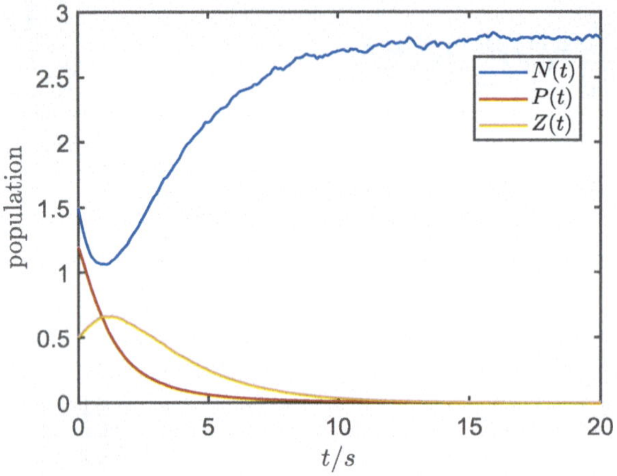

(b)

Fig. 3.3 (Left) The solution of model (3.6) with cell size $x = 1.2$, $y = 1.5$ and (right) $x = 1.2$, $y = 2$ with $\sigma_1 = \sigma_2 = \sigma_3 = 0.05$ and initial condition $(1.5, 1.2, 0.5)$

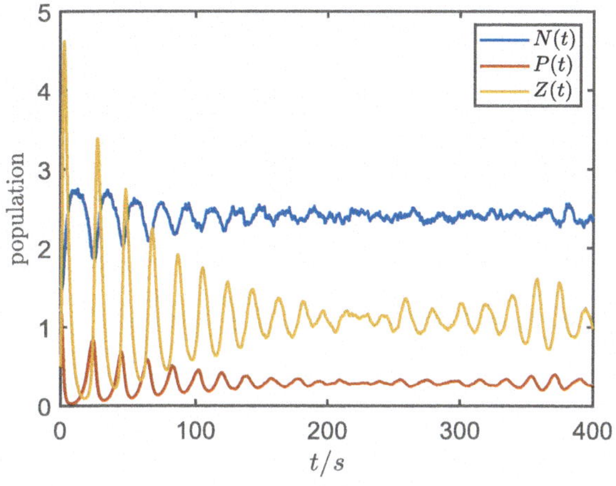

(a)

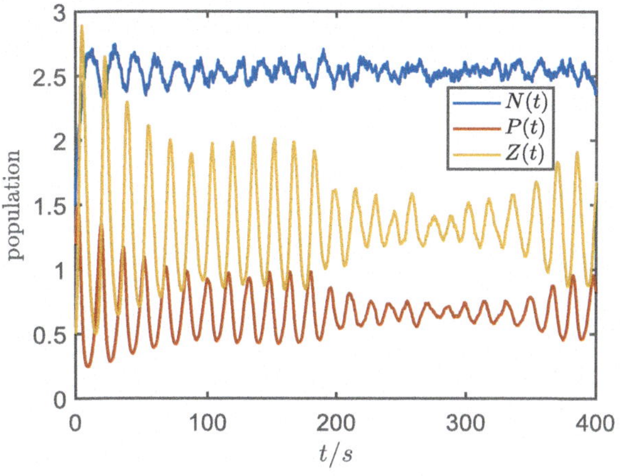

(b)

Fig. 3.4 (Left) The solution of model (3.6) with cell size $x = 0.8$, $y = 0.9$ and (right) $x = 0.5$, $y = 0.9$ with $\sigma_1 = \sigma_2 = \sigma_3 = 0.05$ and initial condition (1.5, 1.2, 0.5)

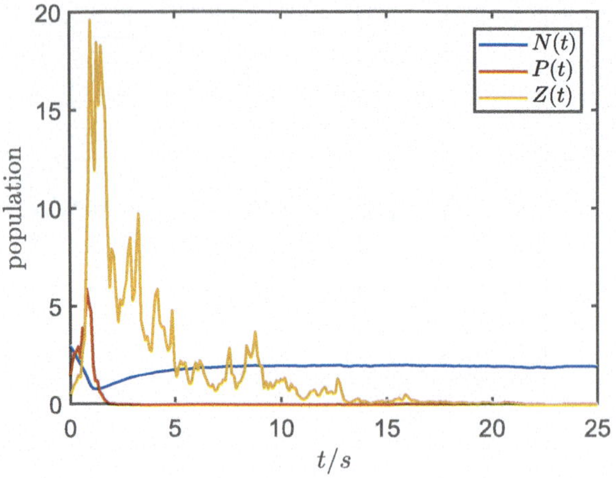

Fig. 3.5 The solution of model (3.6), where $\sigma_1 = 0.05$, $\sigma_2 = 3.05$, $\sigma_3 = 3.25$, $x = 0.8$, $y = 1.5$ and initial condition (1.5, 1.2, 0.5)

Chapter 4
The Effect of Marine Environmental Mutation

Plankton, as an essential producer of the marine ecosystem, like a primary role of a food web, occurs in many applications in material cycle and energy transportation. Moreover, it has applied value in environmental motoring and aquaculture. Based on intrinsic biological discipline, many factors like an ocean current, temperature, nutrients, and perturbation affect the growth and reproduction of plankton, displaying some abrupt changes in a short duration at certain moments. Compared with the whole evolution process, the state of plankton might experience sudden changes according to such fluctuating environment, in which the short duration of these sudden changes can be negligible. Thus these changes can be quantified mathematically by the form of impulse. Furthermore, impulsive differential equations can be used to well-approximate species variation in the plankton ecosystem. With their fast development, more and more accurate mathematical whimsical ecological models exist to dynamically investigate the behavior of biological systems [41–46].

In this chapter, a nonautonomous impulsive plankton model is considered with the mutual help of prey. The relation between solutions of the impulsive system and the corresponding non-impulsive system is used to get sufficient conditions ensuring permanence and global attractivity. Also, the conditions are proposed to study the question of which species of system are driven to extinction. At last, the main results are verified by the numerical simulations.

4.1 Model Description

The plankton ecosystem shows interest as two teams of preys compete with each other by allelopathy in the same habitat or nutrient without a predator. When a predator exists, the two preys will attack the predator, and the damage caused can be seen as a form of help for the other prey in some sense. In the following, the influence of impulse on a plankton ecosystem is presented in which there includes

S. T. Liu and L. Zhang, *Pattern Dynamics of Marine Plankton Behavior*,
https://doi.org/10.1007/978-981-97-5369-7_4

one predator and two kinds of prey cooperating in resisting being caught when the predator is present. An autonomous model without impulse has been applied to study the stability of this type of biosystem based on Tripathis investigation [47]. Then the following nonautonomous impulsive system is obtained considering the varying parameters:

$$\begin{cases} \dot{x}(t) = x(t)(a(t) - a(t)x(t) - a_1(t)y(t) - a_2(t)z(t) + a_3(t)y(t)z(t)), \\ \dot{y}(t) = y(t)(b(t) - b(t)y(t) - b_1(t)x(t) - b_2(t)z(t) + b_3(t)x(t)z(t)), \quad t \neq \tau_k \\ \dot{z}(t) = z(t)(-c(t) - c_1(t)z(t) + c_2(t)x(t) + c_3(t)y(t)), \\ x(\tau_k^+) = h_{1k}x(\tau_k), \\ y(\tau_k^+) = h_{2k}y(\tau_k), \\ z(\tau_k^+) = h_{3k}z(\tau_k), \quad k \in N_+, \end{cases} \tag{4.1}$$

where $x(t),\ y(t),\ z(t)$ stand for population densities of two kinds of preys and one predator at time t; $a(t), b(t), c(t), a_i(t), b_i(t),\ c_i(t)\ (i = 1, 2, 3)$ mean continuous functions bounded by positive constants on $[0, +\infty)$; impulse coefficients h_{ik} $(k = 1, 2, 3 \cdots)$ represent positive; $0 < \tau_1 < \tau_2 < \tau_3 < \cdots < \tau_k < \cdots$ means impulsive points satisfying $\lim_{k \to +\infty} \tau_k = +\infty$; and the initial values $x(0),\ y(0),\ z(0)$ are positive from the view of ecology.

The following is the ecological meaning of system (4.1):

(i) $x(t)(a(t) - a(t)x(t))$ and $y(t)(b(t) - b(t)y(t))$ mean that the preys grow logistically when the predator is absent.
(ii) $a_1(t)$ and $b_1(t)$ represent the competition strength of two preys without predator separately.
(iii) $a_2(t)$ and $b_2(t)$ denote the impact of predation on the two preys, respectively.
(iv) $a_3(t)$ and $b_3(t)$ stand for the benefits of mutual assistance between two preys.
(v) When there is no prey, $c(t)$ expresses the mortality rate of predator and $c_1(t)$ denotes the death of predator due to intraspecific competition.
(vi) $c_2(t)x(t)z(t)$ and $c_3(t)y(t)z(t)$ formulate the benefits of predator from capturing preys.

It is thereby shown that the following ecological factors related with the proposed system (4.1) are the focus in this chapter:

(i) Different rates $a_3(t)$ and $b_3(t)$ express the benefits of co-operation between two preys.
(ii) Assume that the function of help between two preys is weaker than the function of competition and predation.

4.2 Preliminaries

Some useful lemmas and definitions are given in this section.

Assume the given continuous function $f(t)$, and let f^L and f^M denote $\inf\limits_{0\le t<+\infty} f(t)$ and $\sup\limits_{0\le t<+\infty} f(t)$, respectively.

It is in addition to the following lemmas based on the fundamental theories of impulsive differential equations:

Lemma 5 *There is a unique solution*

$$(x(t, x_0), y(t, y_0),$$

$z(t, z_0)) \in PC([0, +\infty), \mathbb{R}^3)$, *for each given initial value* $(x_0, y_0, z_0) \in \mathbb{R}^3$, *where*

$$PC([0, +\infty), \mathbb{R}^3) = \{f : [0, +\infty) \to R^3,$$

$f(t)$ *is a continuous function when* $t \neq \tau_k$, *and* $f(\tau_k^-) = f(\tau_k)$. $f(\tau_k^+)$ *exists,* $k = 1, 2, 3 \cdots \}$.

Consider the following system:

$$\begin{aligned}
\dot{X}(t) &= X(t)(a(t) - \alpha_1(t)X(t) - \alpha_2(t)Y(t) - \alpha_3(t)Z(t) + \alpha_4(t)Y(t)Z(t)),\\
\dot{Y}(t) &= Y(t)(b(t) - \beta_1(t)X(t) - \beta_2(t)Y(t) - \beta_3(t)Z(t) + \beta_4(t)X(t)Z(t)),\\
\dot{Z}(t) &= Z(t)(-c(t) + \gamma_1(t)X(t) + \gamma_2(t)Y(t) - \gamma_3(t)Z(t)).
\end{aligned} \tag{4.2}$$

$$\alpha_1(t) = a(t) \prod_{0<\tau_k<t} h_{1k},\ \beta_1(t) = b_1(t) \prod_{0<\tau_k<t} h_{1k},\ \gamma_1(t) = c_2(t) \prod_{0<\tau_k<t} h_{1k},$$

$$\alpha_2(t) = a_1(t) \prod_{0<\tau_k<t} h_{2k},\ \beta_2(t) = b(t) \prod_{0<\tau_k<t} h_{2k},\ \gamma_2(t) = c_3(t) \prod_{0<\tau_k<t} h_{2k},$$

$$\alpha_3(t) = a_2(t) \prod_{0<\tau_k<t} h_{3k},\ \beta_3(t) = b_2(t) \prod_{0<\tau_k<t} h_{3k},\ \gamma_3(t) = c_1(t) \prod_{0<\tau_k<t} h_{3k},$$

$$\alpha_4(t) = a_3(t) \prod_{0<\tau_k<t} h_{2k}h_{3k},\ \beta_4(t) = b_3(t) \prod_{0<\tau_k<t} h_{1k}h_{3k}.$$

Lemma 6 *[48]* $X(t) > 0$, $Y(t) > 0$,$Z(t) > 0$ *hold for all* $t \ge 0$, *where* $(X(t), Y(t), Z(t))^T$ *is the arbitrary solution of system (4.2) subject to the initial positive values* $X(0)$, $Y(0)$, $Z(0)$.

Proof It yields from system (4.2) that

$$\begin{aligned}&X(t)=X(0)exp\{\int_0^t[a(s)-\alpha_1(s)X(s)-\alpha_2(s)Y(s)-\alpha_3(s)Z(s)\\&+\alpha_4(s)Y(s)Z(s)]ds\},\\&Y(t)=Y(0)exp\{\int_0^t[b(s)-\beta_1(s)X(s)-\beta_2(s)Y(s)-\beta_3(s)Z(s)\\&+\beta_4(s)X(s)Z(s)]ds\},\\&Z(t)=Z(0)exp\{\int_0^t[-c(s)+\gamma_1(s)X(s)+\gamma_2(s)Y(s)-\gamma_3(s)Z(s)]ds\}.\end{aligned}$$

It is obvious that $X(t),\ Y(t),\ Z(t)$ will be positive as $X(0)>0,\ Y(0)>0,\ Z(0)>0$.

This is the end of the proof of Lemma 6. □

Motivated by the biological background of system (4.2), its initial values are assumed to be positive.

Lemma 7 ([48]). *Consider system (4.1) and system (4.2). Then the following conclusions can be drawn:*

(i) $(X(t),\ Y(t),\ Z(t))^T=(\prod_{0<\tau_k<t}h_{1k}^{-1}x(t),\ \prod_{0<\tau_k<t}h_{2k}^{-1}y(t),\ \prod_{0<\tau_k<t}h_{3k}^{-1}z(t))^T$ *is a solution of system (4.2), where* $(x(t),\ y(t),\ z(t))^T$ *is the solution of system (4.1).*

(ii) $(x(t),\ y(t),\ z(t))^T$ $=(\prod_{0<\tau_k<t}h_{1k}X(t),\ \prod_{0<\tau_k<t}h_{2k}Y(t),\ \prod_{0<\tau_k<t}h_{3k}Z(t))^T$ *is a solution of system (4.1), where* $(X(t),\ Y(t),\ Z(t))^T$ *is the solution of system (4.2).*

Proof (i) Define

$$(X(t),\ Y(t),\ Z(t))^T=(\prod_{0<\tau_k<t}h_{1k}^{-1}x(t),\ \prod_{0<\tau_k<t}h_{2k}^{-1}y(t),\ \prod_{0<\tau_k<t}h_{3k}^{-1}z(t))^T.$$

Then substitute

$$(x(t),y(t),z(t))^T=(\prod_{0<\tau_k<t}h_{1k}X(t),\ \prod_{0<\tau_k<t}h_{2k}Y(t),\ \prod_{0<\tau_k<t}h_{3k}Z(t))^T$$

into system (4.1). It is easy to prove that $(X(t),Y(t),Z(t))^T$ is a solution of system (4.2) in the case of $t\neq\tau_k$ $(k\in N_+)$. When $t=\tau_k$,

$$\begin{aligned}X(\tau_k^+)&=\lim_{t\to\tau_k^+}\prod_{0<\tau_k<t}h_{1k}^{-1}x(\tau_k^+)\\&=\prod_{0<\tau_j\le\tau_k}h_{1j}^{-1}x(\tau_k^+)\\&=\prod_{0<\tau_j<\tau_k}h_{1j}^{-1}x(\tau_k)=X(\tau_k),\end{aligned}$$

$$X(\tau_k^-)=\lim_{t\to\tau_k^-}\prod_{0<\tau_k<t}h_{1k}^{-1}x(\tau_k^-)=\prod_{0<\tau_j<\tau_k}h_{1j}^{-1}x(\tau_k)=X(\tau_k).$$

It is obvious to check $Y(\tau_k^+) = Y(\tau_k^-) = Y(\tau_k)$ and $Z(\tau_k^+) = Z(\tau_k^-) = Z(\tau_k)$.

As $X(t)$, $Y(t)$, $Z(t)$ are continuous on $[0, +\infty)$, $(X(t), Y(t), Z(t))^T$ satisfies system (4.2).

(ii) According to the above proofs of (i), it can be easily verified that $(x(t), y(t), z(t))^T$ satisfies the system (4.1) when $t \neq \tau_k$. As $t = \tau_k$, it yields

$$x(\tau_k^+) = \lim_{t\to\tau_k^+} \prod_{0<\tau_k<t} h_{1k}X(t) = h_{1k} \prod_{0<\tau_j<\tau_k} h_{1j}X(\tau_k) = h_{1k}x(\tau_k).$$

$$x(\tau_k^-) = \lim_{t\to\tau_k^-} \prod_{0<\tau_k<t} h_{1k}X(t) = \prod_{0<\tau_j<\tau_k} h_{1j}X(\tau_k) = x(\tau_k).$$

Similarly, $y(\tau_k^+) = h_{2k}y(\tau_k)$, $y(\tau_k^-) = y(\tau_k)$, $z(\tau_k^+) = h_{3k}z(\tau_k)$ and $z(\tau_k^-) = z(\tau_k)$.

Then, it can be proved that $(x(t), y(t), z(t))^T$ is a solution of system (4.1). This is the end of the proof of Lemma 7. □

Lemma 8 ([47]). $\limsup_{t\to\infty} u(t) \leq \frac{q}{p}$ ($\liminf_{t\to\infty} u(t) \geq \frac{q}{p}$), *when* $p > 0$, $q > 0$ *and* $\frac{du}{dt} \leq (\geq) u(t)(q - pu(t))$, $u(t_0) > 0$.

Definition 15 ([49]). The system (4.2) is said to be

(i) Permanent: when there exists positive constants d_i and D_i $(i = 1, 2, 3)$ satisfying

$$\begin{aligned} d_1 &\leq \liminf_{t\to\infty} X(t) \leq \limsup_{t\to+\infty} X(t) \leq D_1, \\ d_2 &\leq \liminf_{t\to\infty} Y(t) \leq \limsup_{t\to+\infty} Y(t) \leq D_2, \\ d_3 &\leq \liminf_{t\to\infty} Z(t) \leq \limsup_{t\to+\infty} Z(t) \leq D_3. \end{aligned}$$

(ii) Global attractive: $\lim_{t\to+\infty}(X_1(t) - X_2(t)) = 0$, $\lim_{t\to+\infty}(Y_1(t) - Y_2(t)) = 0$, $\lim_{t\to+\infty}(Z_1(t) - Z_2(t)) = 0$ hold, for any two solutions $(X_1(t), Y_1(t), Z_1(t))^T$ and $(X_2(t), Y_2(t), Z_2(t))^T$ of system (4.2).

4.3 Permanence and Global Attractivity

The following concerns the permanence and global attractivity as the important long-term behaviors of the above system.

Theorem 46 *Assume that:*

(i) there exists constants $m_i > 0,\ M_i > 0\ (i = 1,\ 2,\ 3)$ *satisfying*

$$m_i \le \prod_{0<\tau_k<t} h_{ik} \le M_i \ for\ all\ t > 0,$$

(ii) $a^L - a_1^M M_2 D_2 - a_2^M M_3 D_3 > 0,\ b^L - b_1^M M_1 D_1 - b_2^M M_3 D_3 > 0$ *and* $-c^M + c_2^L m_1 d_1 + c_3^L m_2 d_2 > 0.$

Then, the following conclusions can be obtained and applied to all solutions $(X(t),\ Y(t),\ Z(t))^T$ *of system (4.2):*

$$\begin{aligned} d_1 &\le \liminf_{t\to\infty} X(t) \le \limsup_{t\to+\infty} X(t) \le D_1, \\ d_2 &\le \liminf_{t\to\infty} Y(t) \le \limsup_{t\to+\infty} Y(t) \le D_2, \\ d_3 &\le \liminf_{t\to\infty} Z(t) \le \limsup_{t\to+\infty} Z(t) \le D_3, \end{aligned}$$

where

$$d_1 = \frac{a^L - a_1^M M_2 D_2 - a_2^M M_3 D_3}{a^M M_1},\ d_2 = \frac{b^L - b_1^M M_1 D_1 - b_2^M M_3 D_3}{b^M M_2},$$

$$d_3 = \frac{-c^M + c_2^L m_1 d_1 + c_3^L m_2 d_2}{c_1^M M_3}, D_1 = \frac{a^M}{a^L m_1},\ D_2 = \frac{b^M}{b^L m_2},$$

$$D_3 = \frac{c_2^M D_1 M_1 + c_3^M D_2 M_2}{c_1^L m_3}.$$

Proof The following conclusions can be drawn:

$$\begin{aligned} \dot{X}(t) &\le X(t)(a(t) - \alpha_1(t)X(t)), \\ &\le X(t)(a^M - a^L m_1 X(t)), \end{aligned}$$

based on system (4.2), Lemma 8, and the assumption that co-operation between two preys has weaker effect than competition and predation. Then it yields $\limsup_{t\to+\infty} X(t) \le \frac{a^M}{a^L m_1} = D_1$.

The following can be obtained by the similar fashion: $\limsup_{t\to+\infty} Y(t) \le \frac{b^M}{b^L m_2} = D_2$. For an arbitrary constant $\varepsilon > 0$, there exists a constant $T_1 > 0$ satisfying $X(t) < D_1 + \varepsilon$ and $Y(t) < D_2 + \varepsilon$ when $t > T_1$. Hence,

$$\begin{aligned}\dot{Z}(t) &\le Z(t)(\gamma_1(t)X(t)+\gamma_2(t)Y(t)-\gamma_3(t)Z(t)),\\ &\le Z(t)(c_2^M M_1(D_1+\varepsilon)+c_3^M M_2(D_2+\varepsilon)-c_1^L m_3 Z(t)).\end{aligned}$$

As $\varepsilon \to 0$, it yields $\limsup_{t\to+\infty} Z(t) \le \frac{c_2^M D_1 M_1 + c_3^M D_2 M_2}{c_1^L m_3} = D_3$.

When $t > T_2$, there also exists a constant $T_2 > 0$ such that $Z(t) < D_3 + \varepsilon$. Choose $T_3 = max\{T_1,\ T_2\}$ and let $t > T_3$. Then it yields $X(t) < D_1 + \varepsilon,\ Y(t) < D_2 + \varepsilon$ and $Z(t) < D_3 + \varepsilon$ hold. Therefore,

$$\begin{aligned}\dot{X}(t) &\ge X(t)(a(t)-\alpha_1(t)X(t)-\alpha_2(t)(D_2+\varepsilon)-\alpha_3(t)(D_3+\varepsilon)),\\ &\ge X(t)(a^L-a^M M_1 X(t)-a_1^M M_2(D_2+\varepsilon)-a_2^M M_3(D_3+\varepsilon)).\end{aligned}$$

When $\varepsilon \to 0$, it follows that $\liminf_{t\to+\infty} X(t) \ge \frac{a^L - a_1^M M_2 D_2 - a_2^M M_3 D_3}{a^M M_1} = d_1$.

Similarly, we can obtain $\liminf_{t\to+\infty} Y(t) \ge \frac{b^L - b_1^M M_1 D_1 - b_2^M M_3 D_3}{b^M M_2} = d_2$. When $t > T_4$, there exists a constant $T_4 > 0$ such that

$$\begin{aligned}\dot{Z}(t) &= Z(t)(-c(t)+\gamma_1(t)X(t)+\gamma_2(t)Y(t)-\gamma_3(t)Z(t)),\\ &\ge Z(t)(-c^M+c_2^L m_1(d_1-\varepsilon)+c_3^L m_2(d_2-\varepsilon)-c_1^M M_3 Z(t)).\end{aligned}$$

Let $\varepsilon \to 0$, it yields $\liminf_{t\to+\infty} Z(t) \ge \frac{-c^M + c_2^L m_1 d_1 + c_3^L m_2 d_2}{c_1^M M_3} = d_3$. This is the end of the proof of Theorem 46. □

Remark 1 From Theorem 46 and Lemma 7, it is obvious that there exists positive constants A and B such that $A < X(t),\ Y(t),\ Z(t) < B$ for all $t > 0$ and the system (4.1) is also permanent on account of the relation between solutions of system (4.1) and (4.2).

Theorem 47 *The system (4.2) is globally attractive when there exists a nonnegative continuous function $g(t)$ satisfying $\int_0^\infty g(t)dt = \infty$ and positive constants h_i^* ($i = 1, 2, 3$) such that:*

$$\begin{aligned}&h_1^*\alpha_1(t)-h_2^*\beta_1(t)-h_2^*\beta_4(t)B-h_3^*\gamma_1(t)\ge g(t),\\ &-h_1^*\alpha_2(t)-h_1^*\alpha_4(t)B+h_2^*\beta_2(t)-h_3^*\gamma_2(t)\ge g(t),\\ &-h_1^*\alpha_3(t)-h_1^*\alpha_4(t)B-h_2^*\beta_3(t)-h_2^*\beta_4(t)B+h_3^*\gamma_3(t)\ge g(t),\end{aligned}$$

with conditions of Theorem 46.

Proof Let $(X_1(t), Y_1(t), Z_1(t))^T$ and $(X_2(t), Y_2(t), Z_2(t))^T$ be two arbitrary solutions of system (4.2). Based on the mean-value theorem, it yields

$$\begin{aligned}&\tfrac{1}{B}|X_1(t)-X_2(t)| \le |lnX_1(t)-lnX_2(t)| \le \tfrac{1}{A}|X_1(t)-X_2(t)|,\\ &\tfrac{1}{B}|Y_1(t)-Y_2(t)| \le |lnY_1(t)-lnY_2(t)| \le \tfrac{1}{A}|Y_1(t)-Y_2(t)|,\\ &\tfrac{1}{B}|Z_1(t)-Z_2(t)| \le |lnZ_1(t)-lnZ_2(t)| \le \tfrac{1}{A}|Z_1(t)-Z_2(t)|.\end{aligned}$$

Assume the Lyapunov function listed below:

$$V(t) = h_1^*|lnX_1(t) - lnX_2(t)| + h_2^*|lnY_1(t) - lnY_2(t)| + h_3^*|lnZ_1(t) - lnZ_2(t)|,$$

where $h_i^* > 0$ $(i = 1, 2, 3)$. After calculating the right derivative $D^+(V)$ and denoting

$$X_1(t) - X_2(t) = \Delta X(t),\ Y_1(t) - Y_2(t) = \Delta Y(t),\ Z_1(t) - Z_2(t) = \Delta Y(t),$$

it yields

$$\begin{aligned}
&D^+(V)\\
&= h_1^* sgn(\Delta X(t))(\tfrac{\dot{X}_1(t)}{X_1(t)} - \tfrac{\dot{X}_2(t)}{X_2(t)}) + h_2^* sgn(\Delta Y(t))(\tfrac{\dot{Y}_1(t)}{Y_1(t)} - \tfrac{\dot{Y}_2(t)}{Y_2(t)})\\
&+h_3^* sgn(\Delta Z(t))(\tfrac{\dot{Z}_1(t)}{Z_1(t)} - \tfrac{\dot{Z}_2(t)}{Z_2(t)}),\\
&= h_1^* sgn(\Delta X(t))L_1(t) + h_2^* sgn(\Delta Y(t))L_2(t) + h_3^* sgn(\Delta Z(t))L_3(t),\\
&\leq h_1^* L_4(t) + h_2^* L_5(t) + h_3^* L_6(t),\\
&\leq (-h_1^*\alpha_1(t) + h_2^*(\beta_1(t) + \beta_4(t)Z_1(t)) + h_3^*\gamma_1(t))|\Delta X(t)| + (h_1^*(\alpha_2(t)\\
&+\alpha_4(t)Z_2(t))\\
&-h_2^*\beta_2(t) + h_3^*\gamma_2(t))|\Delta Y(t)| + (h_1^*(\alpha_3(t) + \alpha_4(t)Y_1(t))\\
&+h_2^*(\beta_3(t) + \beta_4(t)X_2(t)) - h_3^* r_3(t))|\Delta Z(t)|,\\
&\leq -g(t)(|X_1(t) - X_2(t)| + |Y_1(t) - Y_2(t)| + |Z_1(t) - Z_2(t)|),\\
&\leq -\phi g(t)(h_1^*|lnX_1(t) - lnX_2(t)| + h_2^*|lnY_1(t) - lnY_2(t)|\\
&+h_3^*|lnZ_1(t) - lnZ_2(t)|),\\
&\leq -\phi g(t)V,
\end{aligned}$$

where $\phi = \frac{A}{minh_i^*}$ and

$$\begin{aligned}
&L_1(t) = -\alpha_1(t)\Delta X(t) - \alpha_2(t)\Delta Y(t) - \alpha_3(t)\Delta Z(t) + \alpha_4(t)(Y_1(t)\Delta Z(t)\\
&+Z_2(t)\Delta Y(t)),\\
&L_2(t) = -\beta_1(t)\Delta X(t) - \beta_2(t)\Delta Y(t) - \beta_3(t)\Delta Z(t) + \beta_4(t)(Z_1(t)\Delta X(t)\\
&+X_2(t)\Delta Z(t)),\\
&L_3(t) = \gamma_1(t)\Delta X(t) + \gamma_2\Delta Y(t) - \gamma_3(t)\Delta Z(t),\\
&L_4(t) = -\alpha_1(t)|\Delta X(t)| + (\alpha_2(t) + \alpha_4(t)Z_2(t))|\Delta Y(t)|\\
&+(\alpha_3(t) + \alpha_4(t)Y_1(t))|\Delta Z(t)|,\\
&L_5(t) = (\beta_1(t) + \beta_4(t)Z_1(t))|\Delta X(t)| - \beta_2(t)|\Delta Y(t)|\\
&+(\beta_3(t) + \beta_4(t)X_2(t))|\Delta Z(t)|,\\
&L_6(t) = \gamma_1(t)|\Delta X(t)| + \gamma_2(t)|\Delta Y(t)| - \gamma_3(t)|\Delta Z(t)|.
\end{aligned}$$

Then, it is easy to obtain

$$V(t) \leq V(0)exp(-\phi \int_0^t g(s)ds).$$

Therefore as $t \to +\infty$, $V(t) \to 0$, illustrating $\lim_{t\to+\infty}(X_1(t) - X_2(t)) = 0$, $\lim_{t\to+\infty}(Y_1(t) - Y_2(t)) = 0$, $\lim_{t\to+\infty}(Z_1(t) - Z_2(t)) = 0$.

This is the end of the proof of Theorem 47. □

4.4 Extinction

In this section, our main results about extinction of system (4.1) are presented. The following theorem is enlightened by [49].

Theorem 48 *As* $t \to +\infty$, $(x(t), y(t), z(t)) \to (0, 0, 0)$, *meaning that the species will be extinct when there exists positive constants* ω_i $(i = 1, 2, 3)$ *satisfying:*

$$\liminf_{t\to+\infty}\left(\int_t^{t+\omega_1} a(s)ds\right) > 0,\ \limsup_{t\to+\infty}\left(\int_t^{t+\omega_1} a(s)ds + \sum_{t\le\tau_k<t+\omega_1} lnh_{1k}\right) \le 0,$$
$$\liminf_{t\to+\infty}\left(\int_t^{t+\omega_2} b(s)ds\right) > 0,\ \limsup_{t\to+\infty}\left(\int_t^{t+\omega_2} b(s)ds + \sum_{t\le\tau_k<t+\omega_2} lnh_{2k}\right) \le 0,$$
$$\liminf_{t\to+\infty}\left(\int_t^{t+\omega_3} c_1(s)ds\right) > 0,\ \limsup_{t\to+\infty}\left(\int_t^{t+\omega_3} c(s)ds + \sum_{t\le\tau_k<t+\omega_3} lnh_{3k}\right) \le 0,$$

and $h_i(t, \lambda) = |\sum_{t\le\tau_k<t+\lambda} lnh_{ik}| < H_i$ *where* $0 < \lambda < \omega_i$.

Proof According to the above conditions, it is easy to obtain that there exists positive constants δ_1 and T_1 satisfying when $t > T_1$,

$$\int_t^{t+\omega_1}(a(s) - a(s)k_1)ds + \sum_{t\le\tau_k<t+\omega_1} lnh_{1k} < -\delta_1,$$

for an arbitrary constant $k_1 > 0$. Define

$$\begin{aligned} \dot{u}(t) &= u(t)(a(t) - a(t)u(t)) \quad t \ne \tau_k, \\ u(\tau_k^+) &= h_{1k}u(\tau_k) \quad k \in N^+. \end{aligned} \tag{4.3}$$

Then from the comparison theorem $x(t) \le u(t)$ when $\dot{x}(t) \le \dot{u}(t)$.

Next, $u(t) \to 0$ would be proved as $t \to +\infty$ for the following cases:

Case I: There exists a constant $t_1 > T_1^* > 0$ satisfying $u(t) \ge k_1$ for all $t > t_1$.

Case II: There exists a constant $t_1 > T_1^* > 0$ satisfying $u(t) \le k_1$ for all $t > t_1$.

Case III: When $t > T_1^*$, $u(t)$ oscillates around k_1.

For Case I, let $t = t_1 + l_1\omega_1$, l_1 is a positive integer. After integrating system (4.3) from t_1 to t, it yields

$$
\begin{aligned}
u(t) &= u(t_1)exp(\textstyle\int_{t_1}^{t}(a(s) - a(s)u(s))ds + \sum_{t_1 \le \tau_k < t} lnh_{1k}), \\
&\le u(t_1)exp(\textstyle\int_{t_1}^{t_1+\omega_1}(a(s) - a(s)k_1)ds + \sum_{t_1 \le \tau_k < t_1+\omega_1} lnh_{1k} \\
&+ \cdots + \textstyle\int_{t_1+(l_1-1)\omega_1}^{t_1+l_1\omega_1}(a(s) - a(s)k_1)ds + \sum_{t_1+(l_1-1)\omega_1 \le \tau_k < t_1+l_1\omega_1} lnh_{1k}), \\
&\le u(t_1)exp(-l_1\delta_1).
\end{aligned} \tag{4.4}
$$

Obviously when $l_1 \to +\infty$, $u(t) \to 0$ which leads to a contradiction.

For case III, assume two sequences $\{\varphi_n\}$ and $\{\varphi_n^*\}$ satisfying $T_1^* < \varphi_1 < \varphi_1^* < \varphi_2 < \varphi_2^* < \cdots$ and $\lim\limits_{t\to+\infty} \varphi_n = \lim\limits_{t\to+\infty} \varphi_n^* = +\infty$.

$$
\begin{aligned}
&u(\varphi_n) \ge k_1,\ u(\varphi_n^+) \le k_1,\ u(\varphi_n^*) \le k_1,\ u(\varphi_n^{*+}) \ge k_1. \\
&u(t) \le k_1 \text{ for all } t \in (\varphi_n, \varphi_n^*);\ u(t) \ge k_1\ for\ all\ t \in (\varphi_n^*, \varphi_{n+1}).
\end{aligned} \tag{4.5}
$$

There exists positive constants l_2 and λ such that $t = \varphi_n^* + l_2\omega_1 + \lambda(\lambda \in [0, \omega_1))$, for any t satisfying $T_1^* < \varphi_n^* < t < \varphi_{n+1}$. Then it yields

$$
\begin{aligned}
u(t) &= u(\varphi_n^*)exp(\textstyle\int_{\varphi_n^*}^{t}(a(s) - a(s)u(s))ds + \sum_{\varphi_n^* \le \tau_k < t} lnh_{1k}), \\
&\le k_1 exp(\textstyle\int_{\varphi_n^*}^{\varphi_n^*+\omega_1}(a(s) - a(s)k_1)ds + \sum_{\varphi_n^* \le \tau_k < \varphi_n^*+\omega_1} lnh_{1k} \\
&+ \ldots + \textstyle\int_{\varphi_n^*+l_2\omega_1}^{\varphi_n^*+l_2\omega_1+\lambda}(a(s) - a(s)k_1)ds + \sum_{\varphi_n^*+l_2\omega_1 \le \tau_k < \varphi_n^*+l_2\omega_1+\lambda} lnh_{1k}), \\
&\le k_1 exp(\lambda a^M|1 - k_1| + H_1).
\end{aligned} \tag{4.6}
$$

It is obtained that $u(t) \le k_1 \le k_1 exp(\lambda a^M|1 - k_1| + H_1)$ holds when $T_1^* < \varphi_n < t < \varphi_n^*$.

In either case (Case II or Case III), it can always yield $u(t) \le k_1 exp(\lambda a^M|1 - k_1| + H_1)$ for all $t > T_1^*$. Because of the arbitrariness of k_1, we can get $u(t) \to 0$ as $t \to +\infty$.

From $x(t) \le u(t)$, it is easy to get as $x(t) \to 0$ as $t \to +\infty$. In the similar fashion, $y(t) \to 0,\ t \to +\infty$. Hence, there exists a positive constant T_2^* satisfying when $t > T_2^*$, $x(t),\ y(t) < \varepsilon$. for any constant $\varepsilon > 0$. And there are positive constants T_3^* and δ_3 such that for any $t > T_3^*$

$$
\int_t^{t+\omega_3} (c(s) - c_1(s)k_3 + c_2^M\varepsilon + c_3^M\varepsilon)\, ds + \sum_{t \le \tau_k < t+\omega_3} lnh_{3k} < -\delta_3,
$$

for any $k_3 > 0$ when ε is small enough. When $t > T_2^*$, it yields the following inequality:

$$
\begin{aligned}
\dot{z}(t) &\le z(t)(c(t) - c_1(t)z(t) + c_2(t)x(t) + c_3(t)y(t)), \\
&\le z(t)(c(t) - c_1(t)z(t) + c_2^M\varepsilon + c_3^M\varepsilon).
\end{aligned}
$$

When $t \to +\infty$, we can get $z(t) \to 0$ with the similar argument as the above proofs.

Then the three species will be extinct under the conditions of Theorem 48 and this is the end of the proof.

Remark 2 Based on the conditions of Theorem 48, it is easy to get that the species of system may be extinct as the effect of impulse is sharply offensive,.

4.5 Numerical Simulations

In the following, examples of extinctive systems and permanent systems under the conditions are given. Based on the difficulty of obtaining practical data, some hypothetical parametric values illustrate our analytical findings.
In system (4.1), let

$$a(t) = b(t) = c_1(t) = 0.1sint + 0.8,\ c_2(t) = c_3(t) = 0.01sint + 0.4,$$

$$a_3(t) = b_3(t) = 0.01sint + 0.01,\ h_{ik} = 1.7^{\frac{1}{k+1}-\frac{1}{k}}\ (k \in N_+,\ i = 1, 2, 3),$$

$$c(t) = a_1(t) = b_1(t) = a_2(t) = b_2(t) = 0.01sint + 0.1,\ \tau_k = 30 + 5k.$$

From Theorem 46, it is easy to know that

$$m_i = 0.58,\ M_i = 0.77,\ D_1 = D_2 = \frac{a^M}{a^L m_1} = \frac{0.9}{0.7 \times 0.58} = 2.22,$$

$$D_3 = \frac{c_2^M D_1 M_1 + c_3^M D_2 M_2}{c_1^L m_3} = \frac{0.41 \times D_1 \times 0.77 \times 2}{0.58 \times 0.77} = 3.45,$$

$$d_1 = d_2 = \frac{a^L - a_1^M M_2 D_2 - a_2^M M_3 D_3}{a^M M_1} = \frac{0.7 - 0.11 \times 0.77 \times D_1 - 0.11 \times 0.77 \times D_3}{0.9 \times 0.77} = 0.32,$$

$$d_3 = \frac{-c^M + c_2^L m_1 d_1 + c_3^L m_2 d_2}{c_1^M M_3} = \frac{-0.11 + 0.58 \times d_1 \times 0.39 \times 2}{0.9 \times 0.77} = 0.05.$$

From Theorem 47, assume $h_1^* = h_2^* = h_3^* = 1,\ B = 3.5,\ g(t) = 0.12$. Figure 4.1 shows the simulation results. Theorems 46 and 47 tell us that when the intrinsic growth of species x and y and the growth rate of predator z are relatively large while the impulsive effect is relatively small, system (4.1) will probably be permanent and achieve global attractive. Based on the results of Fig. 4.1, all the solutions with different initial values will eventually be same and permanent, where the parameters of system (4.1) satisfy the conditions of Theorems 46 and 47.

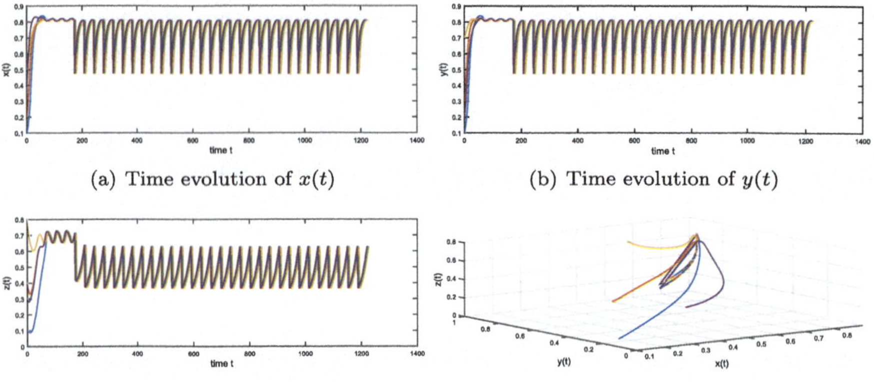

(a) Time evolution of $x(t)$ (b) Time evolution of $y(t)$

(c) Time evolution of $z(t)$ (d) Phase portrait of system (4.1)

Fig. 4.1 Solutions of system (4.1) with the same parameters and the listed initial values (0.1,0.1,0.1), (0.2,0.3,0.4), (0.5,0.7,0.8), (0.4,0.2,0.3)

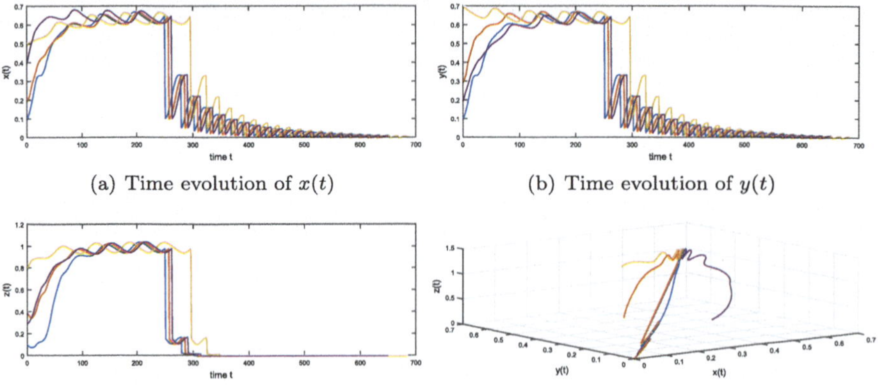

(a) Time evolution of $x(t)$ (b) Time evolution of $y(t)$

(c) Time evolution of $z(t)$ (d) Phase portrait of system (4.1)

Fig. 4.2 Solutions of system (4.1) with the same parameters and the listed initial values (0.1,0.1,0.1), (0.2,0.3,0.4), (0.5,0.7,0.8), (0.4,0.2,0.3)

In system (4.1), let

$$a(t) = b(t) = c(t) = 0.3sint + 0.303, \ \tau_k = 30 + 5k \ (k \in N_+),$$

$$c_1(t) = c_2(t) = c_3(t) = sint + 1.1, \ a_3(t) = b_3(t) = 0.01sint + 0.11,$$

$$a_1(t) = b_1(t) = a_2(t) = b_2(t) = 0.1sint + 0.11, \ w_i = 1, \ h_{ik} = 0.16,$$

$H_i = 2(i = 1, 2, 3)$. Based on Theorem 48, Fig. 4.2 shows the extinction of species of this system. A high impulsive effect to a certain degree means that the plankton ecosystem suffers tremendous impacts such as extensive harvesting, environmental pollution, and loss of habitat. So the system tends toward extinction. Hence, the impulsive effect plays a significant role in the evolution of a system. Based on this result, reasonable maintenance of biological systems can be achieved.

Part II
Stability Analysis of Plankton System

Chapter 5
The Effect of Time Delay

As we all know, phytoplankton can produce harmful algal blooms, seriously affecting many areas' marine ecology, environment, and economy. So to understand the intrinsic mechanism of harmful algal blooms, studying the dynamic behavior of the phytoplankton–zooplankton system becomes significant and urgent [50]. Generally speaking, there are two different approaches for its research: one is from experimental data analysis, and the other is from theoretical analysis [15, 51–56]. From the theoretical view, many functional response functions have been developed to recognize the relationship among species in the phytoplankton–zooplankton system. They include Holling type [57–59], Ivlev type [60], ratio-dependent type [61], Crowley–Martin type [62], Beddington–DeAngelis type [63, 64], and Hassell–Varley type [65]. From the first approach, work [66] based on the experimental data found out that the biological characteristic of the Beddington–DeAngelis-type functional response function is more approximate and also adds the drawback of the singular phenomenon in a low-density state.

The dynamic behavior of phytoplankton–zooplankton system with cell size and time delay is investigated inspired by the recent works [35, 59, 64]. Remarkably, the maximum growth rate relies on the cell size for phytoplankton so that the Beddington–DeAngelis-type functional response function is used to specify the relationship between phytoplankton and zooplankton, making the system more comprehensive and accurate to express the real-world situation. It is obvious that the function is essentially different from those related literature [17, 62, 67]. Since the cell size plays critical role in the growth of phytoplankton–zooplankton system, many works focus on this issue (see, e.g., [35, 36, 68]). However, these works only focus on experimental data analysis instead of theoretical analysis. For its hypothetical case, the positiveness and boundedness of the solution are first derived, and the asymptotic stability of the coexistent equilibrium and Hopf bifurcation is studied using the associated characteristic equation. Finally, the direction of the Hopf bifurcation and the strength of the bifurcated periodic solution are determined.

S. T. Liu and L. Zhang, *Pattern Dynamics of Marine Plankton Behavior*,
https://doi.org/10.1007/978-981-97-5369-7_5

5.1 System Model and Problem Formulation

In the following, the effect of body size on dynamic behavior analysis of phytoplankton–zooplankton system is explored. Let the phytoplankton and zooplankton population be characterized as follows, respectively: population density u_1, cell size x, u_2 and y. This chapter focuses on the effect of body-size-dependent population as the first attempt and the effect of the other abiotic variables influence for phytoplankton and zooplankton growth is not our aim. As a result, the following phytoplankton–zooplankton system is investigated:

$$\begin{cases} \dot{u}_1 = u_1\left(r_1(x) - s(x) - \alpha_1 u_1 - \dfrac{C(x,y)u_2}{a+u_1+bu_2}\right), \\ \dot{u}_2 = -r_2 u_2 + \dfrac{C(x,y)u_1(t-\tau)u_2(t-\tau)}{a+u_1(t-\tau)+bu_2(t-\tau)}, \\ u_1(\theta) = \psi_1(\theta) > 0,\ u_2(\theta) = \psi_2(\theta) > 0,\ \theta \in [-\tau, 0], \end{cases} \tag{5.1}$$

where ψ_1, ψ_2 are initial conditions; $\tau > 0$ symbolizes time delay for gestation of zooplankton; $s(x)$ denotes the sinking rate and $s(x) = \alpha_2 x^2$ [35] $\big(r_1(x) > s(x)\big)$ with sinking rate coefficient α_2; α_1 expresses the crowding effect parameter caused by the intraspecific competition with peers of the same species. The function $\dfrac{C(x,y)u_1}{a+u_1+bu_2}$ stands for the Beddington–DeAngelis functional response; the parameter a signifies a measure of the abundance of phytoplankton and zooplankton relative to the interacted environment and b indicates zooplankton interference [69]; and $C(x,y)$ denotes the zooplankton consumption rate. Moreover, empirical evidence illustrates that zooplankton consumption rate gets maximum when zooplanktons feed on particles with an optimal phytoplankton size ratio [70]. As a result, we formulate $C(x,y)$ as

$$C(x,y) = C_m \exp\left(-\frac{1}{\lambda}(x-\kappa y)^2\right), \tag{5.2}$$

where λ denotes consumption rate coefficient, κ means optimal predator–prey ratio, and C_m is maximum consumption rate. In the following, assume the maximum consumption value is $C(x,y) = C_m$ with $x = \kappa y$. r_2 stands for the natural mortality rate for zooplankton population; $r_1(x)$ specifies the maximum phytoplankton growth rate. And the resource uptake kinetics [74] and empirical observations [71–73] find out a relationship between growth rate and cell size, satisfying as below:

$$r_1(x) = \frac{x}{c_1x^2 + c_2x + c_3}, \tag{5.3}$$

where c_1, c_2 and c_3 are constants.

Next, by considering the effect of time delay and cell size, the dynamic behavior of system (5.1) with (5.2) and (5.3) will be investigated in detail.

5.2 Stability of Equilibrium and Existence of Hopf Bifurcation

The local and global stability of coexisting equilibrium and the existence of Hopf bifurcation are studied in this section. Note that the positiveness and boundedness of the solution of phytoplankton–zooplankton system are essential prerequisites. Consequently, the following lemma is given as below firstly.

Lemma 9 *The solution* $(u_1(t), u_2(t))$ *of system (5.1) is positive and uniformly bounded when considering system (5.1) with (5.2), (5.3) and initial values* $u_1(t_0) > 0$, $u_2(t_0) > 0$.

Proof In view of the right-hand side of system (5.1) as continuous and smooth on $\mathbb{R}^2_+ = \{(u_1, u_2) : u_1, u_2 \geq 0\}$, and initial values $u_1(t_0) > 0$, $u_2(t_0) > 0$, it yields

$$\begin{aligned} u_1(t) &= u_1(t_0)e^{\int_{t_0}^{t}\left(\frac{x}{c_1x^2+c_2x+c_3}-\alpha_2x^2-\alpha_1u_1(s)-\frac{C_mu_2(s)}{a+u_1(s)+bu_2(s)}\right)ds} > 0,\\ u_2(t) &= u_2(t_0)e^{\int_{t_0}^{t}\left(\frac{C_mu_1(s-\tau)u_2(s-\tau)}{u_2(s)(a+u_1(s-\tau)+bu_2(s-\tau))}-r_2\right)ds} > 0. \end{aligned}$$

Utilizing the positiveness of the solution, it produces

$$\dot{u}_1 \leq u_1\left(\frac{x}{c_1x^2+c_2x+c_3} - \alpha_2x^2 - \alpha_1u_1\right).$$

The standard comparison argument displays that

$$\limsup_{t\to\infty} u_1(t) \leq \frac{x}{\alpha_1(c_1x^2+c_2x+c_3)} - \frac{\alpha_2}{\alpha_1}x^2.$$

Assume $\zeta(t) = u_1(t-\tau) + u_2(t)$. Along the solution of system (5.1), it gives

$$\begin{aligned} &\frac{d\zeta(t)}{dt} + l\zeta(t)\\ =&\dot{u}_1(t-\tau) + \dot{u}_2(t) + [rgb]0.00, 0.07, 1.00lu_1(t-\tau) + lu_2(t)\\ =&\frac{xu_1(t-\tau)}{c_1x^2+c_2x+c_3} - \alpha_2x^2u_1(t-\tau) - \alpha_1u_1^2(t-\tau)\\ &-r_2u_2(t) + lu_1(t-\tau) + lu_2(t). \end{aligned}$$

Define $l = r_2$. Then, it provides

$$\begin{aligned} \frac{d\zeta(t)}{dt} + l\zeta(t) =& u_1(t-\tau)\left(\frac{x}{c_1x^2+c_2x+c_3} - \alpha_2x^2 + r_2 - \alpha_1u_1(t-\tau)\right)\\ \leq& \frac{\left(\frac{x}{c_1x^2+c_2x+c_3} - \alpha_2x^2 + r_2\right)^2}{4\alpha_1}, \end{aligned}$$

and

$$0 < \zeta(t) \le \frac{\left(\frac{x}{c_1x^2+c_2x+c_3} - \alpha_2 x^2 + r_2\right)^2}{4\alpha_1 r_2}\left(1-e^{-r_2 t}\right) + \zeta(0)e^{-r_2 t}.$$

Hence, it yields

$$0 < \lim_{t\to\infty} \zeta(t) \le \frac{\left(\frac{x}{c_1x^2+c_2x+c_3} - \alpha_2 x^2 + r_2\right)^2}{4\alpha_1 r_2}.$$

By this and noting $\zeta(t) = u_1(t-\tau) + u_2(t)$, it generates the following region for the solution of system (5.1):

$$B = \{(u_1, u_2) \in \mathbb{R}_+^2 : 0 < u_1 \le \frac{x}{\alpha_1(c_1x^2+c_2x+c_3)} - \frac{\alpha_2}{\alpha_1}x^2,$$
$$0 < u_1 + u_2 \le \frac{\left(\frac{x}{c_1x^2+c_2x+c_3} - \alpha_2 x^2 + r_2\right)^2}{4\alpha_1 r_2}\}.$$

□

Accordingly, it is easy to prove that system (5.1) with (5.2), (5.3) has two boundary equilibria $E^0 = (0,0)$, $E^1 = \left(\frac{x}{\alpha_1(c_1x^2+c_2x+c_3)} - \frac{\alpha_2 x^2}{\alpha_1}, 0\right)$ and a coexistent equilibrium $E^* = (u_1^*, u_2^*)$ as the condition

$$C_m\left(\frac{C_m}{r_2}-1\right)^2 + a\left(\frac{C_m}{r_2}-1\right)\left(\frac{x}{c_1x^2+c_2x+c_3} - \alpha_2x^2 + \frac{r_2}{b} - \frac{C_m}{b}\right)$$
$$> a^2\alpha_1$$

holds, where

$$u_1^* = \frac{r_1(x)-s(x)+\frac{r_2}{b}-\frac{C_m}{b}+\sqrt{\left(r_1(x)-s(x)+\frac{r_2}{b}-\frac{C_m}{b}\right)^2+\frac{4a\alpha_1 r_2}{b}}}{2\alpha_1}$$
$$= \frac{\frac{x}{c_1x^2+c_2x+c_3}-\alpha_2x^2+\frac{r_2}{b}-\frac{C_m}{b}+\sqrt{\left(\frac{x}{c_1x^2+c_2x+c_3}-\alpha_2x^2+\frac{r_2}{b}-\frac{C_m}{b}\right)^2+\frac{4a\alpha_1 r_2}{b}}}{2\alpha_1},$$
$$u_2^* = \frac{u_1^*\left(\frac{C(x,y)}{r_2}-1\right)-a}{b} = \frac{u_1^*\left(\frac{C_m}{r_2}-1\right)-a}{b}.$$

Nevertheless, coexistent equilibrium, representing the coexistent equilibrium state of phytoplankton and zooplankton, is our main focus from the biological viewpoint. As a result, the dynamic behavior of system (5.1) around the coexistent equilibrium E^* is considered.

In the following, the local stability and the existence of Hopf bifurcation, and the global stability at the coexistent equilibrium of system (5.1) are discussed.

5.2.1 *Local Stability and Hopf Bifurcation*

Assume $\overline{u}_1 = u_1 - u_1^*$, $\overline{u}_2 = u_2 - u_2^*$ and define $\overline{u}_1$, $\overline{u}_2$ by u_1, u_2. Using Taylor expansion at the coexistent equilibrium, it yields

$$\begin{cases} \dot{u}_1(t) &= a_{11}u_1(t) + a_{12}u_2(t) + \sum_{i+j\geq 2} f_1^{(ij)} u_1^i(t)u_2^j(t), \\ \dot{u}_2(t) &= a_{22}u_2(t) + b_{21}u_1(t-\tau) + b_{22}u_2(t-\tau) \\ &\quad + \sum_{i+j+k\geq 2} f_2^{(ijk)} u_2^i(t)u_1^j(t-\tau)u_2^k(t-\tau), \end{cases} \tag{5.4}$$

with

$$\begin{cases} a_{11} = -\alpha_1 u_1^* + \frac{C_m u_1^* u_2^*}{(a+u_1^*+bu_2^*)^2}, \quad a_{12} = -\frac{C_m u_1^*(a+u_1^*)}{(a+u_1^*+bu_2^*)^2}, \\ b_{21} = \frac{C_m u_2^*(a+bu_2^*)}{(a+u_1^*+bu_2^*)^2}, \quad a_{22} = -r_2, \quad b_{22} = \frac{C_m u_1^*(a+u_1^*)}{(a+u_1^*+bu_2^*)^2}, \\ f_1 = u_1\left(\frac{x}{c_1x^2+c_2x+c_3} - \alpha_2 x^2 - \alpha_1 u_1 - \frac{C_m u_2}{a+u_1+bu_2}\right), \\ f_2 = -r_2u_2 + \frac{C_m u_1(t-\tau)u_2(t-\tau)}{a+u_1(t-\tau)+bu_2(t-\tau)}, \\ f_1^{(ij)} = \frac{1}{i!j!}\frac{\partial^{i+j} f_1}{\partial u_1^i(t)\partial u_2^j(t)}\Big|_{(u_1^*,\, u_2^*)}, \\ f_2^{(ijk)} = \frac{1}{i!j!k!}\frac{\partial^{i+j+k} f_2}{\partial u_2^i(t)\partial u_1^j(t-\tau)\partial u_2^k(t-\tau)}\Big|_{(u_1^*,\, u_2^*)}. \end{cases} \tag{5.5}$$

From above, the linearized system is produced:

$$\begin{cases} \dot{u}_1 &= a_{11}u_1(t) + a_{12}u_2(t), \\ \dot{u}_2 &= a_{22}u_2(t) + b_{21}u_1(t-\tau) + b_{22}u_2(t-\tau). \end{cases} \tag{5.6}$$

Its characteristic equation at the zero equilibrium (0, 0) is given

$$\lambda^2 - (a_{11} + a_{22})\lambda + a_{11}a_{22} + (-b_{22}\lambda + a_{11}b_{22} - a_{12}b_{21})e^{-\lambda\tau} = 0. \tag{5.7}$$

Now, the local asymptotic stability of system (5.1) in two different cases is supplied.

Case I: $\tau = 0$. Substituting $\tau = 0$, then (5.7) reduces to

$$\lambda^2 - (a_{11} + a_{22} + b_{22})\lambda + a_{11}a_{22} + a_{11}b_{22} - a_{12}b_{21} \;=\; 0. \tag{5.8}$$

Assume λ_1 and λ_2 are two roots of (5.8). Then, it gives

$$\begin{cases} \lambda_1 + \lambda_2 &= a_{11} + a_{22} + b_{22}, \\ \lambda_1\lambda_2 &= a_{11}a_{22} + a_{11}b_{22} - a_{12}b_{21}. \end{cases} \tag{5.9}$$

This together with Routh–Hurwitz criterion provides $\lambda_1 + \lambda_2 < 0$ and $\lambda_1\lambda_2 > 0$ if $a_{11} < 0$, implying that system (5.1) is local stability around the coexistent equilibrium E^* if

$$\frac{C_m u_2^*}{(a + u_1^* + bu_2^*)^2} < \alpha_1. \tag{5.10}$$

Furthermore, it yields

$$a + u_1^* + bu_2^* = \frac{C_m u_2^*}{r_1(x) - s(x) - \alpha_1 u_1^*}. \tag{5.11}$$

Solving above equation for u_2^* has

$$u_2^* = \frac{(a + u_1^*)\big(r_1(x) - s(x) - \alpha_1 u_1^*\big)}{C_m - b\big(r_1(x) - s(x) - \alpha_1 u_1^*\big)}. \tag{5.12}$$

Using (5.12), inequality (5.10) has the form as follows:

$$\big(r_1(x) - s(x) - \alpha_1 u_1^*\big)^2 < \frac{C_m(a + u_1^*)\left(r_1(x) - s(x) - \alpha_1 u_1^*\right)}{C_m - b\left(r_1(x) - s(x) - \alpha_1 u_1^*\right)}. \tag{5.13}$$

Assume $Q = r_1(x) - s(x) + a\alpha_1$. When $C_m < bQ$, it produces

$$\begin{aligned} &b\alpha_1^2 u_1^{*2} + u_1^*\Big(C_m(1 + \alpha_1) - 2b\alpha_1\big(r_1(x) - s(x)\big)\Big) \\ &+b\big(r_1(x) - s(x)\big)^2 - C_m\big(r_1(x) - s(x)\big) + aC_m > 0. \end{aligned} \tag{5.14}$$

The discriminant is

$$\begin{aligned} \Delta = &\Big(C_m(1 + \alpha_1) - 2b\alpha_1\big(r_1(x) - s(x)\big)\Big)^2 \\ &- 4b\alpha_1^2\Big(b\big(r_1(x) - s(x)\big)^2 - C_m\big(r_1(x) - s(x) - a\big)\Big) < 0. \end{aligned} \tag{5.15}$$

After solving the inequality, it earns $C_m < \frac{4b\alpha_1 Q}{(1+\alpha_1)^2}$. For that reason, the following proposition is produced.

The equilibrium E^* of system (5.1) with $\tau = 0$ is locally asymptotically stable when $C_m < \min\left\{bQ, \frac{4b\alpha_1 Q}{(1+\alpha_1)^2}\right\}$.

Case II: $\tau > 0$. Let the root of characteristic equation (5.7) be purely imaginary. Substituting $i\omega$ ($\omega > 0$) in (5.7) and separating real and imaginary parts yields

$$\begin{cases} a_{11}a_{22} + a_{11}b_{22}cos\omega\tau = a_{12}b_{21}cos\omega\tau + \omega^2 + b_{22}\omega sin\omega\tau, \\ (a_{11} + a_{22})\omega + b_{22}\omega cos\omega\tau = a_{12}b_{21}sin\omega\tau - a_{11}b_{22}sin\omega\tau. \end{cases} \tag{5.16}$$

Adding up the squares of the corresponding sides of the above equations produces

$$\omega^4 + m_1\omega^2 + m_2 = 0, \tag{5.17}$$

with

$$m_1 = -b_{22}^2 + a_{11}^2 + a_{22}^2, \quad m_2 = a_{11}^2 a_{22}^2 - a_{11}^2 b_{22}^2 - a_{12}^2 b_{21}^2. \tag{5.18}$$

It gives the roots of (5.17) as follows:

$$\omega_0^2 = -\frac{1}{2}m_1 \pm \frac{1}{2}(m_1^2 - 4m_2)^{\frac{1}{2}}. \tag{5.19}$$

Consequently, the three different scenarios about the roots of (5.17) are discussed as follows:

(1) (5.17) has no positive root when (H1) $m_2 > 0$ and $m_1 > 0$ or $\Delta = m_1^2 - 4m_2 < 0$ hold;

(2) (5.17) has a positive root when (H2) $m_1 < 0,\ \Delta = m_1^2 - 4m_2 = 0$ or $m_2 < 0$ holds;

(3) (5.17) has two positive roots when (H3) $m_1 < 0$ and $m_2 > 0$ hold.

Now, (5.7) has purely imaginary root when (H2) and (H3) hold. From the first equation of (5.16), it is easy to know that the values of τ_j associated with ω_0 satisfy

$$\tau_j = \frac{\arccos\left[\frac{(a_{11}a_{22}-\omega_0^2)(a_{12}b_{21}-a_{11}b_{22})-(a_{11}+a_{22})b_{22}\omega_0^2}{(a_{11}b_{22}-a_{12}b_{21})^2+b_{22}^2\omega_0^2}\right] + \frac{2\pi j}{\omega_0}}{\omega_0}, \tag{5.20}$$

where $j = 0, 1, 2, 3, \cdots$ and $\tau_0 = \min\{\tau_j\}$.

Assume $\lambda = \alpha(\tau) + i\omega(\tau)$ is the root of (5.7) where $\alpha(\tau_j) = 0$ and $\omega(\tau_j) = \omega_0$, as τ is close to τ_j. Differentiation of characteristic equation (5.7) with respect to τ provides

$$\frac{d\lambda}{d\tau} = \frac{\lambda e^{-\lambda\tau}(-b_{22}\lambda + a_{11}b_{22} - a_{12}b_{21})}{2\lambda - a_{11} - a_{22} - (\tau(-b_{22}\lambda + a_{11}b_{22} - a_{12}b_{21}) + b_{22})e^{-\lambda\tau}}, \tag{5.21}$$

and

$$\left(\frac{d\lambda}{d\tau}\right)^{-1} = \frac{2\lambda - (a_{11} + a_{22})}{\lambda e^{-\lambda\tau}(-b_{22}\lambda + a_{11}b_{22} - a_{12}b_{21})} - \frac{\tau}{\lambda} - \frac{b_{22}}{\lambda(-b_{22}\lambda + a_{11}b_{22} - a_{12}b_{21})}. \tag{5.22}$$

Using Rouche's theorem, it gives

$$sign\left\{\frac{d(Re\lambda)}{d\tau}\right\}_{\lambda=i\omega_0} = sign\left\{Re\left(\frac{d(\lambda)}{d\tau}\right)^{-1}\right\}_{\lambda=i\omega_0}.$$

If $\lambda = i\omega_0$, then

$$
\begin{aligned}
&Re\left(\frac{d\lambda}{d\tau}\right)^{-1} \\
&= Re\Bigg(\frac{(2\lambda-(a_{11}+a_{22}))e^{\lambda\tau}}{\lambda(-b_{22}\lambda+a_{11}b_{22}-a_{12}b_{21})}-\frac{\tau}{\lambda} \\
&\quad -\frac{b_{22}}{\lambda(-b_{22}\lambda+a_{11}b_{22}-a_{12}b_{21})}\Bigg)\Bigg|_{\lambda=i\omega_0} \\
&= \frac{(-2\omega_0\sin\omega_0\tau_j-(a_{11}+a_{22})\cos\omega_0\tau_j)b_{22}\omega_0+b_{22}^2\omega_0}{b_{22}^2\omega_0^3+(a_{11}b_{22}-a_{12}b_{21})^2\omega_0} \\
&\quad +\frac{(a_{11}b_{22}-a_{12}b_{21})(2\omega_0\cos\omega_0\tau_j-(a_{11}+a_{22})\sin\omega_0\tau_j)}{b_{22}^2\omega_0^3+(a_{11}b_{22}-a_{12}b_{21})^2\omega_0} \\
&= \frac{-2\omega_0^2b_{22}\sin\omega_0\tau_j-b_{22}\omega_0(a_{11}+a_{22})\cos\omega_0\tau_j}{b_{22}^2\omega_0^3+(a_{11}b_{22}-a_{12}b_{21})^2\omega_0} \\
&\quad +\frac{2\omega_0a_{11}b_{22}\cos\omega_0\tau_j-b_{22}^2\omega_0-2a_{12}b_{21}\omega_0\cos\omega_0\tau_j}{b_{22}^2\omega_0^3+(a_{11}b_{22}-a_{12}b_{21})^2\omega_0} \\
&\quad +\frac{a_{12}b_{21}(a_{11}+a_{22})\sin\omega_0\tau_j-(a_{11}+a_{22})a_{11}b_{22}\sin\omega_0\tau_j}{b_{22}^2\omega_0^3+(a_{11}b_{22}-a_{12}b_{21})^2\omega_0} \\
&= \frac{2\omega_0(\omega_0^2-a_{11}a_{22})+(a_{11}+a_{22})^2\omega_0-b_{22}^2\omega_0}{b_{22}^2\omega_0^3+(a_{11}b_{22}-a_{12}b_{21})^2\omega_0} \\
&= \frac{2\omega_0^2+a_{11}^2+a_{22}^2-b_{22}^2}{b_{22}^2\omega_0^2+(a_{11}b_{22}-a_{12}b_{21})^2}.
\end{aligned}
$$

By the definition of these symbols, it is known that (H4) $2\omega_0^2+a_{11}^2+a_{22}^2-b_{22}^2 > 0$ holds, implying

$$\left.\frac{d(Re\lambda)}{d\tau}\right|_{\lambda=i\omega_0} > 0.$$

Thus, the transversality condition holds and the conditions for Hopf bifurcation are satisfied at $\tau=\tau_j$, $j=0,1,2,\ldots$.

Because of the above analysis, we have the following proposition.

$\left.\frac{d\alpha(\tau)}{d\tau}\right|_{\lambda=i\omega_0} > 0$, if (H4) holds.

Based on the discussion above, the following conclusions on stability and Hopf bifurcation of system (5.1) are obtained.

Theorem 49 *The following asserts hold.*

- *The coexistent equilibrium E^* is locally asymptotically stable for all $\tau \geq 0$, if (H1) holds.*

- *The equilibrium E^* is locally asymptotically stable for $\tau \in [0, \tau_j)$ and unstable for $\tau > \tau_j$, if (H2) or (H3) and (H4) hold.*
- *System (5.1) undergoes a Hopf bifurcation at the equilibrium E^* when $\tau = \tau_j,\ j = 0, 1, 2, \ldots$.*

Remark 3 Ecologically, both phytoplankton and zooplankton can coexist in a certain density if the cell size is within a certain range and there is a small gestation delay. Moreover, the system may have a stable bifurcating periodic solution if gestation delay takes values nearby Hopf bifurcation values.

5.2.2 Global Stability

In what follows, the global stability of system (5.1) is investigated.

Theorem 50 *System (5.1) is globally asymptotically stable around the coexistent equilibrium E^*, if*

$$2P_2^2 b u_1^* - Q_2(a + bu_2^*) - 2u_1^* > 0$$

and

$$4\left(\frac{bP_2u_1^*}{Q_2} - \frac{Q_2(a+bu_2^*)+2u_1^*}{2P_2}\right)\left((\alpha_1 - w)(a + P_1 + bP_2)(a + u_1^* + bu_2^*) - C_m Q_2\right) \geq (a + Q_1)^2$$

hold.

Proof In order to investigate the global stability of system (5.1) around E^*, the set $D = \{(u_1, u_2) : P_1 < u_1(t) < Q_1;\ P_2 < u_2(t) < Q_2\}$ is considered. As D is a compact and bounded region in $\mathbb{R}^2$, there exists certain time T such that the solution of system (5.1) remains in the region D for any $t > T$. Next construct a suitable Lyapunov functional:

$$V = \int_{u_1^*}^{u_1} \left(\frac{u_1 - u_1^*}{u_1}\right) du_1 + \int_{u_2^*}^{u_2} \left(\frac{u_2 - u_2^*}{u_2}\right) du_2 + w \int_{t-\tau}^{t} (u_1(s) - u_1^*)^2 ds, \quad (5.23)$$

where $w = \frac{Q_2 C_m (a+bu_2^*)}{2P_2(a+P_1+bQ_1)(a+u_1^*+bu_2^*)}$.

Now, taking the time derivative of V along the solution of system (5.1) yields

$$
\begin{aligned}
\dot{V} &= \tfrac{u_1-u_1^*}{u_1}\dot{u}_1 + \tfrac{u_2-u_2^*}{u_2}\dot{u}_2 + w\big[(u_1-u_1^*)^2 - (u_1(t-\tau)-u_1^*)^2\big] \\
&= (u_1-u_1^*)\left(r_1(x) - s(x) - \alpha_1 u_1 - \tfrac{C_m u_2}{a+u_1+bu_2}\right) \\
&+w\big[(u_1-u_1^*)^2 - \ (u_1(t-\tau)-u_1^*)^2\big] \\
&+(u_2-u_2^*)\left(\tfrac{C_m u_1(t-\tau)u_2(t-\tau)}{u_2(a+u_1(t-\tau)+bu_2(t-\tau))} - r_2\right) \\
&= -\alpha_1(u_1-u_1^*)^2 + w\big[(u_1-u_1^*)^2 - (u_1(t-\tau)-u_1^*)^2\big] \\
&+(u_1-u_1^*)\left(-\tfrac{C_m u_2}{a+u_1+bu_2} + \tfrac{C_m u_2^*}{a+u_1^*+bu_2^*}\right) \\
&+\tfrac{C_m u_2(t-\tau)u_2(t-\tau)(u_2-u_2^*)}{u_2(a+u_1(t-\tau)+bu_2(t-\tau))} - \tfrac{C_m u_1^*(u_2-u_2^*)}{a+u_1^*+bu_2^*} \\
&= -\alpha_1(u_1-u_1^*)^2 - (\tfrac{\left(aC_m+C_m u_1\right)(u_2-u_2^*)(u_1-u_1^*)}{(a+u_1+bu_2)(a+u_1^*+bu_2^*)} \\
&+\tfrac{C_m u_2(u_1^*-u_1)(u_1-u_1^*)}{(a+u_1+bu_2)(a+u_1^*+bu_2^*)}) \\
&+\tfrac{a(u_2(t-\tau)u_2(t-\tau)-u_1^*u_2)(u_2-u_2^*)C_m}{u_2(a+u_1(t-\tau)+bu_2(t-\tau))(a+u_1^*+bu_2^*)} \\
&+\tfrac{u_1(t-\tau)u_1^*(u_2(t-\tau)-u_2)(u_2-u_2^*)C_m}{u_2(a+u_1(t-\tau)+bu_2(t-\tau))(a+u_1^*+bu_2^*)} \\
&+\tfrac{bu_2(t-\tau)(u_1(t-\tau)u_2^*-u_2u_1^*)(u_2-u_2^*)C_m}{u_2(a+u_1(t-\tau)+bu_2(t-\tau))(a+u_1^*+bu_2^*)} \\
&+w\big[(u_1-u_1^*)^2 - (u_1(t-\tau)-u_1^*)^2\big] \\
&\le \left(-\alpha_1 + w + \tfrac{C_m u_2}{(a+u_1+bu_2)(a+u_1^*+bu_2^*)}\right)(u_1-u_1^*)^2 \\
&-\tfrac{C_m(a+u_1)(u_1-u_1^*)(u_2-u_2^*)}{(a+u_1+bu_2)(a+u_1^*+bu_2^*)} \\
&+C_m(\tfrac{(u_2-u_2^*)^2(Q_2(a+bu_2^*)+2u_1^*)}{2P_2(a+u_1(t-\tau)+bu_2(t-\tau))(a+u_1^*+bu_2^*)} \\
&-\tfrac{bP_2u_1^*(u_2-u_2^*)^2}{Q_2(a+u_1(t-\tau)+bu_2(t-\tau))(a+u_1^*+bu_2^*)}) \\
&\le \left(-\alpha_1 + w + \tfrac{C_m Q_2}{(a+P_1+bP_2)(a+u_1^*+bu_2^*)}\right)(u_1-u_1^*)^2 \\
&-\tfrac{C_m(a+P_2)(u_1-u_1^*)(u_2-u_2^*)}{(a+P_1+bP_2)(a+u_1^*+bu_2^*)} \\
&+C_m(\tfrac{Q_2(a+bu_2^*)+2u_1^*}{2P_2(a+P_1+bP_2)(a+u_1^*+bu_2^*)} \\
&-\tfrac{bP_2u_1^*}{Q_2(a+P_1+bP_2)(a+u_1^*+bu_2^*)})(u_2-u_2^*)^2.
\end{aligned}
\tag{5.24}
$$

Thus $\frac{dV}{dt} < 0$ along all the trajectories except E^*, if

$$2P_2^2bu_1^* - Q_2(a+bu_2^*) - 2u_1^* > 0$$

and

$$4\left(\frac{bP_2u_1^*}{Q_2} - \frac{Q_2(a+bu_2^*)+2u_1^*}{2P_2}\right)\cdot((\alpha_1-w)(a+P_1+bP_2)(a+u_1^*+bu_2^*) - C_mQ_2) \ge (a+Q_1)^2$$

hold. So under the conditions of Theorem 50, E^* is globally asymptotically stable. □

Remark 4 From biological point of view, the cell size, rather than the zooplankton gestation delay, plays the key role in ensuring the global stability of the system. Furthermore, the theoretical result implies that phytoplankton and zooplankton will coexist as time increases to infinity if the parameters are within a certain range.

5.3 Direction and Stability of Bifurcation Limit Cycle

In this section, the explicit formula is derived to determine the Hopf bifurcation at the critical value $\tau = \tau_0$ applying the normal form and the center manifold theory [78, 79].

Let $v_1(t) = u_1(\tau t) - u_1^*$, $v_2(t) = u_2(\tau t) - u_2^*$, $\tau = \mu + \tau_0$. System (5.1) can be rewritten as a function equation in $C = C([-1, 0], \mathbb{R}^2)$:

$$\dot{v}(t) = L_\mu(v_t) + F(\mu, v_t), \tag{5.25}$$

where $v = (v_1, v_2)^T \in \mathbb{R}^2$, $v_t(\theta) = v(t+\theta) \in C$, $\theta \in [-1, 0]$ and $L_\mu : C \to \mathbb{R}^2$, $F : \mathbb{R} \times C \to \mathbb{R}^2$ are provided, respectively, by

$$L_\mu(\phi) = (\mu + \tau_0)[\begin{pmatrix} a_{11} & a_{12} \\ 0 & a_{22} \end{pmatrix} \begin{pmatrix} \phi_1(0) \\ \phi_2(0) \end{pmatrix} + \begin{pmatrix} 0 & 0 \\ b_{21} & b_{22} \end{pmatrix} \begin{pmatrix} \phi_1(-1) \\ \phi_2(-1) \end{pmatrix}], \tag{5.26}$$

$$F(\mu, \phi) = (\mu + \tau_0) \times \begin{pmatrix} h_{11}\phi_1^2(0) + h_{12}\phi_1(0)\phi_2(0) + h_{13}\phi_2^2(0) \\ h_{21}\phi_1^2(-1) + h_{22}\phi_1(-1)\phi_2(-1) + h_{23}\phi_2^2(-1) \end{pmatrix}, \tag{5.27}$$

where $\phi = (\phi_1, \phi_2)^T \in C([-1, 0], \mathbb{R}^2)$ and

$$\begin{cases} h_{11} = -\alpha_1 + \frac{C_m u_2^*(a+bu_2^*)}{(a+u_1^*+bu_2^*)^3}, & h_{13} = \frac{bC_m u_1^*(a+u_1^*)}{(a+u_1^*+bu_2^*)^3}, \\ h_{23} = -\frac{bC_m u_1^*(a+u_1^*)}{(a+u_1^*+bu_2^*)^3}, & h_{21} = -\frac{C_m u_2^*(a+bu_2^*)}{(a+u_1^*+bu_2^*)^3}, \\ h_{12} = -\frac{C_m(a+u_1^*+bu_2^*)^2 - bC_m u_2^*(a+bu_2^*) - C_m u_1^*(a+u_1^*)}{(a+u_1^*+bu_2^*)^3}, \\ h_{22} = \frac{C_m(a+u_1^*+bu_2^*)^2 - bC_m u_2^*(a+bu_2^*) - C_m u_1^*(a+u_1^*)}{(a+u_1^*+bu_2^*)^3}. \end{cases} \tag{5.28}$$

Using the Riesz representation theorem, there is a matrix function $\rho(\theta, \mu)$, whose elements are of bounded variation such that

$$L_\mu(\phi) = \int_{-1}^{0} d\rho(\theta, \mu)\phi(\theta). \tag{5.29}$$

Choose

$$\rho(\theta, \mu) = (\mu + \tau_0) \begin{pmatrix} a_{11} & a_{12} \\ 0 & a_{22} \end{pmatrix} \delta(\theta) - (\mu + \tau_0) \begin{pmatrix} 0 & 0 \\ b_{21} & b_{22} \end{pmatrix} \delta(\theta + 1), \tag{5.30}$$

with $\delta(\theta) = \begin{cases} 0, & \theta \neq 0, \\ 1, & \theta = 0. \end{cases}$

Then (5.29) is satisfied. Let $M(\mu)$ be the infinitesimal generator of the strongly continuous semigroup $T(t)$ $(t \geq 0)$. And corresponding to (5.25), it can be generated by the linear equation $\dot{v} = L(\mu)v_t$. Then $M(\mu)$ and $N(\mu)$ are supplied by

$$M(\mu)\phi(\theta) = \begin{cases} \frac{d\phi(\theta)}{d\theta}, \ \theta \in [-1, 0), \\ \int_{-1}^{0} d\rho(\theta, \mu)\phi(\theta), \ \theta = 0, \end{cases}$$
$$N(\mu)\phi(\theta) = \begin{cases} 0, \ \theta \in [-1, 0), \\ F(\mu, \phi), \ \theta = 0. \end{cases} \quad (5.31)$$

Obviously, (5.25) is equivalent to

$$\dot{v}_t \ = \ M(\mu)v_t + N(\mu)v_t. \quad (5.32)$$

For $\phi \in \ C = C^1([-1, 0], \mathbb{R}^2)$, $\varphi \in C^* = C^1([0, 1], (\mathbb{R}^2)^*)$, an adjoint operator M^* of $M(0)$ is defined by

$$M^*\varphi(s) \ = \ \begin{cases} -\frac{d\varphi(s)}{ds}, \ s \in (0, 1], \\ \int_{-1}^{0} d\rho^T(t, 0)\varphi(-t), \ s = 0, \end{cases} \quad (5.33)$$

and a bilinear inner product is denoted by

$$\langle \varphi(s), \phi(\theta) \rangle \ = \ \bar{\varphi}(0)\phi(0) - \int_{-1}^{0}\int_{\xi=0}^{\theta} \bar{\varphi}(\xi - \theta)d\rho(\theta)\phi(\xi)d\xi. \quad (5.34)$$

From the above analysis, we know it is shown that $\pm i\omega_0\tau_0$ are eigenvalues of $M(0)$ and M^*, respectively. The eigenvectors $\eta(\theta)$ and $\eta^*(s)$ of $M(0)$ and M^* are calculated corresponding to the eigenvalues $i\omega_0\tau_0$ and $-i\omega_0\tau_0$. Hypothesize that $\eta(\theta) = (1, g)^T e^{i\omega_0\tau_0\theta}$. Thus it has $M(0)\eta(\theta) = i\omega_0\tau_0\eta(\theta)$. From the definition of $M(0)$, the following is derived

$$\begin{pmatrix} a_{11} & a_{12} \\ 0 & a_{22} \end{pmatrix} \eta(0) + \begin{pmatrix} 0 & 0 \\ b_{21} & b_{22} \end{pmatrix} \eta(-1) \ = \ i\omega_0\eta(0),$$

and $g = -\frac{i\omega_0 - a_{11}}{a_{12}}$ is also obtained.

Consequently, $M^*\eta^*(s) = -i\omega_0\tau_0\eta^*(s)$ when $\eta^*(s) = E(1, h)e^{i\omega_0\tau_0 s}$. Based on the definition of M^*, it yields

$$\begin{pmatrix} a_{11} & 0 \\ a_{12} & a_{22} \end{pmatrix} \eta^*(0) + \begin{pmatrix} 0 & b_{21} \\ 0 & b_{22} \end{pmatrix} \eta^*(-1) \ = \ -i\omega_0\eta^*(0),$$

representing that $h \ = \ \frac{(-i\omega_0 - a_{11})e^{i\omega_0\tau_0}}{b_{21}}$.

So, it deduces

$$
\begin{aligned}
&\langle \eta^*(s), \eta(\theta)\rangle \\
&= \bar{\eta}^*(0)\eta(0) - \int_{-1}^{0}\int_{\xi=0}^{\theta} \bar{\eta}^*(\xi-\theta)d\rho(\theta)\eta(\xi)d\xi \\
&= \bar{E}(1,\bar{h})(1,g)^T \\
&- \int_{-1}^{0}\int_{\xi=0}^{\theta} \bar{E}(1,\bar{h})e^{-i\omega_0\tau_0(\xi-\theta)}d\rho(\theta)(1,g)^T e^{i\omega_0\tau_0\xi}d\xi \\
&= \bar{E}(1,\bar{h})(1,g)^T - \bar{\eta}^*(0)\int_{-1}^{0}\int_{\xi=0}^{\theta} e^{i\omega_0\tau_0\theta}d\xi d\rho(\theta)\eta(0) \\
&= \bar{E}(1,\bar{h})(1,g)^T - \bar{\eta}^*(0)\int_{-1}^{0} \xi e^{i\omega_0\tau_0\theta}\Big|_{\xi=0}^{\theta} d\rho(\theta)\eta(0) \\
&= \bar{E}(1,\bar{h})(1,g)^T - \bar{\eta}^*(0)\int_{-1}^{0} \theta e^{i\omega_0\tau_0\theta}d\rho(\theta)\eta(0) \\
&= \bar{E}(1,\bar{h})(1,g)^T + \bar{\eta}^*(0)\tau_0\begin{pmatrix} 0 & 0 \\ b_{21} & b_{22}\end{pmatrix} e^{-i\omega_0\tau_0}\eta(0) \\
&= \bar{E}\left(1+\bar{h}g+\tau_0(\bar{h}b_{21}+\bar{h}b_{22}g)e^{-i\omega_0\tau_0}\right).
\end{aligned}
$$

In order to get $\langle \eta^*(s), \eta(\theta)\rangle = 1$, let

$$
\bar{E} = \frac{1}{1+\bar{h}g+\tau_0(\bar{h}b_{21}+\bar{h}b_{22}g)e^{-i\omega_0\tau_0}}.
$$

The coordinates describing center manifold C_0 at $\mu = 0$ are computed. And let v_t be the solution of (5.25) at $\mu = 0$. Then define $z(t) = \langle \eta^*, v_t\rangle$ and $W(t,\theta) = v_t(\theta) - 2Re\{z\eta(\theta)\}$. On the center manifold C_0, it has

$$
W(t,\theta) = W(z,\bar{z},\theta) = W_{20}\frac{z^2}{2} + W_{11}z\bar{z} + W_{02}\frac{\bar{z}^2}{2} + \cdots, \tag{5.35}
$$

where z and $\bar{z}$ stand for local coordinates for center manifold C_0 in the direction of η^* and $\bar{\eta}^*$; W means real if v_t is real. In addition, we shall only consider real solutions. For the solution v_t of (5.25), it yields

$$
\begin{aligned}
\dot{z}(t) &= \langle \eta^*, \dot{v}_t\rangle \\
&= i\omega_0\tau_0 z + \langle \eta^*(0), N(0)(W(z,\bar{z},\theta) + 2Re\{z\eta(\theta)\})\rangle \\
&= i\omega_0\tau_0 z + \bar{\eta}^*(0)F(0)(W(z,\bar{z},0) + 2Re\{z\eta(\theta)\}) \\
&= i\omega_0\tau_0 z + g(z,\bar{z}),
\end{aligned} \tag{5.36}
$$

when $\mu = 0$, where

$$
g(z,\bar{z}) = g_{20}\frac{z^2}{2} + g_{11}z\bar{z} + g_{02}\frac{\bar{z}^2}{2} + g_{21}\frac{z^2\bar{z}}{2} + \cdots. \tag{5.37}
$$

Accordingly, it displays

$$
\begin{aligned}
v_t &= W(t,\theta) + z\eta(\theta) + \bar{z}\bar{\eta}(\theta) \\
&= W(t,\theta) + z(1,g)^T e^{i\theta\omega_0\tau_0} + \bar{z}(1,\bar{g})^T e^{-i\theta\omega_0\tau_0}, \\
v_{1t}(0) &= W_{20}^{(1)}(0)\frac{z^2}{2} + W_{11}^{(1)}(0)z\bar{z} + W_{02}^{(1)}(0)\frac{\bar{z}^2}{2} \\
&\quad + z + \bar{z} + o\Big(|(z,\bar{z})|^3\Big), \\
v_{2t}(0) &= W_{20}^{(2)}(0)\frac{z^2}{2} + W_{11}^{(2)}(0)z\bar{z} + W_{02}^{(2)}(0)\frac{\bar{z}^2}{2} \\
&\quad + gz + \bar{g}\bar{z} + o\Big(|(z,\bar{z})|^3\Big), \\
v_{1t}(-1) &= W_{20}^{(1)}(-1)\frac{z^2}{2} + W_{11}^{(1)}(-1)z\bar{z} + W_{02}^{(1)}(-1)\frac{\bar{z}^2}{2} \\
&\quad + ze^{-i\omega_0\tau_j} + \bar{z}e^{i\omega_0\tau_j} + o\Big(|(z,\bar{z})|^3\Big), \\
v_{2t}(-1) &= W_{20}^{(2)}(-1)\frac{z^2}{2} + W_{11}^{(2)}(-1)z\bar{z} + W_{02}^{(2)}(-1)\frac{\bar{z}^2}{2} \\
&\quad + gze^{-i\omega_0\tau_j} + \bar{g}\bar{z}e^{i\omega_0\tau_j} + o\Big(|(z,\bar{z})|^3\Big).
\end{aligned}
$$

Together with (5.27), the above relationship produces that

$$
\begin{aligned}
& g(z,\bar{z}) \\
&= \bar{\eta}^*(0)F_0(z,\bar{z}) \;=\; \bar{\eta}^*(0)F(0,v_t) \\
&= \tau_0\bar{E}\Big\{h_{11}v_{1t}^2(0) + h_{12}v_{1t}(0)v_{2t}(0) + h_{13}v_{2t}^2(0) \\
&+\bar{h}\Big(h_{21}v_{1t}^2(-1) + h_{22}v_{1t}(-1)v_{2t}(-1) + h_{23}v_{2t}^2(-1)\Big)\Big\} \\
&= \tau_0\bar{E}\Big\{\Big(h_{11} + h_{12}g + h_{13}g^2 + h_{21}\bar{h}e^{-2i\omega_0\tau_0} + h_{22}\bar{h}ge^{-2i\omega_0\tau_0} \\
&+h_{23}\bar{h}g^2e^{-2i\omega_0\tau_0}\Big)z^2 + \Big(h_{11} + h_{12}\bar{g} + h_{13}\bar{g}^2 + h_{21}\bar{h}e^{2i\omega_0\tau_0} \\
&+h_{22}\bar{h}\bar{g}e^{2i\omega_0\tau_0} + h_{23}\bar{h}\bar{g}^2e^{2i\omega_0\tau_0}\Big)\bar{z}^2 + \Big(2h_{11} + h_{12}(g+\bar{g}) \\
&+2h_{13}g\bar{g} + 2h_{21}\bar{h} + h_{22}\bar{h}(g+\bar{g}) + 2h_{23}\bar{h}g\bar{g}\Big)z\bar{z} \\
&+\Big(h_{11}W_{20}^{(1)}(0) + 2h_{11}W_{11}^{(1)}(0) + h_{12}\Big(W_{11}^{(1)}(0)g + W_{20}^{(1)}(0)\tfrac{\bar{g}}{2} \\
&+W_{11}^{(2)}(0) + \tfrac{1}{2}W_{20}^{(2)}(0)\Big) + h_{13}\Big(W_{20}^{(2)}(0)\bar{g} + 2W_{11}^{(2)}(0)g\Big) \\
&+h_{21}\bar{h}\Big(W_{20}^{(1)}(-1)e^{i\omega_0\tau_0} + 2W_{11}^{(1)}(-1)e^{-i\omega_0\tau_0}\Big) \\
&+h_{22}\bar{h}\Big(W_{11}^{(1)}(-1)ge^{-i\omega_0\tau_0} + W_{20}^{(1)}(-1)\tfrac{\bar{g}}{2}e^{i\omega_0\tau_0} \\
&+W_{11}^{(2)}(-1)e^{-i\omega_0\tau_0} + \tfrac{1}{2}W_{20}^{(2)}(-1)e^{i\omega_0\tau_0}\Big) \\
&+h_{23}\bar{h}\Big(W_{20}^{(2)}(-1)\bar{g}e^{i\omega_0\tau_0} + 2W_{11}^{(2)}(-1)ge^{-i\omega_0\tau_0}\Big)\Big)z^2\bar{z} + \cdots\Big\}.
\end{aligned}
$$

By comparing coefficients with (5.37), it provides

$$g_{20} = 2\tau_0 \bar{E}\Big(h_{11} + h_{12}g + h_{13}g^2 + h_{21}\bar{h}e^{-2i\omega_0\tau_0} + h_{22}\bar{h}ge^{-2i\omega_0\tau_0} + h_{23}\bar{h}g^2e^{-2i\omega_0\tau_0}\Big),$$
$$g_{11} = \tau_0 \bar{E}\Big(2h_{11} + h_{12}(g+\bar{g}) + 2h_{13}g\bar{g} + 2h_{21}\bar{h} + h_{22}\bar{h}(g+\bar{g}) + 2h_{23}\bar{h}g\bar{g}\Big),$$
$$g_{02} = 2\tau_0 \bar{E}\Big(h_{11} + h_{12}\bar{g} + h_{13}\bar{g}^2 + h_{21}\bar{h}e^{2i\omega_0\tau_0} + h_{22}\bar{h}\bar{g}e^{2i\omega_0\tau_0} + h_{23}\bar{h}\bar{g}^2e^{2i\omega_0\tau_0}\Big),$$
$$\begin{aligned} g_{21} = 2\tau_0 \bar{E}\Big(& h_{11}W_{20}^{(1)}(0) + 2h_{11}W_{11}^{(1)}(0) + h_{12}\Big(W_{11}^{(1)}(0)g + W_{20}^{(1)}(0)\frac{\bar{g}}{2} + W_{11}^{(2)}(0) \\ &+ \frac{1}{2}W_{20}^{(2)}(0)\Big) + h_{13}\Big(W_{20}^{(2)}(0)\bar{g} + 2W_{11}^{(2)}(0)g\Big) \\ &+ h_{21}\bar{h}\Big(2W_{11}^{(1)}(-1)e^{-i\omega_0\tau_0} + W_{20}^{(1)}(-1)e^{i\omega_0\tau_0}\Big) \\ &+ h_{22}\bar{h}\Big(W_{11}^{(1)}(-1)ge^{-i\omega_0\tau_0} + W_{20}^{(1)}(-1)\frac{\bar{g}}{2}e^{i\omega_0\tau_0} + W_{11}^{(2)}(-1)e^{-i\omega_0\tau_0} \\ &+ \frac{1}{2}W_{20}^{(2)}(-1)e^{i\omega_0\tau_0}\Big) + h_{23}\bar{h}\Big(W_{20}^{(2)}(-1)\bar{g}e^{i\omega_0\tau_0} + 2W_{11}^{(2)}(-1)ge^{-i\omega_0\tau_0}\Big)\Big). \end{aligned} \tag{5.38}$$

Obviously, $W_{20}(\theta)$ and $W_{11}(\theta)$ determine the value of g_{21}. From (5.32) and (5.38), it gives

$$\dot{W}(t,\theta) = \dot{v}_t(\theta) - 2Re\{\dot{z}(t)\eta(\theta)\} = M(0)W + H(z,\bar{z},\theta), \tag{5.39}$$

where

$$\begin{aligned} H(z,\bar{z},\theta) &= -2Re\{g(z,\bar{z})\eta(\theta)\} + N(0)(W + 2Re\{z(t)\eta(\theta)\}) \\ &= H_{20}(\theta)\frac{z^2}{2} + H_{11}(\theta)z\bar{z} + H_{02}(\theta)\frac{\bar{z}^2}{2} + \cdots. \end{aligned} \tag{5.40}$$

From (5.40) and the definition of $N(0)$, we obtain the following:

$$H(z,\bar{z},\theta) = \begin{cases} -2Re\{g(z,\bar{z})\eta(\theta)\} & \theta \in [-1,0), \\ -2Re\{g(z,\bar{z})\eta(\theta)\} + F_0(z,\bar{z}) & \theta = 0. \end{cases} \tag{5.41}$$

On the center manifold C_0 near the origin, it yields

$$\dot{W} = W_z\dot{z} + W_{\bar{z}}\dot{\bar{z}} = W_{20}z\dot{z} + W_{11}\bar{z}\dot{z} + W_{11}z\dot{\bar{z}} + W_{02}\bar{z}\dot{\bar{z}} + \cdots. \tag{5.42}$$

It is then provided:

$$\begin{aligned} (M(0) - 2i\omega_0\tau_0 I)W_{20}(\theta) &= -H_{20}(\theta), \\ M(0)W_{11}(\theta) &= -H_{11}(\theta), \end{aligned} \tag{5.43}$$

after substituting (5.36) into (5.42) and comparing the coefficients with (5.39). Based on (5.40) and (5.41), it follows that

$$H_{20}(\theta) = -g_{20}\eta(\theta) - \bar{g}_{02}\bar{\eta}(\theta), \quad H_{11}(\theta) = -g_{11}\eta(\theta) - \bar{g}_{11}\bar{\eta}(\theta),$$

for $\theta \in [-1, 0)$. It yields

$$\begin{aligned}\dot{W}_{20}(\theta) &= 2i\omega_0\tau_0 W_{20}(\theta) + g_{20}\eta(\theta) + \bar{g}_{20}\bar{\eta}(\theta),\\ \dot{W}_{11}(\theta) &= g_{11}\eta(\theta) + \bar{g}_{11}\bar{\eta}(\theta),\end{aligned}$$

after substituting $H_{20}(\theta)$ and $H_{11}(\theta)$ into (5.43), meaning that

$$\begin{aligned}W_{20}(\theta) &= \frac{ig_{20}\eta(0)}{\omega_0\tau_0}e^{i\omega_0\tau_0\theta} + \frac{i\bar{g}_{02}\bar{\eta}(0)}{3\omega_0\tau_0}e^{-i\omega_0\tau_0\theta} + E_1 e^{2i\omega_0\tau_0\theta},\\ W_{11}(\theta) &= -\frac{ig_{11}\eta(0)}{\omega_0\tau_0}e^{i\omega_0\tau_0\theta} + \frac{i\bar{g}_{11}\bar{\eta}(0)}{\omega_0\tau_0}e^{-i\omega_0\tau_0\theta} + E_2,\end{aligned} \tag{5.44}$$

where $E_1,\ E_2 \in \mathbb{R}^2$ represent constant vectors.

Next, the values of $E_1,\ E_2$ are determined. Note that

$$M(0)W_{20} = \int_{-1}^{0} d\rho(\theta)W_{20}(\theta)$$

for $\theta = 0$ and by (5.43), it supplies

$$\begin{aligned}\int_{-1}^{0} d\eta(\theta)W_{20}(\theta) &= 2i\omega_0\tau_0 W_{20}(0) - H_{20}(0),\\ \int_{-1}^{0} d\eta(\theta)W_{20}(\theta) &= -H_{11}(0).\end{aligned} \tag{5.45}$$

By (5.39)–(5.41), it yields

$$\begin{cases} H_{20}(0) = -g_{20}\eta(0) - \bar{g}_{02}\bar{\eta}(0) \\ +2\tau_0\begin{pmatrix} h_{11} + h_{12}g + h_{13}g^2 \\ h_{21}\bar{h}e^{-2i\omega_0\tau_0} + h_{22}\bar{h}ge^{-2i\omega_0\tau_0} + h_{23}\bar{h}g^2e^{-2i\omega_0\tau_0}\end{pmatrix}, \\ H_{11}(0) = -g_{11}\eta(0) - \bar{g}_{11}\bar{\eta}(0) \\ +\tau_0\begin{pmatrix} 2h_{11} + h_{12}(g+\bar{g}) + 2h_{13}g\bar{g} \\ 2h_{21}\bar{h} + h_{22}\bar{h}(g+\bar{g}) + h_{23}\bar{h}2g\bar{g}\end{pmatrix}. \end{cases} \tag{5.46}$$

Substituting (5.44) in (5.45), it produces

$$\begin{aligned}&\left(2i\omega_0\tau_0 I - \int_{-1}^{0} e^{2i\omega_0\tau_0\theta}d\rho(\theta)\right)E_1\\ &= 2\tau_0\begin{pmatrix} h_{11} + h_{12}g + h_{13}g^2 \\ h_{21}\bar{h}e^{-2i\omega_0\tau_0} + h_{22}\bar{h}ge^{-2i\omega_0\tau_0} + h_{23}\bar{h}g^2e^{-2i\omega_0\tau_0}\end{pmatrix},\end{aligned}$$

and

$$\int_{-1}^{0} d\rho(\theta)E_2 = -\tau_0 \begin{pmatrix} 2h_{11} + h_{12}(g+\bar{g}) + 2h_{13}g\bar{g} \\ 2h_{21}\bar{h} + h_{22}\bar{h}(g+\bar{g}) + 2h_{23}\bar{h}g\bar{g} \end{pmatrix}.$$

It provides

$$E_1 = 2\begin{pmatrix} a_{11} & a_{12} \\ b_{21}e^{-2i\omega_0\tau_0} & a_{22} + b_{22}e^{-2i\omega_0\tau_0} \end{pmatrix}^{-1} \times \begin{pmatrix} h_{11} + h_{12}g + h_{13}g^2 \\ h_{21}\bar{h}e^{-2i\omega_0\tau_0} + h_{22}\bar{h}ge^{-2i\omega_0\tau_0} + h_{23}\bar{h}g^2e^{-2i\omega_0\tau_0} \end{pmatrix},$$

and

$$E_2 = \begin{pmatrix} a_{11} & a_{12} \\ b_{21} & a_{22} + b_{22} \end{pmatrix}^{-1} \begin{pmatrix} 2h_{11} + h_{12}(g+\bar{g}) + 2h_{13}g\bar{g} \\ 2h_{21}\bar{h} + h_{22}\bar{h}(g+\bar{g}) + 2h_{23}\bar{h}g\bar{g} \end{pmatrix},$$

after solving the above linear equations for E_1 and E_2. Accordingly, the parameters of system (5.1) and time delay τ_0 determine each g_{ij}. Hence, these standard results can be computed as given below:

$$\begin{aligned} c(0) &= \frac{i}{2\omega_0\tau_0}\left(g_{20}g_{11} - 2|g_{11}|^2 - \frac{1}{3}|g_{02}|^2\right) + \frac{g_{21}}{2}, \\ \mu_2 &= -\frac{Re\{c(0)\}}{Re\{\lambda'(\tau_0)\}}, \quad \beta_2 = 2Re\{c(0)\}, \\ T_2 &= -\frac{Im\{c(0)\} + \mu_2 Im\{\lambda'(\tau_0)\}}{\omega_0\tau_0}. \end{aligned} \tag{5.47}$$

From [63], it is known that the direction of the Hopf bifurcation is determined by the sign of μ_2: the Hopf bifurcation is supercritical (subcritical) and the bifurcating periodic solution exists for $\tau > \tau_0$ ($\tau < \tau_0$), if $\mu_2 > 0$ ($\mu_2 < 0$). The periodic solution is stable (unstable) if $\beta_2 < 0$ ($\beta_2 > 0$). The period of the bifurcating periodic solution of system increases (decreases) if $T_2 > 0$ ($T_2 < 0$).

5.4 Simulation

The following numerical simulation is given to illustrate the feasibility of our theoretical results by the following system:

$$
\begin{cases}
\dot{u}_1 = \dfrac{u_1 x}{0.01x^2 + 0.02x + 0.3} - 0.04x^2 u_1 - 0.5u_1^2 \\
\qquad - \dfrac{5.5u_2 u_1}{0.8 + u_1 + 0.72u_2}, \\
\dot{u}_2 = -0.9u_2 + \dfrac{5.5u_1(t-\tau)u_2(t-\tau)}{0.8 + u_1(t-\tau) + 0.72u_2(t-\tau)}.
\end{cases} \tag{5.48}
$$

Then a coexistent equilibrium $E^*(0.0831, 0.0426)$ is obtained. Figure 5.1 demonstrates the local stability of system (5.48) without τ around E^*. Figure 5.2 points out that the size of the cell is increasing as the value of x increases in which system (5.48) undergoes stable periodic solution oscillation for $x = 1$ from its stable position. Ecologically, it can give the answer that the increase in cell size of phytoplankton may play significant role in causing planktonic bloom depicted by oscillation.

Let other parameters be the same as those specified in the above, the numerical results of system (5.48) are given for different values of τ. It is easy to get the critical value of delay parameter $\tau = \tau_0 \approx 0.512$ for system (5.48) when the stability behavior changes. Furthermore, Fig. 5.3 reveals that E^* is unstable at $\tau = \tau_0$, and the system exists a limit cycle. In addition, Fig. 5.4 displays that the given system is stable around E^* for $\tau = 0.3$ and Fig. 5.5 points out that the

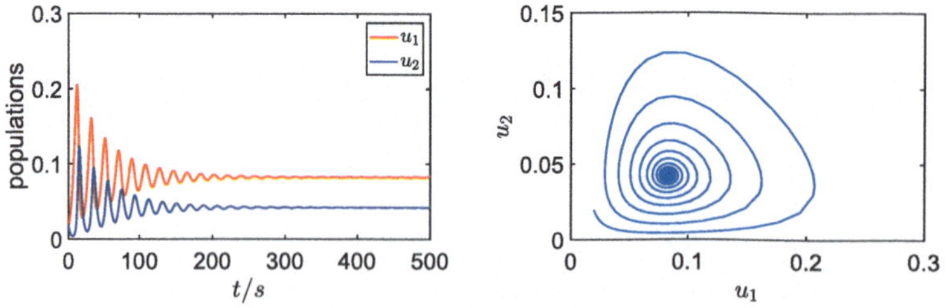

(a) Behaviour of phytoplankton and zooplankton dynamically (b) The phase plot of the system for the locally asymptotically stable condition

Fig. 5.1 In absence of delay ($\tau = 0$), for cell size ($x = 0.09$) and fixed initial condition (0.02, 0.02)

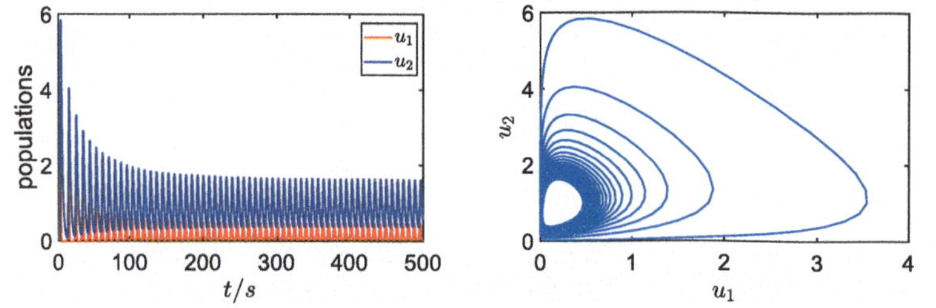

(a) Behaviour of phytoplankton and zooplankton dynamically (b) The phase plot of the system for the cell size changing

Fig. 5.2 In absence of delay ($\tau = 0$), for cell size ($x = 1$) and fixed initial condition (0.02, 0.02)

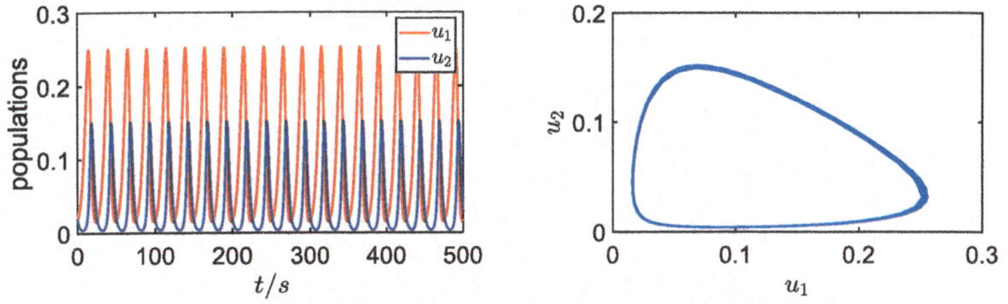

(a) Behaviour of phytoplankton and zooplankton dynamically (b) The phase plot of the system

Fig. 5.3 With delay threshold value ($\tau_0 \approx 0.512$), for cell size ($x = 0.09$) and fixed initial condition (0.02, 0.02)

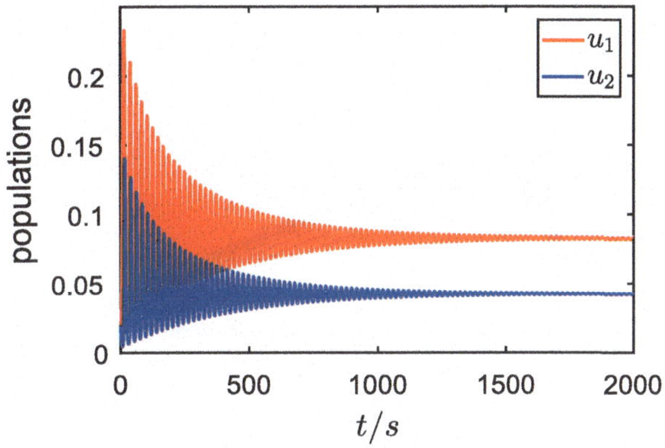

(a)

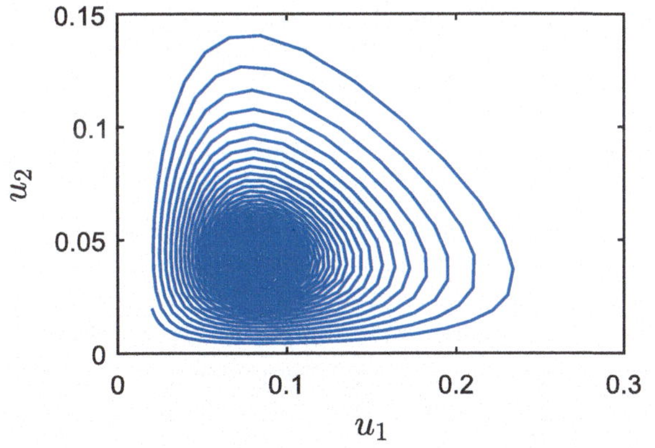

(b)

Fig. 5.4 Behavior of phytoplankton and zooplankton as a function of time (left panel), and the phase plot of the system (right panel) for the locally asymptotically stable condition. The numerical results are obtained with delay ($\tau = 0.3 < \tau_0$), for cell size ($x = 0.09$) and fixed initial condition (0.02, 0.02)

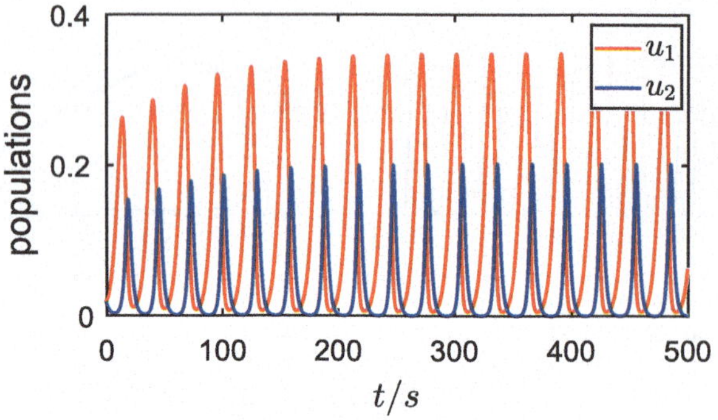

(a)

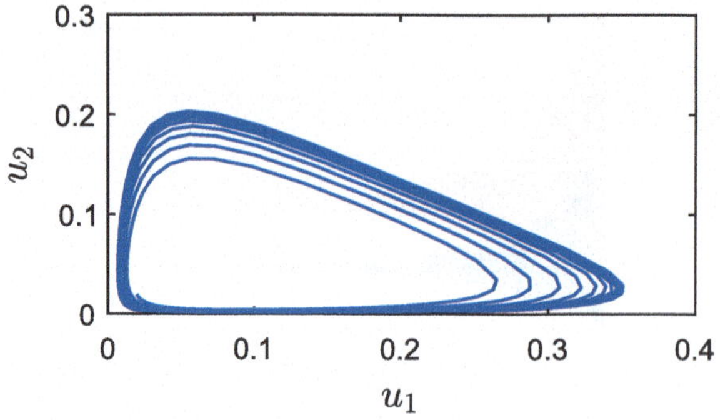

(b)

Fig. 5.5 Behavior of phytoplankton and zooplankton as a function of time (left panel), and the phase plot of the system (right panel) for the unstable condition. The numerical results are obtained with delay ($\tau = 0.7 > \tau_0$), for cell size ($x = 0.09$) and fixed initial condition (0.02, 0.02)

Hopf bifurcation is periodic, respectively. After computation of (5.47), it yields $c(0) = -160.99 + 276.78i$, $\beta_2 = -321.9790$, $\mu_2 = 5126.5$, $T_2 = -89676$. Thus, Fig. 5.5 demonstrates the Hopf bifurcation as supercritical Hopf bifurcation and the bifurcating limit cycle as orbitally stable. So, the period of the bifurcating limit cycle varies with the value of τ from the critical value τ_0 monotonously increasing. Based on the above analysis, it illustrates that the complexity of phytoplankton–zooplankton interaction varies with the delay introduced in zooplankton gestation and the cell size of phytoplankton monotonously increasing. And the possibility of algal bloom is also enhanced.

Chapter 6
The Effect of Water Temperature

Temperature plays an important role in the development of marine environment. Compared with terrestrial organisms, most marine organisms are more sensitive to temperature changes so that they tend to live in relatively stable temperature environment. Temperature changes have significant effects on the growth, metabolism, reproduction, distribution development, and food availability of marine organisms. Hence, the question arises that how to characterize the effect of the changes in temperature on the populations of biological communities. In order to answer this question, many works focus on the dynamics of the water temperature in marine ecosystems from experimental and theoretical point for a few decades, respectively [82–96]. Nowadays experiments have validated the following two major changes: (i) decreasing Cladoceran body size with rising temperature [97] and (ii) variation of critical density with water temperature [98]. Nonetheless, due to the complexity of the mechanism of zooplankton, water temperature indirectly affects its various metabolic activities. Consequently, if we want to change the whole marine food web, we should rely on the effects of temperature on zooplankton to modify its predation ability and induce shifts in the community structure and density of phytoplankton.

Based on the importance of temperature in the marine environment, this chapter explores the influence of temperature on plankton dynamics utilizing temperature-dependent parameters. A coupled plankton–temperature model is considered, and rich dynamics for different temperatures and parameters are also exhibited. Hump-shaped and U-shaped models are used to study the relationships between temperature and parameters of the plankton model in the range of physiologically relevant temperature.

S. T. Liu and L. Zhang, *Pattern Dynamics of Marine Plankton Behavior*,
https://doi.org/10.1007/978-981-97-5369-7_6

6.1 Formulation of Temperature-Dependence Model

Functional response as a significant component of predator–prey relationship displays the strength of predator–prey interactions. These responses, divided into various types by densities of prey and predator, affect the stability of predator–prey dynamics and maintain the dynamic balance of community structure. As the researches worked by C. S. Holling, there are four types for functional responses generally: called as Holling's type I, II, III, and IV. When assume that prey is scarce, a rectangular hyperbola is often used to model the Holling type II functional response. The Holling type III is more appropriate to model the situation that predator actively seeks out large concentrations of prey. Furthermore, the type III functional response can potentially stabilize fluctuations of prey and predator populations, and also reduce prey extinction risk [99, 100]. Therefore, we concentrate on the following phytoplankton–zooplankton model:

$$\begin{cases} \dot{P} &= r\left(1-\frac{P}{K}\right)P - \frac{aP^nZ}{1+ahP^n}, \\ \dot{Z} &= -mZ + e\frac{aP^nZ}{1+ahP^n}, \end{cases} \tag{6.1}$$

where initial conditions $P(0) > 0, Z(0) > 0, n > 0$ mean Holling parameter, $n = 1$ stands for a type II response, and $n > 1$ represents a type III response. P produces the density of phytoplankton population and Z provides the density of zooplankton population. r and K supply the intrinsic growth rate and carrying capacity of phytoplankton, respectively. The parameter a means the capture rate of zooplankton on phytoplankton and e gives the conversion rate of phytoplankton to zooplankton. h and m provide the handling time and the natural mortality rate of zooplankton, respectively.

6.1.1 Temperature Dependence of Model Parameters

The temperature-dependent parameters are considered in order to explore the effect of temperature on the dynamic behavior of plankton with Holling type III function response.

(1) The natural mortality rate of zooplankton increases exponentially with temperature. The following gives the mathematical model from [23]:

$$m(T) = m_0 \exp(-E/(kT)),$$

where m_0 means a constant, E stands for activation energy, k represents Boltzmann's constant ($8.62 \times 10^{-5}\,\text{eVK}^{-1}$), and T means absolute temperature of water. Figure 6.1 and Fig. 1b of [23] show the dynamics of mortality rate function with respect to T, displaying the variation of mortality rate similar as the experimental study [101]. (2) As the intrinsic growth rate of phytoplankton behaviors a hump-shaped relationship with temperature, the following model is given based on the model in [23]:

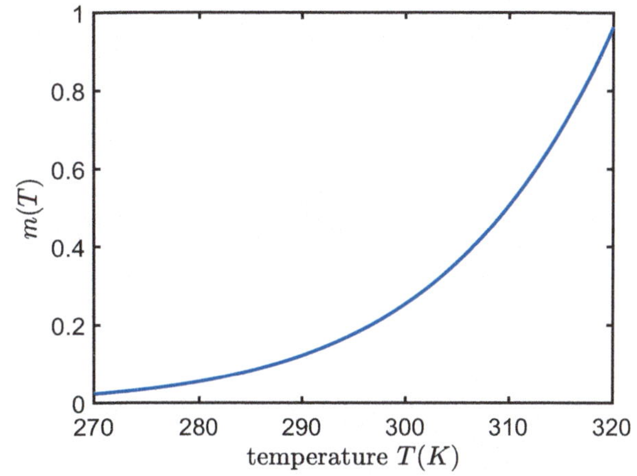

Fig. 6.1 Variation of natural mortality rate of zooplankton with temperature

$$r(T) = r_0 \exp(-(T - T_{opt})^2/(2S^2)),$$

where T_{opt} means optimal Celsius temperature where the intrinsic growth rate r reaches its maximum, S stands for the function breadth, and r_0 represents a constant. Figure 6.2 and Fig. 1a of [23] show the dynamics of intrinsic growth rate function with respect to T in which the variation of intrinsic growth rate is similar to the experimental study [102].

(3) As the relationship that the capture rate of zooplankton is hump shaped with temperature, the following model is deduced from the mathematical model in [23]:

$$a(T) = a_0 \exp(-(T - T_{opt})^2/(2S^2)),$$

where T_{opt} means optimal Celsius temperature where the capture rate a reaches its maximum, S stands for the function breadth, and a_0 represents a constant. Figure 6.3 and Fig. 1c of [23] illustrate the dynamics of capture rate function with respect to T where the variation of capture rate is similar to the experimental study [103].

(4) Based on the relationship that the handling time of zooplankton is U-shaped with temperature, the following model is deduced from the mathematical model in [23]:

$$h(T) = h_0 \exp((T - T_{opt})^2/(2S^2)),$$

where T_{opt} means Celsius optimal temperature where the handling time h reaches its minimum, S stands for the function breadth, and h_0 represents a constant. Figure 6.4 and Fig. 1a of [23] show the dynamics of handling time function with respect to T where the variation of handling time is similar to the experimental study [103].

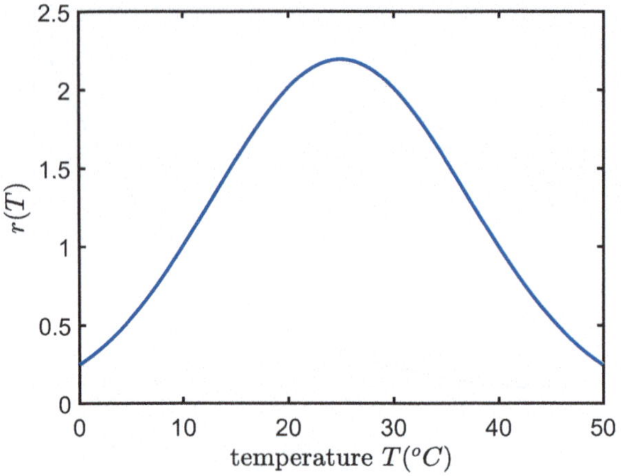

Fig. 6.2 Variation of intrinsic growth rate of phytoplankton with temperature

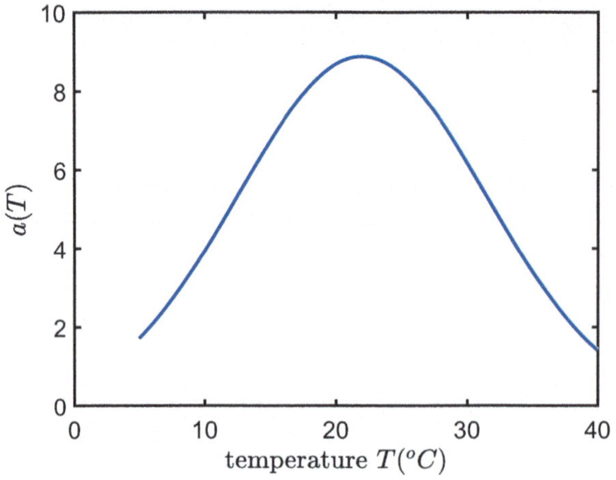

Fig. 6.3 Variation of capture rate of zooplankton with temperature

(5) Based on the relationship that the Holling parameter of functional response is U-shaped with temperature, the following model is deduced from the mathematical model in [23]:

$$n(T) = n_0 \exp((T - T_{opt})^2/(2S^2)),$$

where T_{opt} means Celsius optimal temperature where the Holling parameter n reaches its minimum, S stands for the function breadth, and n_0 represents a constant. Figure 6.5 shows the dynamics of mortality rate function with respect to T where the variation of intrinsic growth rate is similar to Fig. 1e of [23].

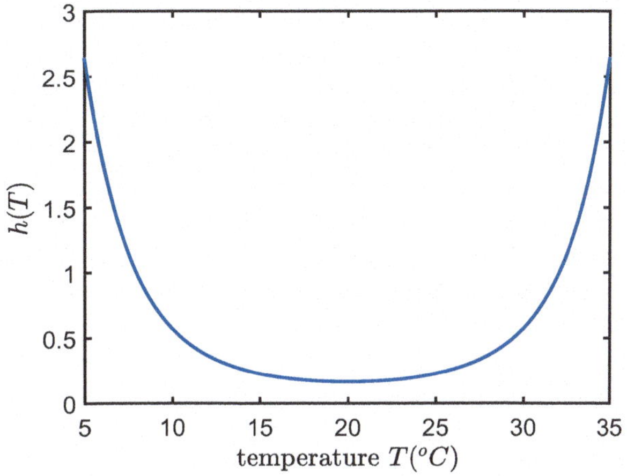

Fig. 6.4 Variation of handling time of zooplankton with temperature

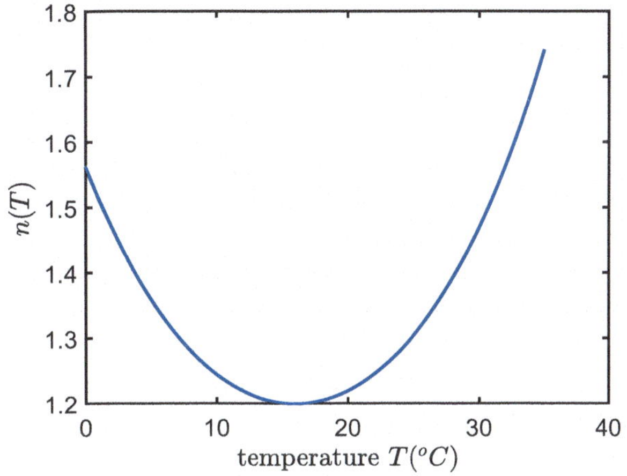

Fig. 6.5 Variation of Holling parameter of functional response with temperature

Accordingly, Fig. 6.5 illustrates that the functional response $p(x) = \frac{ax^n}{1+ahx^n}$ gives Holling type III, where it is a continuously differentiable function defined on $[0, \infty)$ and satisfies

$$p(0) = 0, \quad p^{'}(x) > 0, \quad \lim_{x \to \infty} p(x) = \frac{1}{h} < \infty.$$

Obviously, it displays that $p(x)$ is monotonic. The biological results mean that the phytoplankton population has linearly monotonous effect on predation. But when the number of population increases, the predation will be more intensive. Nevertheless, the nutrient concentration may have an inhibitory effect on the growth rate of the population when the nutrient concentration reaches a high level. Due to the non-monotonic responses, the Holling type III is not suitable when there exists inhibition and defense. As a result, we assume that there exists an exponential relationship between metabolic rates and parameters, and temperature. It enhances our understanding of how temperature variation influences the dynamics of plankton interactions. By these assumptions, it illustrates that the phytoplankton–zooplankton model becomes more closer to the actual situation and also adds complexity of the proposed model.

In [23], a parameterized pelagic producer–predator system is described for green alga Monoraphidium minutum and crustacean Daphnia hyalina. It illustrated that how the temperature dependence of prey carrying capacity is derived and different shapes of this temperature dependence are possible under different resource supply constraints. As the intrinsic growth rate (6.1) shows a plankton growth process with temperature dependence, the interaction between the green alga Monoraphidium minutum and the crustacean Daphnia hyalina can also be described. As a result, to make a detailed analysis of this specific system from a biomathematical perspective is our main aim.

6.2 Qualitative Analysis

In the following, the positivity and boundedness of solutions are firstly given. Then parameter-dependent equilibrium analysis is shown. Additionally, the conditions for local and global stability of model (6.1), and the sufficient conditions for occurrence of Hopf bifurcation are also given.

6.2.1 Positivity and Boundedness

Theorem 51 *Solutions generated by model (6.1) are positive and uniformly bounded.*

Proof Assume $(P(t), Z(t))$ is any solution with positive initial conditions. Because of the right-hand side of model (6.1) as continuous and smooth on $R_+^2 = \{(P, Z) : P, Z > 0\}$, and the initial values $P(0) > 0$, $Z(0) > 0$, it yields

$$\begin{cases} P(t) = P(0)e^{\int_0^t (r(1-\frac{P}{K})-\frac{aP^{n-1}Z}{1+ahP^n})ds} > 0, \\ Z(t) = Z(0)e^{\int_0^t (-m+e\frac{aP^n}{1+ahP^n})ds} > 0. \end{cases}$$

Therefore, the solutions remain in the R_+^2. In view of the first equation of model (6.1), we get

$$\frac{dP}{dt} \leq r\left(1-\frac{P}{K}\right)P.$$

Utilizing comparison theorem, we achieve that $\limsup_{t\to\infty} P(t) \le K$.

Let $\zeta(t) = eP(t) + Z(t)$. Applying the equations of model (6.1), we realize

$$\begin{aligned}\frac{d\zeta(t)}{dt} = e\dot{P}(t) + \dot{Z}(t) =& er\left(1 - \frac{P}{K}\right) P - mZ \\ \le& erP - mZ \\ \le& 2erP - \theta(eP + Z),\end{aligned}$$

where $\theta = \min\{r, m\}$. Hence,

$$\frac{d\zeta(t)}{dt} + \theta\zeta(t) \le 2erK.$$

After calculation, we achieve

$$\limsup_{t\to\infty} Z(t) \le \frac{2erK}{\theta}.$$

Consequently, it is obvious that the solutions of model (6.1) are ultimately bounded. □

Theorem 52 *There exists a unique positive solution* $(N(t), P(t))$ *of model (6.1) for any given initial value* $(P_0, Z_0) \in \mathbb{R}^2_+$.

Proof Let $u(t) = \ln P(t)$ and $v(t) = \ln Z(t)$. Then we get the equations as follows:

$$\begin{cases} \dot{u}(t) &= r\left(1 - \frac{e^{u(t)}}{K}\right) e^{u(t)} - \frac{ae^{u(t)n} e^{v(t)}}{1+ahe^{u(t)n}}, \\ \dot{v}(t) &= -me^{v(t)} + e\frac{ae^{u(t)n} e^{v(t)}}{1+ahe^{u(t)n}}, \end{cases} \tag{6.2}$$

where $t \ge 0$ and the initial values are $u(0) = \ln P_0$, $v(0) = \ln Z_0$. Moreover, the coefficients of (6.2) satisfy local Lipschitz continuous so that there exists a unique solution of (6.2). Hence, $P(t) = e^{u(t)}$, $Z(t) = e^{v(t)}$ is the unique positive solution of (6.1) for initial values $(P_0, Z_0) \in \mathbb{R}^2_+$. □

6.2.2 Parameter-Dependent Equilibrium Analysis

In the following, the influence of parameters is investigated by employing the derivative for the coexisting equilibrium. All biological feasible equilibria of model (6.1) are listed as follows:

(i) The trivial equilibrium $E^0 = (0, 0)$ is always feasible.

(ii) The zooplankton free equilibrium $E^1 = (K, 0)$ is always feasible.

(iii) The coexisting equilibrium $E^* = (P^*, Z^*)$ is feasible if (H1)

$$e > m(T)\left(\frac{1}{K^{n(T)}a(T)} + h(T)\right)$$

holds, where

$$P^* = \left(\frac{m(T)}{ea(T) - a(T)m(T)h(T)}\right)^{\frac{1}{n(T)}},$$

$$Z^* = r(T)\left(1 - \frac{P^*}{K}\right)\frac{1 + a(T)h(T)P^{*n(T)}}{a(T)P^{*(n(T)-1)}},$$

which means that the conversion rate of phytoplankton is greater than the product of the handling time and the natural mortality rate of zooplankton for existence of phytoplankton.

Therefore, the effects of parameters on the coexisting equilibrium are given. Differentiating P^* and Z^* with respect to $a(T)$ and e, we achieve

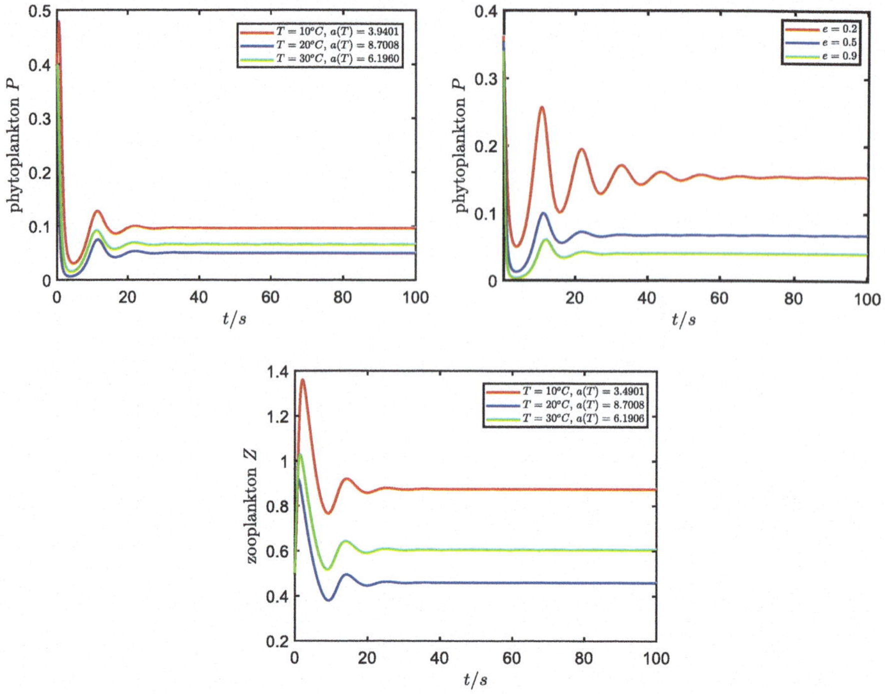

Fig. 6.6 Negative effect of parameters on coexisting equilibrium of model (6.1)

$$\frac{dP^*}{da(T)} = \frac{-m(T)}{n(T)a(T)^2(e - m(T)h(T))}\left(\frac{m(T)}{ea(T) - a(T)m(T)h(T)}\right)^{\frac{1}{n(T)}-1},$$
$$\frac{dP^*}{de} = \frac{-m(T)}{n(T)(e - m(T)h(T))^2}\left(\frac{m(T)}{ea(T) - a(T)m(T)h(T)}\right)^{\frac{1}{n(T)}-1},$$
$$\frac{dZ^*}{da(T)} = \frac{-r(T)}{a(T)^2 P^{*(n(T)-1)}}\left(1 - \frac{P^*}{K}\right).$$

As mentioned above, from Fig. 6.6, we can find that P^* strictly decreases with $a(T)$ and e, meaning that increasing the capture rate of zooplankton and the conversion rate of phytoplankton gets the decreasing effect of coexisting equilibrium density of phytoplankton. In addition, from Fig. 6.6, Z^* also strictly decreases with $a(T)$, meaning that increasing capture rate of zooplankton gets the decreasing effect of coexisting equilibrium density of zooplankton. Again, practising the differential equality, we accomplish

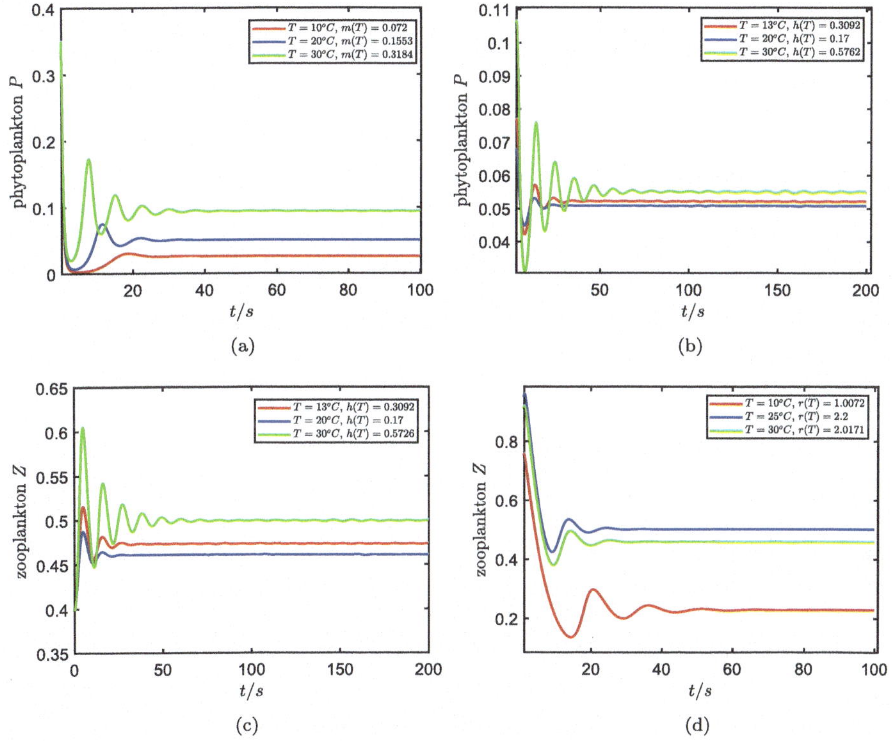

Fig. 6.7 Positive effect of parameters on coexisting equilibrium of model (6.1)

$$\frac{dP^*}{dm(T)} = \left(\frac{m(T)}{ea(T) - a(T)m(T)h(T)}\right)^{\frac{1}{n(T)}-1} \frac{e}{n(T)(e - m(T)h(T))^2},$$

$$\frac{dP^*}{dh(T)} = \left(\frac{m(T)}{ea(T) - a(T)m(T)h(T)}\right)^{\frac{1}{n(T)}-1} \frac{m(T)^2}{n(T)(e - m(T)h(T))^2},$$

$$\frac{dZ^*}{dh(T)} = r(T)\left(1 - \frac{P^*}{K}\right)P^*, \qquad \frac{dZ^*}{dr(T)} = \left(1 - \frac{P^*}{K}\right)\frac{1 + a(T)h(T)P^{*n(T)}}{a(T)P^{*(n(T)-1)}}.$$

The above result and Fig. 6.7 illustrate that P^* varies with $m(T)$ and $h(T)$ increasingly, meaning that, increasing mortality rate, handling time of zooplankton, and growth rate of phytoplankton gets the increasing effect of coexisting equilibrium density of phytoplankton. And Fig. 6.7 illustrates that Z^* varies with $h(T)$ and $r(T)$ increasingly, meaning that, increasing handling time of zooplankton and growth rate of phytoplankton gets the increasing effect of coexisting equilibrium density of zooplankton.

Based on the above results, it is obvious that water temperature and parameters have a significant effect on coexisting equilibrium density of plankton. As for both phytoplankton and zooplankton, increasing and decreasing temperature generally alter the interaction between predator and prey, which may not balance their metabolic demands.

6.2.3 *Stability Analysis*

In the following, the local stability conditions are first derived by applying Routh–Hurwitz criterion. After that, the global asymptotic stability is shown by Lyapunov method. For convenience, the T in the representation of coefficients is dropped in the next subsection.

Theorem 53 *The coexisting equilibrium of model (6.1) is locally asymptotically stable if* (H1) *and* $K\left(\frac{m}{ea-amh}\right)^{\frac{-1}{n}}\left(\frac{e}{e-mh} - n\right) + n < \frac{2e}{e-mh}$ *hold.*

Proof The Jacobian matrix at E^* is given by

$$J(a) = \begin{pmatrix} a_{11}(a) & a_{12}(a) \\ a_{21}(a) & 0 \end{pmatrix},$$

where

$$a_{11}(a) = -\frac{P^*r}{K} + \frac{aZ^*P^{*(n-1)}(1 + ahP^{*n} - n)}{(1 + ahP^{*n})^2},$$

$$a_{12}(a) = -\frac{aP^{*n}}{1 + ahP^{*n}}, \quad a_{21}(a) = \frac{aenZ^*P^{*(n-1)}}{(1 + ahP^{*n})^2}.$$

The characteristic equation of the Jacobian matrix is given as follows:

$$\lambda^2 - a_{11}(a)\lambda - a_{12}(a)a_{21}(a) = 0. \tag{6.3}$$

Let λ_1 and λ_2 be two roots of (6.3). Thus, we obtain

$$\lambda_1 + \lambda_2 = a_{11}(a), \quad \lambda_1\lambda_2 = -a_{12}(a)a_{21}(a). \tag{6.4}$$

This derives $\lambda_1 + \lambda_2 < 0$ and $\lambda_1\lambda_2 > 0$ together with Routh–Hurwitz criterion when $a_{11}(a) < 0$, implying that model (6.1) shows local stability feature around coexisting equilibrium E^* when

$$\frac{aZ^*P^{*(n-2)}(1 + ahP^{*n} - n)}{(1 + ahP^{*n})^2} < \frac{r}{K}.$$

Hence, we get

$$1 + ahP^{*n} = \frac{eaP^{*n}}{m}. \tag{6.5}$$

Utilizing P^*, Z^* and (6.5), inequality (6.5) becomes the form as given below:

$$K\left(\frac{m}{ea - amh}\right)^{\frac{-1}{n}}\left(1 + \frac{hm}{e - mh} - n\right) + n < 2\left(1 + \frac{hm}{e - mh}\right).$$

Therefore it is easy to get that model (6.1) behaves with local stability around coexisting equilibrium E^*. □

Based on the construction method of Lyapunov functional of [105, 106], the global asymptotic stability of coexisting equilibrium can be got.

Theorem 54 *Model (6.1) is globally asymptotically stable around coexisting equilibrium E^* if* (H1) *holds and $n \geq 1 + ahK^n$.*

Proof Let $f(P) = \frac{aP^n}{1+ahP^n}$ and assume a suitable Lyapunov functional listed as follows:

$$V(t) = \int_{P^*}^{P} \frac{ef(u) - m}{f(u)} du + \int_{Z^*}^{Z} \frac{v - Z^*}{v} dv.$$

After calculating the time derivative of $V(t)$ along the solutions of model (6.1), it yields

$$\begin{aligned}
&\tfrac{dV(t)}{dt} = \tfrac{ef(P)-m}{f(P)}\dot{P} + \tfrac{Z-Z^*}{Z}\dot{Z}\\
&= \tfrac{ef(P)-m}{f(P)}\left(r\left(1 - \tfrac{P}{K}\right)P - \tfrac{aP^nZ}{1+ahP^n}\right)\\
&+(Z - Z^*)\left(-m + e\tfrac{aP^n}{1+ahP^n}\right)\\
&= e\,(f(P) - f(P^*))\,(h(P) - h(P^*)),
\end{aligned}$$

where $h(P) = \frac{r(K-P)(1+ahP^n)}{KaP^{n-1}}$ and $h^{'}(P) = \frac{r(ahP^n(K-2P)+(n-1)(P-K)-P)}{KaP^n}$. Then, $f^{'}(P) > 0, h^{'}(P) < 0$ when $n \geq 1 + ahK^n$ for all $P > 0$. Obviously $\dot{V}(t) \leq 0$ and $\dot{V}(t) = 0$ if and only if $P = P^*$. Hence, we get that E^* shows globally asymptotically stable. □

6.2.4 Hopf Bifurcation

Next, we would prove that model (6.1) has periodic solutions about coexisting equilibrium E^*, as the parameter a is larger than the threshold value a^*.

Theorem 55 *Model (6.1) undergoes a Hopf bifurcation at coexisting equilibrium when $a = a^*$, (H1) and $mh < e < mh + \frac{e}{n}$ hold.*

Proof Let

$$\frac{aZ^*P^{*(n-2)}(1 + ahP^{*n} - n)}{(1 + ahP^{*n})^2} = \frac{r}{K}.$$

By P^*, Z^* and (6.5), we can acquire

$$a^* = \frac{m}{e - mh}\left(\frac{nhm + (2-n)e}{K(nhm + (1-n)e)}\right)^n.$$

So, it yields that (6.3) has a pair of purely imaginary roots when $a = a^*$. Again

$$\left.\frac{da_{11}}{da}\right|_{a=a^*} = \frac{r}{Kn}\left(\frac{m}{e-mh}\right)^{\frac{1}{n}} \left.\frac{1}{a^{\frac{1}{n}+1}}\right|_{a=a^*} > 0.$$

Hence it means the transversality condition is satisfied and a simple Hopf bifurcation occurs at $a = a^*$. □

Then, an explicit formula is given to determine the direction of Hopf bifurcation of (6.1) at E^*.

Let translate E^* to the origin by the translation $\bar{x} = P - P^*$, $\bar{y} = Z - Z^*$. The model (6.1) is then transformed into

$$\begin{pmatrix} \dot{\bar{x}} \\ \dot{\bar{y}} \end{pmatrix} = J(a)\begin{pmatrix} \bar{x} \\ \bar{y} \end{pmatrix} + \begin{pmatrix} f(\bar{x}, \bar{y}) \\ h(\bar{x}, \bar{y}) \end{pmatrix}, \tag{6.6}$$

where

$$\begin{aligned} f(\bar{x}, \bar{y}) =& a_1\bar{x}^2 + a_2\bar{x}\bar{y} + a_3\bar{x}^3 + a_4\bar{x}^2\bar{y} + O(|(\bar{x}, \bar{y})|^4), \\ h(\bar{x}, \bar{y}) =& b_1\bar{x}^2 + b_2\bar{x}\bar{y} + b_3\bar{x}^3 + b_4\bar{x}^2\bar{y} + O(|(\bar{x}, \bar{y})|^4), \end{aligned}$$

with

$$a_1 = -\frac{r}{K} - anZ^* P^{*(n-2)} \frac{n-1-ahp^{*n}(n+1)}{2(1+ahP^{*n})^3}, \quad a_2 = \frac{anP^{*(n-1)}}{2(1+ahP^{*n})^2},$$
$$a_3 = -anZ^* P^{*(n-3)} \frac{(n-1)(n-2-2ahP^{*n}(n+1))}{6(1+ahP^{*n})^4}$$
$$+(-anZ^* P^{*(n-3)}) \frac{ah(1+n)P^{*n}(2(1-n)+ahP^{*n}(n+2))}{6(1+ahP^{*n})^4}$$
$$a_4 = -anP^{*(n-2)} \frac{n-1-ahp^{*n}(n+1)}{3(1+ahP^{*n})^3} \bar{x}^2 \bar{y},$$
$$b_1 = aenZ^* P^{*(n-2)} \frac{n-1-ahp^{*n}(n+1)}{2(1+ahP^{*n})^3},$$
$$b_2 = \frac{aenP^{*(n-1)}}{2(1+ahP^{*n})^2}, \quad b_4 = aenP^{*(n-2)} \frac{n-1-ahp^{*n}(n+1)}{3(1+ahP^{*n})^3},$$
$$b_3 = aenZ^* P^{*(n-3)} \frac{(n-1)(n-2-2ahP^{*n}(n+1))}{6(1+ahP^{*n})^4}$$
$$+aenZ^* P^{*(n-3)} \frac{ah(1+n)P^{*n}(2(1-n)+ahP^{*n}(n+2))}{6(1+ahP^{*n})^4}.$$

Assume $\lambda_{1,2}(a) = \alpha(a) \pm i\omega(a)$ with

$$\alpha(a) = \frac{a_{11}(a)}{2}, \omega(a) = \frac{1}{2}\sqrt{-4a_{12}(a)a_{21}(a) - a_{11}(a)^2}.$$

As

$$a = a^*, \lambda_{1,2}(a^*) = \pm i\omega(a^*), \omega(a^*) = \sqrt{-a_{12}(a^*)a_{21}(a^*)},$$

one of the eigenvectors corresponding to the eigenvalues $\lambda_{1,2}(a^*)$ is

$$\xi = (a_{12}(a^*), a_{11}(a^*) - i\omega(a^*))^T.$$

Set

$$B = \begin{pmatrix} a_{12}(a) & 0 \\ a_{11}(a) & -\omega(a) \end{pmatrix}, \quad B^{-1} = \begin{pmatrix} \frac{1}{a_{12}(a)} & 0 \\ \frac{a_{11}(a)}{\omega(a)a_{12}(a)} & -\frac{1}{\omega(a)} \end{pmatrix}.$$

By the transformation $\begin{pmatrix} \tilde{x} \\ \tilde{y} \end{pmatrix} = B \begin{pmatrix} \bar{x} \\ \bar{y} \end{pmatrix}$, (6.6) yields

$$\begin{pmatrix} \dot{\tilde{x}} \\ \dot{\tilde{y}} \end{pmatrix} = \begin{pmatrix} \alpha(a) & -\omega(a) \\ \omega(a) & \alpha(a) \end{pmatrix} \begin{pmatrix} \tilde{x} \\ \tilde{y} \end{pmatrix} + \begin{pmatrix} f_1(\tilde{x}, \tilde{y}) \\ h_1(\tilde{x}, \tilde{y}) \end{pmatrix}, \tag{6.7}$$

where

$$f_1(\tilde{x}, \tilde{y}) = \frac{1}{a_{21}(a)} f(a_{12}(a)\tilde{x}, a_{11}(a)\tilde{x} - \omega(a)\tilde{y}),$$
$$h_1(\tilde{x}, \tilde{y}) = \frac{a_{11}(a)}{\omega(a)a_{12}(a)} f(a_{12}(a)\tilde{x}, a_{11}(a)\tilde{x}$$
$$-\omega(a)\tilde{y}) - \frac{1}{\omega(a)} h(a_{12}(a)\tilde{x}, a_{11}(a)\tilde{x} - \omega(a)\tilde{y}),$$

and

$$\begin{aligned}
&f(a_{12}(a)\tilde{x}, a_{11}(a)\tilde{x} - \omega(a)\tilde{y}) = a_1 a_{12}(a)^2\tilde{x}^2 + a_2 a_{12}(a)\tilde{x}(a_{11}(a)\tilde{x} \\
&-\omega(a)\tilde{y}) + a_3 a_{12}(a)^3\tilde{x}^3 \\
&+a_4 a_{12}(a)^2\tilde{x}^2(a_{11}(a)\tilde{x} - \omega(a)\tilde{y}) \\
&+O(|(a_{12}(a)\tilde{x}, a_{11}(a)\tilde{x} - \omega(a)\tilde{y})|^4), \\
&h(a_{12}(a)\tilde{x}, a_{11}(a)\tilde{x} - \omega(a)\tilde{y}) = b_1 a_{12}(a)^2\tilde{x}^2 + b_2 a_{12}(a)\tilde{x}(a_{11}(a)\tilde{x} \\
&-\omega(a)\tilde{y}) + b_3 a_{12}(a)^3\tilde{x}^3 + b_4 a_{12}(a)^2\tilde{x}^2(a_{11}(a)\tilde{x} - \omega(a)\tilde{y}) \\
&+O(|(a_{12}(a)\tilde{x}, a_{11}(a)\tilde{x} - \omega(a)\tilde{y})|^4).
\end{aligned}$$

Then (6.7) is rewritten into the following polar coordinate form:

$$\begin{cases} \dot{l} = \alpha(a)l + \sigma(a)l^3 + O(|l|^4), \\ \dot{\theta} = \omega(a) + c(a)l^2 + O(|l|^4). \end{cases} \tag{6.8}$$

Utilizing Taylor expansion of (6.8) at $a = a^*$, it yields

$$\begin{cases} \dot{l} = \alpha'(a^*)(a - a^*)l + \sigma(a^*)l^3 + O(|l|^4), \\ \dot{\theta} = \omega(a^*) + \omega'(a^*)(a - a^*) + c(a^*)l^2 + O(|l|^4). \end{cases}$$

Because the sign of $\sigma(a^*)$ determines the stability of bifurcated periodic solutions, it yields

$$\sigma(a^*) = \frac{1}{16}(f_{1\tilde{x}\tilde{x}\tilde{x}} + h_{1\tilde{x}\tilde{x}\tilde{y}}) + \frac{1}{16\omega(a^*)}\Big(f_{1\tilde{x}\tilde{y}} f_{1\tilde{x}\tilde{x}} - h_{1\tilde{x}\tilde{y}} h_{1\tilde{x}\tilde{x}} - f_{1\tilde{x}\tilde{x}} h_{1\tilde{x}\tilde{x}}\Big),$$

where

$$\begin{aligned}
f_{1\tilde{x}\tilde{x}\tilde{x}} &= \tfrac{1}{a_{21}(a^*)}(a_3 a_{12}(a^*)^3 + a_4 a_{12}(a^*)^2 a_{11}(a^*)), \\
h_{1\tilde{x}\tilde{x}\tilde{y}} &= -a_{11}(a^*) a_4 a_{12}(a^*) + b_4 a_{12}(a^*)^2, \\
f_{1\tilde{x}\tilde{y}} &= -\tfrac{1}{a_{21}(a^*)} a_2 a_{12}(a^*)\omega(a^*), \\
f_{1\tilde{x}\tilde{x}} &= \tfrac{1}{a_{21}(a^*)}(a_1 a_{12}(a^*)^2 + a_2 a_{12}(a^*) a_{11}(a^*)), \\
h_{1\tilde{x}\tilde{y}} &= -a_{11}(a^*) a_2 + b_2 a_{12}(a^*), \\
h_{1\tilde{x}\tilde{x}\tilde{y}} &= -a_{11}(a^*) a_4 a_{12}(a^*) + b_4 a_{12}(a^*)^2, \\
h_{1\tilde{x}\tilde{x}} &= \tfrac{a_{11}(a^*)}{\omega(a^*)}(a_1 a_{12}(a^*) + a_2 a_{11}(a^*)) - \tfrac{1}{\omega(a^*)}(b_1 a_{12}(a^*)^2 + b_2 a_{12}(a^*) a_{11}(a^*)).
\end{aligned}$$

Thus, we get

$$\mu_2 = -\frac{\sigma(a^*)}{\alpha'(a^*)}.$$

According to Poincaré–Andronov–Hopf bifurcation theorem, we obtain the following result.

Theorem 56 *The direction of Hopf bifurcation of* (6.1) *at* E^* *is supercritical if* (H1), $mh < e < mh + \frac{e}{n}$ *are satisfied and* $\mu_2 > 0$. *When* $\mu_2 < 0$, *the direction of Hopf bifurcation is subcritical. Furthermore, as* $\sigma(a^*) > 0$, *the bifurcated periodic solutions are unstable while the bifurcated periodic solutions are stable as* $\sigma(a^*) < 0$.

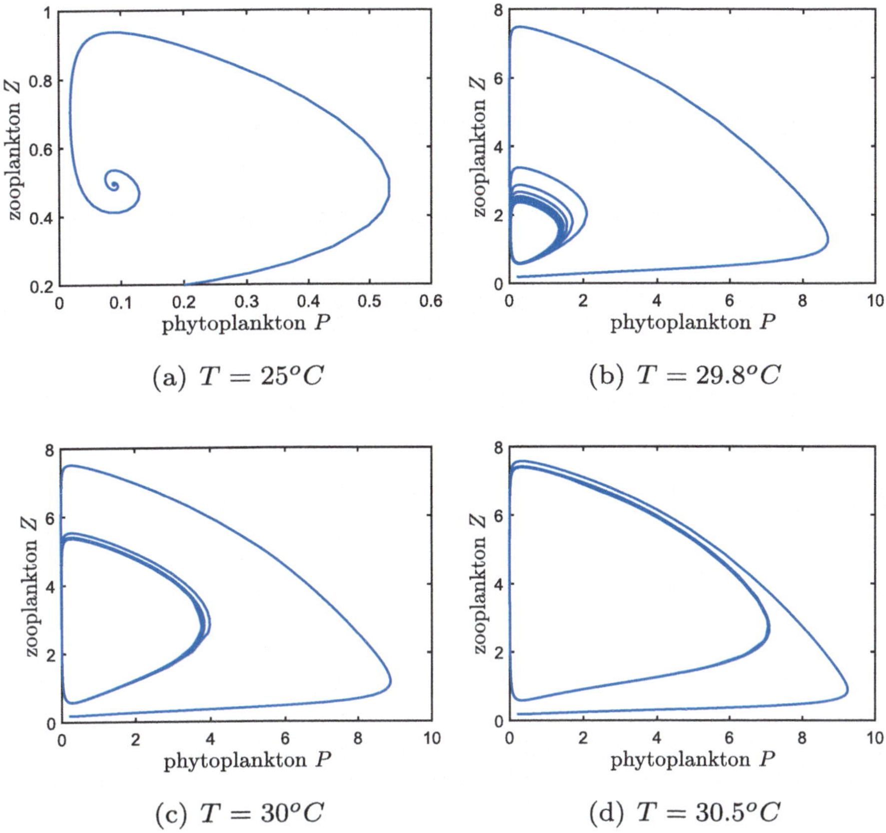

(a) $T = 25^oC$ (b) $T = 29.8^oC$

(c) $T = 30^oC$ (d) $T = 30.5^oC$

Fig. 6.8 Phase plots of model (6.1) with different values of T

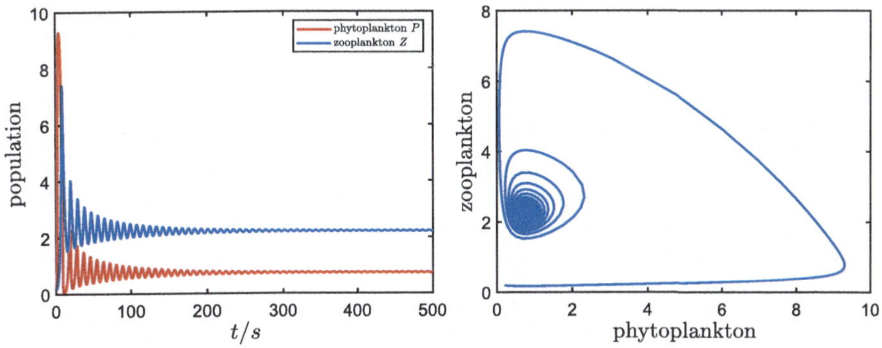

(a) Behavior of plankton (b) The phase plot of model (6.1)

Fig. 6.9 Behavior of plankton and the phase plot of model (6.1) with $a = 1.5 < a^*$ and $T = 30\,°\mathrm{C}$

Table 6.1 Description of the parameters for model (6.1)

Parameters	Description	Values	Reference
r	Intrinsic growth rate of phytoplankton	$r_0 = 2.2$, $S = 12$, $T_{opt} = 25$	[23]
K	Carrying capacity of environment	10	–
h	Handling time	$h_0 = 0.17$, $S = 6.4$ $T_{opt} = 20$	[23]
a	Capture rate of zooplankton	$a_0 = 8.9$, $S = 9.4$ $T_{opt} = 22$	[23]
e	Conversion efficiency	0.5	–
m	Mortality rate of zooplankton	$m_0 = 4.4 \times 10^8$, $E = -0.55$ $k = 8.62 \times 10^{-5}$	[23]
n	Holling parameter of functional response	$n_0 = 1.2$, $S = 22$ $T_{opt} = 16$	[23]
T	Water temperature	–	–

6.3 Numerical Simulations

In the following, numerical simulations are presented to support the analytical results as shown in Table 6.1. Figure 6.8a shows the stability diagram of E^* as taking $T = 25\,°\text{C}$. Figure 6.8b–d illustrates its variation effect of T on the model. Assume the capture rate a of zooplankton as a bifurcation parameter. Taking the values mentioned in Table 6.1 and $T = 30\,°\text{C}$, Fig. 6.9 shows that E^* is locally asymptotically stable when $a = 1.5 < a^*$ and Fig. 6.10 illustrates it is unstable when $a = 2 > a^*$. Obviously the model undergoes a Hopf bifurcation around E^* at $a = a^* = 1.8817$. Then we calculate $\mu_2 > 0$ and $\sigma(a^*) < 0$, meaning that model (6.1) undertakes supercritical Hopf bifurcation at capture rate $a = a^*$ and the bifurcated periodic solutions are stable as shown in Figs. 6.10 and 6.11. When the capture rate is small, the plankton population keeps steady state, which means the sustainable development of our plankton model. However, when the capture rate is relatively high, the ecological balance of our ecosystem will be disrupted. Moreover, Fig. 6.10 shows the appearance of bifurcated periodic solutions of (6.1) is the reason for the fluctuation of the plankton population. It is illustrated clear that even when there exists small amplitude oscillations in our plankton ecosystem, our plankton ecosystem still has the phytoplankton and zooplankton.

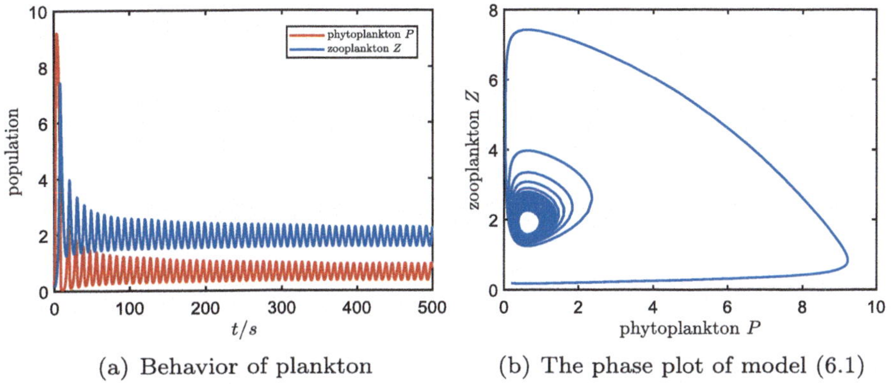

Fig. 6.10 Behavior of plankton and the phase plot of model (6.1) with $a = 2 > a^*$ and $T = 30\,°\mathrm{C}$

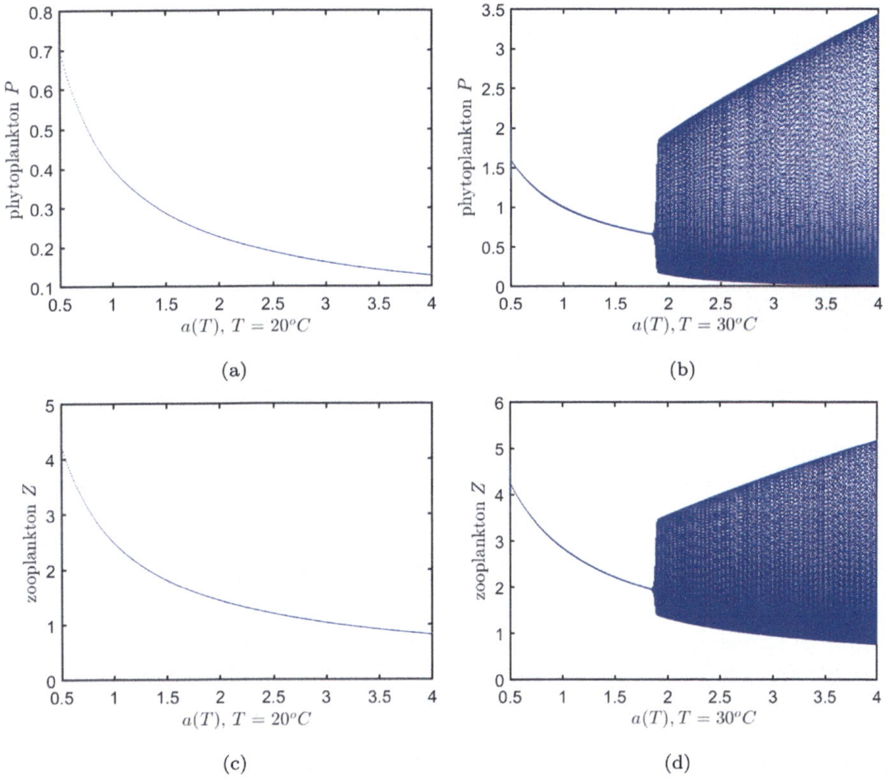

Fig. 6.11 Bifurcation analysis for a function of capture rate a of zooplankton and population densities

Figure 6.11 presents the bifurcation results with respect to capture rate a of zooplankton and other parameters taken as shown in Table 6.1. The left panels of Fig. 6.11 show that, when $T = 20\,°C$, the model remains stable. As shown in the right panels of Fig. 6.11, when $T = 30\,°C$, the model remains stable with a small range of a and has a Hopf bifurcation with larger a. The results show that when the temperature remains high, the ecosystem of phytoplankton and zooplankton becomes unstable according to the increase of the capture rate of zooplankton.

Figure 6.12 presents the bifurcation results with respect to natural mortality rate m of zooplankton and other parameters taken as in Table 6.1. The left panels of Fig. 6.12 show that when $T = 20\,°C$, the model remains stable. The right panels of Fig. 6.12 illustrate that when $T = 28\,°C$, the model remains stable with a small range of m and has a Hopf bifurcation with larger m. The results show that when the temperature remains high, the ecosystem of phytoplankton and zooplankton becomes unstable according to the increase of the natural mortality of zooplankton.

Figure 6.13 presents the bifurcation results with respect to handling time h of zooplankton and other parameters remaining the same as given in Table 6.1. The left panels of Fig. 6.13 show that when $T = 10\,°C$, the model remains stable. The right panels of Fig. 6.13 illustrate that when $T = 20\,°C$, the model remains stable with a small range of h and has a Hopf bifurcation with larger h. The results show that when the temperature remains at appropriate level, the ecosystem of phytoplankton and zooplankton becomes unstable according to the increase of the handling time of zooplankton.

Figure 6.14 presents the bifurcation results with respect to Holling parameter n of functional response and other parameters remaining the same as shown in Table 6.1. The left panels of Fig. 6.14 show that when $T = 20\,°C$, the model remains stable. The right panels of Fig. 6.14 show that when $T = 30\,°C$, the model has a Hopf bifurcation with a small range of n and remains stable with larger n. The results show that when the temperature is high, the ecosystem of phytoplankton and zooplankton becomes unstable according to the decrease of the Holling parameter of type III.

Figure 6.15 presents the bifurcation results with respect to intrinsic growth rate r of phytoplankton and other parameters remaining the same as shown in Table 6.1. The left panels of Fig. 6.15 show that, when $T = 12\,°C$, the model remains stable. The right panels of Fig. 6.15 illustrate that when $T = 29\,°C$, the model remains stable with a small range of r and has a Hopf bifurcation with larger r. The results show that with high temperature, the ecosystem of phytoplankton and zooplankton becomes unstable according to increase of the intrinsic growth rate of phytoplankton.

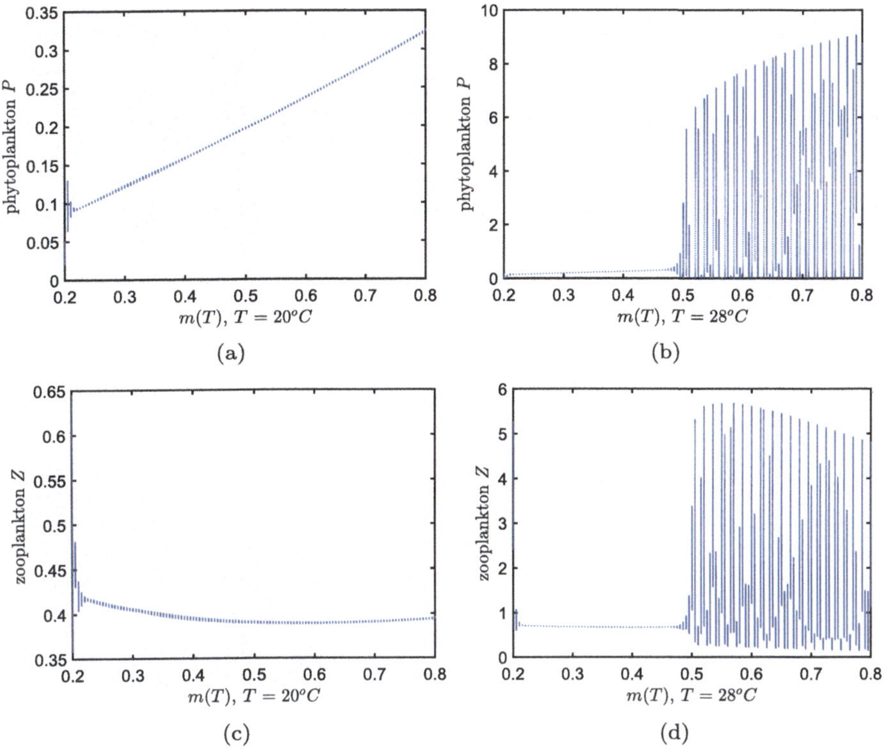

Fig. 6.12 Bifurcation analysis for a function of natural mortality rate m of zooplankton and population densities

Figure 6.16 presents the bifurcation results with respect to water temperature T and other parameters remaining the same as given in Table 6.1. The model has two Hopf bifurcations shown in Fig. 6.16. The results show that if the water temperature is too low or too high, the ecosystem of phytoplankton and zooplankton becomes unstable.

6.4 Sensitivity Analysis

This section focuses on the effects of the relative factors responsible for the variety of biomass in order to understand the dynamical behavior of phytoplankton and zooplankton. The main method is to calculate the sensitivity indices of coexisting equilibrium to the parameters of the model, which is calculated by normalized forward sensitivity method [104].

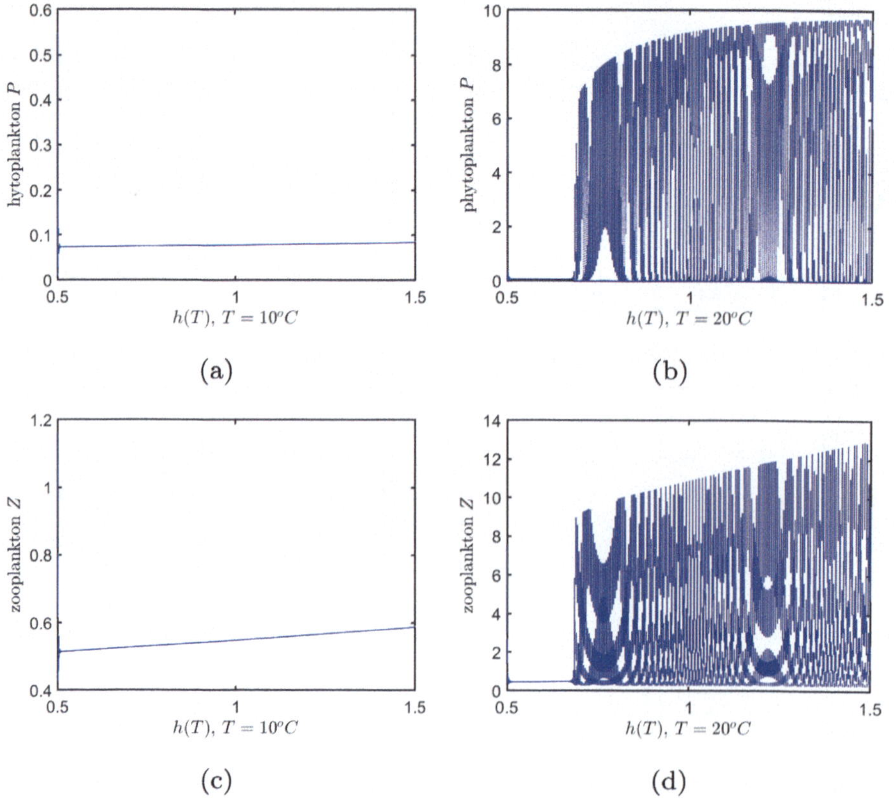

Fig. 6.13 Bifurcation analysis for a function of handling time h of zooplankton and population densities

Definition 16 [104] The normalized forward sensitivity index γ_b^X of a variable X is defined as follows:

$$\gamma_b^X := \frac{\partial X}{\partial b} \times \frac{b}{X}.$$

Table 6.2 shows the sensitivity index of the coexisting equilibrium E^* with temperature $T = 20\,°\mathrm{C}$. These indices illustrate that how crucial each parameter is to plankton biomass. Table 6.2 presents the influence of almost all parameters on the coexisting equilibrium. When the sign of the sensitivity indices of E^* is positive, E^* varies with parameters increasingly. Conversely, E^* varies with parameters decreasingly. It also shows that the equilibrium density of the phytoplankton is most sensitive to the conversion rate e of phytoplankton and the mortality rate $m(T)$ of zooplankton, but it is not sensitive to the growth rate $r(T)$ and the carrying capacity K of phytoplankton. Moreover, the equilibrium density of zooplankton is again most sensitive

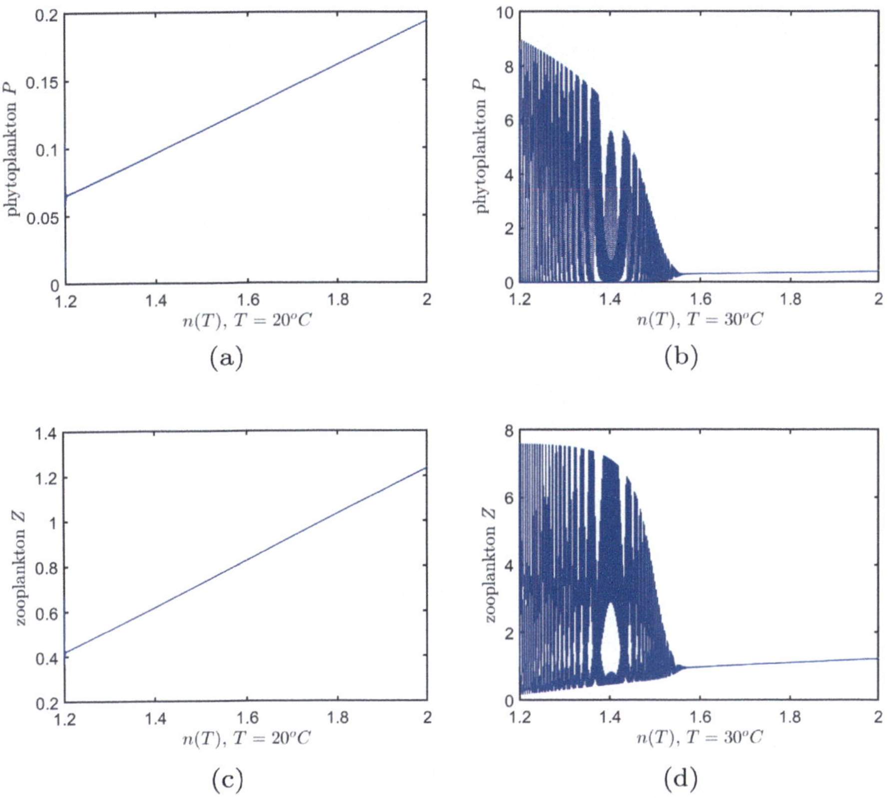

Fig. 6.14 Bifurcation analysis for a function of Holling parameter n of functional response and population densities

Table 6.2 Sensitivity indices of E^* to parameters for model (6.1) with temperature $T = 20\,°\text{C}$

Parameters x_i	$\gamma_{x_i}^{P^*}$	$\gamma_{x_i}^{Z^*}$
r	0	1
K	0	0.0069
h	0.3976	0.0528
a	−0.8197	−0.9472
e	−7.5294	0.1405
m	7.5294	−0.1405
n	−3.9996	2.669

to the Holling parameter $n(T)$ of functional response. Hence, when the temperature remains low level, the decreasing mortality rate of zooplankton and increasing conversion rate would play a significant role in the population growth process of phytoplankton.

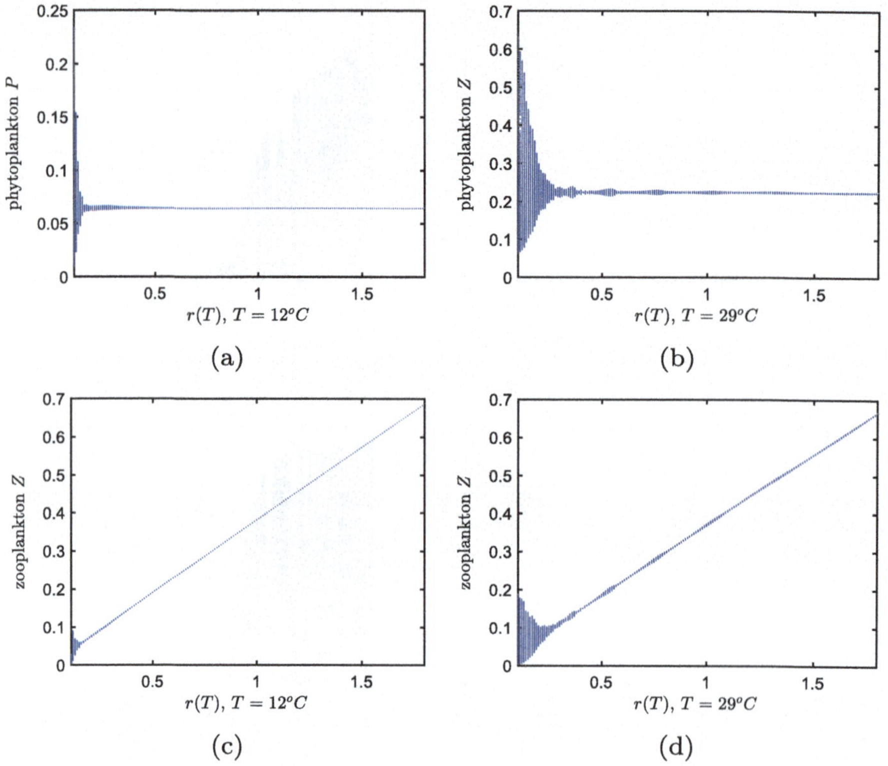

Fig. 6.15 Bifurcation analysis for a function of intrinsic growth rate r of phytoplankton and population densities

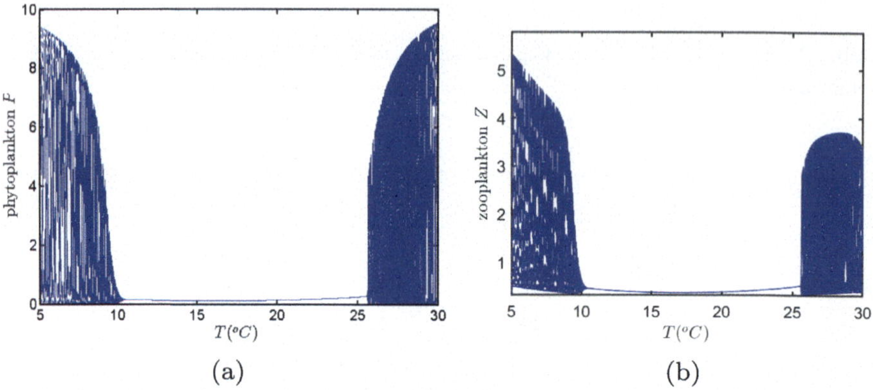

Fig. 6.16 Bifurcation analysis for a function of water temperature T and population densities

Chapter 7
The Effect of Plankton Cell Size

There are numerous models presented to study the dynamical behaviors of phytoplankton and zooplankton [17, 34, 90, 109–113]. Many factors, such as prey-taxis, age structure, allelopathy, prey refuge, gestation time delay, and so on (see [21, 114–119] and the references therein), have been taken into account. The cell size of plankton is significant for marine ecosystem because of its determination to their growth rates, metabolic activity and numerical abundance [36], influence to carbon cycle, species interactions, and community structure [120]. Furthermore, it also affects the densities and success rate of biomass of plankton, like Bythotrephes [121] and larval fishes foraging [122]. However, there exist few results about the trait in plankton system. For example, a nutrient-phytoplankton–zooplankton system with adaptive change of cell size was studied to show the coevolution of zooplankton and phytoplankton leading to oscillation of cell size and ecosystem dynamics [36]. And disruptive selection was found that it can lead to evolutionary branching of phytoplankton and zooplankton using the interaction of trophic and body-size-dependent population growth [35]. Naturally, it is important to determine the influence of cell size on the dynamical behavior of diffusive plankton.

Due to the impact of many factors (food supplies, environment, season, etc.), the distribution of the ocean plankton is inhomogeneous spatially within a fixed bounded domain. Commonly, the diffusion of plankton follows the general process: from an area of high concentration to one of low concentration. So related dynamical models, including the spatiotemporal effect, have been increasingly studied [123, 124]. In the following, a diffusive plankton system about cell size and time delay is presented. Moreover, in order to conquer the drawback about its singular phenomenon in low-density state, we use a Beddington–DeAngelis-type functional response function to describe the relationship between phytoplankton and zooplankton. The gestation time of phytoplankton and capturing time of zooplankton are considered to study its effect on the stability or instability of prey densities for predation. In addition, the influence of cell size on plankton system is investigated by choosing different dependent parameters, e.g., growth rate, sinking rate of phytoplankton, and consumption

S. T. Liu and L. Zhang, *Pattern Dynamics of Marine Plankton Behavior*,
https://doi.org/10.1007/978-981-97-5369-7_7

rate of zooplankton. With this, the influence of diffusion coefficients on the occurrence of Turing instability and time delay on the existence of Hopf bifurcation are also investigated.

7.1 Model Formulation

Some assumptions and related parameters are presented below.

- P means the population density of phytoplankton and z_1 is the cell size, Z and z_2 represent the corresponding population density of zooplankton and cell size, respectively.
- The relationships about cell size, growth rate, and sinking rate are listed by $r_1(z_1) = \frac{z_1}{c_1 z_1^2 + c_2 z_1 + c_3}$ and $s(z_1) = \beta z_1^2$ $(r_1(z_1) > s(z_1))$ [71, 74], respectively.
- The zooplankton consumption rate $C(z_1, z_2) \triangleq C_m \exp\left(-\frac{1}{\lambda}(z_1 - kz_2)^2\right)$ remains maximum when zooplankton lives on particles with an optimal phytoplankton size ratio [70].
- Assume that intraspecific competition with peers of the same species affects the phytoplankton death and a fixed fraction of the zooplankton population represents the zooplankton death rate. After preying upon phytoplankton, the reproduction of zooplankton is not instantaneous, but is mediated by some constant time lag τ for gestation.

These above factors indicate more realistic reproduction behavior of marine life. As stated by the above assumptions, the system in the presence of cell size is considered as follows:

$$\begin{cases} \frac{\partial P(x,t)}{\partial t} - d_1 \Delta P(x,t) = P(x,t)(r_1(z_1) - s(z_1) - \alpha P(x,t)) - \frac{C(z_1,z_2)P(x,t)Z(x,t)}{a+P(x,t)+bZ(x,t)}, \\ x \in \Omega,\ t > 0, \\ \frac{\partial Z(x,t)}{\partial t} - d_2 \Delta Z(x,t) = \frac{C(z_1,z_2)P(x,t-\tau)Z(x,t-\tau)}{a+P(x,t-\tau)+bZ(x,t-\tau)} - r_2 Z(x,t), \\ x \in \Omega,\ t > 0, \\ \frac{\partial P(x,t)}{\partial \nu} = \frac{\partial Z(x,t)}{\partial \nu} = 0, \\ x \in \partial\Omega,\ t > 0, \\ P(x,\theta) = P_0(x,\theta) \geq 0, \quad Z(x,\theta) = Z_0(x,\theta) \geq 0, \\ x \in \bar{\Omega},\ \theta \in [-\tau, 0], \end{cases} \tag{7.1}$$

where Δ expresses the Laplacian operator in $\mathbf{R}^n$, d_1 and d_2 indicate diffusion coefficients of the phytoplankton and zooplankton, respectively. Table 7.1 implies the meaning of each parameter. Suppose $\Omega = (0, l\pi)$ $(l > 0)$ is bounded with smooth boundary $\partial\Omega$. ν shows the outward unit normal vector on $\partial\Omega$. The homogeneous Neumann boundary conditions signify that this system is self-contained with zero population flux across the boundary. In addition, the initial functions $P_0(x, \theta)$ and $Z_0(x, \theta)$ behave nonnegative continuous.

For any solution of (7.1), *it yields*

Table 7.1 Parameter definitions in system (7.1) and their units

Symbol	Parameter definition	Units
z_1	Phytoplankton cell size	μm
z_2	Zooplankton cell size	μm
$r_1(z_1) = \frac{z_1}{c_1 z_1^2 + c_2 z_1 + c_3}$	Maximum phytoplankton growth rate	1/day
(c_1, c_2, c_3)	Coefficients of $r_1(z_1)$	1/μm
$s(z_1) = \beta z_1^2$	Phytoplankton sinking rate	1/day
β	Sinking rate coefficient	1/day
α	Crowding effect parameter	1/day
$C(z_1, z_2) = C_m \exp\left(-\frac{1}{\lambda}(z_1 - kz_2)^2\right)$	Zooplankton consumption rate	L/day
C_m	Maximum consumption rate	L/day
λ	Consumption rate coefficient	μm
k	Consumption rate coefficient	Unitless
a	Measure of the abundance of phytoplankton	1/day
b	Zooplankton interference	1/day
r_2	Zooplankton natural mortality rate	1/day

$$\limsup_{t\to\infty} P(\cdot, t) \le \frac{r_1(z_1) - s(z_1)}{\alpha},$$

$$\limsup_{t\to\infty} \int_\Omega Z(\cdot, t) \le \left(\frac{r_1(z_1) - s(z_1)}{4r_2} + 1\right) \frac{r_1(z_1) - s(z_1)}{\alpha} |\Omega|.$$

Proof Let $(P(x,t), Z(x,t))$ be a nonnegative solution of (7.1). Then the first equation of (7.1) indicates that

$$\begin{cases} \frac{\partial P}{\partial t} \le d_1 \Delta P + P(r_1(z_1) - s(z_1) - \alpha P), & x \in (0, l\pi),\ t > 0, \\ P_x(0,t) = P_x(l\pi, t) = 0, & t > 0. \end{cases}$$

And it is obvious that $P(x,t) \le \frac{r_1(z_1) - s(z_1)}{\alpha}$ by comparison principle. There exists $t_1 > 0$ such that $P(x,t) \le \frac{r_1(z_1) - s(z_1)}{\alpha} + \epsilon_1$ for all $x \in \Omega$ and $t \ge t_1$ and for any $\epsilon_1 > 0$. Let $\int_\Omega P(x,t)dx = U_1(t)$, $\int_\Omega Z(x,t)dx = U_2(t)$. Then by utilizing the Neumann boundary condition, we obtain

$$\frac{dU_1}{dt} = \int_\Omega \frac{\partial P}{\partial t} dx = \int_\Omega \left[P(r_1(z_1) - s(z_1) - \alpha P) - \frac{C(z_1, z_2) P Z}{a + P + bZ}\right] dx,$$
$$\frac{dU_2}{dt} = \int_\Omega \frac{\partial Z}{\partial t} dx = \int_\Omega \left[\frac{C(z_1, z_2) P(x, t-\tau) Z(x, t-\tau)}{a + P(x, t-\tau) + bZ(x, t-\tau)} - r_2 Z\right] dx.$$

Then

$$
\begin{aligned}
&\frac{d(U_1(t)+U_2(t+\tau))}{dt}\\
&=\int_\Omega \frac{\partial P(x,t)}{\partial t}dx+\int_\Omega \frac{\partial Z(x,t+\tau)}{\partial t}dx\\
&=\int_\Omega [P(r_1(z_1)-s(z_1)-\alpha P)-r_2 Z(x,t+\tau)]dx\\
&\leq \frac{(r_1(z_1)-s(z_1))^2}{4\alpha}|\Omega|+r_2\left(\frac{r_1(z_1)-s(z_1)}{\alpha}+\epsilon_1\right)|\Omega|-r_2(U_1(t)+U_2(t+\tau)).
\end{aligned}
$$

Using the comparison principle, we get

$$
\limsup_{t\to\infty}(U_1(t)+U_2(t+\tau))\leq\left(\frac{r_1(z_1)-s(z_1)}{4r_2}+1\right)\frac{r_1(z_1)-s(z_1)}{\alpha}|\Omega|.
$$

From above, we gain that any solution $Z(x,t)$ satisfies an L^1 a priori estimate $J_1=\left(\frac{r_1(z_1)-s(z_1)}{4r_2}+1\right)\frac{r_1(z_1)-s(z_1)}{\alpha}|\Omega|$ for large t, depending on the parameters of system (7.1) and Ω. Moreover, we can apply the L^1 bound to obtain an L^∞ bound c, such that $\limsup_{t\to+\infty} Z(x,t)\leq c$ with c independent of P_0, Z_0, d_1, d_2. Thus,

$$
\limsup_{t\to\infty}\int_\Omega Z(x,t)=\limsup_{t\to\infty}U_2(t)\leq\left(\frac{r_1(z_1)-s(z_1)}{4r_2}+1\right)\frac{r_1(z_1)-s(z_1)}{\alpha}|\Omega|.
$$

□

The following focuses on the conditions of the local stability, Turing instability of system (7.1) for $\tau=0$, and finding the threshold of the occurrence of Hopf bifurcation for $\tau>0$. In order to determining the direction of Hopf bifurcation and the stability of bifurcated periodic solutions, an explicit formula is then obtained using center manifold theory.

7.2 Spatiotemporal Dynamics Analysis

From the viewpoint of ecology, the coexisting equilibrium is our focus. By simple calculation, it yields that there exists coexisting equilibrium $E_*=(P_*,Z_*)$ for system (7.1), where

$$
Z_*=\frac{P_*(r_1(z_1)-s(z_1)-\alpha P_*)}{r_2}=\frac{P_*\left(\frac{z_1}{c_1z_1^2+c_2z_1+c_3}-\beta z_1^2-\alpha P_*\right)}{r_2},
$$

with $P_*<\frac{r_1(z_1)-s(z_1)}{\alpha}$. Then we get

$$h(P) = b\alpha P^2 + (C(z_1, z_2) - r_2 - b(r_1(z_1) - s(z_1)))P - ar_2.$$

Clearly, $h(0) = -ar_2 < 0$ and $h\left(\frac{r_1(z_1)-s(z_1)}{\alpha}\right) = \frac{r_1(z_1)-s(z_1)}{\alpha}(C(z_1, z_2) - r_2) - r_2 a$. When $h\left(\frac{r_1(z_1)-s(z_1)}{\alpha}\right) > 0$, it means that $h(P) = 0$ has a positive root. Consequently, system (7.1) gets a unique coexisting equilibrium E_* if and only if

(H1)

$$(C(z_1, z_2) - r_2)(r_1(z_1) - s(z_1)) > r_2 a\alpha$$

is satisfied, and

$$\begin{aligned} P_* &= \frac{b(r_1(z_1) - s(z_1)) - C(z_1, z_2) + r_2)}{2b\alpha} \\ &+ \frac{b\sqrt{\left(b(r_1(z_1) - s(z_1)) - C(z_1, z_2) + r_2\right)^2 + 4ba\alpha r_2}}{2b\alpha} \\ &= \frac{b\left(\frac{z_1}{c_1 z_1^2 + c_2 z_1 + c_3} - \beta z_1^2\right) - C_m \exp\left(-\frac{1}{\lambda}(z_1 - kz_2)^2\right) + r_2}{2b\alpha} \\ &+ \frac{\sqrt{\left(b\left(\frac{z_1}{c_1 z_1^2 + c_2 z_1 + c_3} - \beta z_1^2\right) - C_m \exp\left(-\frac{1}{\lambda}(z_1 - \kappa z_2)^2\right) + r_2\right)^2 + 4ba\alpha r_2}}{2b\alpha}. \end{aligned}$$

Therefore, let $P_1(x, t) = P(x, t) - P_*$, $Z_1(x, t) = Z(x, t) - Z_*$ and the linearized system of (7.1) is assumed as

$$\begin{pmatrix} \frac{\partial P_1}{\partial t} \\ \frac{\partial Z_1}{\partial t} \end{pmatrix} = D\Delta \begin{pmatrix} P_1(x, t) \\ Z_1(x, t) \end{pmatrix} + L_1 \begin{pmatrix} P_1(x, t) \\ Z_1(x, t) \end{pmatrix} + L_2 \begin{pmatrix} P_1(x, t - \tau) \\ Z_1(x, t - \tau) \end{pmatrix}, \tag{7.2}$$

where

$$D = \begin{pmatrix} d_1 & 0 \\ 0 & d_2 \end{pmatrix}, \quad L_1 = \begin{pmatrix} a_{11} & C(z_1, z_2)a_{12} \\ 0 & a_{22} \end{pmatrix}, \quad L_2 = \begin{pmatrix} 0 & 0 \\ C(z_1, z_2)b_{21} & C(z_1, z_2)b_{22} \end{pmatrix}, \tag{7.3}$$

and

$$\begin{aligned} a_{11} &= -\alpha P_* + \frac{C(z_1, z_2)P_* Z_*}{(a + P_* + bZ_*)^2}, \\ a_{12} &= \frac{-P_*(a + P_*)}{(a + P_* + bZ_*)^2}, \\ a_{22} &= -r_2, \end{aligned}$$

$$\begin{aligned} b_{21} &= \frac{Z_*(a + bZ_*)}{(a + P_* + bZ_*)^2}, \\ b_{22} &= \frac{-P_*(a + P_*)}{(a + P_* + bZ_*)^2}. \end{aligned}$$

The characteristic equation of (7.2) is

$$\det(\lambda I - M_n - L_1 - L_2 e^{-\lambda\tau}) = 0, \tag{7.4}$$

where I indicates the 2×2 identity matrix and $M_n = -\frac{n^2}{l^2}D, n \in N_0$. Consequently, we realize

$$\begin{aligned} &\lambda^2 + \left((d_1 + d_2)\tfrac{n^2}{l^2} - a_{11} - a_{22}\right)\lambda + d_1 d_2 \tfrac{n^4}{l^4} - (d_1 a_{22} + d_2 a_{11})\tfrac{n^2}{l^2} + a_{11}a_{22} \\ &- C(z_1, z_2)\left(C(z_1, z_2)a_{12}b_{21} + d_1 \tfrac{n^2}{l^2} b_{22} - a_{11}b_{22} + \lambda b_{22}\right)e^{-\lambda\tau} = 0. \end{aligned} \tag{7.5}$$

Next, we examine the local stability, Turing instability of coexisting equilibrium E_*, and the existence of Hopf bifurcation. For then, we assume that condition (H1) holds in the next parts.

7.2.1 *Stability and Turing Instability Without Delay*

In this part, the effects of diffusion and cell size on the dynamics of system (7.1) are investigated.

As $\tau = 0$, (7.5) is equivalent to

$$\lambda^2 - tr_n + \Delta_n = 0, \tag{7.6}$$

with

$$\begin{cases} tr_n = -(d_1 + d_2)\frac{n^2}{l^2} + a_{11} + a_{22} + C(z_1, z_2)b_{22}, \\ \Delta_n = d_1 d_2 \frac{n^4}{l^4} - (d_1 a_{22} + d_2 a_{11} + C(z_1, z_2)d_1 b_{22})\frac{n^2}{l^2} \\ \quad + a_{11}a_{22} - C(z_1, z_2)^2 a_{12}b_{21} + C(z_1, z_2)a_{11}b_{22}, \end{cases} \tag{7.7}$$

we can prove the stability of coexisting equilibrium E_* as the following three cases.

(i) When $d_1 = d_2 = 0$ holds and Routh–Hurwitz criterion yields

$(H2)\ tr_0 = a_{11} + a_{22} + C(z_1, z_2)b_{22} < 0,$
that is $C(z_1, z_2) > -(a_{11} + a_{22})/b_{22}$ *and* $a_{11} > 0,$
$(H3)\ \Delta_0 = a_{11}a_{22} - C(z_1, z_2)^2 a_{12}b_{21} + C(z_1, z_2)a_{11}b_{22} > 0,$ *that is*

$$C(z_1, z_2) > \frac{-a_{11}b_{22} + \sqrt{a_{11}^2 b_{22}^2 + 4a_{12}b_{21}a_{11}a_{22}}}{-2a_{12}b_{21}},$$

E_* remains locally asymptotically stable without diffusion and time delay.

(ii) Assume $d_1 > 0$, $d_2 > 0$ hold. Then it gives $tr_n < tr_0 < 0$ and $\Delta_n > \Delta_0 > 0$ for $n \in N_0$ with conditions (H2), (H3) and $d_1 a_{22} + d_2 a_{11} + d_1 C(z_1, z_2)b_{22} < 0$, implying that all eigenvalues of (7.6) have negative real parts. Then E_* remains locally asymptotically stable without time delay.

(iii) Turing instability occurs when (7.6) has a positive eigenvalue under conditions (H2) and (H3), where the conditions mean that there exists $n \in N$ such that $tr_n > 0$ or $\Delta_n < 0$. It is obvious that if (H2) holds and from the definition of tr_n, we get $tr_n < 0$ for any $n \in N$. Thus, it yields that Turing instability occurs when there exists $n \in N$ such that $\Delta_n < 0$. For the sign of the function listed as given below:

$$\begin{aligned} G(y) = & d_1 d_2 y^2 - (d_1 a_{22} + d_2 a_{11} + d_1 C(z_1, z_2) b_{22}) y \\ & + a_{11} a_{22} - C(z_1, z_2)^2 a_{12} b_{21} + C(z_1, z_2) a_{11} b_{22}, \end{aligned}$$

where $G(y) = \Delta_n$, $y = \frac{n^2}{l^2}$. Then, we get $G(0) = a_{11}a_{22} - C(z_1, z_2)^2 a_{12} b_{21} + C(z_1, z_2) a_{11} b_{22} > 0$ under (H3). The sufficient conditions for the occurrence of Turing instability are presented by

$$\begin{aligned} & d_1 a_{22} + d_2 a_{11} + d_1 C(z_1, z_2) b_{22} > 0, \\ & (d_1 a_{22} + d_2 a_{11} + d_1 C(z_1, z_2) b_{22})^2 - 4 d_1 d_2 (a_{11} a_{22} - C(z_1, z_2)^2 a_{12} b_{21} \\ & + C(z_1, z_2) a_{11} b_{22}) > 0. \end{aligned}$$

After calculation, we gain

$$\begin{cases} C(z_1, z_2) < -\frac{d_1 a_{22} + d_2 a_{11}}{d_1 b_{22}}, \\ d_1 b_{22}^2 + 4 d_2 a_{12} b_{21} < 0, \\ 0 < C(z_1, z_2) < C_+, \end{cases}$$

where $C_+ = \frac{(d_2 a_{11} - d_1 a_{22})(d_1 b_{22} - 2\sqrt{-d_1 d_2 a_{12} b_{21}})}{d_1^2 b_{22}^2 + 4 d_1 d_2 a_{12} b_{21}}$. Summarizing the above analysis, the following theorem is obtained.

Theorem 57 *Suppose* (H2) *and* (H3) *hold and* $\tau = 0$. *Then for system (7.1), the following statements remain true:*

(i) *If* $d_1 = d_2 = 0$, *coexisting equilibrium* E_* *is locally asymptotically stable.*

(ii) *If* $C(z_1, z_2) > \frac{d_1 a_{22} + d_2 a_{11}}{-d_1 b_{22}}$ *holds and* $d_1 > 0$, $d_2 > 0$, *coexisting equilibrium* E_* *is locally asymptotically stable.*

(iii) *If* $C(z_1, z_2) < \frac{d_1 a_{22} + d_2 a_{11}}{-d_1 b_{22}}$, $C(z_1, z_2) > C_+$, *and* $d_1 b_{22}^2 + 4 d_2 a_{12} b_{21} < 0$ *hold,* $d_1 > 0$ *and* $d_2 > 0$, *coexisting equilibrium* E_* *is locally asymptotically stable.*

(iv) *If* $C(z_1, z_2) < \frac{d_1 a_{22} + d_2 a_{11}}{-d_1 b_{22}}$, $C(z_1, z_2) < C_+$, *and* $d_1 b_{22}^2 + 4 d_2 a_{12} b_{21} < 0$ *hold,* $d_1 > 0$ *and* $d_2 > 0$, *coexisting equilibrium* E_* *is Turing unstable.*

7.2.2 Hopf Bifurcation Induced by Time Delay

In this subsection, the effects of time delay and cell size on the dynamics of system (7.1) are considered.

Suppose $\tau > 0$, $d_1 > 0$, $d_2 > 0$, and $i\omega$ ($\omega > 0$) mean a root of (7.5). Then ω assures

$$
\begin{aligned}
&-\omega^2 + i\omega\left((d_1+d_2)\tfrac{n^2}{l^2} - a_{11} - a_{22}\right) + d_1 d_2 \tfrac{n^4}{l^4} - (a_{11}d_2 + a_{22}d_1)\tfrac{n^2}{l^2} + a_{11}a_{22} \\
&-C(z_1,z_2)\left(C(z_1,z_2)a_{12}b_{21} + d_1\tfrac{n^2}{l^2}b_{22} - a_{11}b_{22} + i\omega b_{22}\right)(\cos\omega\tau - i\sin\omega\tau) = 0.
\end{aligned}
\tag{7.8}
$$

After separating real and imaginary parts of (7.8), we get the equations as given below:

$$
\begin{aligned}
&-\omega^2 + d_1 d_2\tfrac{n^4}{l^4} - (d_1 a_{22} + d_2 a_{11})\tfrac{n^2}{l^2} + a_{11}a_{22} \\
&= C(z_1,z_2)\left(C(z_1,z_2)a_{12}b_{21} + d_1 b_{22}\tfrac{n^2}{l^2} - a_{11}b_{22}\right)\cos\omega\tau + \omega C(z_1,z_2)b_{22}\sin\omega\tau, \\
&\left((d_1+d_2)\tfrac{n^2}{l^2} - a_{11} - a_{22}\right)\omega \\
&+C(z_1,z_2)(C(z_1,z_2)a_{12}b_{21} + d_1 b_{22}\tfrac{n^2}{l^2} - a_{11}b_{22})\sin\omega\tau \\
&= \omega C(z_1,z_2)b_{22}\cos\omega\tau.
\end{aligned}
\tag{7.9}
$$

Adding up the squares of the above equations gives

$$
\omega^4 + A_n\omega^2 + B_n = 0, \tag{7.10}
$$

with

$$
\begin{aligned}
A_n &= (d_1^2 + d_2^2)\tfrac{n^4}{l^4} + a_{11}^2 + a_{22}^2 - 2(d_1 a_{11} + d_2 a_{22})\tfrac{n^2}{l^2} - C(z_1,z_2)^2 b_{22}^2, \\
B_n &= \left(d_1 d_2\tfrac{n^4}{l^4} - (d_1 a_{22} + d_2 a_{11})\tfrac{n^2}{l^2} + a_{11}a_{22}\right)^2 \\
&-C(z_1,z_2)^2\left(C(z_1,z_2)a_{12}b_{21} + d_1 b_{22}\tfrac{n^2}{l^2} - a_{11}b_{22}\right)^2.
\end{aligned}
$$

The following gives the roots of (7.10) by

$$
\omega_n^{2\pm} = -\frac{1}{2}A_n \pm \frac{1}{2}(A_n^2 - 4B_n)^{\frac{1}{2}}. \tag{7.11}
$$

As stated by the definition of A_n and B_n, it is easy to get that $A_n \to \infty$, $B_n \to \infty$ $(n \to \infty)$, meaning that there exists at most a finite number of n such that $A_n < 0$ or $B_n < 0$. Therefore, the three different scenarios about the roots of (7.10) are discussed.

(i) When (H4) (i1) $B_n > 0$ and (i2) either $A_n > 0$ or $\Delta = A_n^2 - 4B_n < 0$ hold, there is no positive root for (7.10).

(ii) When there exist $n = 0, 1, \cdots, N_1$, $N_1 \in N_0$, (H5) either $A_n < 0$, $\Delta = A_n^2 - 4B_n = 0$ or $B_n < 0$ holds, there is a positive root for (7.10).

(iii) When there exist $n = 0, 1, \cdots, N_2$, $N_2 \in N_0$, (H6) $A_n < 0$, $B_n > 0$ and $\Delta = A_n^2 - 4B_n > 0$ hold, there are two positive roots for (7.10).

Now, when (H5) holds, there is a pair of purely imaginary roots $\pm i\omega_n$ $(0 \le n \le N_1)$ for (7.5) at $\tau_n^j = \tau_n^0 + \frac{2j\pi}{\omega_n}$, where

$$
\tau_n^0 = \frac{\arccos\left(\frac{d_1 d_2\frac{n^4}{l^4} - \omega_n^2 - (d_1 a_{22} + d_2 a_{11})\frac{n^2}{l^2} + a_{11}a_{22} - \omega_n^2 C(z_1,z_2)b_{22}\left((d_1+d_2)\frac{n^2}{l^2} - a_{11} - a_{22}\right)}{C(z_1,z_2)^2(C(z_1,z_2)a_{12}b_{21} + d_1 b_{22}\frac{n^2}{l^2} - a_{11}b_{22})^2 + \omega_n^2 C(z_1,z_2)^2 b_{22}^2}\right)}{\omega_n}, \tag{7.12}
$$

and $\omega_n = \omega^+$. When (H6) holds, there are two pairs of purely imaginary roots $\pm i\omega_n^{\pm}$ $(0 \le n \le N_2)$ for (7.5) at $\tau_n^{j,\pm} = \tau_n^{0,\pm} + \frac{2j\pi}{\omega_n^{\pm}}$, $j \in N_0$, where

$$\tau_n^{0,\pm} = \frac{\arccos\left(\frac{d_1 d_2 \frac{n^4}{l^4} - (\omega_n^{\pm})^2 - (d_1 a_{22} + d_2 a_{11})\frac{n^2}{l^2} + a_{11}a_{22} - (\omega_n^{\pm})^2 C(z_1,z_2) b_{22}\left((d_1+d_2)\frac{n^2}{l^2} - a_{11} - a_{22}\right)}{C(z_1,z_2)^2 (C(z_1,z_2)a_{12}b_{21} + d_1 b_{22}\frac{n^2}{l^2} - a_{11}b_{22})^2 + (\omega_n^{\pm})^2 C(z_1,z_2)^2 b_{22}^2}\right)}{\omega_n^{\pm}}, \tag{7.13}$$

and $\omega_n^{\pm} = \sqrt{\omega_n^{2\pm}}$.

Let $\lambda_n = \alpha_n(\tau) + i\omega_n(\tau)$ be the root of (7.5) satisfying $\alpha_n(\tau_n^{j,\pm}) = 0$ and $\omega_n(\tau_n^{j,\pm}) = \omega_n^{\pm}$. Then, we get

$$\begin{aligned}\left(\frac{d\lambda}{d\tau}\right)^{-1} &= \frac{-e^{\lambda\tau}\left(2\lambda + (d_1+d_2)\frac{n^2}{l^2} - a_{11} - a_{22}\right)}{\lambda C(z_1,z_2)\left(C(z_1,z_2)a_{12}b_{21} + d_1\frac{n^2}{l^2}b_{22} - a_{11}b_{22} + \lambda b_{22}\right)} \\ &\quad + \frac{b_{22}}{\lambda\left(C(z_1,z_2)a_{12}b_{21} + d_1\frac{n^2}{l^2}b_{22} - a_{11}b_{22} + \lambda b_{22}\right)} - \frac{\tau}{\lambda},\end{aligned} \tag{7.14}$$

after differentiating (7.5) with respect to τ. If $\tau = \tau_n^{j,\pm}$, we obtain

$$\begin{aligned}
&Re\left(\frac{d\lambda}{d\tau}\right)^{-1} \\
&= Re\left(\frac{e^{\lambda\tau}(2\lambda + (d_1+d_2)\frac{n^2}{l^2} - a_{11} - a_{22}) - b_{22}C(z_1,z_2)}{-\lambda C(z_1,z_2)(C(z_1,z_2)a_{12}b_{21} + d_1\frac{n^2}{l^2}b_{22} - a_{11}b_{22} + \lambda b_{22})} - \frac{\tau}{\lambda}\right)\Bigg|_{\tau=\tau_n^{j,\pm}} \\
&= \frac{-2(\cos\omega_n^{\pm}\tau_n^{j,\pm}(C(z_1,z_2)a_{12}b_{21} + d_1\frac{n^2}{l^2}b_{22} - a_{11}b_{22}) + \omega_n^{\pm}\sin\omega_n^{\pm}\tau_n^{j,\pm}b_{22})}{C(z_1,z_2)\omega_n^{\pm 2}b_{22}^2 + C(z_1,z_2)(C(z_1,z_2)a_{12}b_{21} + d_1\frac{n^2}{l^2}b_{22} - a_{11}b_{22})^2} \\
&+ \frac{((d_1+d_2)\frac{n^2}{l^2} - a_{11} - a_{22})\omega_n^{\pm}\cos\omega_n^{\pm}\tau_n^{j,\pm}b_{22}}{C(z_1,z_2)\omega_n^{\pm}(\omega_n^{\pm 2}b_{22}^2 + (C(z_1,z_2)a_{12}b_{21} + d_1\frac{n^2}{l^2}b_{22} - a_{11}b_{22})^2)} \\
&+ \frac{((d_1+d_2)\frac{n^2}{l^2} - a_{11} - a_{22})(-\sin\omega_n^{\pm}\tau_n^{j,\pm}(C(z_1,z_2)a_{12}b_{21} + d_1\frac{n^2}{l^2}b_{22} - a_{11}b_{22}))}{C(z_1,z_2)\omega_n^{\pm}(\omega_n^{\pm 2}b_{22}^2 + (C(z_1,z_2)a_{12}b_{21} + d_1\frac{n^2}{l^2}b_{22} - a_{11}b_{22})^2)} \\
&+ \frac{-b_{22}^2}{b_{22}^2\omega_n^{\pm 2} + (C(z_1,z_2)a_{12}b_{21} + d_1\frac{n^2}{l^2}b_{22} - a_{11}b_{22})^2} \\
&= \pm\frac{2\omega_n^{\pm 2} + A_n}{C(z_1,z_2)^2\omega_n^{\pm 2}b_{22}^2 + C(z_1,z_2)^2(C(z_1,z_2)a_{12}b_{21} + d_1\frac{n^2}{l^2}b_{22} - a_{11}b_{22})^2} \\
&= \pm\frac{\sqrt{A_n^2 - 4B_n}}{C(z_1,z_2)^2\omega_n^{\pm 2}b_{22}^2 + C(z_1,z_2)^2(C(z_1,z_2)a_{12}b_{21} + d_1\frac{n^2}{l^2}b_{22} - a_{11}b_{22})^2}.
\end{aligned}$$

Hence, we get

(i) $\alpha_n'(\tau_n^{j,\pm}) = 0$ when $A_n^2 - 4B_n = 0$ holds;

(ii) $\alpha_n'(\tau_n^{j,+}) > 0$, $\alpha_n'(\tau_n^{j,-}) < 0$, when $A_n^2 - 4B_n > 0$ holds.

From (7.12) and (7.13), we get $\tau_n^0 < \tau_n^j$ and $\tau_n^{0,\pm} < \tau_n^{j,\pm}$. Denote $\tau^0 = \min\{\tau_n^{0,\pm} \text{ or } \tau_n^0\}$. According to the above analysis, the following theorem is obtained.

Theorem 58 *Assume (H2), (H3) and (ii) or (iii) of Theorem 57 hold. The assertions as below hold.*

- *E_* remains locally asymptotically stable for all $\tau > 0$ as* (H4) *holds.*
- *E_* remains locally asymptotically stable for $\tau \in [0, \tau^0)$ and unstable for $\tau > \tau^0$ as* (H5) *or* (H6) *holds.*
- *System (7.1) undergoes a Hopf bifurcation at E_* when (H5) or (H6) holds, $\tau = \tau_n^{j,+}$, $(\tau = \tau_n^{j,-})$, $j = 0, 1, 2, \ldots$.*

Next, using the normal form and the center manifold theory in [79, 125], the explicit formula is derived to determine the Hopf bifurcation at the critical value $\tau = \tau_n^j$.

For simplicity, let $\tau_0 = \tau_n^j$, $v_1(x,t) = P(x, \tau t) - P_*$, $v_2(x,t) = Z(x, \tau t) - Z_*$, and $x \in (0, l\pi)$ $\tau = \mu + \tau_0$. Then we can rewrite system (7.1) as the following form in the phase space $\mathcal{C} = C([-1, 0], X)$:

$$\dot{v}(t) = \tau_0 D \Delta v(t) + L_\mu(v_t) + F(\mu, v_t), \tag{7.15}$$

where $v = (v_1(\cdot, t), v_2(\cdot, t))^T$, $v_t(\theta) = v(t+\theta) \in \mathcal{C}$, and $L : \mathbb{R} \times \mathcal{C} \to X$, $F : \mathcal{C} \times \mathbb{R} \to X$ are given by

$$L_\mu(\phi) = (\mu + \tau_0) \begin{pmatrix} a_{11}\phi_1(0) + C(z_1, z_2) a_{12}\phi_2(0) \\ a_{22}\phi_2(0) + C(z_1, z_2) b_{21}\phi_1(-1) + C(z_1, z_2) b_{22}\phi_2(-1) \end{pmatrix}, \tag{7.16}$$

$$F(\mu, \phi) = (\mu + \tau_0) \begin{pmatrix} F_1 \\ F_2 \end{pmatrix}, \tag{7.17}$$

and $\phi = (\phi_1, \phi_2)^T \in \mathcal{C}([-1, 0], \mathbb{R}^2)$

$$\begin{aligned} F_1 =& h_{v_1v_1}\phi_1^2(0) + h_{v_1v_2}\phi_1(0)\phi_2(0) + h_{v_2v_2}\phi_2^2(0) + h_{v_1v_1v_1}\phi_1^3(0) \\ &+ h_{v_1v_2v_2}\phi_1(0)\phi_2^2(0) + h_{v_2v_2v_1}\phi_2^2(0)\phi_1(0) + h_{v_2v_2v_2}\phi_2^3(0), \\ F_2 =& f_{v_1v_1}\phi_1^2(-1) + f_{v_1v_2}\phi_1(-1)\phi_2(-1) + f_{v_2v_2}\phi_2^2(-1) + f_{v_1v_1v_1}\phi_1^3(-1) \\ &+ f_{v_1v_2v_2}\phi_1(-1)\phi_2^2(-1) + f_{v_2v_2v_1}\phi_2^2(-1)\phi_1(-1) + f_{v_2v_2v_2}\phi_2^3(-1), \end{aligned}$$

where the coefficients of F_1 and F_2 are supplied in Appendix C.

As $\mu = 0$ means a Hopf bifurcation value of (7.15), the linear part is given as

$$\dot{v}(t) = -\tau_0 D \frac{n^2}{l^2} \Delta v(t) + L_\mu(v_t). \tag{7.18}$$

Based on the results in previous section, it is obvious that $\pm i\omega_n\tau_0$ are a pair of purely imaginary eigenvalues of (7.18). After applying the Riesz representation theorem, we prove that there is a matrix function $\rho(\theta, \mu)$, such that

$$-\tau_0 D\frac{n^2}{l^2}\phi(0) + L_\mu(\phi) = \int_{-1}^{0} d\rho(\theta, \tau)\phi(\theta). \tag{7.19}$$

Choose

$$\rho(\theta, \tau) = \begin{cases} (\tau_0 + \mu)\begin{pmatrix} a_{11} - d_1\frac{n^2}{l^2} & C(z_1, z_2)a_{12} \\ 0 & a_{22} - d_2\frac{n^2}{l^2} \end{pmatrix} & \theta = 0, \\ O & \theta \in (-1, 0), \\ -(\tau_0 + \mu)\begin{pmatrix} 0 & 0 \\ C(z_1, z_2)b_{21} & C(z_1, z_2)b_{22} \end{pmatrix} & \theta = -1, \end{cases} \tag{7.20}$$

where O stands for a 2×2 zero matrix. Hence (7.19) is satisfied. Assume that $M(\mu)$ is the infinitesimal generator of semigroup and

$$M(\mu)\phi(\theta) = \begin{cases} \frac{d\phi(\theta)}{d\theta} & \theta \in [-1, 0), \\ \int_{-1}^{0} d\rho(\theta, \mu)\phi(\theta) & \theta = 0. \end{cases} \tag{7.21}$$

For $\phi \in C = C^1([-1, 0], \mathbf{R}^2)$, $\varphi \in C^* = C^1([0, 1], (\mathbf{R}^2)^*)$, the adjoint operator M^* of M is defined by

$$M^*\varphi(s) = \begin{cases} -\frac{d\varphi(s)}{ds} & s \in (0, 1], \\ \int_{-1}^{0} d\rho^T(s, \tau_0)\varphi(-s) & s = 0, \end{cases} \tag{7.22}$$

and a bilinear pairing product is denoted by

$$\begin{aligned} \langle\varphi(s), \phi(\theta)\rangle &= \varphi(0)\phi(0) - \int_{-1}^{0}\int_{\xi=0}^{\theta} \varphi(\xi - \theta)d\rho(\theta)\phi(\xi)d\xi \\ &= \varphi(0)\phi(0) + \tau_0\int_{-1}^{0} \varphi(\xi + 1)F\phi(\xi)d\xi. \end{aligned} \tag{7.23}$$

Because of the above analysis, it is easy to get that $\pm i\omega_n\tau_0$ are eigenvalues of M^*. Then the eigenfunction $\eta(\theta)$ of $M(0)$ corresponding to $i\omega_n\tau_0$ and the eigenfunction $\eta^*(s)$ of M^* corresponding to $-i\omega_n\tau_0$ are calculated. Let $\eta(\theta) = \binom{1}{g} e^{i\omega_n\tau_0\theta}$. Thus we get $M(0)\eta(\theta) = i\omega_n\tau_0\eta(\theta)$. From the definition of $M(0)$, it yields

$$\begin{pmatrix} a_{11} - d_1\frac{n^2}{l^2} & C(z_1, z_2)a_{12} \\ 0 & a_{22} - d_2\frac{n^2}{l^2} \end{pmatrix}\eta(0) + \begin{pmatrix} 0 & 0 \\ C(z_1, z_2)b_{21} & C(z_1, z_2)b_{22} \end{pmatrix}\eta(-1) = i\omega_n\eta(0),$$

thus, we get $g = -\frac{i\omega_n - a_{11} + d_1\frac{n^2}{l^2}}{C(z_1, z_2)a_{12}}$.

If $\eta^*(s) = (1\ h)e^{-i\omega_n\tau_0 s}$, we get $M^*\eta^*(s) = -i\omega_n\tau_0\eta^*(s)$. From the definition of M^*, we have

$$\begin{pmatrix} a_{11} - d_1\frac{n^2}{l^2} & 0 \\ C(z_1, z_2)a_{12} & a_{22} - d_2\frac{n^2}{l^2} \end{pmatrix}\eta^*(0) + \begin{pmatrix} 0 & C(z_1, z_2)b_{21} \\ 0 & C(z_1, z_2)b_{22} \end{pmatrix}\eta^*(-1) = -i\omega_n\eta^*(0),$$

which means that $h = \frac{(-i\omega_n - a_{11} + d_1\frac{n^2}{l^2})e^{i\omega_n\tau_0}}{C(z_1,z_2)b_{21}}$.

Moreover, let $\Phi = (\Phi_1, \Phi_2)$, $\Phi^* = (\Phi_1^*, \Phi_2^*)^T$, and

$$\Phi_1(\theta) = \frac{\eta(\theta) + \bar{\eta}(\theta)}{2} = \begin{pmatrix} Re\{e^{i\omega_n\tau_0\theta}\} \\ Re\{ge^{i\omega_n\tau_0\theta}\} \end{pmatrix}$$

$$= \begin{pmatrix} \cos\omega_n\tau_0\theta \\ \frac{\left(a_{11} - d_1\frac{n^2}{l^2}\right)\cos\omega_n\tau_0\theta + \omega_n\sin\omega_n\tau_0\theta}{C(z_1,z_2)a_{12}} \end{pmatrix},$$

$$\Phi_2(\theta) = \frac{\eta(\theta) - \bar{\eta}(\theta)}{2i} = \begin{pmatrix} Im\{e^{i\omega_n\tau_0\theta}\} \\ Im\{ge^{i\omega_n\tau_0\theta}\} \end{pmatrix}$$

$$= \begin{pmatrix} \sin\omega_n\tau_0\theta \\ \frac{\left(a_{11} - d_1\frac{n^2}{l^2}\right)\sin\omega_n\tau_0\theta - \omega_n\cos\omega_n\tau_0\theta}{C(z_1,z_2)a_{12}} \end{pmatrix},$$

$$\Phi_1^*(s) = \frac{\eta^*(s) + \bar{\eta}^*(s)}{2} = \begin{pmatrix} Re\{e^{-i\omega_n\tau_0 s}\} \\ Re\{he^{-i\omega_n\tau_0 s}\} \end{pmatrix}^T$$

$$= \begin{pmatrix} \cos\omega_n\tau_0 s \\ \frac{\left(d_1\frac{n^2}{l^2} - a_{11}\right)\cos(1+s)\omega_n\tau_0 + \omega_n\sin(1+s)\omega_n\tau_0}{C(z_1,z_2)b_{21}} \end{pmatrix}^T,$$

$$\Phi_2^*(s) = \frac{\eta^*(s) - \bar{\eta}^*(s)}{2i} = \begin{pmatrix} Im\{e^{-i\omega_n\tau_0 s}\} \\ Im\{he^{-i\omega_n\tau_0 s}\} \end{pmatrix}^T$$

$$= \begin{pmatrix} -\sin\omega_n\tau_0 s \\ \frac{\left(d_1\frac{n^2}{l^2} - a_{11}\right)\sin(1+s)\omega_n\tau_0 - \omega_n\cos(1+s)\omega_n\tau_0}{C(z_1,z_2)b_{21}} \end{pmatrix}^T.$$

Then the following is got by (7.23)

$$\langle \Phi_1^*(s), \Phi_1(\theta)\rangle = 1 + \frac{(a_{11}-d_1\frac{n^2}{l^2})\omega_n \sin\omega_n\tau_0-(a_{11}-d_1\frac{n^2}{l^2})^2 \cos\omega_n\tau_0}{C(z_1,z_2)^2 b_{21}a_{12}}$$
$$+\frac{\tau_0}{2}\Bigg[\Big(d_1\frac{n^2}{l^2}-a_{11}\Big)\Big(1+\frac{\sin 2\omega_n\tau_0}{2\omega_n\tau_0}\Big)$$
$$-\frac{\sin^2\omega_n\tau_0}{\tau_0}$$
$$+\frac{b_{22}}{C(z_1,z_2)a_{12}b_{21}}\Big(\omega_n^2-(a_{11}-d_1\frac{n^2}{l^2})^2\Big)\Big(1+\frac{\sin 2\omega_n\tau_0}{2\omega_n\tau_0}\Big)\Bigg],$$
$$\langle \Phi_1^*(s), \Phi_2(\theta)\rangle = \frac{\Big(a_{11}-d_1\frac{n^2}{l^2}\Big)\omega_n \cos\omega_n\tau_0-\omega_n^2 \sin\omega_n\tau_0}{C(z_1,z_2)^2 b_{21}a_{12}}$$
$$+\frac{\tau_0}{2}\Bigg[\Big(d_1\frac{n^2}{l^2}-a_{11}\Big)\frac{\sin^2\omega_n\tau_0}{\omega_n\tau_0}$$
$$-\omega_n\Big(1+\frac{\sin 2\omega_n\tau_0}{2\omega_n\tau_0}\Big)+\frac{b_{22}}{C(z_1,z_2)a_{12}b_{21}}\Big(\frac{1}{2\omega_n\tau_0}-\frac{\sin^2\omega_n\tau_0}{\omega_n\tau_0}\big(\omega_n^2+(a_{11}-d_1\frac{n^2}{l^2})^2\big)\Big)\Bigg],$$
$$\langle \Phi_2^*(s), \Phi_1(\theta)\rangle = \frac{\Big(d_1\frac{n^2}{l^2}-a_{11}\Big)\omega_n \cos\omega_n\tau_0-\Big(a_{11}-d_1\frac{n^2}{l^2}\Big)^2 \sin\omega_n\tau_0}{C(z_1,z_2)^2 b_{21}a_{12}}$$
$$+\frac{\tau_0}{2}\Bigg[\Big(d_1\frac{n^2}{l^2}-a_{11}\Big)\frac{\sin^2\omega_n\tau_0}{\omega_n\tau_0}$$
$$-\frac{\sin 2\omega_n\tau_0}{2\omega_n\tau_0}+\omega_n+\frac{b_{22}}{C(z_1,z_2)a_{12}b_{21}}\Big(-(a_{11}-d_1\frac{n^2}{l^2})^2\frac{\sin^2\omega_n\tau_0}{\omega_n\tau_0}+2\omega_n(a_{11}-d_1\frac{n^2}{l^2})$$
$$-\frac{\omega_n\sin^2\omega_n\tau_0}{\tau_0}\Big)\Bigg],$$
$$\langle \Phi_2^*(s), \Phi_2(\theta)\rangle = \frac{\Big(a_{11}-d_1\frac{n^2}{l^2}\Big)\omega_n \sin\omega_n\tau_0+\omega_n^2 \cos\omega_n\tau_0}{C(z_1,z_2)^2 b_{21}a_{12}}$$
$$+\frac{\tau_0}{2}\Bigg[\Big(a_{11}-d_1\frac{n^2}{l^2}\Big)\Big(1+\frac{\sin 2\omega_n\tau_0}{2\omega_n\tau_0}\Big)$$
$$-\frac{\sin^2\omega_n\tau_0}{\tau_0}+\frac{b_{22}}{C(z_1,z_2)a_{12}b_{21}}\Big(\Big(a_{11}-d_1\frac{n^2}{l^2}\Big)^2\Big(1+\frac{\sin 2\omega_n\tau_0}{2\omega_n\tau_0}\Big)+\omega_n^2\Big(\frac{\sin 2\omega_n\tau_0}{2\omega_n\tau_0}-1\Big)\Big)\Bigg].$$

And let's denote $\langle \Phi^*, \Phi\rangle = \langle \Phi_j^*, \Phi_k\rangle$ $(j,k=1,2)$ and construct a new basis $\Psi = \langle \Psi_1, \Psi_2\rangle^T = \langle \Phi^*, \Phi\rangle^{-1}\Phi^*$ with $\langle \Psi, \Phi\rangle = I$. Then $f_n = (\beta_n^1, \beta_n^2)$ and $c\cdot f_n = c_1\beta_n^1 + c_2\beta_n^2$ for $c \in C([-1,0], X)$ are also defined, where

$$\beta_n^1 = \begin{pmatrix}\cos\frac{n}{l}x\\ 0\end{pmatrix},\quad \beta_n^2 = \begin{pmatrix}0\\ \cos\frac{n}{l}x\end{pmatrix}.$$

Hence the center subspace of (7.18) is provided by $P_{CN}\mathcal{C} \oplus P_S\mathcal{C}$ where $P_S\mathcal{C}$ means the complement subspace of $P_{CN}\mathcal{C}$ in $\mathcal{C}$,

$$P_{CN}\mathcal{C}(\phi) = \Phi(\Psi, \langle\phi, f_n\rangle)f_n, \tag{7.24}$$

where $\langle\phi, f_n\rangle = (\langle\phi,\beta_n^1\rangle, \langle\phi,\beta_n^2\rangle)^T$, $\langle u, v\rangle = \frac{1}{l\pi}\int_0^{l\pi} u_1\bar{v}_1 dx + \frac{1}{l\pi}\int_0^{l\pi} u_2\bar{v}_2 dx$, for $u=(u_1,u_2)^T$, $v=(v_1,v_2)^T$.

By the above results, it follows that (7.15) is equivalent to

$$\dot{v}_t = M(\mu)v_t + N(\mu)v_t, \tag{7.25}$$

with $N(\mu)\phi(\theta) = \begin{cases} 0 & \theta \in [-1, 0), \\ F(\mu, \phi) & \theta = 0. \end{cases}$
After using $\mathcal{C} = P_{CN}\mathcal{C} \oplus P_S\mathcal{C}$ and (7.23), we get the solution at $\mu = 0$ as

$$v_t = \Phi(x_1, x_2)^T f_n + h(x_1, x_2, 0), \tag{7.26}$$

where $(x_1, x_2)^T = (\Psi, \langle\phi, f_n\rangle)$. Let $z = x_1 - ix_2$ and $\eta = \Phi_1 + i\Phi_2$. Then, (7.26) is

$$v_t = \frac{1}{2}(\eta z + \bar{\eta}\bar{z}) f_n + h\left(\frac{z + \bar{z}}{2}, \frac{(z - \bar{z})i}{2}, 0\right) = \frac{1}{2}(\eta z + \bar{\eta}\bar{z}) f_n + W(z, \bar{z}). \tag{7.27}$$

On the center manifold C_0, following the approach of [79, 125], we obtain

$$W(z, \bar{z}) = W_{20}\frac{z^2}{2} + W_{11}z\bar{z} + W_{02}\frac{\bar{z}^2}{2} + \cdots, \tag{7.28}$$

$$g(z, \bar{z}) = g_{20}\frac{z^2}{2} + g_{11}z\bar{z} + g_{02}\frac{\bar{z}^2}{2} + g_{21}\frac{z^2\bar{z}}{2} + \cdots, \tag{7.29}$$

$$\dot{z}(t) = i\omega_n\tau_0 z + g(z, \bar{z}) = i\omega_n\tau_0 z + (\Psi_1(0) - i\Psi_2(0))\langle F(0, v_t), f_n\rangle. \tag{7.30}$$

Thus, from (7.27) and (7.28) we have

$$\begin{aligned}
&v_{1t}(0) = W_{20}^{(1)}(0)\tfrac{z^2}{2} + W_{11}^{(1)}(0)z\bar{z} + W_{02}^{(1)}(0)\tfrac{\bar{z}^2}{2} \\
&+\tfrac{1}{2}(z + \bar{z})\cos(\tfrac{nx}{l}) + o\Big(|(z, \bar{z})|^3\Big), \\
&v_{2t}(0) = W_{20}^{(2)}(0)\tfrac{z^2}{2} + W_{11}^{(2)}(0)z\bar{z} + W_{02}^{(2)}(0)\tfrac{\bar{z}^2}{2} \\
&+\tfrac{1}{2}(gz + \bar{g}\bar{z})\cos(\tfrac{nx}{l}) + o\Big(|(z, \bar{z})|^3\Big), \\
&v_{1t}(-1) = W_{20}^{(1)}(-1)\tfrac{z^2}{2} + W_{11}^{(1)}(-1)z\bar{z} + W_{02}^{(1)}(-1)\tfrac{\bar{z}^2}{2} \\
&\qquad +\tfrac{1}{2}(ze^{-i\omega_n\tau_0} + \bar{z}e^{i\omega_n\tau_0})\cos(\tfrac{nx}{l}) + o\Big(|(z, \bar{z})|^3\Big), \\
&v_{2t}(-1) = W_{20}^{(2)}(-1)\tfrac{z^2}{2} + W_{11}^{(2)}(-1)z\bar{z} + W_{02}^{(2)}(-1)\tfrac{\bar{z}^2}{2} \\
&\qquad +\tfrac{1}{2}(gze^{-i\omega_n\tau_0} + \bar{g}\bar{z}e^{i\omega_n\tau_0})\cos(\tfrac{nx}{l}) + o\Big(|(z, \bar{z})|^3\Big).
\end{aligned} \tag{7.31}$$

This, together with (7.17), suggests that

$$\begin{aligned}
&\bar{F}_1(0, v_t) = \tfrac{F_1}{\tau_0} = \cos^2\left(\tfrac{nx}{l}\right)\left(\tfrac{z^2}{2}\chi_{20} + z\bar{z}\chi_{11} + \tfrac{\bar{z}^2}{2}\bar{\chi}_{20}\right) \\
&+\tfrac{z^2\bar{z}}{2}\left(\chi_1\cos\left(\tfrac{nx}{l}\right) + \chi_2\cos^3\left(\tfrac{nx}{l}\right)\right) + \cdots, \\
&\bar{F}_2(0, v_t) = \tfrac{F_2}{\tau_0} = \cos^2\left(\tfrac{nx}{l}\right)\left(\tfrac{z^2}{2}\varsigma_{20} + z\bar{z}\varsigma_{11} + \tfrac{\bar{z}^2}{2}\bar{\varsigma}_{20}\right) \\
&+\tfrac{z^2\bar{z}}{2}\left(\varsigma_1\cos\left(\tfrac{nx}{l}\right) + \varsigma_2\cos^3\left(\tfrac{nx}{l}\right)\right) + \cdots,
\end{aligned} \tag{7.32}$$

and

$$\begin{aligned}\langle F(0, v_t), f_n\rangle &= \tau_0(\bar{F}_1(0, v_t) f_n^1 + \bar{F}_2(0, v_t) f_n^2)\\ &= \tfrac{z^2}{2}\tau_0\begin{pmatrix}\chi_{20}\\ \varsigma_{20}\end{pmatrix}\tfrac{1}{\pi}\int_0^{l\pi}\cos^3\left(\tfrac{nx}{l}\right)dx + z\bar{z}\tau_0\begin{pmatrix}\chi_{11}\\ \varsigma_{11}\end{pmatrix}\tfrac{1}{\pi}\int_0^{l\pi}\cos^3\left(\tfrac{nx}{l}\right)dx\\ &+\tfrac{\bar{z}^2}{2}\tau_0\begin{pmatrix}\bar{\chi}_{20}\\ \bar{\varsigma}_{20}\end{pmatrix}\tfrac{1}{\pi}\int_0^{l\pi}\cos^3\left(\tfrac{nx}{l}\right)dx + \tfrac{z^2\bar{z}}{2}\tau_0\begin{pmatrix}\kappa_1\\ \kappa_2\end{pmatrix} + \cdots,\end{aligned} \tag{7.33}$$

where the coefficients are listed in Appendix D.
When $\Psi_1(0) - i\Psi_2(0) = (\gamma_1, \gamma_2)$, we can have

$$\begin{aligned}&(\Psi_1(0) - i\Psi_2(0))\langle F(0, v_t), f_n\rangle\\ &= \tfrac{\tau_0}{\pi}\int_0^{l\pi}\cos^3\left(\tfrac{nx}{l}\right)dx\Big(\tfrac{z^2}{2}(\gamma_1\chi_{20} + \gamma_2\varsigma_{20}) + z\bar{z}(\gamma_1\chi_{11} + \gamma_2\varsigma_{11})\\ &+\tfrac{\bar{z}^2}{2}(\gamma_1\bar{\chi}_{20} + \gamma_2\bar{\varsigma}_{20})\Big) + \tfrac{z^2\bar{z}}{2}\tau_0(\gamma_1\kappa_1 + \gamma_2\kappa_2) + \cdots.\end{aligned} \tag{7.34}$$

It follows from (7.29), (7.30), and (7.34),

$$\begin{aligned}&n \in N,\ \ g_{20} = g_{11} = g_{02} = 0;\\ &n = 0,\ \ g_{20} = \gamma_1\tau_0\chi_{20} + \gamma_2\tau_0\varsigma_{20},\ \ g_{11} = \gamma_1\tau_0\chi_{11} + \gamma_2\tau_0\varsigma_{11},\\ &g_{02} = \gamma_1\tau_0\bar{\chi}_{20} + \gamma_2\tau_0\bar{\varsigma}_{20};\\ &n \in N_0,\ \ g_{21} = \tau_0(\gamma_1\kappa_1 + \gamma_2\kappa_2).\end{aligned}$$

It is obvious that the value of g_{21} is determined by $W_{20}(\theta)$ and $W_{11}(\theta)$. From (7.25), we get

$$\dot{W}(t, \theta) = M(0)W + H(z, \bar{z}, \theta), \tag{7.35}$$

where

$$\begin{aligned}H(z, \bar{z}, \theta) &= H_{20}(\theta)\frac{z^2}{2} + H_{11}(\theta)z\bar{z} + H_{02}(\theta)\frac{\bar{z}^2}{2} + \cdots\\ &= X_0F(0, v_t) - \Phi(\Psi, \langle X_0F(0, v_t), f_n\rangle f_n).\end{aligned} \tag{7.36}$$

Substituting (7.30) into the derivative of (7.28) and comparing the coefficients with (7.35) and (7.36), we have

$$(M(0) - 2i\omega_n\tau_0 I)W_{20}(\theta) = -H_{20}(\theta),\quad M(0)W_{11}(\theta) = -H_{11}(\theta). \tag{7.37}$$

It follows from (7.34) that for $\theta \in [-1, 0)$,

$$\begin{aligned}H(z,\bar{z},\theta) &= -\Phi(\theta)\Psi(0)\langle F(0,v_t), f_n\rangle f_n\\ &= -\left(\tfrac{\eta(\theta)+\bar{\eta}(\theta)}{2}, \tfrac{\eta(\theta)-\bar{\eta}(\theta)}{2i}\right)(\Psi_1(0), \Psi_2(0))\langle F(0,v_t), f_n\rangle f_n\\ &= -\tfrac{1}{2}(\eta(\theta)g_{20}+\bar{\eta}(\theta)\bar{g}_{02})\tfrac{z^2}{2} - \tfrac{1}{2}(\eta(\theta)g_{11}+\bar{\eta}(\theta)\bar{g}_{11})z\bar{z}\\ &\quad -\tfrac{1}{2}(\eta(\theta)g_{02}+\bar{\eta}(\theta)\bar{g}_{20})\tfrac{\bar{z}^2}{2} + \cdots.\end{aligned}$$

Hence, for $\theta \in [-1, 0)$

$$H_{20}(\theta) = \begin{cases} 0 & n \in N,\\ -\frac{1}{2}(\eta(\theta)g_{20}+\bar{\eta}(\theta)\bar{g}_{02})f_0 & n = 0,\end{cases}$$

$$H_{11}(\theta) = \begin{cases} 0 & n \in N,\\ -\frac{1}{2}(\eta(\theta)g_{11}+\bar{\eta}(\theta)\bar{g}_{11})f_0 & n = 0,\end{cases}$$

and for $\theta = 0$, $H(z,\bar{z})(0) = f(0,v_t) - \Phi(\Psi, \langle f(0,v_t), f_n\rangle f_n)$, we get

$$\begin{aligned}H_{20}(0) &= \begin{cases} \tau_0\begin{pmatrix}\chi_{20}\\ \varsigma_{20}\end{pmatrix}\cos^2(\frac{nx}{l}) & n \in N,\\ \tau_0\begin{pmatrix}\chi_{20}\\ \varsigma_{20}\end{pmatrix} - \frac{1}{2}(\eta(0)g_{20}+\bar{\eta}(0)\bar{g}_{02})f_0 & n = 0,\end{cases}\\ H_{11}(0) &= \begin{cases} \tau_0\begin{pmatrix}\chi_{11}\\ \varsigma_{11}\end{pmatrix}\cos^2(\frac{nx}{l}) & n \in N,\\ \tau_0\begin{pmatrix}\chi_{11}\\ \varsigma_{11}\end{pmatrix} - \frac{1}{2}(\eta(0)g_{11}+\bar{\eta}(0)\bar{g}_{11})f_0 & n = 0,\end{cases}\end{aligned} \tag{7.38}$$

By (7.37), we get

$$\begin{aligned}\dot{W}_{20}(\theta) &= 2i\omega_n\tau_0 W_{20}(\theta) + \frac{1}{2}\left(\eta(\theta)g_{20}+\bar{\eta}(\theta)\bar{g}_{02}\right)f_n,\\ \dot{W}_{11}(\theta) &= -\frac{i}{2\omega_n\tau_0}(g_{11}\eta(\theta)+\bar{g}_{11}\bar{\eta}(\theta))f_n, \quad \theta \in [-1, 0),\end{aligned}$$

which indicates that

$$\begin{aligned}W_{20}(\theta) &= \frac{i}{2i\omega_n\tau_0}\left(g_{20}\eta(\theta)+\frac{\bar{g}_{02}}{3}\bar{\eta}(\theta)\right)f_n + E_1 e^{2i\omega_n\tau_0\theta},\\ W_{11}(\theta) &= \frac{i}{2i\omega_n\tau_0}(\bar{g}_{11}\bar{\eta}(\theta) - g_{11}\eta(\theta)) + E_2,\end{aligned} \tag{7.39}$$

where $E_1,\ E_2 \in \mathbf{R}^2$ stands for constant vectors.

In the following, the values of $E_1,\ E_2$ are determined. By (7.37)–(7.39), it yields that for $\theta \in [-1, 0)$

$$\begin{cases} -(g_{20}\eta(0) + \frac{\bar{g}_{02}}{3}\bar{\eta}(0))f_0 + E_1 2i\omega_n\tau_0 - M(0)\frac{i}{2\omega_n\tau_0}(g_{20}\eta(0) + \frac{\bar{g}_{02}}{3}\bar{\eta}(0))f_0 \\ -M(0)E_1 \\ -L(\frac{i}{2\omega_n\tau_0}(g_{20}\eta(0) + \frac{\bar{g}_{02}}{3}\bar{\eta}(0))f_0 + E_1 e^{2i\omega_n\tau_0\theta}) = \tau_0\begin{pmatrix}\chi_{20}\\ \varsigma_{20}\end{pmatrix} - \frac{1}{2}(\eta(0)g_{20} \\ +\bar{\eta}(0)\bar{g}_{02})f_0, \\ -M(0)\frac{i}{2\omega_n\tau_0}(\bar{g}_{11}\bar{\eta}(0) - g_{11}\eta(0)) - M(0)E_2 \\ -L(\frac{i}{2\omega_n\tau_0}(\bar{g}_{11}\bar{\eta}(0) - g_{11}\eta(0)) + E_2) \\ = \tau_0\begin{pmatrix}\chi_{11}\\ \varsigma_{11}\end{pmatrix} - \frac{1}{2}(\bar{g}_{11}\bar{\eta}(0) - g_{11}\eta(0))f_0. \end{cases} \tag{7.40}$$

As

$$M(0)\eta(0) + L(\eta(0)\cdot f_0) = i\omega_n\eta(0)f_0, \quad M(0)\bar{\eta}(0) + L(\bar{\eta}(0)\cdot f_0) = -i\omega_n\bar{\eta}(0)f_0,$$

we get

$$E_1 2i\omega_n - M(0)E_1 - LE_1e^{2i\omega_n} = \tau_0\begin{pmatrix}\chi_{20}\\ \varsigma_{20}\end{pmatrix}\cos^2\left(\frac{nx}{l}\right),$$
$$-M(0)E_2 - LE_2 = \tau_0\begin{pmatrix}\chi_{11}\\ \varsigma_{11}\end{pmatrix}\cos^2\left(\frac{nx}{l}\right).$$

Thus

$$E_1 = \tau_0 E\begin{pmatrix}\chi_{20}\\ \varsigma_{20}\end{pmatrix}\cos^2\left(\frac{nx}{l}\right), \quad E_2 = \tau_0 E^*\begin{pmatrix}\chi_{11}\\ \varsigma_{11}\end{pmatrix}\cos^2\left(\frac{nx}{l}\right),$$

where

$$E = \begin{pmatrix} 2i\omega_n\tau_0 + d_1\frac{n^2}{l^2} - a_{11} & -C(z_1, z_2)a_{12} \\ -C(z_1, z_2)b_{21}e^{-2i\omega_n\tau_0} & 2i\omega_n\tau_0 + d_2\frac{n^2}{l^2} - a_{22} - C(z_1, z_2)b_{22}e^{-2i\omega_n\tau_0} \end{pmatrix}^{-1},$$

and

$$E^* = \begin{pmatrix} d_1\frac{n^2}{l^2} - a_{11} & -C(z_1, z_2)a_{12} \\ -C(z_1, z_2)b_{21} & d_2\frac{n^2}{l^2} - a_{22} - C(z_1, z_2)b_{22} \end{pmatrix}^{-1}.$$

Hence, each g_{ij} is determined by the parameters of system (7.1). The following values can be evaluated:

$$c(0) = \frac{i}{2\omega_n\tau_0}\left(g_{20}g_{11} - 2|g_{11}|^2 - \frac{1}{3}|g_{02}|^2\right) + \frac{g_{21}}{2}, \quad \mu_2 = -\frac{Re\{c(0)\}}{Re\{\lambda'(\tau_n^j)\}},$$
$$\beta_2 = 2Re\{c(0)\}, \quad T_2 = -\frac{Im\{c(0)\} + \mu_2 Im\{\lambda'(\tau_n^j)\}}{\omega_n\tau_0}. \tag{7.41}$$

Based on [79], the properties of Hopf bifurcation are determined by these values. When $\mu_2 > 0$ ($\mu_2 < 0$), the Hopf bifurcation is supercritical (subcritical) and there exists the bifurcating periodic solutions for $\tau > \tau_0$ ($\tau < \tau_0$). If $\beta_2 < 0$ ($\beta_2 > 0$), then the periodic solutions are stable (unstable). When $T_2 > 0$ ($T_2 < 0$), the period of bifurcating periodic solutions increases (decreases).

7.3 Numerical Simulation

In this part, some numerical simulations are carried out to experimentally confirm our theoretical results. Assume $a = 0.35, b = 0.8, r_2 = 0.5, C_m = 3, \alpha = 0.6, \beta = 0.6, c_1 = 0.08, c_2 = 0.2, c_3 = 0.02, k = 2, \lambda = 4, l = 3$. Then system (7.1) can be written as follows:

$$\begin{cases} \frac{\partial P}{\partial t} - d_1 \Delta P = P\left(\frac{z_1}{0.08z_1^2+0.2z_1+0.02} - 0.6z_1^2 - 0.6P\right) - \frac{3e^{-\frac{1}{4}(z_1-2z_2)^2} PZ}{0.35+P+0.8Z}, \\ x \in (0, 3\pi),\ t > 0, \\ \frac{\partial Z}{\partial t} - d_2 \Delta Z = \frac{3e^{-\frac{1}{4}(z_1-2z_2)^2} P(x,t-\tau)Z(x,t-\tau)}{0.35+P(x,t-\tau)+0.8Z(x,t-\tau)} - 0.5Z, \\ x \in (0, 3\pi),\ t > 0, \\ P_x(0,t) = Z_x(0,t) = P_x(3\pi,t) = Z_x(3\pi,t) = 0, \\ t > 0, \\ P(x,\theta) = P_0(x,\theta) \geq 0, \quad Z(x,\theta) = Z_0(x,\theta) \geq 0, \\ x \in [0, 3\pi],\ \theta \in [-\tau, 0]. \end{cases} \tag{7.42}$$

Based on condition (H1), we can obtain that there exists a unique coexisting equilibrium $E_* = (0.2850, 1.2961)$ for system (7.42) with the cell size of phytoplankton $z_1 = 0.1$ and the cell size of zooplankton $z_2 = 0.2$. Next different diffusive coefficients are chosen to verify the effects of diffusion and cell size on the dynamics of system (7.42).

When $\tau = 0$, $d_1 = d_2 = 0$, we get (H2) $tr_0 = a_{11} + a_{22} + C(z_1, z_2)b_{22} = -0.0935 < 0$ and (H3) $\Delta_0 = a_{11}a_{22} - C(z_1, z_2)^2 a_{12}b_{21} + C(z_1, z_2)a_{11}b_{22} = 0.2088 > 0$. According to (i) of Theorem 57, we obtain that E_* is asymptotically stable as shown in Fig. 7.1.

When $\tau = 0$, $d_1 = d_2 = 0.06$, and condition (ii) $d_1 a_{22} + d_2 a_{11} + d_1 C(z_1, z_2)b_{22} = -0.0284 < 0$ of Theorem 57 is satisfied, we get that the numerical simulation of the dynamics of system (7.42) is illustrated in Fig. 7.2.

When $\tau = 0$, $d_1 = 0.06$, $d_2 = 2$, we have $d_1 a_{22} + d_2 a_{11} + d_1 C(z_1, z_2)b_{22} = 0.3919 > 0$, $C(z_1, z_2) - C(z_1, z_2)_+ = 2.6284 > 0$, $d_1 b_{22}^2 + 4d_2 a_{12}b_{21} = -0.3328 < 0$. Then we get equilibrium E_* as locally asymptotically stable by (iii) of Theorem 57 as shown in Fig. 7.3. When $\tau = 0$, $d_1 = 0.12$, $d_2 = 10$, $n = 1$,

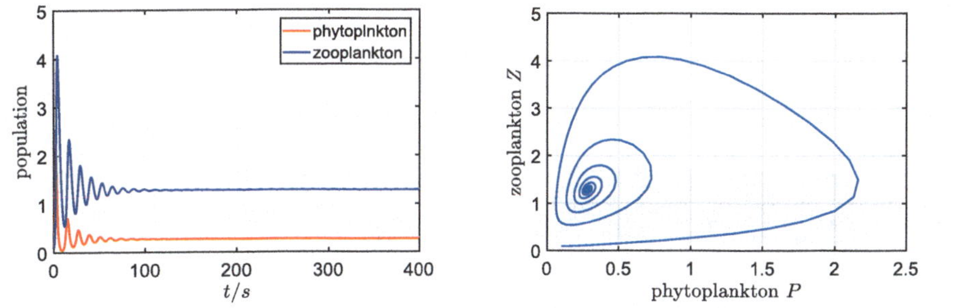

(a) Behavior of plankton

(b) the phase plot of the system (right panel) for stable conditions

Fig. 7.1 Behavior of plankton and the phase plot of the system for stable conditions, where $\tau = 0$, $d_1 = d_2 = 0$, $z_1 = 0.1$, $z_2 = 0.2$ and initial condition $(0.1, 0.1)$

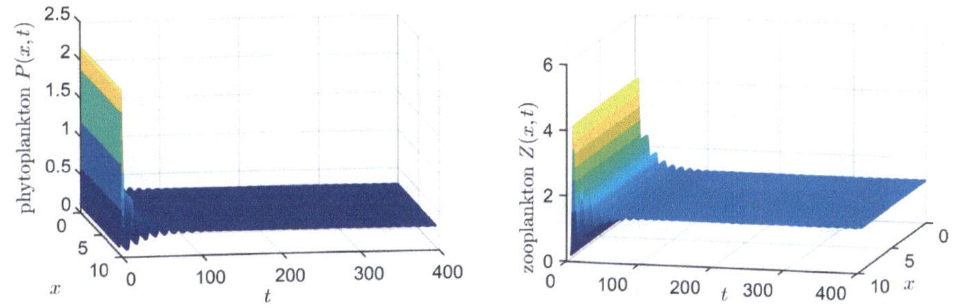

(a)

(b)

Fig. 7.2 Behavior of plankton for stable conditions, where $\tau = 0$, $d_1 = d_2 = 0.06$, $z_1 = 0.1$, $z_2 = 0.2$ and initial condition $(0.1, 0.1)$

we get $d_1 a_{22} + d_2 a_{11} + d_1 C(z_1, z_2) b_{22} = 2.0835 > 0$, $C(z_1, z_2) - C(z_1, z_2)_+ = -2.1349 < 0$ and $d_1 b_{22}^2 + 4 d_2 a_{12} b_{21} = -1.6649 < 0$. Then by (iv) of Theorem 57, we get that E_* is Turing unstable shown in Fig. 7.4.

When $\tau = 0$, $d_1 = 0.06$, $d_2 = 2$, and increasing the value of the maximum consumption rate to $C_m = 6$, we get that equilibrium E_* of system (7.42) becomes unstable as shown in Fig. 7.5.

When $\tau = 0$, $d_1 = 0.12$, $d_2 = 10$, and cell size is changed as $z_1 = 0.15$ and $z_2 = 0.2$, we get that E_* is local asymptotically stable as shown in Fig. 7.6.

From the above numerical simulations, it is easy to illustrate the effect of cell size on the spatiotemporal dynamics of plankton system. When other coefficients and living environment remain fixed, the plankton densities fluctuate with time because of large zooplankton consumption rate. Then planktonic bloom could emerge as shown in Figs. 7.3 and 7.5. As for the case of Turing instability caused by larger diffusion parameters, we find that with the increasing of the cell size of plank-

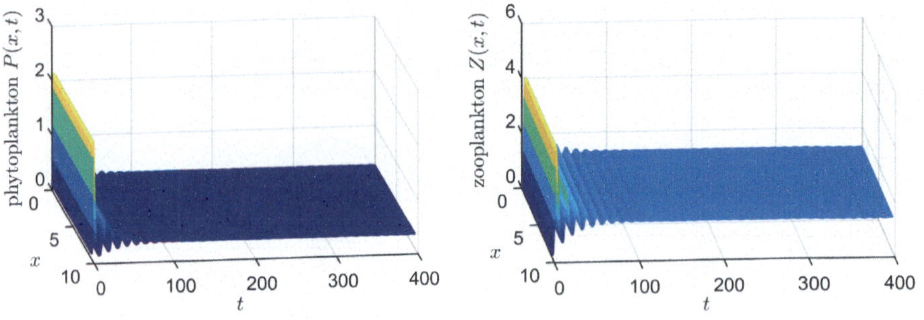

(a) (b)

Fig. 7.3 Behavior of plankton for stable conditions, where $\tau = 0$, $d_1 = 0.06$, $d_2 = 2$, $z_1 = 0.1$, $z_2 = 0.2$ and initial condition $(0.1, 0.1)$

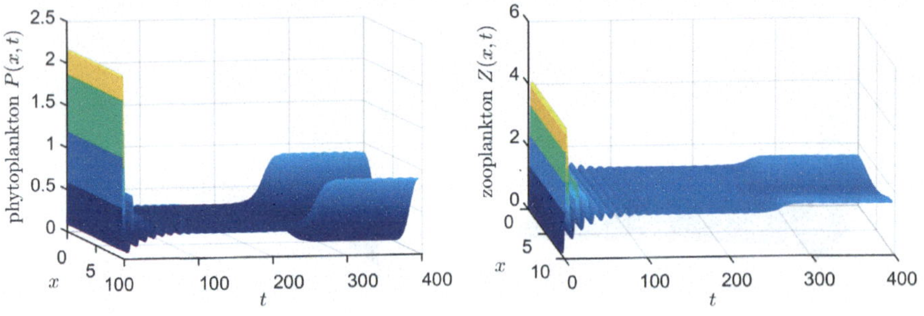

(a) (b)

Fig. 7.4 Behavior of plankton for Turing unstable conditions, where $\tau = 0$, $d_1 = 0.12$, $d_2 = 10$, $z_1 = 0.1$, $z_2 = 0.2$ and initial condition $(0.1, 0.1)$

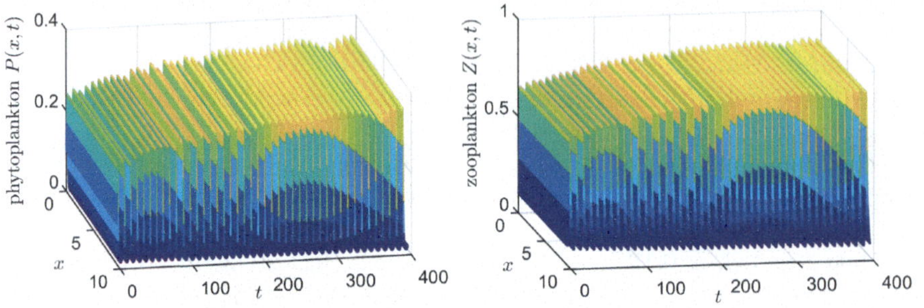

(a) (b)

Fig. 7.5 Behavior of plankton for the change of consumption rate, where $\tau = 0, d_1 = 0.06, d_2 = 2$, $z_1 = 0.1$, $z_2 = 0.2$, $C_m = 6$ and initial condition $(0.1, 0.1)$

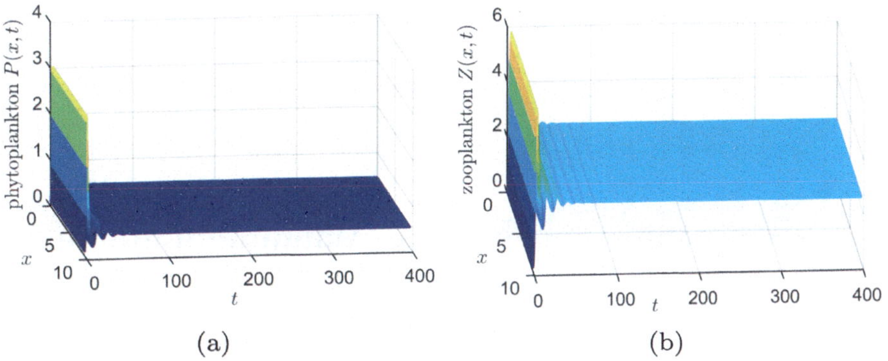

Fig. 7.6 Behavior of plankton for the change of cell size, where $\tau = 0$, $d_1 = 0.12$, $d_2 = 10$, $z_1 = 0.15$, $z_2 = 0.2$ and initial condition $(0.1, 0.1)$

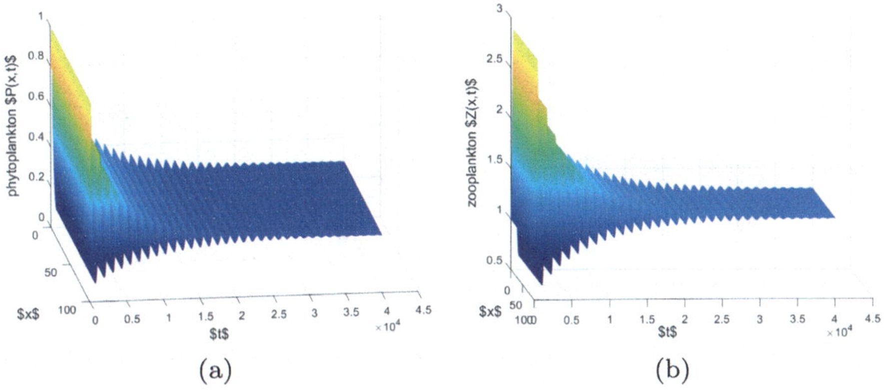

Fig. 7.7 Behavior of plankton for stable conditions, where $\tau = 0.16 < \tau_*$, $d_1 = 0.06$, $d_2 = 0.06$, $z_1 = 0.1$, $z_2 = 0.2$ and initial condition $(0.8, 1)$

ton, the behaviors for phytoplankton and zooplankton change from Turing instability to asymptotic stability shown in Figs. 7.4 and 7.6.

Next, we also show the results of the numerical simulation of system (7.42) for different values of τ with $d_1 = 0.06$, $d_2 = 0.06$, $z_1 = 0.1$, $z_2 = 0.2$. By direct computation, we have $\tau_* = \tau_0^0 \approx 0.3$. By Theorem 58, we know that if $\tau \in [0, \tau_*)$, then equilibrium E_* is stable (see Fig. 7.7) and when τ crosses through the critical τ_*, E_* loses its stability and Hopf bifurcation occurs (see Fig. 7.8). After computation of (7.41), we can obtain $\beta_2 \approx -32.7812$, $\mu_2 \approx 49.59$, $T_2 \approx 20.5689$. Hence, the Hopf bifurcation is known as supercritical Hopf bifurcation and the bifurcating limit cycle is orbitally stable, which can be depicted in Fig. 7.8. In addition, if we increase the cell size of plankton to $z_1 = 0.2$, $z_2 = 0.3$ and set the time delay to $\tau = 0.33$ (see Fig. 7.9), the phytoplankton and zooplankton will coexist and tend to the coexisting equilibrium.

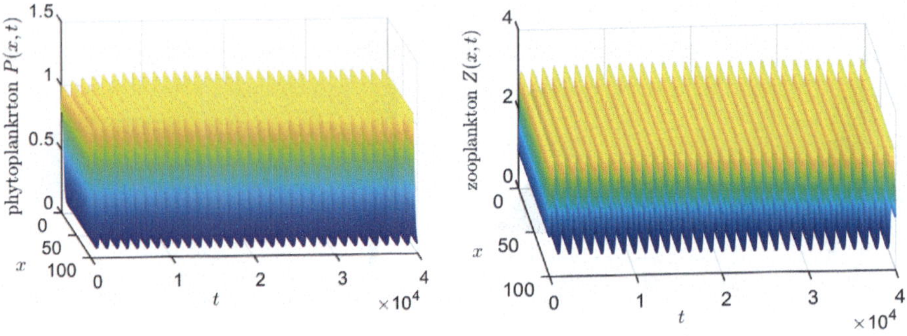

(a) (b)

Fig. 7.8 Behavior of plankton for unstable conditions, where $\tau = 0.33 > \tau_*, d_1 = 0.06, d_2 = 0.06,$ $z_1 = 0.1$, $z_2 = 0.2$ and initial condition (0.8, 1)

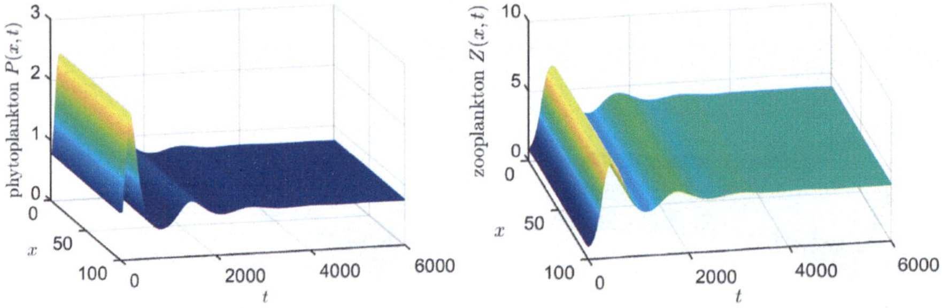

(a) (b)

Fig. 7.9 Behavior of plankton for the change of cell size, where $\tau = 0.33 > \tau_*$, $d_1 = 0.06$, $d_2 = 0.06$, $z_1 = 0.2$, $z_2 = 0.3$ and initial condition (0.8, 1)

Chapter 8
The Effect of Defensive and Aggressive Behavior

Complex interaction between chemical–physical conditions and biological processes plays significant role in the dynamic behavior of plankton [17, 34, 126, 127]. There exists a direct effect of some factors like pH [128], turbulence [129], and temperature [23] on the marine ecosystem. Moreover, various factors also affect the biological processes including growth, mortality, competition, and predation [21, 114–118]. For instance, a phytoplankton–zooplankton-fish model including nonlinear fish harvesting and taxation is investigated in Ref. [130]. In Ref. [15], the local Hopf bifurcation of a phytoplankton–zooplankton model with gestational delays was addressed, while in Ref. [114] a nutrient-phytoplankton–zooplankton model was presented with spatiotemporal dynamics under the effects of prey-taxis and intratrophic predation. There are many factors affecting parameters of plankton ecosystem which exhibit stochastic disturbances such as fluctuations of environment like earthquakes, carrying capacity, epidemics and tsunami, and the birth and death rates. Thus, stochastic disturbances were introduced into these deterministic models to reveal the effect of environmental variability on the population dynamics [25, 131–133]. Specially, the environmental noise has negative effect on the Listeria monocytogenes mean-value concentration [132] and stochastic model can also reproduce the phytoplankton distributions observed in real data [133].

Moreover, it should be noted that the population dynamics can be affected by the adaptive changes of plankton like genetic evolution, behavioral selection, and phenotypic plasticity [134]. There are two main adaptive changes: one is the defense of phytoplankton and the other is the offense of zooplankton. By producing the effect of defense, zooplankton is often the main drivers for phytoplankton to adapt to environment, while increasing defense capability of phytoplankton, it may yield the occurrence of offense [135]. There are several defense factors including morphology (e.g., silica shell, colony formation), physiology (e.g., toxicity, bioluminescence), and behavior (e.g., escape response) [136]. And there exists some monotone linear relationships: digestion time of predator, predator's motivation to search prey by producing toxic substances, and the encounter rate with a predator [137]. As for the

S. T. Liu and L. Zhang, *Pattern Dynamics of Marine Plankton Behavior*,
https://doi.org/10.1007/978-981-97-5369-7_8

Table 8.1 Description of the parameters for model (8.1)

Parameters	Description
$r_1(u)$	Intrinsic growth rate of phytoplankton
r_0	Maximum growth rate
c_1	Costliness of defense
K	Carrying capacity of environment
b	Handling time
$a(u, v)$	Capture rate of zooplankton
a_0	Maximum capture rate
θ	Sensitivity of capture rate to the difference between the trait values
$\epsilon(v)$	Conversion efficiency
ϵ_0	Maximum conversion efficiency
c_2	Costliness of offense
r_2	Mortality rate of zooplankton

defense of phytoplankton, the offense of zooplankton varies positively with immunity against prey toxins and encounter area and negatively with handling time.

This chapter presents a diffusive plankton model to study the effects of defense of phytoplankton and offense of zooplankton. Compared with the other models, our model focuses on the intrinsic growth rate of phytoplankton and the conversion rate of zooplankton, respectively, depending on defense and offense. In the meantime, there exists an extendibility for the construction conditions of biological mathematical model in which the capture rate of zooplankton to phytoplankton is shared by defense and offense. Furthermore, the single compartment model is studied and its spatially extended version is also proposed to choose capture rate as the control parameter.

8.1 Formulation of Diffusive Plankton Model

The phytoplankton P and zooplankton Z with defense u and offense v is considered to model the defensive and offensive effects as given below:

$$\begin{cases} \dot{P} = P\left(r_1(u)\left(1-\frac{P}{K}\right) - \dfrac{a(u,v)Z}{1+ba(u,v)P}\right), \\ \dot{Z} = Z\left(\dfrac{\epsilon(v)a(u,v)P}{1+ba(u,v)P} - r_2 Z\right). \end{cases} \tag{8.1}$$

Table 8.1 describes the biological significance of the parameters of the model and the functions $r_1(u)$, $a(u, v)$, and $\epsilon(v)$ are assumed as follows:

- A trade-off between defense u and the intrinsic growth rate r_1 of phytoplankton is assumed to be $r_1(u) = r_0 e^{-c_1 u^2}$ [138]. Hence the r_1 on the dynamics has negative impact as prey biomass approaches the carrying capacity. As for the prey biomass, it has the negative linear effect on the cost of the same defense level.
- The levels of defense and offense determine the capture rate of the predator to the prey which is represented by $a(u, v) = \frac{a_0}{1+e^{\theta(u-v)}}$ [138, 139]. Hence, the capture rate has negative relationship with defense and offense.
- ϵ, the trade-off between offense v and the conversion efficiency, is defined as $\epsilon(v) = \epsilon_0 e^{-c_2 v^2}$ [140].

Generally, the plankton distributes inhomogeneously and moves from areas with higher density to areas with smaller density because of diffusion and current. Hence, by introducing diffusion terms into model (8.1), it can be modified to better present the real-world situation.

$$\begin{cases} \dfrac{\partial P}{\partial t} - d_1 \Delta P = P\left(r_1(u)\left(1 - \frac{P}{K}\right) - \frac{a(u,v)Z}{1+ba(u,v)P}\right), & x \in (0, l\pi),\ t > 0, \\ \dfrac{\partial Z}{\partial t} - d_2 \Delta Z = Z\left(\frac{\epsilon(v)a(u,v)P}{1+ba(u,v)P} - r_2 Z\right), & x \in (0, l\pi),\ t > 0, \\ P_x(0,t) = P_x(l\pi, t) = 0, \quad Z_x(0,t) = Z_x(l\pi, t) = 0, & t > 0, \\ P(x,0) = \phi(x,0) > 0, \quad Z(x,0) = \varphi(x,0) > 0, & x \in (0, l\pi), \end{cases} \tag{8.2}$$

where d_1 and d_2 mean diffusive coefficients of phytoplankton and zooplankton, respectively. The boundary conditions remain Neumann boundary conditions (the region is closed), meaning that there is no phytoplankton and zooplankton species entering and leaving the region at the boundary. The Laplacian operator $\Delta = \frac{\partial^2}{\partial x^2} \in \mathbb{R}$ represents that plankton is spatially heterogeneous and migrates from regions of higher density to regions of lower density.

8.2 Dynamics Analysis of Single Compartment Model

Next, the stability and bifurcation analyses are performed to prove the coexisting equilibrium by calculating the eigenvalues of the Jacobian matrix of model (8.1). And the stability of the bifurcated periodic solutions is also proved. Firstly, we investigate the stability of the coexisting equilibrium.

Apparently, we assume that (i) the extinction equilibrium $E_0 = (0, 0)$ remains unstable; (ii) the boundary equilibrium $E_1 = (K, 0)$ satisfies a saddle point.

From the viewpoint of ecology, the coexisting equilibrium $E_* = (P_*, Z_*)$ is our interested aim from the following equations:

$$\begin{cases} r_1(u)\left(1-\frac{P}{K}\right)-\dfrac{a(u,v)Z}{1+ba(u,v)P}=0, \\ \dfrac{\epsilon(v)a(u,v)P}{1+ba(u,v)P}-r_2Z=0. \end{cases}$$

It is obvious to obtain that

$$Z_*=\frac{r_1(u)}{a(u,v)}\left(1-\frac{P_*}{K}\right)(1+ba(u,v)P_*),$$

where $Z_*>0$ as $P_*<K$ and P_* represents a positive root of the equation as given below:

$$S(P)=b^2a(u,v)^2P^3+P^2ba(u,v)(2-Kba(u,v)) \\ +P\left(1-2Kba(u,v)+\frac{K\epsilon(v)a(u,v)^2}{r_2r_1(u)}\right)-K=0.$$

Clearly, $S(0)=-K<0$, $S(K)=\frac{\epsilon(v)a(u,v)^2K^2}{r_2r_1(u)}>0$. Therefore it is easy to obtain the proof about the existence of the coexisting equilibrium E_*. Now, we turn our attention to the dynamics of model (8.1) with E_*. Then we give the Jacobian matrix at E_* as

$$J=\begin{pmatrix} a_{11} & b_{11} \\ a(u,v)a_{22} & a(u,v)b_{22} \end{pmatrix},$$

with

$$a_{11}=-\frac{r_1(u)}{K}P_*+\frac{ba(u,v)^2Z_*P_*}{(1+ba(u,v)P_*)^2},\quad b_{11}=-\frac{P_*a(u,v)}{1+ba(u,v)P_*},$$
$$a_{22}=\frac{\epsilon(v)Z_*}{(1+ba(u,v)P_*)^2},\quad b_{22}=-\frac{r_2Z_*}{a(u,v)}.$$

The characteristic equation displays

$$\lambda^2-Tr(J)\lambda+Det(J)=0, \tag{8.3}$$

where

$$Tr(J)=a_{11}+a(u,v)b_{22},\qquad Det(J)=a(u,v)a_{11}b_{22}-a(u,v)b_{11}a_{22}.$$

The roots of (8.3) are obtained by

$$\lambda_{1,2}=\alpha(a(u,v))\pm i\omega(a(u,v))=\frac{Tr(J)\pm\sqrt{Tr(J)^2-4Det(J)}}{2}.$$

Thus, it yields that the roots of (8.3) get negative real parts if and only if $a_{11}+a(u,v)b_{22}<0$, $\left(a(u,v)<\frac{Q}{\sqrt{bP_*Z_*-bP_*Q}},\ Q=\left(\frac{r_1(u)P_*}{K}+r_2Z_*\right)^{\frac{1}{2}}\right)$ and

$$a_{11}b_{22} - b_{11}a_{22} > 0 \qquad (H1)$$

hold.

Moreover, as $a(u, v) = a = \frac{Q}{\sqrt{bP_*Z_*} - bP_*Q}$, we get $Tr(J) = 0$. Because of $Det(J) > 0$ under (H1), (8.3) obtains a pair of purely imaginary roots. After choosing $a(u, v)$ as perturbation parameter, we get

$$\left\{\frac{d(Re\lambda(a(u,v)))}{da(u,v)}\right\}_{\lambda=i\omega,\ a(u,v)=a} = \frac{baZ_*P_*}{(1+baP_*)^3} > 0.$$

This displays that the transversality condition holds. Thus, we get the result of model (8.1) as below immediately.

Theorem 59 *Let (H1) hold and $Q = (\frac{r_1(u)P_*}{K} + r_2Z_*)^{\frac{1}{2}}$. For model (8.1),*

(i) the equilibrium E_* is locally asymptotically stable if $a_{11} < 0$ holds;

(ii) the equilibrium E_* is locally asymptotically stable if $a_{11} > 0$ holds, for $a(u, v) < \frac{Q}{\sqrt{bP_*Z_*} - bP_*Q}$;

(iii) model (8.1) undergoes Hopf bifurcation at E_* if $a_{11} > 0$ holds and $a(u, v)$ crosses $a = \frac{Q}{\sqrt{bP_*Z_*} - bP_*Q}$.

In the following, the stability of bifurcated periodic solution is given. To this end, E_* is translated to the origin by the translation $\bar{x} = P - P_*$, $\bar{y} = Z - Z_*$. Then, model (8.1) becomes

$$\begin{cases} \dot{\bar{x}} = (\bar{x} + P_*)\left(r_1(u)\left(1 - \frac{(\bar{x}+P_*)}{K}\right) - \frac{a(u,v)(\bar{y}+Z_*)}{1+ba(u,v)(\bar{x}+P_*)}\right), \\ \dot{\bar{y}} = (\bar{y} + Z_*)\left(\frac{\epsilon(v)a(u,v)(\bar{x}+P_*)}{1+ba(u,v)(\bar{x}+P_*)} - r_2(\bar{y} + Z_*)\right). \end{cases} \tag{8.4}$$

Thus (8.4) is rewritten into the following form:

$$\begin{pmatrix} \dot{\bar{x}} \\ \dot{\bar{y}} \end{pmatrix} = J\begin{pmatrix} \bar{x} \\ \bar{y} \end{pmatrix} + \begin{pmatrix} f(\bar{x}, \bar{y}, a(u,v)) \\ h(\bar{x}, \bar{y}, a(u,v)) \end{pmatrix}, \tag{8.5}$$

where

$$\begin{aligned} f(\bar{x}, \bar{y}, a(u,v)) &= \frac{\bar{x}^2}{2}\left(-\frac{r_1(u)}{K} + \frac{ba(u,v)^2Z_*(1-ha(u,v)P_*)}{(1+ba(u,v)P_*)^3}\right) + \frac{(ba(u,v)P_* - 1)a(u,v)\bar{x}\bar{y}}{2(1+ba(u,v)P_*)^2} \\ &- \frac{b^2a(u,v)^3Z_*(2-ba(u,v)P_*)\bar{x}^3}{3(1+ba(u,v)P_*)^4} + \frac{ba(u,v)^2(1-ba(u,v)P_*)\bar{x}^2\bar{y}}{6(1+ba(u,v)P_*)^4} \\ &+ \frac{ba(u,v)^2(1-ba(u,v)P_*)\bar{x}\bar{y}\bar{x}}{6(1+ba(u,v)P_*)^3} + \frac{ba(u,v)^2\bar{y}\bar{x}^2}{3(1+ba(u,v)P_*)^3} + O(|(\bar{x}, \bar{y})|^4), \\ h(\bar{x}, \bar{y}, a(u,v)) &= -\frac{\epsilon(v)ba(u,v)^2Z_*\bar{x}^2}{(1+ba(u,v)P_*)^3} + \frac{\epsilon(v)a(u,v)\bar{x}\bar{y}}{2(1+ba(u,v)P_*)^2} - \frac{r_2\bar{y}^2}{2} - \frac{\epsilon(v)ba(u,v)^2\bar{x}^2\bar{y}}{3(1+ba(u,v)P_*)^3} \\ &+ \frac{\epsilon(v)b^2a(u,v)^3Z_*\bar{x}^3}{(1+ba(u,v)P_*)^4} - \frac{\epsilon(v)ba(u,v)^2\bar{x}\bar{y}\bar{x}}{3(1+ba(u,v)P_*)^3} + O(|(\bar{x}, \bar{y})|^4). \end{aligned}$$

As $a(u, v) = a$, $\lambda_{1,2}(a) = \pm i\omega$, $\omega = \sqrt{a(a_{11}b_{22} - b_{11}a_{22})}$, it yields the one eigenvector corresponding to the eigenvalues $\lambda_{1,2}(a)$: where $\xi = (-b_{11}, a_{11} - i\omega)^T$. Assume

$$B = \begin{pmatrix} -b_{11} & 0 \\ a_{11} & -\omega \end{pmatrix}, \quad B^{-1} = \begin{pmatrix} -\frac{1}{b_{11}} & 0 \\ -\frac{a_{11}}{\omega b_{11}} & -\frac{1}{\omega} \end{pmatrix}.$$

Utilizing the transformation $\begin{pmatrix} \tilde{x} \\ \tilde{y} \end{pmatrix} = B \begin{pmatrix} \bar{x} \\ \bar{y} \end{pmatrix}$, we get the following form based on (8.5)

$$\begin{pmatrix} \dot{\tilde{x}} \\ \dot{\tilde{y}} \end{pmatrix} = \begin{pmatrix} 0 & -\omega \\ \omega & 0 \end{pmatrix} \begin{pmatrix} \tilde{x} \\ \tilde{y} \end{pmatrix} + \begin{pmatrix} f_1(\tilde{x}, \tilde{y}) \\ h_1(\tilde{x}, \tilde{y}) \end{pmatrix}, \tag{8.6}$$

where

$$f_1(\tilde{x}, \tilde{y}) = -\frac{1}{b_{11}} f(-b_{11}\tilde{x}, a_{11}\tilde{x} - \omega\tilde{y}, a(u, v)),$$
$$h_1(\tilde{x}, \tilde{y}) = -\frac{1}{\omega} h(-b_{11}\tilde{x}, a_{11}\tilde{x} - \omega\tilde{y}, a(u, v)) - \frac{a_{11}}{\omega b_{11}} f(-b_{11}\tilde{x}, a_{11}\tilde{x} - \omega\tilde{y}, a(u, v)),$$

$$\begin{aligned}
&f(-b_{11}\tilde{x}, a_{11}\tilde{x} - \omega\tilde{y}, a(u, v)) \\
&= \frac{1}{2}\left(-\frac{r_1(u)}{K} + \frac{ba(u,v)^2 Z_*(1 - ba(u,v)P_*)}{(1+ba(u,v)P_*)^3}\right) b_{11}^2 \tilde{x}^2 \\
&- \frac{b_{11}\tilde{x} a(u, v)(ba(u, v)P_* - 1)(a_{11}\tilde{x} - \omega\tilde{y})}{2(1 + ba(u, v)P_*)^2} - \frac{ba(u, v)^3 Z_*(2 - ba(u, v)P_*)(-b_{11}\tilde{x})^3}{3(1 + ba(u, v)P_*)^4} \\
&- \frac{(b_{11}\tilde{x})^2 ba(u, v)^2 (1 - ba(u, v)P_*)(a_{11}\tilde{x} - \omega\tilde{y})}{6(1 + ba(u, v)P_*)^4} \\
&- \frac{(b_{11}\tilde{x})^2 ba(u, v)^2 (3 - ba(u, v)P_*)(a_{11}\tilde{x} - \omega\tilde{y})}{6(1 + ba(u, v)P_*)^3} + O(|(-b_{11}\tilde{x}, a_{11}\tilde{x} - \omega\tilde{y})|^4), \\
&h(-b_{11}\tilde{x}, a_{11}\tilde{x} - \omega\tilde{y}, a(u, v)) \\
&= -\frac{\epsilon(v)ba(u,v)^2 Z_*(b_{11}\tilde{x})^2}{(1+ba(u,v)P_*)^3} - \frac{b_{11}\tilde{x}\epsilon(v)a(u, v)(a_{11}\tilde{x} - \omega\tilde{y})}{2(1 + ba(u, v)P_*)^2} - \frac{r_2(a_{11}\tilde{x} - \omega\tilde{y})^2}{2} \\
&- \frac{\epsilon(v)ba(u, v)^2 (b_{11}\tilde{x})^2 (a_{11}\tilde{x} - \omega\tilde{y})}{3(1 + ba(u, v)P_*)^3} - \frac{\epsilon(v)b^2 a(u, v)^3 Z_*(b_{11}\tilde{x})^3}{(1 + ba(u, v)P_*)^4} \\
&- \frac{\epsilon(v)ba(u, v)^2 (b_{11}\tilde{x})^2 (a_{11}\tilde{x} - \omega\tilde{y})}{3(1 + ba(u, v)P_*)^3} + O(|(-b_{11}\tilde{x}, a_{11}\tilde{x} - \omega\tilde{y})|^4).
\end{aligned}$$

Equation (8.6) is rewritten into the polar coordinate form as given below:

$$\begin{cases} \dot{r} = \alpha(a(u, v))r + \sigma(a(u, v))r^3 + O(|r|^4), \\ \dot{\theta} = \omega(a(u, v)) + c(a(u, v))r^2 + O(|r|^4). \end{cases} \tag{8.7}$$

Using Taylor expansion of (8.7) at $a(u, v) = a$, we get

$$\begin{cases} \dot{r} = \alpha'(a)(a(u, v) - a)r + \sigma(a)r^3 + O(|r|^4), \\ \dot{\theta} = \omega(a) + \omega'(a)(a(u, v) - a) + c(a)r^2 + O(|r|^4). \end{cases}$$

The sign of $\sigma(a)$ determines the stability of bifurcated periodic solution so that we obtain

$$\sigma(a)=\frac{1}{16}(f_{1\tilde{x}\tilde{x}\tilde{x}}+h_{1\tilde{x}\tilde{x}\tilde{y}})+\frac{1}{16\omega(a)}\Big(f_{1\tilde{x}\tilde{y}}f_{1\tilde{x}\tilde{x}}-h_{1\tilde{x}\tilde{y}}(h_{1\tilde{x}\tilde{x}}+r_2\omega)-f_{1\tilde{x}\tilde{x}}h_{1\tilde{x}\tilde{x}}\Big),$$

where

$$\begin{aligned}
f_{1\tilde{x}\tilde{x}\tilde{x}}&=-\Big(2\frac{b^2a(u,v)^3Z_*(2-ba(u,v)P_*)b_{11}^2}{(1+ba(u,v)P_*)^4}+\frac{b_{11}ba(u,v)^2(1-ba(u,v)P_*)a_{11}}{3(1+ba(u,v)P_*)^4}\\
&+\frac{2b_{11}ba(u,v)^2(1-ba(u,v)P_*)a_{11}}{3(1+ba(u,v)P_*)^3}+\frac{2ba(u,v)^2a_{11}b_{11}}{3(1+ba(u,v)P_*)^3}\Big),\\
h_{1\tilde{x}\tilde{x}\tilde{y}}&=-a_{11}\Big(\frac{b_{11}ba(u,v)^2(1-ba(u,v)P_*)}{3(1+ba(u,v)P_*)^4}+\frac{b_{11}ba(u,v)^2(1-ba(u,v)P_*)}{3(1+ba(u,v)P_*)^3}\\
&+\frac{2ba(u,v)^2b_{11}}{3(1+ba(u,v)P_*)^3}\Big)-\frac{4\epsilon(v)ba(u,v)^2b_{11}^2}{3(1+ba(u,v)P_*)^3},\\
f_{1\tilde{x}\tilde{y}}&=-\frac{ba(u,v)^2P_*\omega}{2(1+ba(u,v)P_*)^2}+\frac{a(u,v)\omega}{2(1+ba(u,v)P_*)^2},\\
f_{1\tilde{x}\tilde{x}}&=-b_1\left(-\frac{r_1(u)}{K}+\frac{ba(u,v)^2Z_*(1-ba(u,v)P_*)}{(1+ba(u,v)P_*)^3}\right)+\frac{ba(u,v)^2P_*a_{11}}{(1+ba(u,v)P_*)^2}\\
&-\frac{a(u,v)a_{11}\tilde{x}}{(1+ba(u,v)P_*)^2},\\
h_{1\tilde{x}\tilde{y}}&=-a_{11}\Big(\frac{ba(u,v)^2P_*}{2(1+ba(u,v)P_*)^2}-\frac{a(u,v)}{2(1+ba(u,v)P_*)^2}\Big)\\
&-\Big(\frac{b_{11}\epsilon(v)a(u,v)}{2(1+ba(u,v)P_*)^2}+r_2a_{11}\Big),\\
h_{1\tilde{x}\tilde{x}}&=-\frac{a_{11}}{\omega}\Big(\Big(-\frac{r_1(u)}{K}+\frac{ba(u,v)^2Z_*(1-ba(u,v)P_*)}{(1+ba(u,v)P_*)^3}\Big)b_{11}\\
&+\frac{a(u,v)a_{11}(1-ba(u,v)P_*)}{(1+ba(u,v)P_*)^2}\Big)\\
&+\frac{1}{\omega}\Big(\frac{2\epsilon(v)ba(u,v)^2Z_*b_{11}^2}{(1+ba(u,v)P_*)^3}+\frac{b_{11}\epsilon(v)a(u,v)a_{11}}{(1+ba(u,v)P_*)^2}+r_2a_{11}^2\Big).
\end{aligned}$$

From the results in Ref. [141], it is obvious that the bifurcated periodic solution remains stable when $\sigma<0$. On the contrary, if $\sigma>0$, the bifurcated periodic solution remains unstable.

8.3 Dynamics Analysis of Spatially Extended Model

In the following, the diffusion-driven instability and the bifurcation analysis for model (8.2) is generated under condition (H1).

8.3.1 Turing Instability

Next, sufficient conditions are derived for occurrence of Turing instability for model (8.2). A constant solution is Turing unstable when it is stable in the absence of diffusion, but becomes unstable when there exists diffusion [142].

We get the linearization of (8.2):

$$\begin{pmatrix} \frac{\partial P}{\partial x} \\ \frac{\partial Z}{\partial x} \end{pmatrix} = \begin{pmatrix} a_{11} + d_1 \frac{\partial^2}{\partial x^2} & b_{11} \\ a(u,v)a_{22} & a(u,v)b_{22} + d_2 \frac{\partial^2}{\partial x^2} \end{pmatrix} \begin{pmatrix} P \\ Z \end{pmatrix}. \tag{8.8}$$

Subject to the Neumann boundary conditions, the characteristic equation of (8.8) is displayed

$$\lambda^2 - Tr_n \lambda + Det_n = 0, \quad n = 0, 1, 2, \ldots, \tag{8.9}$$

where

$$\begin{cases} Tr_n = -(d_1 + d_2)\frac{n^2}{l^2} + a_{11} + a(u,v)b_{22} = Tr(J) - (d_1 + d_2)\frac{n^2}{l^2}, \\ Det_n = d_1 d_2 \frac{n^4}{l^4} - (d_1 b_{22} a(u,v) + a_{11} d_2)\frac{n^2}{l^2} + a(u,v)a_{11}b_{22} - a(u,v)a_{22}b_{11} \\ \quad = \left(\frac{n^2}{l^2} - \frac{a_{11}}{d_1}\right) d_1 d_2 \frac{n^2}{l^2} - a(u,v)d_1 b_{22}\frac{n^2}{l^2} + Det(J), \end{cases} \tag{8.10}$$

and the eigenvalues are obtained by

$$\lambda_{1,2}^n = \frac{Tr_n \pm \sqrt{Tr_n^2 - 4Det_n}}{2}.$$

Under (H1) and (ii) of Theorem 59, we get $Tr_n < Tr(J) < 0$ and $Det(J) > 0$. Hence, Turing instability occurs as it exists $n \in N$ such that $Det_n < 0$. Define

$$d_m = \min_{1 \le n \le m} \frac{Det(J) - a(u,v)d_1 b_{22}\frac{n^2}{l^2}}{\left(a_{11} - d_1 \frac{n^2}{l^2}\right)\frac{n^2}{l^2}}.$$

Then, we get $\frac{n^2}{l^2} - \frac{a_{11}}{d_1} \ge \frac{a_{11}}{d_1}(n^2 - 1) \ge 0$ when $\frac{a_{11} l^2}{d_1} \le 1$ holds. And when $m^2 \le \frac{a_{11} l^2}{d_1} \le (m+1)^2$ and $1 \le n \le m$ hold, we get $0 < a_{11}\left(1 - \frac{n^2}{(m+1)^2}\right) \le a_{11} - d_1 \frac{n^2}{l^2} \le a_{11}\left(1 - \frac{n^2}{m^2}\right)$ is satisfied.

According to the above analysis, we obtain the theorem as follows.

Theorem 60 *When (H1) and (ii) of Theorem 59 hold,*

(i) The coexisting equilibrium E_* of model (8.2) gets locally asymptotically stable as (H2) $\frac{a_{11} l^2}{d_1} \le 1$ or (H3) $m^2 \le \frac{a_{11} l^2}{d_1} \le (m+1)^2$, $d_2 < d_m$ holds.

(ii) The coexisting equilibrium E_* of model (8.2) gets Turing unstable if $m^2 \le \frac{a_{11} l^2}{d_1} \le (m+1)^2$ and $d_2 > d_m$ hold.

8.3.2 Hopf Bifurcation

Next, the occurrence of Hopf bifurcation is considered when $a(u, v)$ is identified as bifurcation parameter. Applying the presentation of eigenvalues, it yields that (8.9) obtains purely imaginary roots when

$$a(u, v) = a_n = \frac{a_{11} - (d_1 + d_2)\frac{n^2}{l^2}}{-b_{22}},$$

and $Det_n > 0$. Hence, $Tr_n(a_n) = 0$, $Tr_j(a_n) \neq 0$, for $j \neq n$. Meanwhile,

$$Det_n(a_n) > -(d_1 b_{22} a(u, v) + a_{11} d_2)\frac{n^2}{l^2} > \left((d_1 - d_2)a_{11} - d_1(d_1 + d_2)\frac{n^2}{l^2}\right)\frac{n^2}{l^2}.$$

If there exists some $n \in N$ such that $Det_n(a_n) > 0$ holds, we obtain $a_1 > \frac{d_1(d_1+d_2)\frac{n^2}{l^2}}{d_1-d_2}$ and $d_1 > d_2$. Moreover, the complex eigenvalues $\alpha_n(a(u, v)) \pm i\omega_n(a(u, v))$ satisfy

$$\alpha_n^{'}(a_n) = \frac{ba_n Z_* P_*}{(1 + ba_n P_*)^3} > 0,$$

implying that it satisfies the transversal condition.

Then, the following result is obtained.

Theorem 61 *Assume that condition (H1) and (ii) of Theorem 59 hold. Model (8.2) undergoes the Hopf bifurcation at $a(u, v) = a_n$ if there exists some $n \in N$ such that the conditions $a_{11} > \frac{d_1(d_1+d_2)\frac{n^2}{l^2}}{d_1-d_2}$ and $d_1 > d_2$ are satisfied.*

Moreover:

(i) When it coincides with the periodic solution of model (8.1), the bifurcated periodic solution from $a(u, v) = a$ is spatially homogeneous.

(ii) The bifurcated periodic solutions from $a(u, v) = a_n$ are spatially nonhomogeneous, where $(n \geq 1)$.

Next, the stability of the bifurcated periodic solutions would be studied by utilizing center manifold theorem and normal form theorem of partial differential equations [79, 143]. Let $a(u, v) = a$ $(n = 0)$ be the bifurcation parameter, $\bar{P} = P - P_*$ and $\bar{Z} = Z - Z_*$. For convenience, model (8.2) can be written as

$$U_t = \begin{pmatrix} d_1\frac{\partial^2}{\partial x^2} & 0 \\ 0 & d_2\frac{\partial^2}{\partial x^2} \end{pmatrix} U + JU + F(U, a), \tag{8.11}$$

dropping the bar, with $U = (P, Z)^T$ and $F(U, a) = (f(U, a), h(U, a))^T$, f, h are defined in (8.5). After the straightforward computation, we get

$$f_{PP} = -\frac{r_1(u)}{K} + \frac{ba^2 Z_*(1-haP_*)}{(1+baP_*)^3}, \quad f_{PZ} = \frac{ba^2 P_*}{(1+baP_*)^2},$$
$$f_{ZP} = -\frac{a}{(1+baP_*)^2}, \; f_{PPP} = -\frac{2b^2a^3 Z_*(2-baP_*)}{(1+baP_*)^4},$$
$$f_{PPZ} = \frac{ba^2(1-baP_*)}{(1+baP_*)^4}, \quad f_{PZP} = \frac{ba^2(1-baP_*)}{(1+baP_*)^3},$$
$$f_{ZPP} = \frac{2ba^2}{(1+baP_*)^3}, \quad h_{PP} = -\frac{2\epsilon(v)ha^2 Z_*}{(1+baP_*)^3},$$
$$h_{PZ} = \frac{\epsilon(v)a}{(1+baP_*)^2}, \quad h_{ZZ} = -r_2,$$
$$h_{PPZ} = -\frac{2\epsilon(a)ba^2}{(1+baP_*)^3}, \quad h_{PPP} = \frac{6\epsilon(v)b^2a^3 Z_*}{(1+baP_*)^4},$$
$$h_{PZP} = -\frac{2\epsilon(a)ba^2}{(1+baP_*)^3},$$

and we also get the other partial derivatives as 0. The linear operator $L = L(a)$ of (8.8) at $a(u, v) = a$ is defined as

$$LU = \begin{pmatrix} d_1\frac{\partial^2}{\partial x^2} + a_{11} & b_{11} \\ aa_{22} & d_2\frac{\partial^2}{\partial x^2} + ab_{22} \end{pmatrix} U,$$

and the adjoint operator of L is as

$$L^*U = \begin{pmatrix} d_1\frac{\partial^2}{\partial x^2} + a_{11} & -aa_{22} \\ -b_{11} & d_2\frac{\partial^2}{\partial x^2} + ab_{22} \end{pmatrix} U.$$

Obviously, $\langle L^*U, V\rangle = \langle U, LV\rangle$ for any $U = (P_1, P_2)^T, V = (Z_1, Z_2)^T \in \mathbf{H}^2([0, l\pi]) \times \mathbf{H}^2([0, l\pi])$ and the inner product in $\mathbf{H}^2([0, l\pi]) \times \mathbf{H}^2([0, l\pi])$ is $\langle U, V\rangle = \frac{1}{l\pi}\int_0^{l\pi}(P_1\bar{Z}_1 + P_2\bar{Z}_2)dx$. Then, the linearization equation of (8.11) at (0, 0) is obtained as follows:

$$\dot{U}_t = LU. \tag{8.12}$$

It's easy to get that the characteristic equation of (8.12) has a pair of purely imaginary roots $\Lambda = \{i\omega, -i\omega\}$ as $a(u, v) = a$. Assume that q and q^* are the eigenfunctions of L and L^* corresponding to $i\omega$ and $-i\omega$:

$$q = (q_1, q_2)^T = \left(\frac{1}{b_{11}}, -\frac{i\omega + a_{11}}{\omega^2 + a_{11}^2}\right)^T, \quad q^* = (q_1^*, q_2^*)^T = \left(\frac{1}{aa_{22}}, \frac{a_{11} - i\omega}{\omega^2 + a_{11}^2}\right)^T.$$

Thus $\langle q^*, q\rangle = 1$, $\langle q^*, \bar{q}\rangle = 0$.

At $a(u, v) = a$, the coordinates describing center manifold C are computed. Then we rewrite the solution of (8.11) as

$$U = zq + \bar{z}\bar{q} + w(z, \bar{z}), \tag{8.13}$$

where $w = (w_1, w_2)^T$, $z = \langle q^*, U\rangle$. On the center manifold C, we get

$$w = w_{20}\frac{z^2}{2} + w_{11}z\bar{z} + w_{02}\frac{\bar{z}^2}{2} + O(|z|^3).$$

Then, (8.11) is rewritten into the (z, w) coordinate form

$$\begin{aligned} \dot{z} &= i\omega z + \langle q^*, F(U, a)\rangle = i\omega z + g(z, \bar{z}), \\ \dot{w} &= \dot{U} - \dot{z}q - \dot{\bar{z}}\bar{q} = Lw + H(z, \bar{z}), \end{aligned} \tag{8.14}$$

where

$$\begin{aligned} g(z, \bar{z}) &= g_{20}\frac{z^2}{2} + g_{11}\bar{z}z + g_{02}\frac{\bar{z}^2}{2} + g_{21}\frac{z^2\bar{z}}{2} + O(|z|^3, |z|, |w|, |w|^3), \\ H(z, \bar{z}) &= H_{20}\frac{z^2}{2} + H_{11}z\bar{z} + H_{02}\frac{\bar{z}^2 z}{2} + O(|z|^3, |z|, |w|, |w|^3). \end{aligned}$$

Thus, we get

$$\begin{aligned} &\langle q^*, F(U, a)\rangle = \tfrac{1}{2}\langle q^*, C(q, q)\rangle z^2 + \langle q^*, C(q, \bar{q})\rangle z\bar{z} + \tfrac{1}{2}\langle q^*, C(\bar{q}, \bar{q})\rangle \bar{z}^2 \\ &+O(|z|^3, |z|, |w|, |w|^3), \\ &H(z, \bar{z}) = F(U, a) - \langle \bar{q}^*, F(U, a)\rangle q - \langle \bar{q}^*, F(U, a)\rangle \bar{q}, \end{aligned}$$

$$C(q, q) = \begin{pmatrix} f_{PP}q_1^2 + f_{PZ}q_1q_2 + f_{ZP}q_2q_1 + f_{ZZ}q_2^2 \\ h_{PP}q_1^2 + h_{PZ}q_1q_2 + h_{ZP}q_2q_1 + h_{ZZ}q_2^2 \end{pmatrix},$$

$$\begin{aligned} E(q, q, \bar{q}) &= \begin{pmatrix} f_{PPP}q_1^2\bar{q}_1 + f_{PPZ}q_1^2\bar{q}_2 + f_{PZP}q_1q_2\bar{q}_1 + f_{PZZ}q_1q_2\bar{q}_2 \\ h_{PPP}q_1^2\bar{q}_1 + h_{PPZ}q_1^2\bar{q}_2 + h_{PZP}q_1q_2\bar{q}_1 + h_{PZZ}q_1q_2\bar{q}_2 \end{pmatrix} \\ &+ \begin{pmatrix} f_{ZPP}\bar{q}_1q_2q_1 + f_{ZPZ}q_1q_2\bar{q}_2 + f_{ZZP}q_2^2\bar{q}_1 + f_{ZZZ}q_2^2\bar{q}_2 \\ h_{ZPP}\bar{q}_1q_2q_1 + h_{ZPZ}q_1q_2\bar{q}_2 + h_{ZZP}q_2^2\bar{q}_1 + h_{ZZZ}q_2^2\bar{q}_2 \end{pmatrix}. \end{aligned}$$

This, together with (8.14), indicates that

$$\begin{aligned} &g_{20} = \langle q^*, C(q, q)\rangle, \quad h_{11} = \langle q^*, C(q, \bar{q})\rangle, \quad g_{02} = \langle q^*, C(\bar{q}, \bar{q})\rangle, \\ &g_{21} = 2\langle q^*, C(w_{11}, q)\rangle + \langle q^*, C(w_{20}, \bar{q})\rangle + \langle q^*, E(q, q, \bar{q})\rangle, \\ &H_{20} = C(q, q) - \langle q^*, C(q, q)\rangle q - \langle \bar{q}^*, C(q, q)\rangle \bar{q}, \\ &H_{11} = C(q, \bar{q}) - \langle q^*, C(q, \bar{q})\rangle q - \langle \bar{q}^*, C(q, \bar{q})\rangle \bar{q}, \\ &H_{02} = C(\bar{q}, \bar{q}) - \langle q^*, C(\bar{q}, \bar{q})\rangle q - \langle \bar{q}^*, C(\bar{q}, \bar{q})\rangle \bar{q}. \end{aligned}$$

Obviously, we obtain that the values of H_{20}, H_{11}, H_{02} are 0.

On the center manifold $\mathcal{C}$ near the origin, we get

$$\dot{w} = w_z\dot{z} + w_{\bar{z}}\dot{\bar{z}} = w_{20}z\dot{z} + w_{11}\bar{z}\dot{z} + w_{11}z\dot{\bar{z}} + w_{02}\bar{z}\dot{\bar{z}} + \cdots. \tag{8.15}$$

Substituting $\dot{z}$ of (8.14) into (8.15) and comparing the coefficients with $\dot{w}$ of (8.14), we get

$$w_{20} = (2i\omega - L)^{-1}H_{20} = (0,0)^T, \quad w_{11} = -L^{-1}H_{11} = (0,0)^T,$$
$$w_{02} = (-2i\omega - L)^{-1}H_{02} = (0,0)^T,$$

then $g_{21} = \langle q^*, E(q, q, \bar{q})\rangle$.

Consequently, the parameters of model (8.2) and a determine g_{ij} $(0 \leq i, j \leq 2)$. Then, we get the standard result as

$$\delta_1(a) = \frac{g_{20}g_{11}(3\alpha(a) + i\omega(a))}{2(\alpha^2(a) + \omega^2(a))} + \frac{|g_{11}|^2}{\alpha(a) + i\omega(a)} + \frac{|g_{02}|^2}{2(\alpha(a) + 3i\omega(a))} + \frac{g_{21}}{2},$$
$$\beta_2 = Im(\delta_1(a)) - \frac{Re(\delta_1(a))}{\alpha'(a)}\omega'(a).$$

According to the properties of Hopf bifurcation [144], the bifurcated periodic solutions are stable (unstable) when $Re(\delta_1(a)) < 0\,(Re(\delta_1(a)) > 0)$. The period of bifurcated periodic solutions increases (decreases) if $\beta_2 > 0$ $(\beta_2 < 0)$.

8.4 Numerical Simulations

In this section, some numerical simulations are carried out to verify our theoretical results. Let parametric values be taken as

$$r_0 = 0.95, \ c_1 = 0.4, \ K = 6, \ b = 0.6, \ a_0 = 1.85, \ \theta = 5,$$
$$c_2 = 0.8, \ \epsilon_0 = 0.5, \ r_2 = 0.1, u = 0.2, v = 0.15.$$

Then we get that model (8.1) has a coexisting equilibrium $E_* = (0.3819, 1.2814)$. In order to verify the effects of defense of phytoplankton and offense of zooplankton on the dynamics of (8.1), we choose different capture rate $a(u, v)$ of zooplankton.

Figures 8.1 and 8.2 show the locally asymptotically stable and bifurcated periodic stable, where $a(u, v)$ is the bifurcation parameter and the Hopf bifurcation threshold is taken as $a(u, v) = a = 0.9787$. When $a(u, v) = 0.81 < a$, that is $u = 0.2$, $v = 0.15$, Fig. 8.1 shows the coexisting equilibrium E_* as locally asymptotically stable. And when $a(u, v) = 1.5125 > a$, that is, $u = 0.2$, $v = 0.5$, Fig. 8.2 illustrates the loses stability. It implies that different defense of phytoplankton and offense of zooplankton have different effect on the stability of coexisting equilibrium. We can also get that stable plankton population density starts decaying with the reduced defense of phytoplankton or increased offense of zooplankton. This indicates that appropriate increase of phytoplankton defense and decrease zooplankton offense might be beneficial for coexistence of phytoplankton and zooplankton under certain circumstances in deterministic environment. Moreover, we get $\sigma = -82.7657 < 0$. Thus, Fig. 8.2 shows the bifurcated periodic solution.

According to the importance of stochastic disturbance, numerical simulations are performed to illustrate the effect of Brownian motion on dynamic behavior of model (8.1). Any nonlinear system $\dot{x}(t) = f(x(t), t)$ in $\mathbb{R}^d$ can be stabilized by Brownian

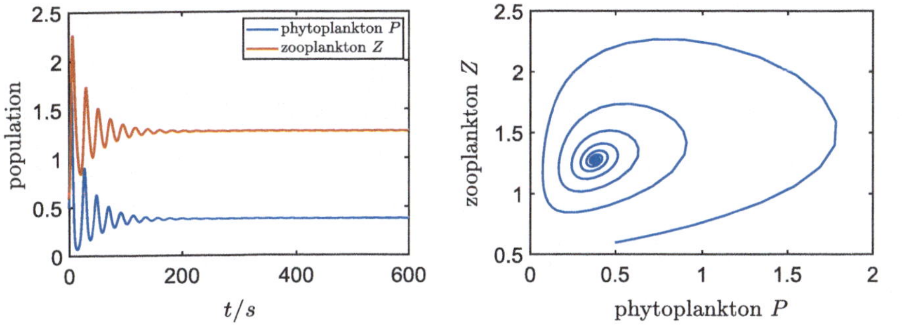

(a) Behavior of plankton

(b) Phase plot of model (8.1) for stable conditions

Fig. 8.1 Behavior of plankton and phase plot of model (8.1) for stable conditions, where $d_1 = d_2 = 0$, $u = 0.2$, $v = 0.15$ and initial condition (0.5, 0.6)

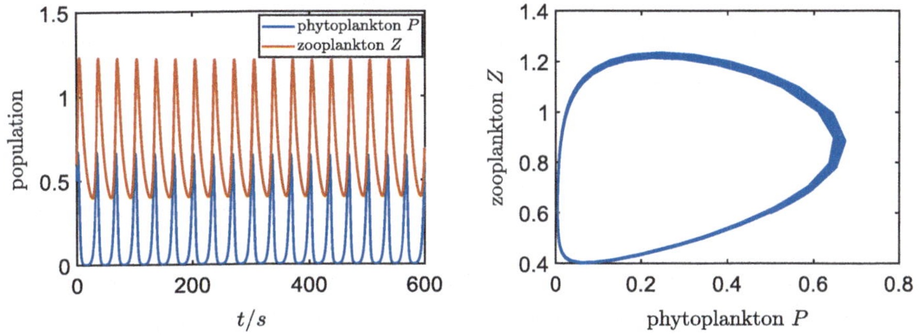

(a) Behavior of plankton

(b) Phase plot of model (8.1)

Fig. 8.2 Behavior of plankton and phase plot of model (8.1) for different values of defense and offense, where $d_1 = d_2 = 0$, $u = 0.2$, $v = 0.5$ and initial condition (0.5, 0.6)

motion provided $|f(x,t)| \leq K|x|$ for some $K > 0$ [145]. On the other hand, it can also be destabilized by Brownian motion when the dimension $d > 2$. Based on Theorems 59 and 60 of [145], the left panel of Fig. 8.3 shows that stochastic disturbance causes extinction of phytoplankton in a short time for (8.4). And Fig. 8.1 illustrates the choosing matrix of intensity of white noise $\begin{pmatrix} 0 & 1.6 \\ -1.6 & 0 \end{pmatrix}$ for the same defense and offense ($u = 0.2$, $v = 0.15$). Moreover, the right panel of Fig. 8.3 shows that model (8.1) can be transformed from instability to stability by the white noise, meaning that white noise makes the densities of phytoplankton and zooplankton move up and down randomly in a small neighborhood, where $\begin{pmatrix} 1.2 & 0 \\ 0 & 1.2 \end{pmatrix}$ for the same defense and offense ($u = 0.2$, $v = 0.5$) as shown in Fig. 8.2. The numerical simulations above indicate that stable coexistence of both populations can be kept by different stochastic disturbances so that there may be the extinctions of phytoplankton and zooplankton

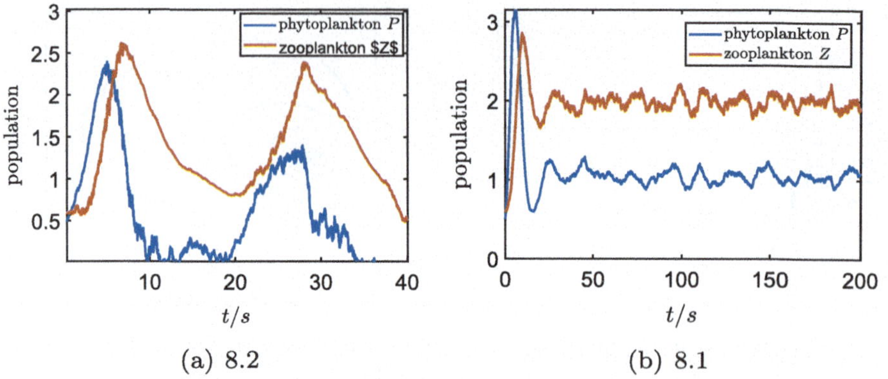

Fig. 8.3 Behavior of plankton with intensity of white noise, where $u = 0.2$, $v = 0.15$ and $u = 0.2$, $v = 0.5$, respectively

Next, the combined effects of defense and offense in diffusive model (8.2) are obtained. Assume that other coefficients are the same as (8.4), $l = 3$, $u = 0.2$, and $v = 0.15$.

According to the simple calculations, Figs. 8.4 and 8.5 show that the conditions (H2) and (H3) of Theorem 60 hold as $d_1 = 0.6$, $d_2 = 0.2$ and $d_1 = 0.05$, $d_2 = 0.2$, respectively. Figures 8.4 and 8.5 illustrate that E_* remains locally asymptotically stable for different diffusive coefficient of phytoplankton. Especially, phytoplankton and zooplankton can coexist when the diffusion of phytoplankton is small.

Figure 8.6 shows that Turing instability occurs when the value of diffusive coefficient d_2 gradually increases when the condition (ii) $m^2 \leq \frac{a_{11}l^2}{d_1} = 13.9567 \leq (m+1)^2$ and $d_2 > d_m = 13.4907$ of Theorem 61 hold for $d_1 = 0.05$, $d_2 = 15$, $m = 3$, $n = 1$. This displays that the accelerated diffusive movement of zooplank-

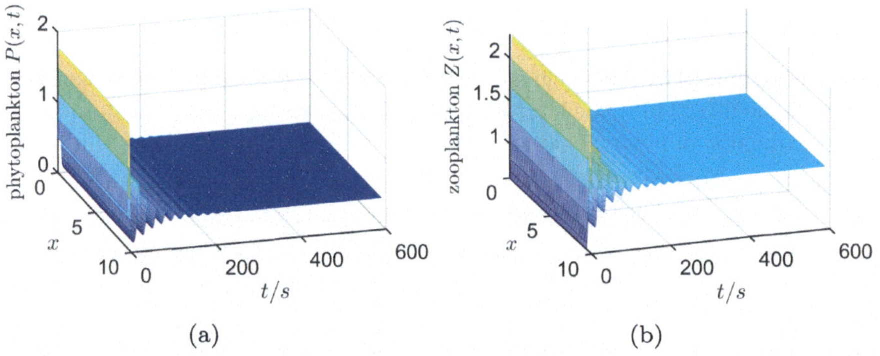

Fig. 8.4 Behavior of plankton with stable conditions (model (8.2)), where $d_1 = 0.6$, $d_2 = 0.2$, $u = 0.2$, $v = 0.15$ and initial condition (0.5, 0.6)

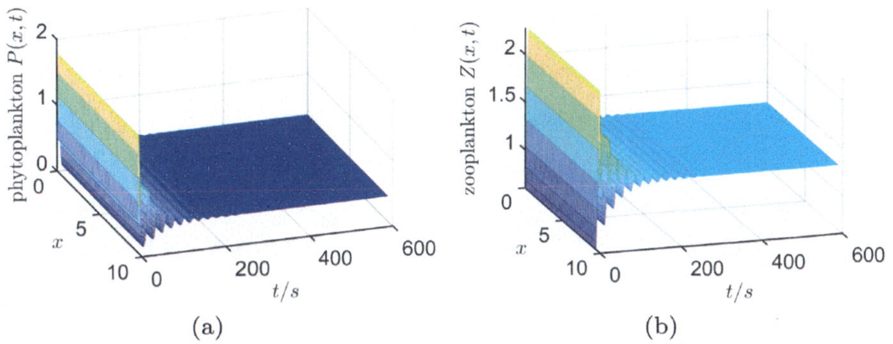

Fig. 8.5 Behavior of plankton with stable conditions (model (8.2)), where $d_1 = 0.05$, $d_2 = 0.2$, $u = 0.2$, $v = 0.15$ and initial condition (0.5, 0.6)

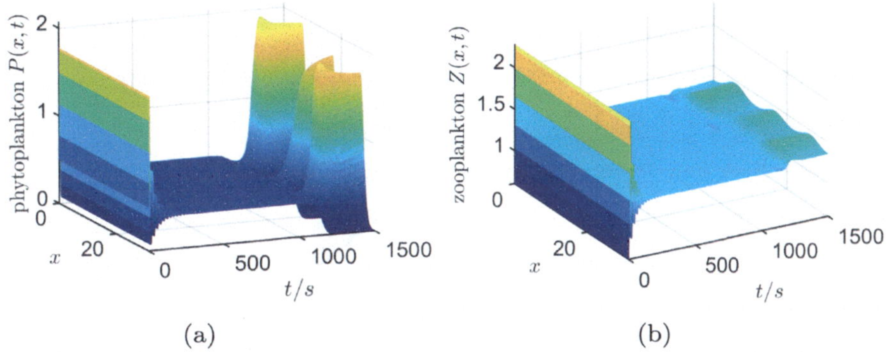

Fig. 8.6 Behavior of plankton with Turing instable conditions (model (8.2)), where $d_1 = 0.05$, $d_2 = 15$, $u = 0.2$, $v = 0.15$ and initial condition (0.5, 0.6)

ton monitors the formation of the pattern. Biologically, the formation of the pattern indicates that zooplankton moves quickly for the successful predation.

Let $n = 2$, $u = 0.42$, $v = 0.65$, $d_1 = 0.15$, and $d_2 = 0.08$. Taking $a(u, v)$ as the bifurcation parameter and after solving numerically the equations, we get that $a(u, v) = 1.4051 < a_n = 1.8983$ and $a_{11} = 0.2753 > \frac{d_1(d_1+d_2)\frac{n^2}{l^2}}{d_1-d_2} = 0.2190$, thus Fig. 8.7 shows that Hopf bifurcation occurs. It illustrates that increase of defense of phytoplankton and offense of zooplankton play important role in the stability of coexisting equilibrium with small diffusive coefficient of zooplankton. Furthermore, when $Re(\delta_1(a)) = -124.5601 < 0$, Fig. 8.7 shows the stability of the bifurcated periodic solutions and the monotone increasing effect on the period of bifurcated periodic solutions for $\beta_2 = 53.6071 > 0$.

Let $u = 0.6$, $v = 1.5$ and the other parameters remain the same as in Fig. 8.7. Figure 8.8 indicates the locally asymptotically stability of the coexisting equilibrium of (8.2). When effects of defense and offense increase, transition of the bifurcated periodic solutions into locally asymptotically stable is observed. Hence, in the

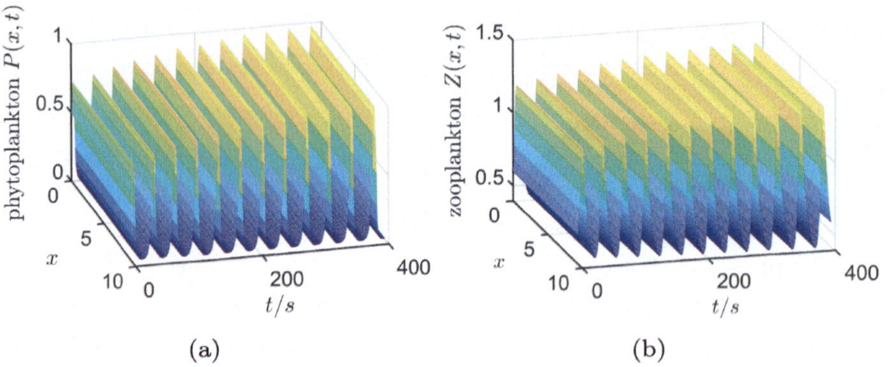

Fig. 8.7 Behavior of plankton with unstable conditions (model (8.2)), where $d_1 = 0.15$, $d_2 = 0.08$, $u = 0.42$, $v = 0.65$ and initial condition $(0.5, 0.6)$

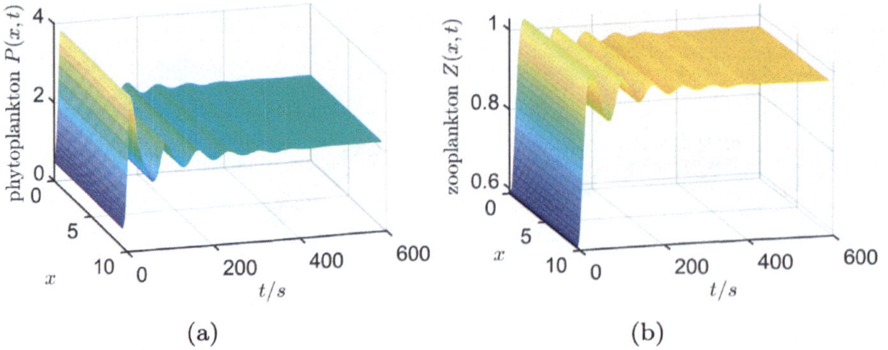

Fig. 8.8 Behavior of plankton for different values of defense and offense, where $d_1 = 0.15$, $d_2 = 0.08$, $u = 0.6$, $v = 1.5$ and initial condition $(0.5, 0.6)$

plankton ecosystem, defense and offense play great role in the development of plankton population.

Part III
Pattern Formation of Plankton Ecosystem

Chapter 9
Effect of Schooling Behavior and Harvest

In real world, heterogeneous distribution of species is usually derived by the random movement of prey and predator in a wide area usually move randomly in a wide area. This heterogeneity, as the driver of biodiversity, shows the significant ecological implication. There are still hot topics about searching for its possible unifying mechanisms to explain its spatial patterns and recognize the reason of pattern formation, the spatial organization, or both. Turing used reaction–diffusion theory to study pattern formations firstly [146]. As a result of diffusion, an initially uniform distribution of reacting components could become unstable, whereby there arises a stationary spatial pattern, which is called as diffusion-driven instability or Turing instability. Subsequently, it was suggested that ecological interactions can generate diffusive instabilities [75]. Following this, there are extensive studies focusing on the pattern dynamics in biological systems [147–156].

As the pioneering work of Lotka [157] and Volterra [158], mathematical models advance deeply the understanding of mechanisms behind predation. Based on the basic Lotka-Volterra model, many researchers made refinements and modifications to explore the more accurate characterizations of ecological scenarios [159–166, 169–173]. For example, there has been insight about various ecological factors such as herding behavior [174–176], alternative food [178], infection [177], cannibalism [179], mutualism [180], Allee effects [182], and migration [181]. As it's widespread in mammals, amphibians, fish, birds, and invertebrates, the population's collective behavior becomes hot topics [183, 184]. Schooling behavior is an indicator that individuals of population move in the same direction in a coordinated fashion to foraging, avoiding predation, etc. So predator–prey model where only prey populations show collective behavior has been the current concentration. Importantly, like many fish species with important economic, both the prey and predator live in group [185–187].

S. T. Liu and L. Zhang, *Pattern Dynamics of Marine Plankton Behavior*,
https://doi.org/10.1007/978-981-97-5369-7_9

9.1 Model Description

Motivated by the previous works, the schooling behavior of a reaction–diffusion predator–prey model is focused on. Recently, in order to consider the interaction phenomena that prey populations exhibit strong herd structure so that the predator interacts with prey along the outer corridor of groups [174], the following basic model is established introducing interaction terms as square root of prey populations rather than simply the prey populations:

$$\begin{aligned} \frac{dX}{dt} &= rX\left(1-\frac{X}{K}\right)-\frac{\alpha\sqrt{X}Y}{1+t_h\alpha\sqrt{X}}, \\ \frac{dY}{dt} &= -sY+\frac{c\alpha\sqrt{X}Y}{1+t_h\alpha\sqrt{X}}, \end{aligned} \tag{9.1}$$

where $X(t)$ and $Y(t)$ display the densities of prey and predator at time t, respectively. The parameter r denotes the growth rate of prey, K mean the carrying capacity, s stands for the death rate of predator, α represents the search efficiency of predator for prey, c implies the biomass conversion or consumption rate, and t_h means the average handing time. For model (9.1), Braza used a typical Lotka-Volterra interaction term to compare and contrast the dynamics of square root system with predator–prey system. Some subtle and interesting results about solution behavior were also given. Since then, based on this basic model, there are models modified from it to obtain richer dynamical behavior. For the phenomena that both prey and predator exhibit schooling behavior, the predator–prey model (9.2) for exploited fish populations was introduced and a number of interesting results about the stability behavior, extinction and optimal harvest policy were established [189].

$$\begin{aligned} \frac{dx}{dt} &= rx\left(1-\frac{x}{K}\right)-b\sqrt{x}\sqrt{y}-q_1Ex, \\ \frac{dy}{dt} &= c\sqrt{x}\sqrt{y}-dy-q_2Ey, \end{aligned} \tag{9.2}$$

where x and y means the densities of prey and predator populations, respectively. Table 9.1 lists the interpretations of parameters. It is our interest that spatial dynamics of the diffusive version of model (9.2) have remained unuttered before. Equation (9.3) is obtained by modifying system (9.2) by introducing diffusion terms. Then using linear stability analysis and amplitude equations, we can capture the corresponding pattern dynamics. To the best of our knowledge, there are few literature involved in this topic up to now and relevant results are expected to enrich the existing studies.

Table 9.1 The description of parameters in model (9.3)

Parameter	Description	Value
r	Intrinsic growth rate of prey	0.92
K	Carrying capacity of environment	7.05
b	Consumption rate of predator	0.70
c	Conversion rate of predator	0.49
d	Death rate of predator	0.39
E	Combined harvesting effort	1.89
q_1	Catchability coefficient of prey	0.18
q_2	Catchability coefficient of predator	0.14
d_1	Self-diffusion coefficient of prey	0.09
d_2	Self-diffusion coefficient of predator	3.28
a	Cross-diffusion coefficient of prey	2.40

$$
\begin{aligned}
\frac{\partial x}{\partial t} &= d_1 \nabla^2 x + a \nabla^2 y + rx\left(1 - \frac{x}{K}\right) \\
&\quad - b\sqrt{x}\sqrt{y} - q_1 Ex, \\
\frac{\partial y}{\partial t} &= d_2 \nabla^2 y + c\sqrt{x}\sqrt{y} - dy - q_2 Ey,
\end{aligned}
\tag{9.3}
$$

where

$$
\nabla^2 = \frac{\partial^2}{\partial \xi_1^2} + \frac{\partial^2}{\partial \xi_2^2}, \quad (\xi_1, \xi_2) \in \Omega = [0, \mathbb{R}] \times [0, \mathbb{R}],
$$

with ∇^2 representing the *Laplacian Operator* and Ω denoting a connected domain with smooth boundary $\partial\Omega$, x and y stand for the densities of prey and predator populations respectively. Its non-zero initial conditions are listed as below

$$
x(\xi_1, \xi_2, 0) > 0, \ \ y(\xi_1, \xi_2, 0) > 0,
$$

and the zero-flux boundary conditions are listed as following

$$
\frac{\partial x}{\partial \boldsymbol{n}} = \frac{\partial y}{\partial \boldsymbol{n}} = 0,
$$

where $\boldsymbol{n}$ is the outward unit normal vector of boundary $\partial\Omega$. Table 9.1 interprets its parameters. As our focus is self-organization of pattern, the zero-flux boundary condition is chosen. In addition, we assume that the populations of prey and predator move randomly from a higher to a lower concentration region, so that their self-diffusion can be expressed by $d_1\nabla^2 x$ and $d_2\nabla^2 y$ respectively. Besides, because of the moving in groups of individuals of predator, it leads to the risk that the prey will be more inclined to diffuse toward the domain of lower concentration of predator.

This phenomenon that the fluxes of one specie induced by the density gradient of another specie can be characterized by cross-diffusion term $a \bigtriangledown^2 y$. This distinctive feature is more significant in current model compared with some individualistic systems. It is noteworthy that cross-diffusion plays an important role in pattern formation [190–192]. Hence, how the pattern dynamics develop under the interplay of herd behavior, cross-diffusion and if there exist some interactions between self-diffusion and cross-diffusion becomes our focus. Therefore, due to the presence of predator, we choose the cross-diffusion of prey. In the following, model (9.3) shows that at least different cross-diffusion coefficient values can lead to pattern transitions between spots and stripes.

9.2 Linear Stability Analysis

The following conditions are for the nonnegative equilibria of the non-spatial model (9.2)
(i) $E_0 = (0, 0)$is feasible;
(ii) $E_b = (K(1 - \frac{q_1 E}{r}), 0)$is feasible when $r - q_1 E > 0$;
(iii) $E^*(x^*, y^*) = (K(1 - \frac{q_1 E}{r}) - \frac{Kbc}{r(d+q_2 E)}, (\frac{c\sqrt{x^*}}{d+q_2 E})^2)$ is feasible when $q_1 q_2 E^2 + (q_1 d - r q_2)E + bc - rd < 0$.
Manna proved that the model cannot be linearizable about E_0 and E_b [76]. Thus, the coexistence equilibrium E^* is our focus. The linear form of model (9.3) at the equilibrium E^* can be written as

$$\begin{aligned}\frac{\partial \hat{x}}{\partial t} &= d_1 \nabla^2 \hat{x} + a \nabla^2 \hat{y} + J_{11}\hat{x} + J_{12}\hat{y},\\ \frac{\partial \hat{y}}{\partial t} &= d_2 \nabla^2 \hat{y} + J_{21}\hat{x} + J_{22}\hat{y},\end{aligned} \tag{9.4}$$

where

$$J_{11} = \frac{3bc}{2(d+q_2 E)} - r + q_1 E,\ J_{12} = -\frac{b(d+q_2 E)}{2c} < 0,$$

$$J_{21} = \frac{c^2}{2(d+q_2 E)} > 0,\ J_{22} = -\frac{d+q_2 E}{2} < 0,$$

with $(\hat{x}, \hat{y})$ displaying small perturbations around E^*. For simplicity, the hats of $\hat{x}$ and $\hat{y}$ are removed. In view of nonuniform perturbations, we get the form of the solution of model (9.4) expanding the form of [193, 194]

$$\begin{pmatrix} x \\ y \end{pmatrix} = \sum_k \begin{pmatrix} z_k^1 \\ z_k^2 \end{pmatrix} e^{\lambda t + i\mathbf{k}\cdot\mathbf{r}} \tag{9.5}$$

where λ means the growth rate of perturbations at time t, i stands for the imaginary unit, k represents the wave number with $\mathbf{k} \cdot \mathbf{k} = k^2$ and $\mathbf{r}$ is the directional vector. Then substituting Eq. (9.5) into model (9.4), we get the dispersion equation as below:

$$\lambda^2 - T_k\lambda + \Delta_k = 0$$

where

$$T_k = J_{11} + J_{22} - (d_1 + d_2)k^2,$$

$$\Delta_k = d_1 d_2 k^4 + (J_{21}a - J_{11}d_2 - J_{22}d_1)k^2 + J_{11}J_{22} - J_{12}J_{21}.$$

After simple calculations, we obtain that

$$\lambda = \frac{T_k \pm \sqrt{T_k^2 - 4\Delta_k}}{2},$$

where eigenvalue λ represents the function of wave number k. It yields Hopf bifurcation as

$$Re(\lambda) = 0, \; Im(\lambda) \neq 0, \; at \; k = 0,$$

i.e., the critical condition is

$$(2q_1q_2 - q_2^2)E^2 + 2(dq_1 - rq_2 - dq_2)E + 3bc - 2rd - d^2 = 0.$$

Turing bifurcation occurs when

$$Re(\lambda) = 0, \; Im(\lambda) = 0, \; at \; k = k_T \neq 0$$

with $k_T^2 = -\frac{J_{21}a - J_{11}d_2 - J_{22}d_1}{2d_1d_2}$. Hence, we get the necessary conditions for the occurrence of Turing instability:

$$(i) \; J_{11} + J_{22} < 0,$$

$$(ii) \; J_{21}a - J_{11}d_2 - J_{22}d_1 < 0,$$

$$(iii) \; 2\sqrt{d_1d_2(J_{11}J_{22} - J_{12}J_{21})} < J_{11}d_2 + J_{22}d_1 - J_{21}a.$$

Then, after setting the cross-diffusion coefficient a as bifurcation parameter, we obtain the critical value of Turing bifurcation:

$$a_T = \frac{J_{11}d_2 + J_{22}d_1 - 2\sqrt{d_1d_2(J_{11}J_{22} - J_{12}J_{21})}}{J_{21}}.$$

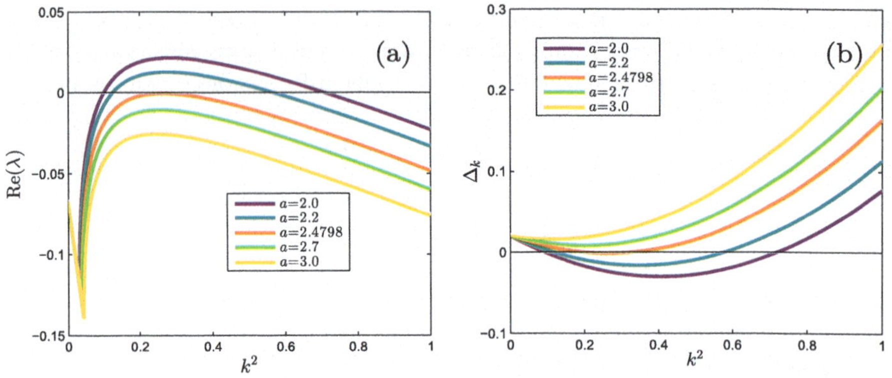

Fig. 9.1 Solutions of system (9.3) with the above parameters and the listed initial values (0.1,0.1,0.1), (0.2,0.3,0.4), (0.5,0.7,0.8), (0.4,0.2,0.3)

Mathematically, diffusion-driven instability or named as Turing instability occurs when there exists no diffusion so that the equilibrium point becomes asymptotically stable, and it is unstable with diffusion. Figure 9.1 illustrates the existence of Turing instability by plotting the graphs of Δ_k and the largest real part of λ w.r.t. k^2. Table 9.1 shows the parameter values except a and we calculate the corresponding critical value as $a_T = 2.4798$. Assume a in the two sides of a_T as $a = 2.0,\ 2.2,\ 2.4798,\ 2.7,\ 3.0$. Figure 9.1a displays that $Re(\lambda)$ remains positive because of the increasing values of a leading to the gradual shrink of the range of k^2. If values lager than a_T, $Re(\lambda)$ becomes completely negative, meaning the stationary of the uniform steady state E^* timely and spacely, which stands for that Turing instability cannot occur. Figure 9.1b shows the same relations between Δ_k and k^2. Notably, when there is no cross-diffusion, and self-diffusivity ratio of prey and predator becomes unity, i.e., $a = 0$ and $d_1 = d_2$, the necessary conditions yield that Turing instability cannot be satisfied simultaneously, meaning that Turing instability cannot occur. The following section focuses on the effect of cross-diffusion on Turing instability together with pattern formations.

9.3 Amplitude Equations of Turing Patterns

Distinct Turing patterns are explored employing multiple-scale analysis, which take the cross-diffusion coefficient a as control parameter to derive amplitude equations near the threshold of Turing bifurcation. The detailed process is divided into three steps briefly [194]. If a near the critical point a_T, we get the solutions of model (9.3) as below

$$\begin{pmatrix} x \\ y \end{pmatrix} = \begin{pmatrix} x^* \\ y^* \end{pmatrix} + \sum_{j=1}^{3} \begin{pmatrix} A_j^x \\ A_j^y \end{pmatrix} exp(i\boldsymbol{k}_j \cdot \boldsymbol{r}) + c.c.$$

where $\begin{pmatrix} A_j^x \\ A_j^y \end{pmatrix}$ represents for the amplitude related to the wave vector $\boldsymbol{k}_j$, and c.c. stands for the conjugate terms [195]. A_j^x and A_j^y is the original parameters in model (9.3) needed to be defined.

Step I: After expanding model (9.3) around the equilibrium $\begin{pmatrix} x^* \\ y^* \end{pmatrix}$ by Taylor expansion, we obtain:

$$\frac{\partial \mathbf{X}}{\partial t} = \mathbf{L}\mathbf{X} + \mathbf{N} \tag{9.6}$$

where

$$\mathbf{X} = \begin{pmatrix} \hat{x} \\ \hat{y} \end{pmatrix}, \ \mathbf{L} = \begin{pmatrix} J_{11} + d_1\nabla^2 & J_{12} + a\nabla^2 \\ J_{21} & J_{22} + d_2\nabla^2 \end{pmatrix},$$

$$\mathbf{N} = \begin{pmatrix} \frac{1}{2} f_{xx}\hat{x}^2 + \frac{1}{2} f_{yy}\hat{y}^2 + f_{xy}\hat{x}\hat{y} + \frac{1}{6} f_{xxx}\hat{x}^3) \\ \frac{1}{2} g_{xx}\hat{x}^2 + \frac{1}{2} g_{yy}\hat{y}^2 + g_{xy}\hat{x}\hat{y} + \frac{1}{6} g_{xxx}\hat{x}^3) \end{pmatrix}$$

$$+ \begin{pmatrix} \frac{1}{2}\frac{1}{6} f_{yyy}\hat{y}^3 + \frac{1}{2} f_{xxy}\hat{x}^2\hat{y} + \frac{1}{2} f_{xyy}\hat{x}\hat{y}^2 + o(\rho^3) \\ \frac{1}{2}\frac{1}{6} g_{yyy}\hat{y}^3 + \frac{1}{2} g_{xxy}\hat{x}^2\hat{y} + \frac{1}{2} g_{xyy}\hat{x}\hat{y}^2 + o(\rho^3) \end{pmatrix},$$

with

$$\hat{x} = x - x^*, \ \hat{y} = y - y^*,$$

and

$$f_{xx} = -\frac{2r}{K} + \frac{b}{4} x^{*-\frac{3}{2}} y^{*\frac{1}{2}}, \ f_{xxx} = -\frac{3b}{8} x^{*-\frac{5}{2}} y^{*\frac{1}{2}},$$

$$f_{xy} = -\frac{b}{4} x^{*-\frac{1}{2}} y^{*-\frac{1}{2}}, \ f_{xxy} = \frac{b}{8} x^{*-\frac{3}{2}} y^{*-\frac{1}{2}},$$

$$f_{yy} = \frac{b}{4} x^{*\frac{1}{2}} y^{*-\frac{3}{2}}, \ f_{yyy} = -\frac{3b}{8} x^{*\frac{1}{2}} y^{*-\frac{5}{2}},$$

$$f_{xyy} = \frac{b}{8} x^{*-\frac{1}{2}} y^{*-\frac{3}{2}}, \ g_{xx} = -\frac{c}{4} x^{*-\frac{3}{2}} y^{*\frac{1}{2}},$$

$$g_{xxx} = \frac{3c}{8} x^{*-\frac{5}{2}} y^{*\frac{1}{2}}, \ g_{xy} = \frac{c}{4} x^{*-\frac{1}{2}} y^{*-\frac{1}{2}},$$

$$g_{xxy} = -\frac{c}{8} x^{*-\frac{3}{2}} y^{*-\frac{1}{2}}, \ g_{yy} = -\frac{c}{4} x^{*\frac{1}{2}} y^{*-\frac{3}{2}},$$

$$g_{yyy} = \frac{3c}{8} x^{*\frac{1}{2}} y^{*-\frac{5}{2}}, \ g_{yyx} = -\frac{c}{8} x^{*-\frac{1}{2}} y^{*-\frac{3}{2}}.$$

The hat of $\hat{x}$ and $\hat{y}$ can be removed for simplicity. Next, decompose the linear operator $\boldsymbol{L}$ into

$$\mathbf{L} = \mathbf{L}_T + (a_T - a)\mathbf{M} \tag{9.7}$$

where

$$\mathbf{L}_T = \begin{pmatrix} J_{11} + d_1\nabla^2 & J_{12} + a_T\nabla^2 \\ J_{21} & J_{22} + d_2\nabla^2 \end{pmatrix},$$

$$\mathbf{M} = \begin{pmatrix} 0 & -\nabla^2 \\ 0 & 0 \end{pmatrix}.$$

Step II: Derive amplitude equations under the first-order and the second-order perturbations, respectively.

Let ϵ be a sufficiently small parameter. Then, the control parameter a, the variable $\mathbf{X}$ and the nonlinear term $\mathbf{N}$ w.r.t. ϵ can be expanded as follows:

$$a_T - a = \epsilon a_1 + \epsilon^2 a_2 + \epsilon^3 a_3 + o(\epsilon^3), \tag{9.8}$$

X

$$\mathbf{X} = \epsilon \begin{pmatrix} x_1 \\ y_1 \end{pmatrix} + \epsilon^2 \begin{pmatrix} x_2 \\ y_2 \end{pmatrix} + \epsilon^3 \begin{pmatrix} x_3 \\ y_3 \end{pmatrix} + o(\epsilon^3), \tag{9.9}$$

$$\mathbf{N} = \epsilon^2 \mathbf{h}_2 + \epsilon^3 \mathbf{h}_3 + o(\epsilon^3), \tag{9.10}$$

where

$$\mathbf{h}_2 = \begin{pmatrix} \frac{1}{2} f_{xx} x_1^2 + f_{xy} x_1 y_1 + \frac{1}{2} f_{yy} y_1^2 \\ \frac{1}{2} g_{xx} x_1^2 + g_{xy} x_1 y_1 + \frac{1}{2} g_{yy} y_1^2 \end{pmatrix}, \tag{9.11}$$

$$\begin{aligned} \mathbf{h}_3 = & \begin{pmatrix} f_{xx} x_1 x_2 + f_{yy} y_1 y_2 + f_{xy} x_1 y_2 + f_{xy} x_2 y_1 \\ g_{xx} x_1 x_2 + g_{yy} y_1 y_2 + g_{xy} x_1 y_2 + g_{xy} x_2 y_1 \end{pmatrix} \\ & + \begin{pmatrix} \frac{1}{6} f_{xxx} x_1^3 + \frac{1}{6} f_{yyy} y_1^3 + \frac{1}{2} f_{xxy} x_1^2 y_1 + \frac{1}{2} f_{xyy} x_1 y_1^2 \\ \frac{1}{6} g_{xxx} x_1^3 + \frac{1}{6} g_{yyy} y_1^3 + \frac{1}{2} g_{xxy} x_1^2 y_1 + \frac{1}{2} g_{xyy} x_1 y_1^2 \end{pmatrix}. \end{aligned} \tag{9.12}$$

Separating time scales of system upon ϵ into independent terms $T_1 = \epsilon t,\ T_2 = \epsilon^2 t,\ T_3 = \epsilon^3 t$, we can get the time derivative expressed as

$$\frac{\partial}{\partial t} = \epsilon \frac{\partial}{\partial T_1} + \epsilon^2 \frac{\partial}{\partial T_2} + \epsilon^3 \frac{\partial}{\partial T_3} + o(\epsilon^3). \tag{9.13}$$

Hence, substituting Eqs. (9.7)–(9.13) into Eq. (9.6), we get the following equations according to the order of ϵ:

The first order of ϵ:

$$\mathbf{L}_T \begin{pmatrix} x_1 \\ y_1 \end{pmatrix} = \mathbf{0}; \tag{9.14}$$

The second order of ϵ:

$$\mathbf{L}_T \begin{pmatrix} x_2 \\ y_2 \end{pmatrix} = \frac{\partial}{\partial T_1} \begin{pmatrix} x_1 \\ y_1 \end{pmatrix} - a_1 \mathbf{M} \begin{pmatrix} x_1 \\ y_1 \end{pmatrix} - \mathbf{h}_2; \tag{9.15}$$

The third order of ϵ:

$$\mathbf{L}_T \begin{pmatrix} x_3 \\ y_3 \end{pmatrix} = \frac{\partial}{\partial T_1} \begin{pmatrix} x_2 \\ y_2 \end{pmatrix} + \frac{\partial}{\partial T_2} \begin{pmatrix} x_1 \\ y_1 \end{pmatrix} - a_2 \mathbf{M} \begin{pmatrix} x_1 \\ y_1 \end{pmatrix} - a_1 \mathbf{M} \begin{pmatrix} x_2 \\ y_2 \end{pmatrix} - \mathbf{h}_3. \tag{9.16}$$

From Eq. (9.14) and since $\boldsymbol{L}_T$ is the linear operator, it yields that $\begin{pmatrix} x_1 \\ y_1 \end{pmatrix}$ is the linear combinations of eigenvectors corresponding to zero eigenvalue. Then, we get the solution $\begin{pmatrix} x_1 \\ y_1 \end{pmatrix}$ as

$$\begin{pmatrix} x_1 \\ y_1 \end{pmatrix} = \begin{pmatrix} \varphi \\ 1 \end{pmatrix} (\sum_{j=1}^{3} W_j exp(i\boldsymbol{k}_j \cdot \boldsymbol{r})) + c.c., \quad j = 1, 2, 3, \tag{9.17}$$

where W_j means the amplitude corresponding to $e^{i\boldsymbol{k}_j \cdot \boldsymbol{r}}$ as system is under the first-order perturbations, and

$$\varphi = -\frac{a_T \hat{k}_T^2 - J_{12}}{d_1 \hat{k}_T^2 - J_{11}}, \quad |\boldsymbol{k}_j| = \hat{k}_T, \; \hat{k}_T^2 = k_T^2(a_T).$$

Based on the Fredholm solubility conditions, the right side of Eqs. (9.14)–(9.16) need to be orthogonal with the zero eigenvectors of adjoint operator of $\boldsymbol{L}_T$. So the existence of nontrivial solutions of Eqs. (9.14)–(9.16) can be guaranteed. Assume $\boldsymbol{L}_T^*$ is the adjoint operator of $\boldsymbol{L}_T$ and the zero eigenvector of $\boldsymbol{L}_T^*$ can be expressed as:

$$\begin{pmatrix} 1 \\ \psi \end{pmatrix} exp(-i\boldsymbol{k}_j \cdot \boldsymbol{r}) + c.c., \qquad j = 1, 2, 3, \tag{9.18}$$

with $\psi = \frac{d_1 \hat{k}_T^2 - J_{11}}{J_{21}}$. For Eq. (9.15), utilizing the orthogonality conditions, we get the amplitude equations with respect to the first-order perturbations as follows:

$$\begin{cases} (\varphi+\psi)\frac{\partial W_1}{\partial T_1} = a_1\hat{k}_T^2 W_1 + 2(l_1+\psi l_2)\bar{W}_2\bar{W}_3, \\ (\varphi+\psi)\frac{\partial W_2}{\partial T_1} = a_1\hat{k}_T^2 W_2 + 2(l_1+\psi l_2)\bar{W}_1\bar{W}_3, \\ (\varphi+\psi)\frac{\partial W_3}{\partial T_1} = a_1\hat{k}_T^2 W_3 + 2(l_1+\psi l_2)\bar{W}_1\bar{W}_2, \end{cases} \tag{9.19}$$

with

$$\begin{cases} l_1 = \frac{1}{2}f_{xx}\varphi^2 + f_{xy}\varphi + \frac{1}{2}f_{yy}, \\ l_2 = \frac{1}{2}g_{xx}\varphi^2 + g_{xy}\varphi + \frac{1}{2}g_{yy}. \end{cases}$$

Next, Eq. (9.15) can be solved to get

$$\begin{aligned} \begin{pmatrix} x_2 \\ y_2 \end{pmatrix} = \begin{pmatrix} X_0 \\ Y_0 \end{pmatrix} + \sum_{j=1}^{3}\begin{pmatrix} X_j \\ Y_j \end{pmatrix} e^{i\boldsymbol{k}_j\cdot\boldsymbol{r}} + \sum_{j=1}^{3}\begin{pmatrix} X_{jj} \\ Y_{jj} \end{pmatrix} e^{i2\boldsymbol{k}_j\cdot\boldsymbol{r}} + \begin{pmatrix} X_{12} \\ Y_{12} \end{pmatrix} e^{i(\boldsymbol{k}_1-\boldsymbol{k}_2)\cdot\boldsymbol{r}} \\ + \begin{pmatrix} X_{23} \\ Y_{23} \end{pmatrix} e^{i(\boldsymbol{k}_2-\boldsymbol{k}_3)\cdot\boldsymbol{r}} + \begin{pmatrix} X_{31} \\ Y_{31} \end{pmatrix} e^{i(\boldsymbol{k}_3-\boldsymbol{k}_1)\cdot\boldsymbol{r}} + c.c. \end{aligned} \tag{9.20}$$

where

$$\begin{pmatrix} X_0 \\ Y_0 \end{pmatrix} = \begin{pmatrix} x_{00} \\ y_{00} \end{pmatrix}(|W_1|^2+|W_2|^2+|W_3|^2),\ X_j = \varphi Y_j,$$

$$\begin{pmatrix} X_{jj} \\ Y_{jj} \end{pmatrix} = \begin{pmatrix} x_{11} \\ y_{11} \end{pmatrix} W_j^2,\ \begin{pmatrix} X_{jl} \\ Y_{jl} \end{pmatrix} = \begin{pmatrix} x_* \\ y_* \end{pmatrix} W_j\bar{W}_l$$

with

$$\begin{pmatrix} x_{00} \\ y_{00} \end{pmatrix} = \frac{-2}{J_{11}J_{22}-J_{21}J_{12}}\begin{pmatrix} J_{22}l_1 - J_{12}l_2 \\ J_{11}l_2 - J_{21}l_1 \end{pmatrix},$$

$$\begin{pmatrix} x_{11} \\ y_{11} \end{pmatrix} = \frac{-1}{(J_{11}-4d_1\hat{k}_T^2)(J_{22}-4d_2\hat{k}_T^2)-J_{21}(J_{12}-4a_T\hat{k}_T^2)}\begin{pmatrix} (J_{22}-4d_2\hat{k}_T^2)l_1-(J_{12}-4a_T\hat{k}_T^2)l_2 \\ (J_{11}-4d_1\hat{k}_T^2)l_2-J_{21}l_1 \end{pmatrix},$$

$$\begin{pmatrix} x_* \\ y_* \end{pmatrix} = \frac{-2}{(J_{11}-3d_1\hat{k}_T^2)(J_{22}-3d_2\hat{k}_T^2)-J_{21}(J_{12}-3a_T\hat{k}_T^2)}\begin{pmatrix} (J_{22}-3d_2\hat{k}_T^2)l_1-(J_{12}-3a_T\hat{k}_T^2)l_2 \\ (J_{11}-3d_1\hat{k}_T^2)l_2-J_{21}l_1 \end{pmatrix}.$$

Substituting Eqs. (9.17), (9.20) into Eq. (9.16), corresponding to the second-order perturbations and applying orthogonality conditions again, we can get the amplitude equations:

$$
\begin{cases}
(\varphi+\psi)(\frac{\partial Y_1}{\partial T_1}+\frac{\partial W_1}{\partial T_2}) = \hat{k}_T^2(a_2W_1+a_1Y_1) - [(I_1+\psi S_1)|W_1|^2 \\
+(I_2+\psi S_2)(|W_2|^2+|W_3|^2)]W_1 \qquad + 2(l_1+\psi l_2)(\bar{W_2}\bar{Y_3}+\bar{W_3}\bar{Y_2}), \\
(\varphi+\psi)(\frac{\partial Y_2}{\partial T_1}+\frac{\partial W_2}{\partial T_2}) = \hat{k}_T^2(a_2W_2+a_1Y_2) - [(I_1+\psi S_1)|W_2|^2 \\
+(I_2+\psi S_2)(|W_1|^2+|W_3|^2)]W_2 \\
\qquad + 2(l_1+\psi l_2)(\bar{W_1}\bar{Y_3}+\bar{W_3}\bar{Y_1}), \\
(\varphi+\psi)(\frac{\partial Y_3}{\partial T_1}+\frac{\partial W_3}{\partial T_2}) = \hat{k}_T^2(a_2W_3+a_1Y_3) - [(I_1+\psi S_1)|W_3|^2 \\
+(I_2+\psi S_2)(|W_2|^2+|W_1|^2)]W_3 \\
\qquad + 2(l_1+\psi l_2)(\bar{W_2}\bar{Y_1}+\bar{W_1}\bar{Y_2}).
\end{cases}
\tag{9.21}
$$

$$
\begin{aligned}
I_1 &= -(\varphi f_{xx}+f_{xy})(x_{00}+x_{11}) - (\varphi f_{xy}+f_{yy})(y_{00}+y_{11}) \\
&\quad -\tfrac{1}{2}f_{xxx}\varphi^3 - \tfrac{1}{2}f_{yyy} - \tfrac{3}{2}f_{xxy}\varphi^2 - \tfrac{3}{2}f_{xyy}\varphi, \\
I_2 &= -(\varphi f_{xx}+f_{xy})(x_{00}+x_*) - (\varphi f_{xy}+f_{yy})(y_{00}+y_*) \\
&\quad - f_{xxx}\varphi^3 - f_{yyy} - 3f_{xxy}\varphi^2 - 3f_{xyy}\varphi, \\
S_1 &= -(\varphi g_{xx}+g_{xy})(x_{00}+x_{11}) - (\varphi g_{xy}+g_{yy})(y_{00}+y_{11}) \\
&\quad -\tfrac{1}{2}g_{xxx}\varphi^3 - \tfrac{1}{2}g_{yyy} - \tfrac{3}{2}g_{xxy}\varphi^2 - \tfrac{3}{2}g_{xyy}\varphi, \\
S_2 &= -(\varphi g_{xx}+g_{xy})(x_{00}+x_*) - (\varphi g_{xy}+g_{yy})(y_{00}+y_*) \\
&\quad - g_{xxx}\varphi^3 - g_{yyy} - 3g_{xxy}\varphi^2 - 3g_{xyy}\varphi.
\end{aligned}
$$

Step III: Write the final representation of amplitude equations.

The coefficients of $exp(i\boldsymbol{k}_j \cdot \boldsymbol{r})$ can then be given on account of the above derivation as:

$$
\begin{pmatrix} A_j^x \\ A_j^y \end{pmatrix} = \epsilon \begin{pmatrix} \varphi \\ 1 \end{pmatrix} W_j + \epsilon^2 \begin{pmatrix} \varphi \\ 1 \end{pmatrix} Y_j + \cdots, \quad j = 1,\ 2,\ 3. \tag{9.22}
$$

Let A_j mean the amplitude A_j^y. Then, we can write the derivative of amplitude A_j as

$$
\frac{\partial A_j}{\partial t} = \epsilon \frac{\partial A_j}{\partial T_1} + \epsilon^2 \frac{\partial A_j}{\partial T_2} + \cdots. \tag{9.23}
$$

Together with Eqs. (9.19), (9.21), (9.22), and (9.23), we obtain the following results:

$$
\begin{cases}
\tau_0 \frac{\partial A_1}{\partial t} = \mu A_1 + h\bar{A}_2\bar{A}_3 - [g_1|A_1|^2 + g_2(|A_2|^2+|A_3|^2)]A_1, \\
\tau_0 \frac{\partial A_2}{\partial t} = \mu A_2 + h\bar{A}_1\bar{A}_3 - [g_1|A_2|^2 + g_2(|A_1|^2+|A_3|^2)]A_2, \\
\tau_0 \frac{\partial A_3}{\partial t} = \mu A_3 + h\bar{A}_1\bar{A}_2 - [g_1|A_3|^2 + g_2(|A_1|^2+|A_2|^2)]A_3,
\end{cases}
\tag{9.24}
$$

where

$$\tau_0 = \frac{\varphi + \psi}{a_T \hat{k}_T^2},\ \mu = \frac{a_T - a}{a_T},\ h = \frac{2(l_1 + \psi l_2)}{a_T \hat{k}_T^2},\ g_1 = \frac{I_1 + \psi S_1}{a_T \hat{k}_T^2},\ g_2 = \frac{I_2 + \psi S_2}{a_T \hat{k}_T^2}.$$

9.4 Stability Analysis of Pattern Formation

As stable Turing patterns correspond to stable steady states, it is easy to know that each amplitude A_j can be decomposed into $A_j = \rho_j e^{i\phi_j}$ where ϕ_j means the phase angle and $\rho_j = |A_j|$. Substituting $A_j = \rho_j e^{i\phi_j}$ into Eq. (9.24), we get the equations w.r.t. real variables ρ_j and ϕ_j as follows:

$$\begin{cases} \tau_0 \frac{\partial \phi}{\partial t} = -h \frac{\rho_1^2\rho_2^2 + \rho_2^2\rho_3^2 + \rho_1^2\rho_3^2}{\rho_1\rho_2\rho_3} sin\phi, \\ \tau_0 \frac{\partial \rho_1}{\partial t} = \mu\rho_1 + |h|\rho_2\rho_3 - [g_1\rho_1^2 + g_2(\rho_2^2 + \rho_3^2)]\rho_1, \\ \tau_0 \frac{\partial \rho_2}{\partial t} = \mu\rho_2 + |h|\rho_1\rho_3 - [g_1\rho_2^2 + g_2(\rho_1^2 + \rho_3^2)]\rho_2, \\ \tau_0 \frac{\partial \rho_3}{\partial t} = \mu\rho_3 + |h|\rho_1\rho_2 - [g_1\rho_3^2 + g_2(\rho_1^2 + \rho_2^2)]\rho_3, \end{cases} \tag{9.25}$$

with

$$\phi = \phi_1 + \phi_2 + \phi_3.$$

There are five kinds of solutions for system (9.25) summarized as follows:
(1) The stationary state (**O**):

$$\rho_1 = \rho_2 = \rho_3 = 0,$$

is stable as $\mu < \mu_2 = 0$ and unstable as $\mu > \mu_2$.
(2) Stripe patterns (**S**):

$$\rho_1 = \sqrt{\frac{\mu}{g_1}} \neq 0,\ \rho_2 = \rho_3 = 0,$$

is stable as $\mu > \mu_3 = \frac{h^2 g_1}{(g_1 - g_2)^2}$ and unstable as $\mu < \mu_3$.
(3) Hexagon pattern ($\mathbf{H}_0,\ \mathbf{H}_\pi$):

$$\rho_1 = \rho_2 = \rho_3 = \frac{|h| \pm \sqrt{h^2 + 4(g_1 + 2g_2)\mu}}{2(g_1 + 2g_2)}$$

with $\phi = 0$ or $\phi = \pi$ are feasible when $\mu > \mu_1 = -\frac{h^2}{4(g_1+2g_2)}$. As $\mu < \mu_4 = \frac{2g_1+g_2}{(g_2-g_1)^2}h^2$, the solution $\rho^+ = \frac{|h|+\sqrt{h^2+4(g_1+2g_2)\mu}}{2(g_1+2g_2)}$ is stable. Oppositely, the solution $\rho^- = \frac{|h|-\sqrt{h^2+4(g_1+2g_2)\mu}}{2(g_1+2g_2)}$ is always unstable.

(4) The mixed states (**MS**):

$$\rho_1 = \frac{|h|}{g_2 - g_1}, \quad \rho_2 = \rho_3 = \sqrt{\frac{\mu - g_1\rho_1^2}{g_1 + g_2}}$$

is feasible when $\mu > g_1\rho_1^2$, $g_2 > g_1$ and this constant solution is always unstable.

Based on the above stability analysis, it is easy to get that $\mu_1 < \mu_2 < \mu_3 < \mu_4$ with $h^2 \neq 0$. The relations between μ and μ_i $(i = 1, 2, 3, 4)$ control the selection and stability of pattern formation. For example, as μ increases to $\mu_2 = 0$, the stationary state becomes losing stability. And instead, there exist hexagon patterns. Here, when $h > 0$, hexagon patterns are $\boldsymbol{H}_0$ $(\phi = 0)$ and always unstable, otherwise the patterns are $\boldsymbol{H}_\pi$ $(\phi = \pi)$ and always unstable. Subcritical bifurcation can induce the emergence of hexagon patterns, i.e., there exists a bistable region (μ_1, μ_2) where the stationary state and hexagon patterns become all stable. Furthermore, hexagon patterns achieve stable until $\mu > \mu_4$ and stripe patterns remain stability when $\mu > \mu_3$. Therefore, it yields that (μ_3, μ_4) is a bistable region for hexagon patterns and stripe patterns.

9.5 Numerical Simulations

The following consider particular examples and compare the predictions with the numerical simulations. In these simulations, the forward finite difference method with a time step size $\Delta t = 0.01$ is employed and a space step size $\Delta h = 1$ is taken in two-dimensional domain which is uniformly divided into 200×200 lattice sites for model (9.3). In addition, we use the Neumann conditions as boundary conditions and a small random perturbation around the equilibrium $E^*(x^*, y^*)$ as the initial data. Notably, we assume that the characteristics of final patterns have no time effect. Because the pattern formations of prey and predator behave the same type, for simplicity, the patterns of prey are given as an example.

9.5.1 Effect of a on Pattern Formation

Table 9.1 lists the different values of a near the Turing bifurcation threshold and other parameters. These parameters satisfy the conditions of Turing instability. Thus, we can assume that $a_T = 2.4798$, $h = -2.3154$, $g_1 = 141.8966$, $g_2 = 332.9359$, $\mu_1 = -0.0016593$, $\mu_3 = 0.020844$, $\mu_4 = 0.090597$.

When a is chosen to be 2.45, we can get $\mu = 0.012 \in (\mu_2, \mu_3)$. Figure 9.2 shows the hexagon patterns: the system goes through an irregular state in the domain firstly;

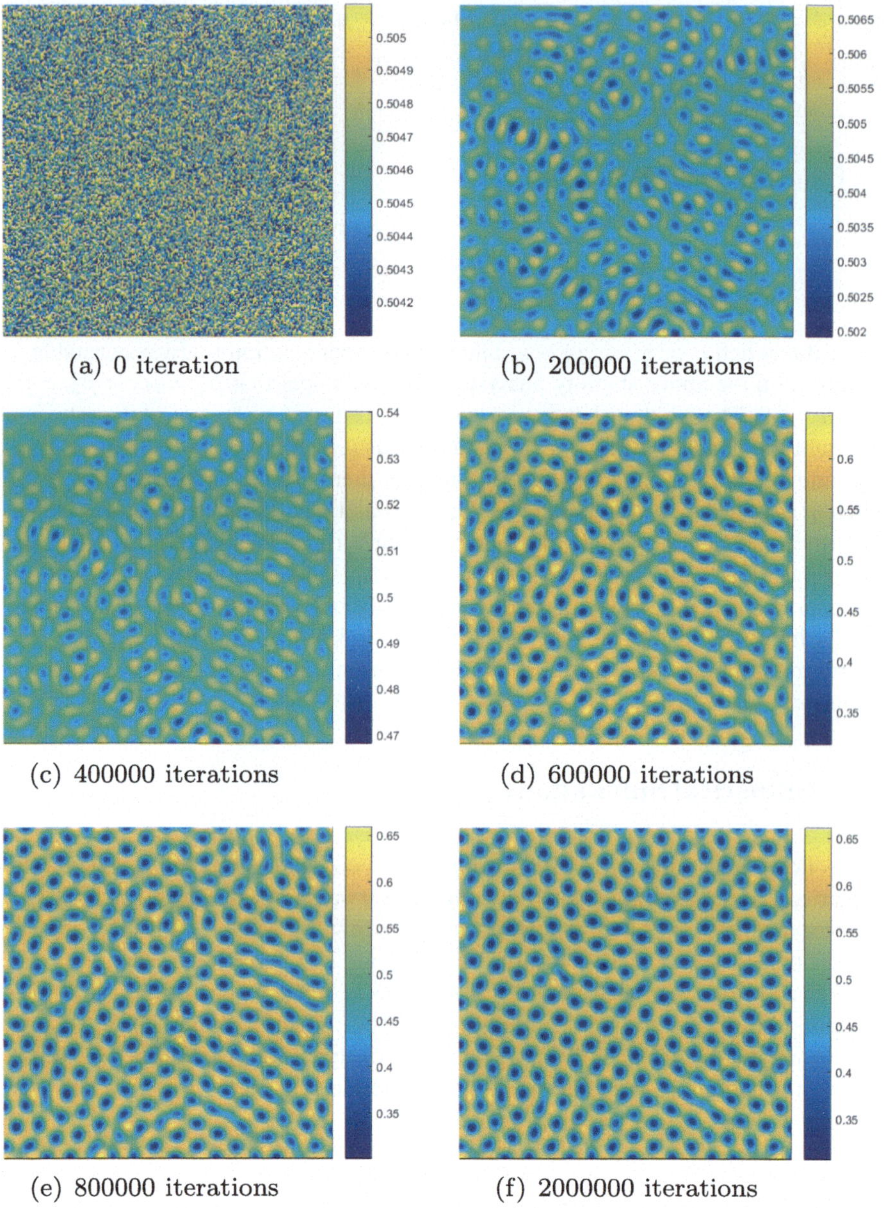

Fig. 9.2 Snapshots of contour pictures for the time evolution of the prey with $a = 2.45$, where other parameters are the same in Table 9.1

then, there appear the mixture of hexagon patterns and stripe patterns gradually and clearly. With the evolutions, there exists the stripe patterns in the hexagons state and then there are only hexagon patterns left after 2000000 iterations. So, we get the simulation results according to the theoretical prediction.

When $a = 2.4$ and μ: $\mu = 0.0322 \in (\mu_3, \mu_4)$, Fig. 9.3 illustrates that hexagons and stripes take shape little by little after the initial irregular pattern. In this evolution process, some hexagons are joined together to be smooth stripes and others become

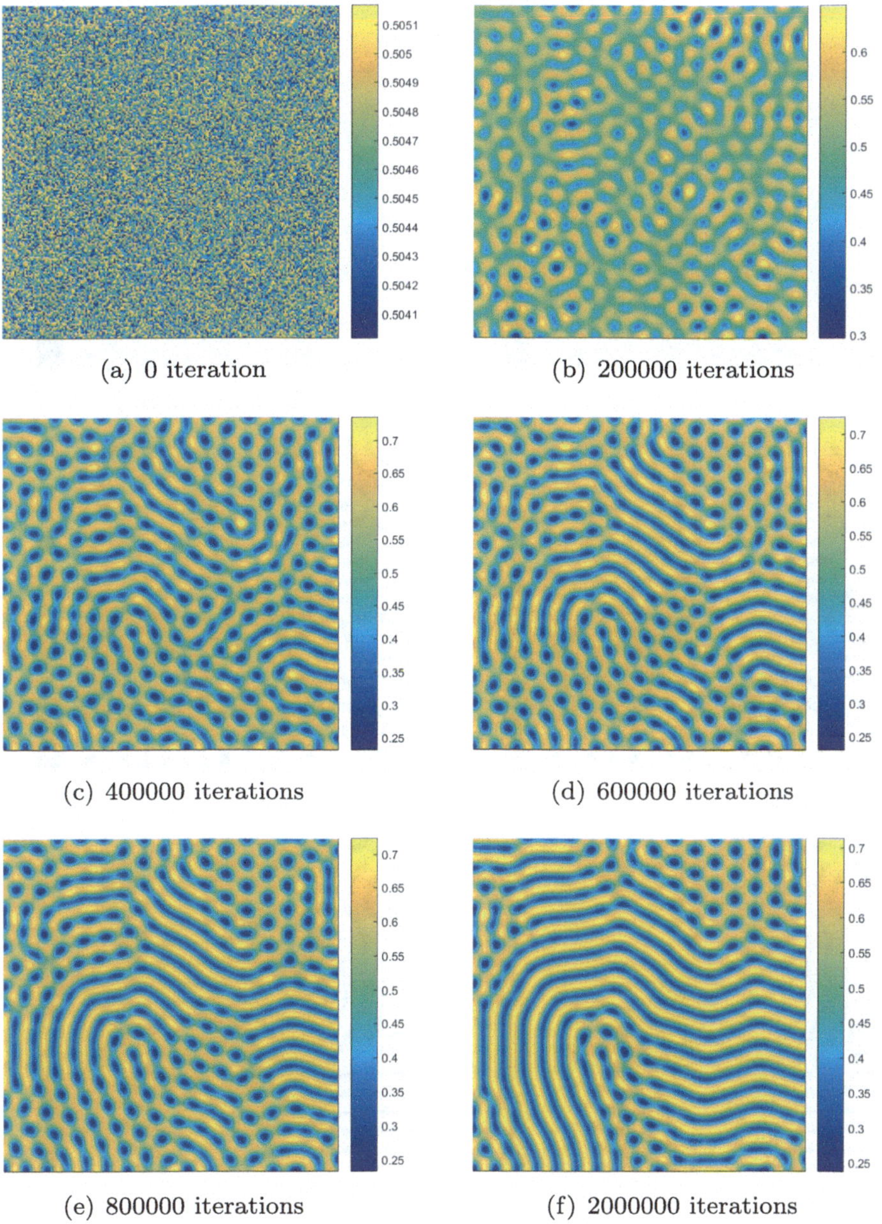

(a) 0 iteration (b) 200000 iterations

(c) 400000 iterations (d) 600000 iterations

(e) 800000 iterations (f) 2000000 iterations

Fig. 9.3 Snapshots of contour pictures for the time evolution of the prey with $a = 2.4$, where other parameter values are the same as in Table 9.1

time independent leading to the concurrence of hexagons and stripes at last. Hence, the simulations also agree well with the analytical results.

When $a = 2.2$ and $\mu = 0.1128 > \mu_4$, Fig. 9.4 illustrates the formation of the stationary stripe patterns after 0, 50000, 120000, 180000, 400000, 1000000 iterations.

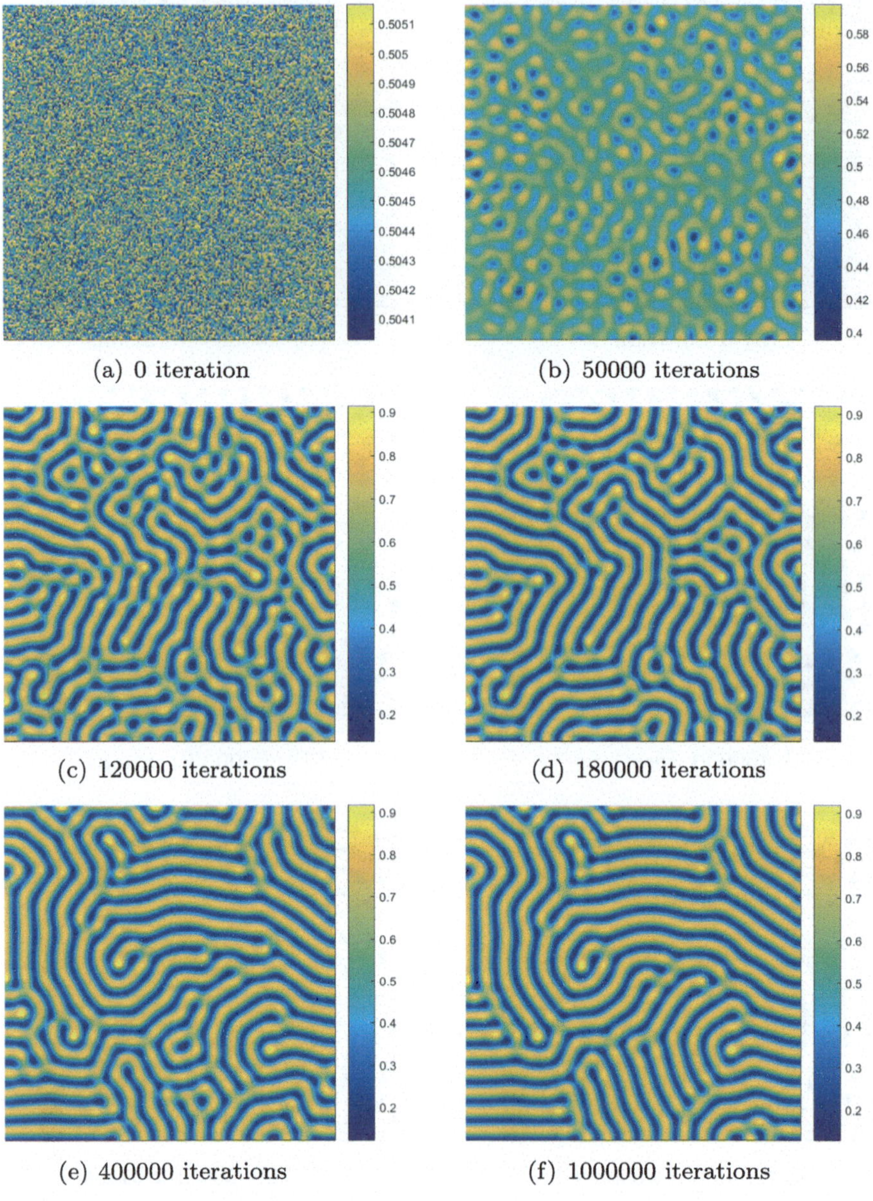

Fig. 9.4 Snapshots of contour pictures for the time evolution of the prey with $a = 2.2$, where other parameter values are the same as in Table 9.1

At first, there appears an irregular density distribution of prey. Then it is replaced by some regular hexagons and stripes subsequently. With the variation of density, hexagons fade away gradually and stripes well-distributed in shapes and colors prevail the whole domain in the end. This scenario is also consistent with the theoretical analysis.

9.5.2 Effect of Catchability Coefficients on Pattern Formation

Subsequently, Fig. 9.5 illustrates the impact of harvesting on pattern selection for a series of pattern formations corresponding to varied q_1 and q_2. First, the catchability coefficient of prey is fixed at $q_1 = 0.13$ and predator's catchability coefficient is varied. At $q_2 = 0.078$, Fig. 9.5a shows that the spot patterns prevail the whole domain. Further increasing to $q_2 = 0.09$, the mixture of spots and stripes are obtained shown in Fig. 9.5b. While for larger $q_2 = 0.098$, Fig. 9.5c shows that hole replication replaces the stripe and spot patterns. That is to say that increasing the catchability coefficient of predator can obtain pattern sequence following "spots-mixture of stripes and spots-holes". On the other hand, when the catchability coefficient of predator is constant ($q_2 = 0.34$), Fig. 9.5c–f shows the opposite transition "holes-stripes-spots" as the increase of catchability intensity of prey ($q_1 = 0.298, 0.299, 0.305$), implying that the prey moves from the high density regions surround isolated low ones to the opposite. From the biological point of view, it also indicates a switch of spatial density distribution dominance between prey and predator populations. Strikingly, pattern transitions remain sensitive to the changes in harvesting intensity.

When cross-diffusion coefficient varies, we can get different pattern structures such as spots and stripes, indicating the vital effect of cross-diffusion on pattern formation. In view of population dynamics, these pattern structures are actually representative of the distribution of populations spatially corresponding to different population predominant state. For instance, spot pattern was recognized as the representation of the formation of patches where high density area of population is surrounded by the low [196]. It is worth noting that under all random initial conditions tested, and for each set of parameter, there are no observed results about qualitative differences. Obviously, we use random initials here to provide spatially inhomogeneous perturbations around the uniform steady state to which can break the symmetry and induce the emergence of spacial pattern. Whereas when there is spatially homogeneous initial concentration, distribution of population becomes homogeneous all the time [197]. Besides the choice of initial data also has different vital effect on the spatiotemporal dynamical behavior. For example, the following modes are all dependent on the diversity in initial values: collective behavior in neural network [200], the type of spatiotemporal patterns arising from the interaction of Hopf and Turing modes in Brusselator model [198], and synchronization stability for memristive system [199]. In these fields, there should be more work on the application of different initial conditions.

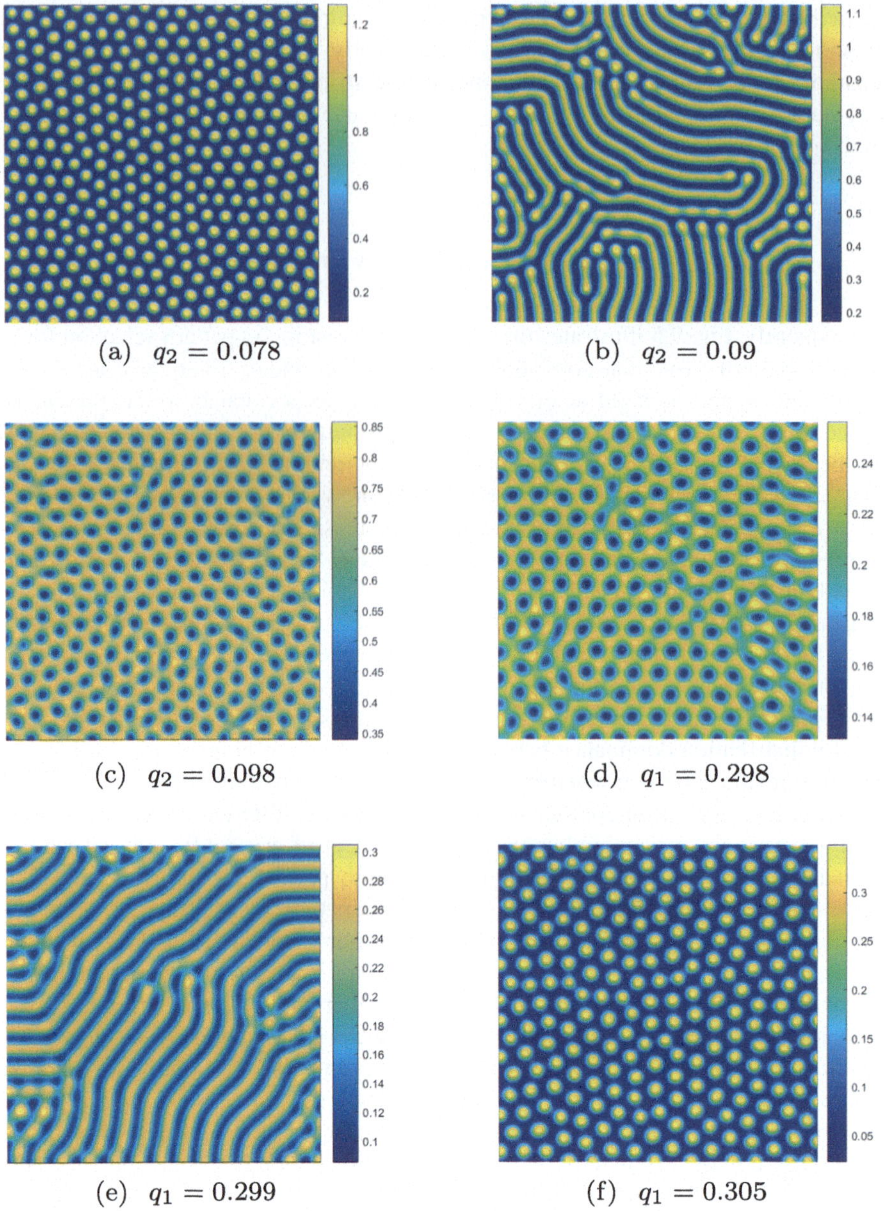

Fig. 9.5 Pattern formations for the prey developed at **a–c** $q_1 = 0.13$ and q_2, **d–f** $q_2 = 0.34$ and q_1, where other parameter values can be found in Table 9.1

Besides, many other pattern formations also attract much more attention, e.g., chemical [202] and biological systems [188], and spiral waves which are often encountered in physical [201]. There appear two research themes of significant biological implications in recent years. One is the stability and break-up mechanism behind spiral wave and researchers believe that it is associated with the occurrence of arrhythmias and tachycardias in cardiac tissue [203]. The other is pattern stability in neural networks because it is known that the study of pattern formation and selection in network of neurons helps deeply understanding the potential emergence mechanism for disease in nervous system [204–207].

As the amplitude equations utilizing multiple-scale analysis are only valid near the threshold, how to predict the selection and stability of steady patterns away from the bifurcation point becomes attractive. Furthermore, pattern formations induced by Turing instability are our focus in this chapter. While, the pattern dynamic generated by the interplay of different instabilities is also a hot topic [208, 209] such as the Turing-Hopf bifurcation. The chaotic behavior [210], spatiotemporal dynamics [212], and the forced pattern [211] have drawn extensive attention. Moreover, there exist some other significant researches about pattern dynamics like defining spatial correlation functions to estimate the regularity for distinct patterns. Readers can refer details in Refs. [200, 213]. These topics can be our future topics.

Chapter 10
Effect of Toxin and Cross-Diffusion

The reaction–diffusion theory was introduced in 1952 by Turing [146] to explain the regulated pattern formation occurring in the developing animal embryo. Subsequently, pattern formation theory has attracted considerable attention in chemistry [214, 215], physics [216], and ecology [217, 218]. Following the search for diffusive instabilities in ecological interactions [219], and the proposal of a mechanism [220] for generating a planktonic patchiness, there have increasing contributions in the area of pattern formation in ecosystems [221]. In view of the population dynamics, different plankton patterns such as spots and stripes are associated with different distributions of the plankton density. Therefore, it also represents different dominant states of the species in the spatial domain [196]. For instance, "spots" pattern indicates abundant population in isolated zones and the remaining space corresponds to low densities [222]. In this case, controlling the distribution of the population remains easier for comparing certain complex patterns such as stripes [196]. Hence, to understand the formation of patterns at different plankton interactions and consider reasonable methods to regulate the spread of plankton in space become significant.

Various factors such as time delay [152], mortality type [151], herd behavior [175], and functional response [223] can affect the pattern dynamics in an ecosystem. Specifically, it is common that a plankton model containing effect like phytoplankton release toxins and cause damage to zooplankton. Though the advantages of toxic phytoplankton on species diversity have been revealed by some literatures [224], an excessive amount of toxins can also lead to the depletion of zooplankton and possibly cause the over-growth of phytoplankton. When toxins from phytoplankton strongly damage zooplankton, it will look for target alternative food for survival. Hence, the supply of additional food has important effect on the persistence of zooplankton like further enhancing the consumption of phytoplankton. In this case, it is useful to provide additional food for zooplankton regarded as a biological control and also avoid algal bloom and eradicate certain harmful phytoplankton [225, 226]. Usually, there exist tools such as toxin liberation and additional food simultaneously. However, to the best of our knowledge, few works focus on the effect of the combined effects

S. T. Liu and L. Zhang, *Pattern Dynamics of Marine Plankton Behavior*,
https://doi.org/10.1007/978-981-97-5369-7_10

of toxin liberation and additional food on pattern dynamics of a plankton model. To fill this gap, we present the investigation of the pattern dynamics in a diffusive plankton model with the two factors mentioned above in this chapter.

10.1 Mathematical Model and Assumptions

Enlightened by [226–228], we establish a toxin-producing phytoplankton–zooplankton model in the presence of additional food for zooplankton as follows:

$$\begin{aligned} \frac{dP}{dT} &= rP(1-\frac{P}{K}) - \frac{e_1 PZ}{1+e_1h_1P+e_2h_2A}, \\ \frac{dZ}{dT} &= \frac{n_1e_1PZ+n_2e_2AZ}{1+e_1h_1P+e_2h_2A} - \tau Z - \sigma\frac{PZ}{\beta+P}, \end{aligned} \tag{10.1}$$

where Table 10.1 gives the descriptions of the parameters. Assume P is the density of the toxic phytoplankton population and Z is the density of the zooplankton population at time T. For model (10.1), the following assumptions are satisfied:
(i) Term $rP(1-\frac{P}{K})$ exists because the toxic phytoplankton grow logistically in the absence of zooplankton.
(ii) The reduction in phytoplankton due to predation is regarded as

$$\frac{e_1 PZ}{1+e_1h_1P+e_2h_2A}$$

with the presence of additional food for zooplankton.
(iii) The predation functional response adopted here is represented by

$$\frac{n_1e_1PZ+n_2e_2AZ}{1+e_1h_1P+e_2h_2A}$$

which is modified Holling Type II when the zooplankton feeds on the the additional food and toxic phytoplankton. Empirical results [164] suggest Holling Type II functional response assuming that the predators do not interfere with the activities of one another. Hence, competitions among zooplankton only occur when there exists a shortage of food.
(iv) The term τZ denotes the mortality of zooplankton because of starvation, natural death competition or higher predation, etc.
(v) $\sigma\frac{PZ}{\beta+P}$ is present when the toxic effect of phytoplankton causes the death of zooplankton.

Next, we rewritten the system (10.1) in a nondimensionalized form for simplicity. Let

$$c=\frac{1}{h_1},\ b=n_1c,\ a=\frac{1}{e_1h_1},\ \eta=\frac{n_2e_2}{n_1e_1},\ \alpha=\frac{n_1h_2}{n_2h_1},$$

Table 10.1 Descriptions and values of the parameters in the model (10.1)

Parameter	Description	Value
r	Intrinsic growth rate of TPP	1.00
K	Carrying capacity of TPP	2.00
h_1	Handling time of zooplankton per unit quantity of TPP	1.00
h_2	Handling time of zooplankton per unit quantity of additional food	0.24
e_1	Ability of zooplankton to detect TPP	1.00
e_2	Ability of zooplankton to detect additional food	2.08
n_1	Nutritive value of TPP	4.16
n_2	Nutritive value of additional food	2.00
A	Biomass of additional food for zooplankton	0.71
τ	Mortality of zooplankton	2.64
σ	Rate of toxin liberation by TPP.	0.80
β	Half saturation constant	9.25

1 *Notes* (i) TPP refers Toxic phytoplankton population.
2 (ii) We assume that the biomass of additional food for zooplankton is distributed uniformly in the habitat.
3 (iii) We assume the parameters given here all positive considering the biological significance.

$$x = \frac{P}{a},\ y = \frac{cZ}{ar},\ t = rT,\ \gamma = \frac{K}{a},\ \delta = \frac{b}{r},\ \xi = \frac{\eta A}{a},\ m = \frac{\tau}{r},\ n = \frac{\sigma}{r},\ h = \frac{\beta}{a},$$

and get the system (10.1) as below

$$\begin{aligned} \frac{dx}{dt} &= x(1 - \frac{x}{\gamma}) - \frac{xy}{1 + \alpha\xi + x}, \\ \frac{dy}{dt} &= \frac{\delta(x + \xi)y}{1 + \alpha\xi + x} - my - n\frac{xy}{h + x}. \end{aligned} \tag{10.2}$$

We introduce diffusion terms in the model (10.2) to express the spread of plankton in the spatial domain. Specifically, it is known that the tendency of predator to chase its prey is common in nature so that the velocity of predator may be bound up with the diffusive velocity of prey. This phenomenon named as cross-diffusion was first proposed by Kerner [229]. In this chapter, the zooplankton is subject to cross-diffusion as follows considering the motion of zooplankton toward phytoplankton:

$$\begin{aligned} \frac{\partial x}{\partial t} &= d_1\nabla^2 x + x(1 - \frac{x}{\gamma}) - \frac{xy}{1 + \alpha\xi + x}, \\ \frac{\partial y}{\partial t} &= d\nabla^2 x + d_2\nabla^2 y + \frac{\delta(x + \xi)y}{1 + \alpha\xi + x} - my - n\frac{xy}{h + x}, \end{aligned} \tag{10.3}$$

where

$$\nabla^2 = \frac{\partial^2}{\partial \zeta_1^2} + \frac{\partial^2}{\partial \zeta_2^2}, \ (\zeta_1, \zeta_2) \in \Omega = [0, R] \times [0, R],$$

∇^2 means the *Laplacian Operator* in two-dimensional space, Ω stands for a square bounded connected domain with smooth boundary $\partial\Omega$, d_1 and d_2 represent self-diffusion coefficients of toxic phytoplankton and zooplankton, respectively. They describe the random motion of individuals from a higher to a lower concentration region. Due to the presence of TPP, d means the cross-diffusion efficient expressing the population fluxes of zooplankton. The non-zero initial condition is given in this reaction–diffusion model (10.3),

$$x(\zeta_1, \zeta_2, 0) > 0, \ y(\zeta_1, \zeta_2, 0) > 0,$$

and zero-flux (or Neumann) boundary condition is displayed

$$\frac{\partial x}{\partial \vartheta} = \frac{\partial y}{\partial \vartheta} = 0, \ (\zeta_1, \zeta_2) \in \partial\Omega,$$

where ϑ means the outward unit normal vector on $\partial\Omega$. The boundary condition indicates that there exists no external input of the system (10.3) and the pattern formations concentrated on the self-organized cases.

10.2 Stability Analysis of the Non-spatial Model

In the following, we discuss the stability of the equilibria and Hopf bifurcation to investigate the temporal dynamics of the non-spatial model (10.2).

10.2.1 Existence of Nonnegative Equilibria

Considering the following case for system (10.2) which implies the interaction of phytoplankton and zooplankton nullclines

$$\begin{cases} x(1 - \frac{x}{\gamma}) - \frac{xy}{1+\alpha\xi+x} = 0, \\ \frac{\delta(x+\xi)y}{1+\alpha\xi+x} - my - n\frac{xy}{h+x} = 0, \end{cases}$$

we can get its nonnegative equilibrium points as below:
(i) $E_0 = (0, 0)$, which remains feasible;
(ii) $E_1 = (\gamma, 0)$, which remains feasible;
(iii) $E^* = (x^*, \ y^*)$, where $y^* = (1 + \alpha\xi + x^*)(1 - \frac{x^*}{\gamma})$ and x^* means the positive

solution of the quadratic equation as follows:

$$a_0x^2 + a_1x + a_2 = 0,$$

where

$$a_0 = \delta - m - n,\ a_2 = \delta\xi h - mh - m\alpha\xi h,$$

$$a_1 = \delta\xi + \delta h - m\alpha\xi - mh - n\alpha\xi - n - m.$$

Then, we denote the expressions of the equilibrium points E^* as

$$\begin{cases} x^*_{1,2} = \dfrac{-(\delta\xi + \delta h - m\alpha\xi - mh - n\alpha\xi - n - m)}{2(\delta - m - n)}, \\ + \dfrac{\pm\sqrt{(\delta\xi + \delta h - m\alpha\xi - mh - n\alpha\xi - n - m)^2 - 4(\delta - m - n)(\delta\xi h - mh - m\alpha\xi h)}}{2(\delta - m - n)}, \\ y^*_{1,2} = (1 + \alpha\xi + x^*)(1 - \frac{x^*}{\gamma}). \end{cases}$$

Obviously, the only one positive coexistence equilibrium point E^* for system (10.2) is obtained, providing $\frac{a_2}{a_0} < 0$ and $0 < x^* < \gamma$. Subsequently, the dynamics of the unique positive equilibrium case will be our focus.

10.2.2 Stability and Hopf Bifurcation Analysis

When the eigenvalues of the Jacobian matrix evaluated at equilibrium point E_0 of the system (10.2) are 1 and $\frac{\delta\xi}{1+\alpha\xi} - m$, the equilibrium point E_0 becomes unstable.

And when the eigenvalues of equilibrium point E_1 are -1 and $\frac{\delta(\gamma+\xi)}{1+\alpha\xi+\gamma} - m - \frac{n\gamma}{h+\gamma}$, we get that this equilibrium is stable as $\frac{\delta(\gamma+\xi)}{1+\alpha\xi+\gamma} - m - \frac{n\gamma}{h+\gamma} < 0$.

Let's write the Jacobian matrix corresponding to E^* as

$$J \triangleq \begin{pmatrix} J_{11} & J_{12} \\ J_{21} & J_{22} \end{pmatrix} = \begin{pmatrix} 1 - \frac{2x^*}{\gamma} - \frac{\alpha\xi y^* + y^*}{(1+\alpha\xi+x^*)^2} & -\frac{x^*}{1+\alpha\xi+x^*} \\ \frac{\delta y^*(1+\alpha\xi-\xi)}{(1+\alpha\xi+x^*)^2} - \frac{nhy^*}{(h+x^*)^2} & 0 \end{pmatrix}, \tag{10.4}$$

and the corresponding characteristic equation as

$$\lambda^2 + D\lambda + B = 0, \tag{10.5}$$

where

$$D = -J_{11},\ B = -J_{12}J_{21}.$$

Based on the Routh-Hurwitz criterion and conditions: $J_{11} < 0$ and $J_{12}J_{21} < 0$, we get that equilibrium E^* will be locally asymptotically stable. When we take parameter

γ as the bifurcation parameter, it is obvious that Hopf bifurcation of E^* occurs as $D(\gamma) = 0$ and $B(\gamma) > 0$. Then we obtain its transversality condition. We then get the critical value of Hopf bifurcation γ_H after solving the following equations:

$$\begin{cases} D(\gamma_H) = 0, \\ B(\gamma_H) > 0. \end{cases}$$

Assume $\lambda(\gamma) = u(\gamma) + iv(\gamma)$ as the solution of Eq. (10.5). We can get the following equations after substituting $\lambda(\gamma)$ into Eq. (10.5) and separating the real and imaginary parts:

$$\begin{cases} u^2 - v^2 + Du + B = 0, \\ 2uv + Dv = 0. \end{cases} \tag{10.6}$$

Differentiating Eq. (10.6) w.r.t γ yields

$$\begin{cases} 2uu_\gamma - 2vv_\gamma + D_\gamma u + B_\gamma + Du_\gamma = 0, \\ 2u_\gamma v + 2uv_\gamma + D_\gamma v + Dv_\gamma = 0. \end{cases} \tag{10.7}$$

Solving Eq. (10.7) at $\gamma = \gamma_H$, we obtain

$$\frac{d(Re\lambda(\gamma))}{d\gamma}|_{\gamma_H} = \frac{du(\gamma)}{d\gamma}|_{\gamma_H} = \frac{-2D_\gamma v^2 - DB_\gamma}{D^2 + 4v^2}|_{\gamma_H}.$$

Thus, we conclude the conditions for the occurrence of Hopf bifurcation from E^* as below:

Theorem 62 *There exists the Hopf bifurcation from equilibrium E^* when there is $\gamma = \gamma_H$ such that,*

$$(i)\ D(\gamma_H) = 0,\ \ B(\gamma_H) > 0,$$

$$(ii)\ \frac{d(Re\lambda(\gamma))}{d\gamma}|_{\gamma_H} = \frac{du(\gamma)}{d\gamma}|_{\gamma_H} = \frac{-2D_\gamma v^2 - DB_\gamma}{D^2 + 4v^2}|_{\gamma_H} \neq 0.$$

We can see the switch feature for the stability of equilibrium E^* as γ passes through γ_H. Assigned the values of γ in the two sides of γ_H, Figs. 10.1 and 10.2 show the different dynamical behaviors such as asymptotical stability and oscillation, respectively. Here, the parameters of the model (10.2) are chosen as $\alpha = 0.5,\ \delta = 4.2,\ \xi = 0.7,\ m = 2.6,\ h = 9.3,\ n = 0.8$ and $\gamma = 2.5$ (Fig. 10.1), 1.95 (Fig. 10.2). The corresponding γ_H is calculated to be 2.2771.

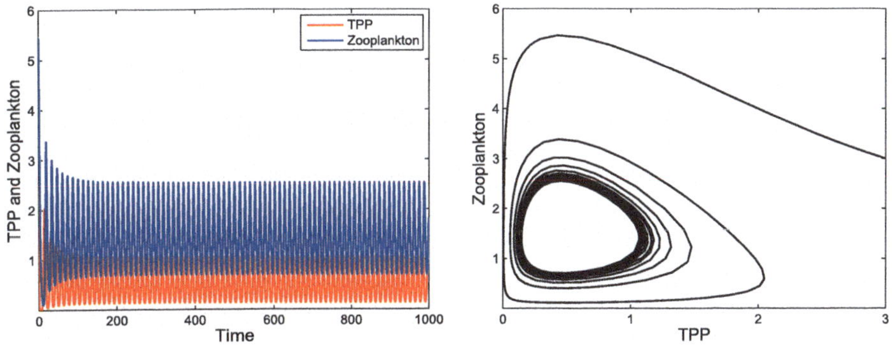

(a) Time evolution of TPP and zooplankton

(b) Phase portrait of the system (10.2)

Fig. 10.1 The oscillatory behavior for the system (10.2) around equilibrium E^*, where $\gamma = 2.5$ is larger than the critical valve γ_H, and other parameters are chosen as $\alpha = 0.5,\ \delta = 4.2,\ \xi = 0.7,\ m = 2.6,\ h = 9.3$ and $n = 0.8$

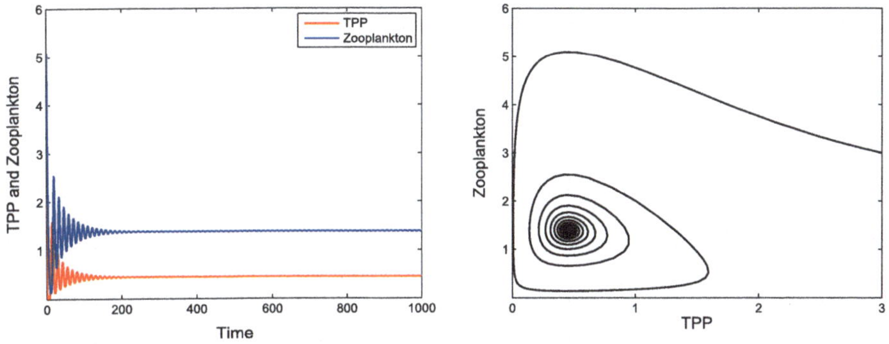

(a) Time evolution of TPP and zooplankton

(b) Phase portrait of the system (10.2)

Fig. 10.2 The asymptotically stable behavior for equilibrium E^* of the system (10.2), where $\gamma = 1.95$ is less than the critical valve γ_H, and other parameters are chosen as $\alpha = 0.5,\ \delta = 4.2,\ \xi = 0.7,\ m = 2.6,\ h = 9.3$ and $n = 0.8$

10.3 Spatial Dynamics of the Reaction–Diffusion Model

10.3.1 Turing Instability Analysis

It is well-known that the interaction of the reaction and diffusion process can generate the self-organization phenomenon in the reaction–diffusion system [146]. Specifically, the system may get patterns through non-equilibrium phase transition as the random distribution triggers off the stability loss of the uniform steady state while it is stable in the absence of diffusion [146]. In order to investigate this mechanism of pattern formation, perturbation analysis was employed to study the different types of spatio-temporal symmetry-breaking [43]. Actually, the main method in this perturbation analysis remains exactly linear stability analysis. Besides, a practical way to analyze the stability of the nonlinear system can be investigated by the corresponding linearization form usually according to the linear stability theorem and the condition that the perturbation is small [43]. Overall, in the following, the occurrence of Turing instability is our focus because that it indicates the loss of the stability for the uniform steady state owing to the diffusion term.

First, we linearize the system (10.3) at the interior equilibrium E^* as below:

$$\begin{aligned}\frac{\partial \hat{x}}{\partial t} &= d_1 \nabla^2 \hat{x} + J_{11}\hat{x} + J_{12}\hat{y},\\ \frac{\partial \hat{y}}{\partial t} &= d\nabla^2 \hat{x} + d_2 \nabla^2 \hat{y} + J_{21}\hat{x},\end{aligned} \tag{10.8}$$

where $x = \hat{x} + x^*,\ \ y = \hat{y} + y^*$ and $(\hat{x}, \hat{y})$ stands for a small perturbation of (x, y) around $E^*(x^*, y^*)$. For simplicity, we remove the hat of $(\hat{x}, \hat{y})$ and get the model as follows:

$$\begin{aligned}\frac{\partial x}{\partial t} &= d_1 \nabla^2 x + J_{11}x + J_{12}y,\\ \frac{\partial y}{\partial t} &= d\nabla^2 x + d_2 \nabla^2 y + J_{21}x.\end{aligned} \tag{10.9}$$

Next, we consider the following form of the solution for the system (10.9)

$$\begin{pmatrix} x \\ y \end{pmatrix} = \begin{pmatrix} c_k^1 \\ c_k^2 \end{pmatrix} e^{\hat{\lambda} t + i\boldsymbol{k}\cdot\boldsymbol{r}}. \tag{10.10}$$

Here, $\hat{\lambda}$ means the growth rate of perturbations, i indicates the imaginary unit where $i^2 = -1$, k means the wave number and $\boldsymbol{k} \cdot \boldsymbol{k} = k^2$, $\boldsymbol{r}$ represents the directional vector, and $c.c.$ indicates the complex conjugate terms. Hence, we obtain the following dispersion relation after substituting Eqs. (10.10) into (10.9):

$$\hat{\lambda}^2 + \hat{D}\hat{\lambda} + \hat{B} = 0 \tag{10.11}$$

where

$$\hat{D} = (d_1 + d_2)k^2 + D,$$

$$\hat{B} = d_1 d_2 k^4 + (J_{12}d - J_{11}d_2)k^2 + B.$$

It is true that Turing instability occurs when there exists no diffusion term for the uniform steady state as stable, whereas it remains unstable including the diffusion term. Thus, let $D > 0$ and $B > 0$ which ensure the local stability of the equilibrium E^* for the ordinary differential equations (10.2). Obviously, $\hat{D} > 0$ owing to the positivity of D and $\hat{B} < 0$ means the only case leading to the diffusive instability. Therefore, we can write the necessary condition for the Turing instability as

$$\hat{B} = H = d_1 d_2 k^4 + (J_{12} d - J_{11} d_2) k^2 + B < 0. \tag{10.12}$$

Here, H means a quadratic function w.r.t. k^2. Then, we obtain the minimum value of $H(k^2)$ at $k_T^2 = -\frac{J_{12}d - J_{11}d_2}{2d_1 d_2}$ as below:

$$H_{min} = \frac{4d_1 d_2 B - (J_{11} d_2 - J_{12} d)^2}{4d_1 d_2}.$$

When there exists certain $k^2 > 0$ such that $H(k^2) < 0$, implying that the minimum value of $H(k^2)$ is negative, there would generate Turing instability. And from $H_{min} = 0$, we can obtain the critical value γ_T of Turing bifurcation. Obviously, we will not give the accurate expression of γ_T because of its complexity. According to the above analysis, we can propose the following theorem:

Theorem 63 *The necessary conditions are listed as below for the presence of Turing instability*

$$(i)\ D > 0 \ and \ B > 0,$$

$$(ii)\ 2\sqrt{d_1 d_2 B} < J_{11} d_2 - J_{12} d.$$

Remark When $d = 0$ of the system (10.3), the expression of $\hat{B}$ modifies to $H(k^2) = d_1 d_2 k^4 - J_{11} d_2 k^2 + B$. Clearly, there is no positive k^2 to satisfy $H(k^2) < 0$. Hence, if there is no cross-diffusion term in the plankton system (10.3), pattern formation cannot occur. It indicates that the cross-diffusion term has a crucial effect on the occurrence of Turing instability. Under the constraint of temporal stability conditions and without the cross-diffusion term, the positive of k^2 can't be guaranteed, which means that the conditions of Turing instability can't hold simultaneously. Furthermore, the diffusive instability cannot occur also unless $\frac{J_{12}}{J_{11}} > \frac{d_2}{d}$ considering the positive of k_T^2.

The necessary condition (10.12) of the Turing instability indicates that diffusion has loss effect on the stability with the perturbation of certain wave numbers. Moreover, Fig. 10.3 illustrates the existence of Turing instability by plotting $H(k^2)$ and the largest real part of $\hat{\lambda}$ versus k^2. The parameters used here in the model (10.3) are taken as $\alpha = 0.5$, $\delta = 4.2$, $\xi = 0.7$, $m = 2.6$, $h = 9.3$ and $n = 0.8$ with varied $\gamma = 1.71,\ 1.83,\ 1.91068,\ 2.00$ and 2.13. And we calculate the critical value of Turing instability, γ_T, as 1.91068. Clearly, Fig. **??**a shows that the parameter γ has monotone increasing relationship with the interval of k^2 which makes $H(k^2)$ negative. Here, a wider range of k^2 indicates a higher possibility for the occurrence of Turing instability. Graph Fig. **??**b reflects the same tendency as Fig. **??**a concerning the largest real part of $\hat{\lambda}$. When $\gamma < \gamma_T$, the real part of $\hat{\lambda}$ remains completely neg-

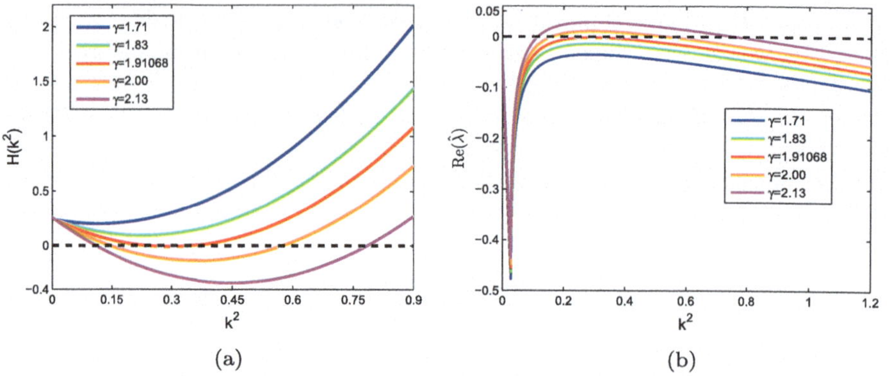

Fig. 10.3 Curves of the functions **a** $H(k^2)$ and **b** $\text{Re}(\hat{\lambda})$ with respect to k^2 for different values of γ, where other parameters are fixed at $\alpha = 0.5,\ \delta = 4.2,\ \xi = 0.7,\ m = 2.6,\ h = 9.3$ and $n = 0.8$. And the black dashed lines indicate **a** $H(k^2) = 0$ and **b** $\text{Re}(\hat{\lambda}) = 0$

ative, indicating that the equilibrium E^* becomes stable in the spatial model (10.3). In this way, it yields no Turing instability and no pattern formation.

10.3.2 Combined Effects of Additional Food and Toxins on Stability

The combined role of additional food and toxins in the stability of the model (10.3) is our aim in this subsection. First, Fig. 10.4 shows the change of the stability as parameters n and ξ varying together. Here, because that n and ξ are propositional to σ and A respectively, we assume the variation in parameters n and ξ as consistent with σ (the rate of toxin liberation by TPP) and A (the biomass of additional food); Table 10.1 lists all other parameters, except σ and A, as fixed. Region I of Fig. 10.4 represents the zooplankton extinct region. Its Region II indicates the stable coexistence range of phytoplankton and zooplankton irrespective of the model with the diffusion term. Turing instability will be cased by n and ξ located in Regions III and IV. The values of Region V as the Turing–Hopf region, can lead to oscillations in the temporal and spatial dynamics. Notably, Region IV induces the Turing instability which relies on the interaction between additional food and toxin liberation since the temporal stability will disappear without toxin liberation.

Figure 10.4 shows that the presence of a small amount of additional food for zooplankton, phytoplankton, and zooplankton makes its stability despite some toxin liberation. As toxin liberation exceeds some critical value, the balance between these two species would be broken, and then the zooplankton vanishes. Next, with the increasing of the supply of additional food, phytoplankton and zooplankton become coexistence again and the toxin liberation hardly affects it. As the additional food

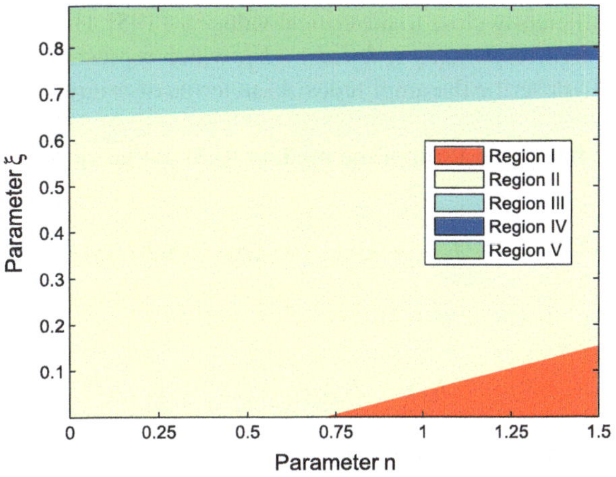

Fig. 10.4 Different stability regions as a function of ξ and n, where Region I stands for the zooplankton extinction region, Region II means the phytoplankton and zooplankton coexistence region, Regions III and IV indicate the Turing instability region, Region V represents the Turing–Hopf Region and other parameters are the same as in Table 10.1

remains significant, Turing instability occurs, i.e., we can see the heterogeneity in the spatial distribution and different patterns. Interestingly, when there emerges a certain amount of additional food (such as $\xi = 0.75$), we can observe the Turing instability regardless of the value of toxin liberation. Region V shows that inhomogeneous stationary patterns may be initiated for very large ξ.

10.3.3 Amplitude Equations

In the following, the amplitude equations are derived utilizing a multiple-scale method near the critical value of Turing instability γ_T. The multiple-scale method indicates that eigenvalues evaluated at γ around γ_T are associated with the active modes determining the critical dynamics and enslaving other rapidly changing modes. Hence, the analysis of the whole dynamics can be simplified into the ones only concerning the active modes by the multiple-scale method [42]. As the amplitude equations can be used to reveal the dynamics of the active mode, the deduction of amplitude equations is discussed in brief in follows. The readers can refer to [43] for detailed derivation process.

Overall, we expand the solutions of the model (10.3) as below

$$\begin{pmatrix} x \\ y \end{pmatrix} = \begin{pmatrix} x^* \\ y^* \end{pmatrix} + \sum_{j=1}^{3} A_j exp(i\boldsymbol{k}_j \cdot \boldsymbol{r}) + \sum_{j=1}^{3} \overline{A}_j exp(-i\boldsymbol{k}_j \cdot \boldsymbol{r})$$

when γ means sufficiently close to the critical value γ_T [195]. Here, A_j and $\overline{A}_j$ stand for the amplitudes corresponding to $\boldsymbol{k}_j$ and $-\boldsymbol{k}_j$, which represents the wave vectors with $|\boldsymbol{k}_j| = k_T$. To describe the amplitudes A_j in terms of original parameters is our aim.

We obtain the linearized form of the model (10.3) at the equilibrium $E^*(x^*, y^*)$ as following,

$$\begin{cases} \frac{\partial x}{\partial t} = d_1\nabla^2 x + J_{11}x + J_{12}y + \frac{1}{2}b_{20}x^2 + \frac{1}{2}b_{02}y^2 + b_{11}xy + \frac{1}{6}b_{30}x^3 \\ \quad + \frac{1}{6}b_{03}y^3 + \frac{1}{2}b_{21}x^2y + \frac{1}{2}b_{12}xy^2, \\ \frac{\partial y}{\partial t} = d\nabla^2 x + d_2\nabla^2 y + J_{21}x + J_{22}y + \frac{1}{2}c_{20}x^2 + \frac{1}{2}c_{02}y^2 + c_{11}xy + \frac{1}{6}c_{30}x^3 \\ \quad + \frac{1}{6}c_{03}y^3 + \frac{1}{2}c_{21}x^2y + \frac{1}{2}c_{12}xy^2, \end{cases} \tag{10.13}$$

where

$$b_{20} = -\frac{2}{\gamma} + \frac{2(1+\alpha\xi)(1-\frac{x^*}{\gamma})}{(1+\alpha\xi+x^*)^2},\ b_{02} = 0,\ b_{11} = -\frac{1+\alpha\xi}{(1+\alpha\xi+x^*)^2},$$

$$b_{30} = -\frac{6(1+\alpha\xi)(1-\frac{x^*}{\gamma})}{(1+\alpha\xi+x^*)^3},\ b_{03} = 0,\ b_{21} = \frac{2\alpha\xi+2}{(1+\alpha\xi+x^*)^3},\ b_{12} = 0,$$

$$c_{20} = \frac{-2\delta(1+\alpha\xi-\xi)(1-\frac{x^*}{\gamma})}{(1+\alpha\xi+x^*)^2} + \frac{2nhy^*}{(h+x^*)^3},\ c_{02} = 0,$$

$$c_{11} = \frac{(1+\alpha\xi-\xi)\delta}{(1+\alpha\xi+x^*)^2} - \frac{nh}{(h+x^*)^2},$$

$$c_{30} = \frac{6\delta(1-\frac{x}{\gamma})(1+\alpha\xi-\xi)}{(1+\alpha\xi+x^*)^3} - \frac{6nhy^*}{(h+x^*)^4},\ c_{03} = 0,$$

$$c_{21} = \frac{-2\delta(1+\alpha\xi-\xi)}{(1+\alpha\xi+x^*)^3} + \frac{2nh}{(h+x^*)^3},\ c_{12} = 0,$$

perturbation $(x - x^*, y - y^*)^T$ is still denoted by $(x, y)^T$. Let $\mathbf{X} = (x, y)^T$. Then we convert model (10.13) into

$$\frac{\partial \mathbf{X}}{\partial t} = \mathbf{LX} + \mathbf{N} \tag{10.14}$$

where

$$\mathbf{L} = \begin{pmatrix} d_1\nabla^2 + J_{11} & J_{12} \\ d\nabla^2 + J_{21} & d_2\nabla^2 + J_{22} \end{pmatrix}, \tag{10.15}$$

$$\mathbf{N} = \begin{pmatrix} \frac{1}{2}b_{20}x^2 + \frac{1}{2}b_{02}y^2 + b_{11}xy + \frac{1}{6}b_{30}x^3 + \frac{1}{6}b_{03}y^3 + \frac{1}{2}b_{21}x^2y + \frac{1}{2}b_{12}xy^2 \\ \frac{1}{2}c_{20}x^2 + \frac{1}{2}c_{02}y^2 + c_{11}xy + \frac{1}{6}c_{30}x^3 + \frac{1}{6}c_{03}y^3 + \frac{1}{2}c_{21}x^2y + \frac{1}{2}c_{12}xy^2 \end{pmatrix}.$$

Operator $\mathbf{L}$ is decomposed into two components as

$$\mathbf{L} = \mathbf{L}_T + (\gamma - \gamma_T)\mathbf{M} \tag{10.16}$$

where

$$\mathbf{L}_T = \begin{pmatrix} d_1\nabla^2 + J_{11}^T & J_{12}^T \\ d\nabla^2 + J_{21}^T & d_2\nabla^2 + J_{22}^T \end{pmatrix}, \; J_{ij}^T = J_{ij}(\gamma_T), \; i, j = 1, 2,$$

$$\mathbf{M} = \begin{pmatrix} -(\frac{1+\alpha\xi}{1+\alpha\xi+x^*} - 2)\frac{x}{\gamma_T^2} & 0 \\ \frac{x^*}{\gamma_T^2}[\frac{\delta(1+\alpha\xi-\xi)}{1+\alpha\xi+x^*} - \frac{nh(1+\alpha\xi+x^*)}{(h+x)^2}] & 0 \end{pmatrix} \triangleq \begin{pmatrix} m_{11} & m_{12} \\ m_{21} & m_{22} \end{pmatrix}.$$

Next, we get the expanding of the parameter γ, solution $\mathbf{X}$ and nonlinear term $\mathbf{N}$ respectively by a sufficiently small parameter ϵ, as follows:

$$\gamma - \gamma_T = \epsilon\gamma_1 + \epsilon^2\gamma_2 + \epsilon^3\gamma_3 + o(\epsilon^3), \tag{10.17}$$

$$\mathbf{X} = \epsilon\begin{pmatrix} x_1 \\ y_1 \end{pmatrix} + \epsilon^2\begin{pmatrix} x_2 \\ y_2 \end{pmatrix} + \epsilon^3\begin{pmatrix} x_3 \\ y_3 \end{pmatrix} + o(\epsilon^3), \tag{10.18}$$

$$\mathbf{N} = \epsilon^2\mathbf{h}_2 + \epsilon^3\mathbf{h}_3 + o(\epsilon^3), \tag{10.19}$$

where

$$\mathbf{h}_2 = \begin{pmatrix} \frac{1}{2}b_{20}x_1^2 + b_{11}x_1y_1 \\ \frac{1}{2}c_{20}x_1^2 + c_{11}x_1y_1 \end{pmatrix}, \tag{10.20}$$

$$\mathbf{h}_3 = \begin{pmatrix} b_{20}x_1x_2 + b_{11}x_1y_2 + b_{11}x_2y_1 + \frac{1}{6}b_{30}x_1^3 + \frac{1}{2}b_{21}x_1^2y_1 \\ c_{20}x_1x_2 + c_{11}x_1y_2 + c_{11}x_2y_1 + \frac{1}{6}c_{30}x_1^3 + \frac{1}{2}c_{21}x_1^2y_1 \end{pmatrix}. \tag{10.21}$$

Based on the multiple-scale method, the time scale of the system is separated into $T_1 = \epsilon t, \; T_2 = \epsilon^2 t, \; T_3 = \epsilon^3 t$, which remains independent of each other. Hence, we get the expression of the time derivative w.r.t t as

$$\frac{\partial}{\partial t} = \epsilon\frac{\partial}{\partial T_1} + \epsilon^2\frac{\partial}{\partial T_2} + \epsilon^3\frac{\partial}{\partial T_3} + o(\epsilon^3). \tag{10.22}$$

Substituting Eqs. (10.16)–(10.22) into Eq. (10.14), we can obtain the equations for different orders of ϵ as below:

The first order of ϵ:

$$\epsilon : \mathbf{L}_T \begin{pmatrix} x_1 \\ y_1 \end{pmatrix} = \mathbf{0}; \tag{10.23}$$

The second order of ϵ:

$$\epsilon^2 : \mathbf{L}_T \begin{pmatrix} x_2 \\ y_2 \end{pmatrix} = \frac{\partial}{\partial T_1} \begin{pmatrix} x_1 \\ y_1 \end{pmatrix} - \gamma_1 \mathbf{M} \begin{pmatrix} x_1 \\ y_1 \end{pmatrix} - \mathbf{h}_2; \tag{10.24}$$

The third order of ϵ:

$$\epsilon^3 : \mathbf{L}_T \begin{pmatrix} x_3 \\ y_3 \end{pmatrix} = \frac{\partial}{\partial T_1} \begin{pmatrix} x_2 \\ y_2 \end{pmatrix} + \frac{\partial}{\partial T_2} \begin{pmatrix} x_1 \\ y_1 \end{pmatrix} - \gamma_2 \mathbf{M} \begin{pmatrix} x_1 \\ y_1 \end{pmatrix} - \gamma_1 \mathbf{M} \begin{pmatrix} x_2 \\ y_2 \end{pmatrix} - \mathbf{h}_3. \tag{10.25}$$

We get $\mathbf{L}_T$ as a linear operator obtained at the onset γ_T, so that solutions of Eq. (10.23) can be described as

$$\begin{pmatrix} x_1 \\ y_1 \end{pmatrix} = \begin{pmatrix} \varphi \\ 1 \end{pmatrix} (\sum_{j=1}^{3} W_j exp(i\boldsymbol{k_j} \cdot \boldsymbol{r}) + c.c.), \qquad j = 1, 2, 3, \tag{10.26}$$

where $\varphi = \frac{-J_{12}^T}{J_{11}^T - d_1 k_T^{*2}}$, $|\boldsymbol{k_j}| = k_T^*$, $k_T^* = k_T(\gamma_T)$ and W_j means the amplitude related to the mode of $exp(i\boldsymbol{k_j} \cdot \boldsymbol{r})$ with the first-order perturbations. The higher-order perturbations determine the concrete expression of W_j . $c.c.$ represents the conjugate terms. Next, according to the existence of nontrivial solutions of Eq. (10.24) and the Fredholm solubility conditions, we can get the orthogonality for the right side of Eq. (10.24) with the zero eigenvectors of $\mathbf{L}_T^*$ as the adjoint operator of $\mathbf{L}_T$. And we also obtain the expression of the zero eigenvectors of $\mathbf{L}_T^*$ as

$$\begin{pmatrix} 1 \\ \psi \end{pmatrix} exp(-i\boldsymbol{k_j} \cdot \boldsymbol{r}) + c.c., \qquad j = 1, 2, 3, \tag{10.27}$$

with $\psi = -\frac{J_{11}^T - d_1 k_T^{*2}}{J_{21}^T - d k_T^{*2}}$. Moreover, the orthogonality condition can be written as

$$(1, \psi) \begin{pmatrix} F_x^j \\ F_y^j \end{pmatrix} = 0, \qquad j = 1, 2, 3. \tag{10.28}$$

Here, F_x^j and F_y^j mean the coefficients of $\exp(i\boldsymbol{k_j} \cdot \boldsymbol{r})$ in F_x and F_y where

$$\begin{pmatrix} F_x \\ F_y \end{pmatrix} \triangleq \frac{\partial}{\partial T_1}\begin{pmatrix} x_1 \\ y_1 \end{pmatrix} - \gamma_1 \mathbf{M}\begin{pmatrix} x_1 \\ y_1 \end{pmatrix} - \mathbf{h}_2.$$

After utilizing Eq. (10.28) and with regard to the first-order perturbations, we obtain the amplitude equations as below:

$$\begin{cases} (\varphi+\psi)\frac{\partial W_1}{\partial T_1} = \gamma_1[(\varphi m_{11}+m_{12})+\psi(\varphi m_{21}+m_{22})]W_1 + 2(l_1+\psi l_2)\bar{W}_2\bar{W}_3, \\ (\varphi+\psi)\frac{\partial W_2}{\partial T_1} = \gamma_1[(\varphi m_{11}+m_{12})+\psi(\varphi m_{21}+m_{22})]W_2 + 2(l_1+\psi l_2)\bar{W}_1\bar{W}_3, \\ (\varphi+\psi)\frac{\partial W_3}{\partial T_1} = \gamma_1[(\varphi m_{11}+m_{12})+\psi(\varphi m_{21}+m_{22})]W_3 + 2(l_1+\psi l_2)\bar{W}_1\bar{W}_2, \end{cases} \tag{10.29}$$

with

$$\begin{cases} l_1 = \frac{1}{2}b_{20}\varphi^2 + b_{11}\varphi, \\ l_2 = \frac{1}{2}c_{20}\varphi^2 + c_{11}\varphi. \end{cases}$$

Next, we introduce the higher-order perturbation terms to ensure the convergence. After solving Eq. (10.24), we obtain

$$\begin{aligned} \begin{pmatrix} x_2 \\ y_2 \end{pmatrix} = \begin{pmatrix} X_0 \\ Y_0 \end{pmatrix} &+ \sum_{j=1}^{3}\begin{pmatrix} X_j \\ Y_j \end{pmatrix} e^{i\boldsymbol{k}_j\cdot\boldsymbol{r}} + \sum_{j=1}^{3}\begin{pmatrix} X_{jj} \\ Y_{jj} \end{pmatrix} e^{2i\boldsymbol{k}_j\cdot\boldsymbol{r}} + \begin{pmatrix} X_{12} \\ Y_{12} \end{pmatrix} e^{i(\boldsymbol{k}_1-\boldsymbol{k}_2)\cdot\boldsymbol{r}} \\ &+ \begin{pmatrix} X_{23} \\ Y_{23} \end{pmatrix} e^{i(\boldsymbol{k}_2-\boldsymbol{k}_3)\cdot\boldsymbol{r}} + \begin{pmatrix} X_{31} \\ Y_{31} \end{pmatrix} e^{i(\boldsymbol{k}_3-\boldsymbol{k}_1)\cdot\boldsymbol{r}} + c.c., \end{aligned} \tag{10.30}$$

with

$$\begin{pmatrix} X_0 \\ Y_0 \end{pmatrix} = \begin{pmatrix} x_{00} \\ y_{00} \end{pmatrix}(|W_1|^2+|W_2|^2+|W_3|^2),\ X_j = \varphi Y_j,$$

$$\begin{pmatrix} X_{jj} \\ Y_{jj} \end{pmatrix} = \begin{pmatrix} x_{11} \\ y_{11} \end{pmatrix} W_j^2,\ \begin{pmatrix} X_{jl} \\ Y_{jl} \end{pmatrix} = \begin{pmatrix} x_* \\ y_* \end{pmatrix} W_j\bar{W}_l,$$

where

$$\begin{pmatrix} x_{00} \\ y_{00} \end{pmatrix} = \frac{-2}{J_{11}^T J_{22}^T - J_{21}^T J_{12}^T}\begin{pmatrix} J_{22}^T l_1 - J_{12}^T l_2 \\ J_{11}^T l_2 - J_{21}^T l_1 \end{pmatrix},$$

$$\begin{pmatrix} x_{11} \\ y_{11} \end{pmatrix} = \frac{-1}{(J_{11}^T - 4d_1k_T^{*2})(J_{22}^T - 4d_2k_T^{*2}) - J_{12}^T(J_{21}^T - 4dk_T^{*2})}\begin{pmatrix} (J_{22}^T - 4d_2k_T^{*2})l_1 - J_{12}^T l_2 \\ (J_{11}^T - 4d_1k_T^{*2})l_2 - (J_{21}^T - 4dk_T^{*2})l_1 \end{pmatrix},$$

$$\begin{pmatrix} x_* \\ y_* \end{pmatrix} = \frac{-2}{(J_{11}^T - 3d_1k_T^{*2})(J_{22}^T - 3d_2k_T^{*2}) - J_{12}^T(J_{21}^T - 3dk_T^{*2})}\begin{pmatrix} (J_{22}^T - 3d_2k_T^{*2})l_1 - J_{12}^T l_2 \\ (J_{11}^T - 3d_1k_T^{*2})l_2 - (J_{21}^T - 3dk_T^{*2})l_1 \end{pmatrix}.$$

We utilize the Fredholm solubility conditions again for Eq. (10.25) and get

$$\begin{cases} (\varphi+\psi)(\frac{\partial Y_1}{\partial T_1}+\frac{\partial W_1}{\partial T_2}) = [(\varphi m_{11}+m_{12})+\psi(\varphi m_{21}+m_{22})](\gamma_2 W_1+\gamma_1 Y_1) \\ -[(I_1+\psi S_1)|W_1|^2 \\ \qquad + (I_2+\psi S_2)(|W_2|^2+|W_3|^2)]W_1 \\ +(b_{20}\varphi^2+2b_{11}\varphi+\psi\varphi^2 c_{20} \qquad + 2\varphi\psi c_{11})(\bar{W}_2\bar{Y}_3+\bar{W}_3\bar{Y}_2), \\ \\ (\varphi+\psi)(\frac{\partial Y_2}{\partial T_1}+\frac{\partial W_2}{\partial T_2}) = [(\varphi m_{11}+m_{12})+\psi(\varphi m_{21}+m_{22})](\gamma_2 W_2+\gamma_1 Y_2) \\ -[(I_1+\psi S_1)|W_2|^2 \\ \qquad + (I_2+\psi S_2)(|W_1|^2+|W_3|^2)]W_2 \\ +(b_{20}\varphi^2+2b_{11}\varphi+\psi\varphi^2 c_{20} \\ \qquad + 2\varphi\psi c_{11})(\bar{W}_1\bar{Y}_3+\bar{W}_3\bar{Y}_1), \\ \\ (\varphi+\psi)(\frac{\partial Y_3}{\partial T_1}+\frac{\partial W_3}{\partial T_2}) = [(\varphi m_{11}+m_{12})+\psi(\varphi m_{21}+m_{22})](\gamma_2 W_3+\gamma_1 Y_3) \\ -[(I_1+\psi S_1)|W_3|^2 \qquad + (I_2+\psi S_2)(|W_2|^2+|W_1|^2)]W_3 \\ +(b_{20}\varphi^2+2b_{11}\varphi+\psi\varphi^2 c_{20} \qquad + 2\varphi\psi c_{11})(\bar{W}_2\bar{Y}_1+\bar{W}_1\bar{Y}_2), \end{cases} \tag{10.31}$$

where

$$I_1 = -(\varphi b_{20}+b_{11})(x_{00}+x_{11}) - \varphi b_{11}(y_{00}+y_{11}) - \frac{1}{2}b_{30}\varphi^3 - \frac{3}{2}b_{21}\varphi^2,$$

$$I_2 = -(\varphi b_{20}+b_{11})(x_{00}+x_*) - \varphi b_{11}(y_{00}+y_*) - b_{30}\varphi^3 - 3b_{21}\varphi^2,$$

$$S_1 = -(\varphi c_{20}+c_{11})(x_{00}+x_{11}) - \varphi c_{11}(y_{00}+y_{11}) - \frac{1}{2}c_{30}\varphi^3 - \frac{3}{2}c_{21}\varphi^2,$$

$$S_2 = -(\varphi c_{20}+c_{11})(x_{00}+x_*) - \varphi c_{11}(y_{00}+y_*) - c_{30}\varphi^3 - 3c_{21}\varphi^2.$$

Finally, we expand the amplitude A_j as

$$A_j = \epsilon W_j + \epsilon^2 Y_j + o(\epsilon^3) \tag{10.32}$$

and also get the derivative of amplitude A_j w.r.t. t as

$$\frac{\partial A_j}{\partial t} = \epsilon\frac{\partial A_j}{\partial T_1} + \epsilon^2\frac{\partial A_j}{\partial T_2} + \cdots. \tag{10.33}$$

Substituting Eqs. (10.29), (10.31) into (10.32) and (10.33), we get the results about A_j as below:

$$\begin{cases} \tau_0\frac{\partial A_1}{\partial t} = \mu A_1 + h\bar{A}_2\bar{A}_3 - [g_1|A_1|^2 + g_2(|A_2|^2+|A_3|^2)]A_1, \\ \tau_0\frac{\partial A_2}{\partial t} = \mu A_2 + h\bar{A}_1\bar{A}_3 - [g_1|A_2|^2 + g_2(|A_1|^2+|A_3|^2)]A_2, \\ \tau_0\frac{\partial A_3}{\partial t} = \mu A_3 + h\bar{A}_1\bar{A}_2 - [g_1|A_3|^2 + g_2(|A_1|^2+|A_2|^2)]A_3, \end{cases} \tag{10.34}$$

where

$$\tau_0 = \frac{\varphi + \psi}{\gamma_T[(\varphi m_{11} + m_{12}) + \psi(\varphi m_{21} + m_{22})]}, \quad \mu = \frac{\gamma - \gamma_T}{\gamma_T},$$

$$h = \frac{b_{20}\varphi^2 + 2b_{11}\varphi + c_{20}\varphi^2\psi + 2c_{11}\varphi\psi}{\gamma_T[(\varphi m_{11} + m_{12}) + \psi(\varphi m_{21} + m_{22})]},$$

$$g_1 = \frac{I_1 + \psi S_1}{\gamma_T[(\varphi m_{11} + m_{12}) + \psi(\varphi m_{21} + m_{22})]},$$

$$g_2 = \frac{I_2 + \psi S_2}{\gamma_T[(\varphi m_{11} + m_{12}) + \psi(\varphi m_{21} + m_{22})]}.$$

10.3.4 Amplitude Instability

In the following, the selection and stability of pattern formations will be investigated on account of the amplitude Eq. (10.34). At first, we decompose each amplitude into $\rho_j = |A_j|$, correspond phase angle θ_i, and denote A_j by $\rho_j exp(i\theta_j)$. Then we substitute $A_j = \rho_j exp(i\theta_j)$ in Eq. (10.34) and obtain the equations about real variables as follows:

$$\begin{cases} \tau_0 \frac{\partial\theta}{\partial t} = -h\frac{\rho_2^2\rho_3^2 + \rho_1^2\rho_3^2 + \rho_1^2\rho_2^2}{\rho_1\rho_2\rho_3} sin\theta, \\ \tau_0 \frac{\partial\rho_1}{\partial t} = \mu\rho_1 + h\rho_2\rho_3 cos\theta - [g_1\rho_1^2 + g_2(\rho_2^2 + \rho_3^2)]\rho_1, \\ \tau_0 \frac{\partial\rho_2}{\partial t} = \mu\rho_2 + h\rho_1\rho_3 cos\theta - [g_1\rho_2^2 + g_2(\rho_1^2 + \rho_3^2)]\rho_2, \\ \tau_0 \frac{\partial\rho_3}{\partial t} = \mu\rho_3 + h\rho_1\rho_2 cos\theta - [g_1\rho_3^2 + g_2(\rho_1^2 + \rho_2^2)]\rho_3, \end{cases} \tag{10.35}$$

where $\theta = \theta_1 + \theta_2 + \theta_3$. System (10.35) possesses the five types of solutions [43] as following:
(1) The stationary state (**O**): associated constant solution

$$\rho_1 = \rho_2 = \rho_3 = 0$$

becomes stable as $\mu < \mu_2 = 0$ and unstable as $\mu > \mu_2$.
(2) Stripe patterns (**S**): associated constant solution

$$\rho_1 = \sqrt{\frac{\mu}{g_1}} \neq 0, \ \rho_2 = \rho_3 = 0$$

becomes stable as $\mu > \mu_3 = \frac{h^2 g_1}{(g_1-g_2)^2}$ and unstable as $\mu < \mu_3$.
(3) Hexagon pattern ($\mathbf{H}_0$, $\mathbf{H}_\pi$): associated constant solutions

$$\rho_1 = \rho_2 = \rho_3 = \frac{|h| \pm \sqrt{h^2 + 4(g_1 + 2g_2)\mu}}{2(g_1 + 2g_2)}$$

with $\theta = 0$ or $\theta = \pi$ become feasible as $\mu > \mu_1 = -\frac{h^2}{4(g_1+2g_2)}$. When $\mu < \mu_4 = \frac{2g_1+g_2}{(g_2-g_1)^2}h^2$, the solution $\rho^+ = \frac{|h|+\sqrt{h^2+4(g_1+2g_2)\mu}}{2(g_1+2g_2)}$ becomes stable, whereas the other solution $\rho^- = \frac{|h|-\sqrt{h^2+4(g_1+2g_2)\mu}}{2(g_1+2g_2)}$ remains always unstable.

(4) The mixed states (**MS**): associated solution

$$\rho_1 = \frac{|h|}{g_2 - g_1}, \quad \rho_2 = \rho_3 = \sqrt{\frac{\mu - g_1\rho_1^2}{g_1 + g_2}}$$

becomes feasible as $\mu > \mu_3$, $g_2 > g_1$ and becomes always unstable.

According to the above results, we can get $\mu_1 < \mu_2 < \mu_3 < \mu_4$ as $h^2 \neq 0$. The relations between μ and μ_i $(i = 1, 2, 3, 4)$ play important role in determining the existence and stability of different patterns. For instance, the stationary state will lose its stability and there will emerge the hexagon patterns when μ increases to $\mu_2 = 0$. Here, if $h > 0$ we get the hexagon patterns as H_0 $(\theta = 0)$ and also the unstability of the other hexagon patterns H_π $(\theta = \pi)$. Otherwise, compared with the former case, the stabilities of H_0 $(\theta = 0)$ and H_π $(\theta = \pi)$ remain completely contrary when $h < 0$. It is noteworthy that (μ_1, μ_2) becomes bistable where there exist two stability for the hexagon patterns and the stationary state. The hexagon patterns will remain stable until $\mu > \mu_4$. While for the stripe patterns induced by supercritical bifurcation, they will remain stable only for $\mu > \mu_3$. Thus, if we take control parameter μ in (μ_3, μ_4), we can get the stability for the hexagon patterns and stripe patterns simultaneously. This can be named as the other bistable region. Finally, we highlight that the stripe patterns vary as below: they can change into the hexagon patterns as μ decreases below μ_3, whereas they can change towards the opposite direction for μ beyond μ_4.

10.4 Numerical Simulation

In this section, numerical simulations of the pattern formations for system (10.3) are performed to illustrate the above analytical results. In order to illustrate the distribution of plankton in two-dimensional space, we use a forward finite difference method for the system (10.3) numerically with the space step $\Delta h = 1$ and the time step $\Delta t = 0.001$. Here, we restrict the distribution range uniformly divided by 200×200 lattice sites and we employ the Neumann condition as the boundary condition. We also take the initial data from a uniformly distributed random perturbation around the interior equilibrium point $E^*(x^*, y^*)$. It is known that when we take the initial data as homogeneous distribution, we can hardly expect its pattern variation because of its homogeneity [230]. It is to be noted that we guarantee the patterns fixed further according to the simulation time. Because we assume that the distributions of phytoplankton are the same type as that of zooplankton, we only display the distribution of phytoplankton for the system (10.3) as an instance. In what follows, different conditions are examined by varying the parameter γ and the effect of additional food and toxin liberation proportional to ξ and n, respectively, are also studied.

10.4.1 Effect of Varied γ on the Pattern Formations

In this subsection, we carry out different patterns of model (10.3) due to varied γ. For the following simulations, we choose the parameters to satisfy the conditions of Turing instability as $d_1 = 0.1,\ d_2 = 29.7,\ d = 12.6,\ \alpha = 0.5,\ \delta = 4.2,\ \xi = 0.7,\ m = 2.6,\ h = 9.3,\ n = 0.8$ with $\gamma = 1.942,\ 2.003$, and 2.130. Figure 10.5 shows a clear interpretation about the curves of corresponding μ and μ_i $(i = 2, 3, 4)$ w.r.t. γ and certain critical values about the pattern formations. In order to generate different patterns, we choose specific values of γ from the regions according to the different relations of μ and μ_i. Table 10.2 lists the values of γ and the corresponding calculated coefficients.

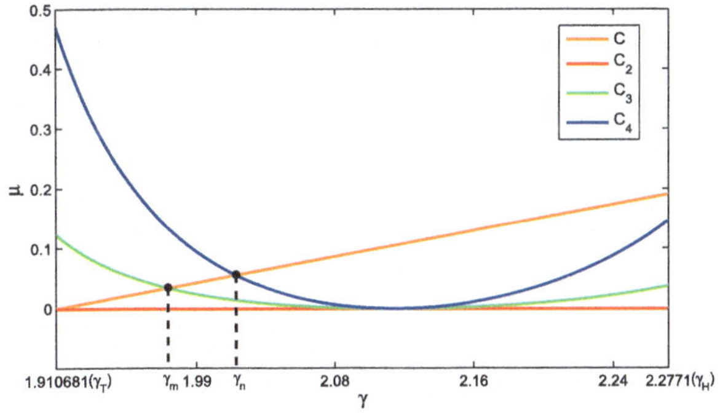

Fig. 10.5 Graphs of μ and μ_i $(i = 2, 3, 4)$ w.r.t. γ, where Curves $C,\ C_2,\ C_3$ and C_4 correspond to the functions $\mu,\ \mu_2,\ \mu_3$, and μ_4, respectively. In addition, $\gamma_T = 1.91068$ means the critical values for the occurrence of Turing instability, $\gamma_n = 2.0178$ indicates the spot patterns, $\gamma_m = 1,9768$ stands for the stripe patterns and $\gamma_H = 2.2771$ represents the Hopf bifurcation. And other parameters of model (10.3) are fixed at $d_1 = 0.1,\ d_2 = 29.7,\ d = 12.6,\ \alpha = 0.5,\ \delta = 4.2,\ \xi = 0.7,\ m = 2.6,\ h = 9.3$ and $n = 0.8$

Table 10.2 Values of γ and the corresponding calculated coefficients

γ	μ	μ_1	μ_3	μ_4	h	g_1	g_2
1.942	0.016	−0.003	0.068	0.261	1.115	24.852	46.211
2.003	0.048	−0.001	0.020	0.079	0.696	26.812	52.300
2.130	0.115	−0.00002	0.0004	0.001	−0.100	28.328	56.613

1 *Note* other parameters in model (10.3) are fixed at $d_1 = 0.1,\ d_2 = 29.7,\ d = 12.6,\ \alpha = 0.5,\ \delta = 4.2,\ \xi = 0.7,\ m = 2.6,\ h = 9.3$ and $n = 0.8$

When we take $\gamma = 1.942$ and other parameters of model (10.3) fixed, we can see clear the evolution of the spatial patterns of phytoplankton after 0, 800000, 1600000, 2000000, 2400000, and 3000000 iterations in Fig. 10.6. As presented in Table 10.2, μ satisfies $\mu_2 < \mu < \mu_3$ and h is taken as 1.115. Thus, according to the theoretical results, we can expect only the hexagon patterns to be present in the end. Initially, we can see the system (10.3) in an irregular state from the simulations. As the perturbation spreads, there emerges a mixture of hexagon and stripe patterns, and the outline of those patterns gradually takes form. Over time, more and more hexagon patterns can be produced with the stripes separating into hexagon patterns. After 3000000 iterations, there yields a time-independent stationary state for the patterns in which the hexagon patterns occupy the entire domain. Hence, the simulations testify the theoretical results.

The formation of the mixed hexagon and stripe patterns of phytoplankton is displayed in Fig. 10.7 after 0, 500000, 800000, 1600000, 1900000 and 3000000 iterations with $\gamma = 2.003$ and $\mu_3 < \mu < \mu_4$. Initially, there presents an irregular transient pattern due to a random perturbation. Afterwards, the irregular state is replaced by mixed hexagon and stripe patterns with the variation in the density of phytoplankton. We can see the gradual aggregation for some of the stripes and hexagons, and also see some mixture separating into clear hexagons. After 3000000 iterations, we can see the simulations in agreement with the theoretical results that there yield obvious time-independent mixed hexagon and stripe patterns, which are well-distributed in their shapes and colors.

The formation of the stripe patterns of phytoplankton is depicted in Fig. 10.8 after 0, 100000, 300000, 800000, 1600000, and 3000000 iterations. In the simulation, we assume $\gamma = 2.13$ and that the corresponding μ is larger than μ_4 with $h = -0.1$. At first, it is clear that the random perturbation can generate an irregular distribution. Subsequently, there yields variation of the irregular state and the presence of the regular stripe patterns instead. After the gradual evolution, we can see clearly the aggregation of the stripe patterns and entwines with each other smoothly. Finally, it is obvious that the distribution of phytoplankton remains unchanged and the domination pattern in the entire range is the stripe patterns. Therefore, the simulations are also in accordance with the theoretical results.

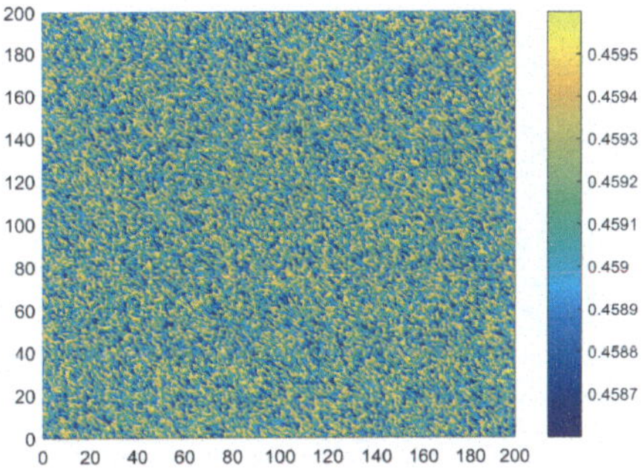

(a) 0 iteration

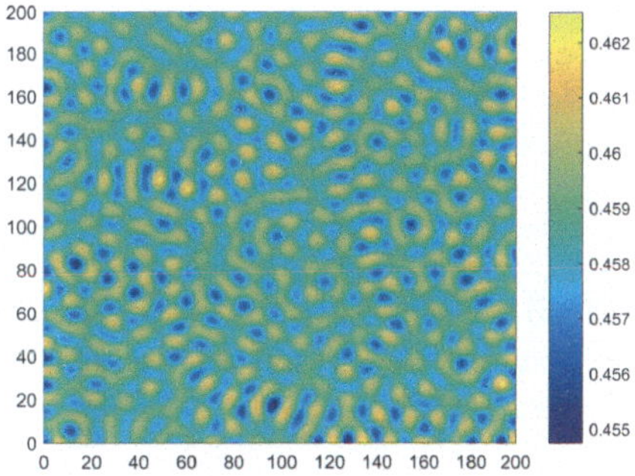

(b) 800000 iterations

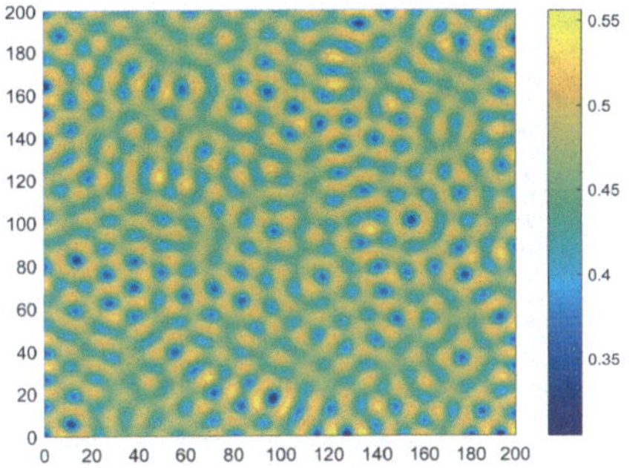

(c) 1600000 iterations

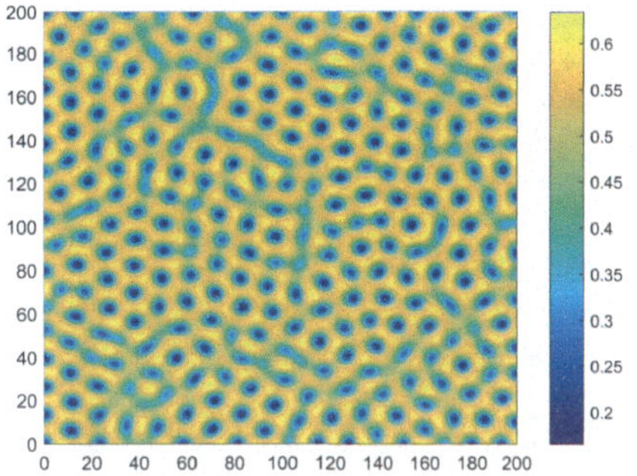

(d) 2000000 iterations

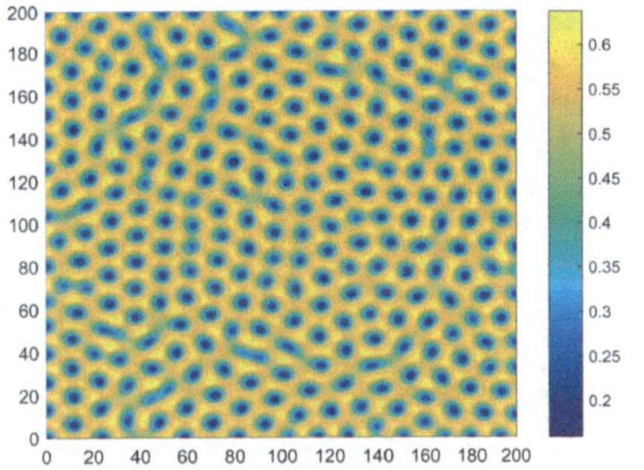

(e) 2400000 iterations

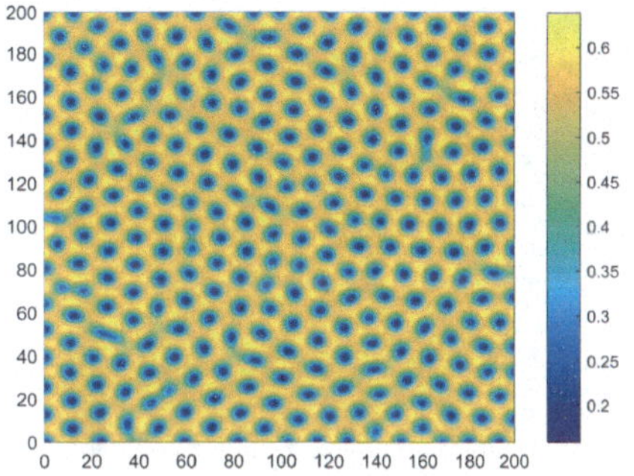

(f) 3000000 iterations

Fig. 10.6 Pattern structures of phytoplankton, where $d_1 = 0.1$, $d_2 = 29.7$, $d = 12.6$, $\alpha = 0.5$, $\delta = 4.2$, $\xi = 0.7$, $m = 2.6$, $h = 9.3$, $n = 0.8$, and $\gamma = 1.942$

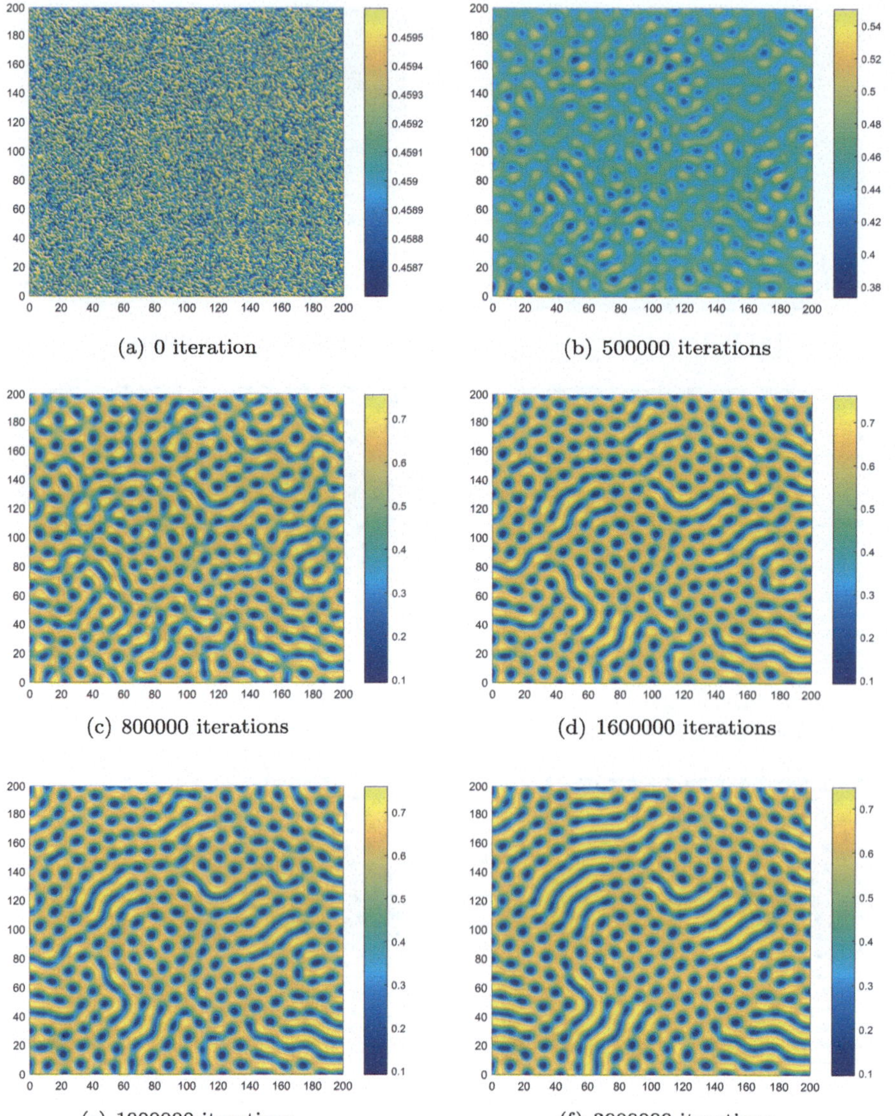

(a) 0 iteration

(b) 500000 iterations

(c) 800000 iterations

(d) 1600000 iterations

(e) 1900000 iterations

(f) 3000000 iterations

Fig. 10.7 Pattern structures of phytoplankton, where $d_1 = 0.1$, $d_2 = 29.7$, $d = 12.6$, $\alpha = 0.5$, $\delta = 4.2$, $\xi = 0.7$, $m = 2.6$, $h = 9.3$, $n = 0.8$, and $\gamma = 2.003$

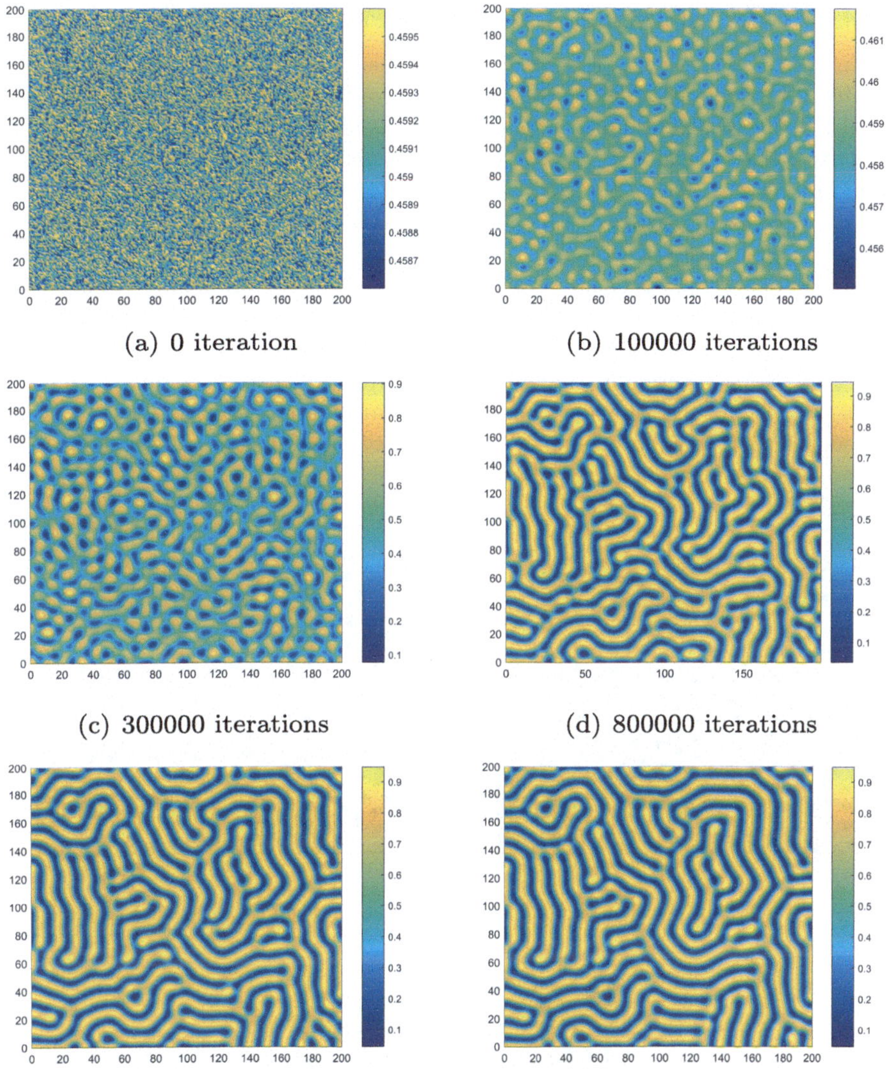

(a) 0 iteration (b) 100000 iterations

(c) 300000 iterations (d) 800000 iterations

(e) 1600000 iterations (f) 3000000 iterations

Fig. 10.8 Pattern structures of phytoplankton, where $d_1 = 0.1$, $d_2 = 29.7$, $d = 12.6$, $\alpha = 0.5$, $\delta = 4.2$, $\xi = 0.7$, $m = 2.6$, $h = 9.3$, $n = 0.8$, and $\gamma = 2.13$

10.4.2 Effect of Variation in ξ on the Pattern Formations

In this following, according to the variation of ξ as an important reflecting term for the biomass of additional food for zooplankton, we observe different spatial patterns of phytoplankton by varying ξ. Here, we assume the parameters of the model (10.3) as $d_1 = 0.1$, $d_2 = 29.7$, $d = 12.6$, $\alpha = 0.5$, $\delta = 4.2$, $\gamma = 2$, $m = 2.6$, $h = 9.3$, $n = 0.7$ and $\xi = 0.7$, 0.712, 0.743, respectively. Figure 10.9 illustrates the variations of the pattern sequence: "holes $\rightarrow$ stripe-hole mixtures $\rightarrow$ stripes". For little additional food, hole patterns generated from the distribution of phytoplankton means the phytoplankton dominance of the spatial domain and also represents that the population is isolated with low density and surrounded by high quantity [196]. However, we can investigate a transition from the hole patterns to the mixed hole and stripe patterns by increasing the quantity of additional food. Furthermore, increasing the additional food supply generates the disappearance of hole patterns and there exist only stripe patterns present in the domain.

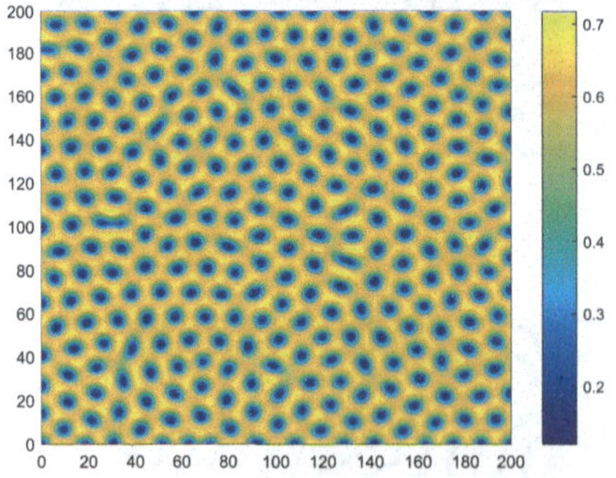

(a) $\xi = 0.7$

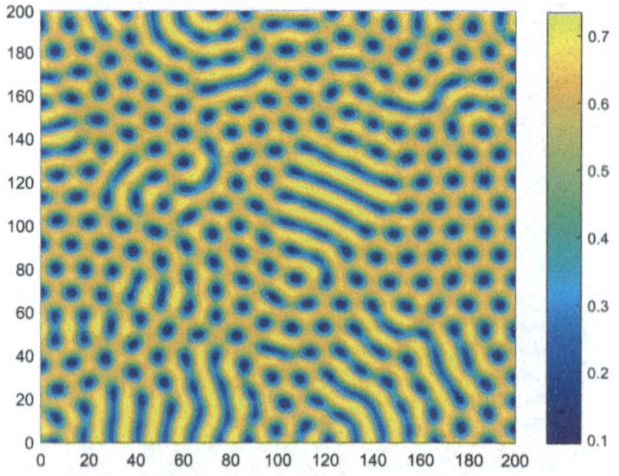

(b) $\xi = 0.712$

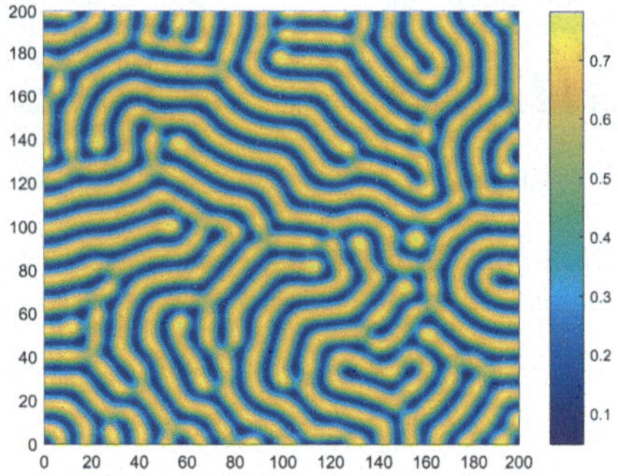

(c) $\xi = 0.743$

Fig. 10.9 Turing patterns of phytoplankton developed of model (10.3), where $d_1 = 0.1$, $d_2 = 29.7$, $d = 12.6$, $\alpha = 0.5$, $\delta = 4.2$, $\gamma = 2$, $m = 2.6$, $h = 9.3$, and $n = 0.7$

10.4.3 Effect of Variation in n on the Pattern Formations

In this following, the investigation of different patterns will be discussed by the variation of n reflecting the toxin liberation of phytoplankton. Here, we assume the parameters of the model (10.3) as $d_1 = 0.1,\ d_2 = 29.7,\ d = 12.6,\ \alpha = 0.5,\ \delta = 4.2,\ \gamma = 2,\ m = 2.6,\ h = 9.3,\ \xi = 0.73$, and $n = 0.1,\ 0.9,\ 1.3$, respectively. Figure 10.10 illustrates that a transition from stripe patterns to a mixture of hole and stripe patterns is exhibited for the spatial patterns. It is found that when the density distribution of the phytoplankton has peaks and troughs, small values of n can generate stripe patterns. And for intermediate values of n, it is clear that the formation of stripes still is prevalent. As the toxin liberation remains high, there emerges a mixture of stripe and hole patterns displaying that the majority of spatial region is dominated gradually by a high-density phytoplankton.

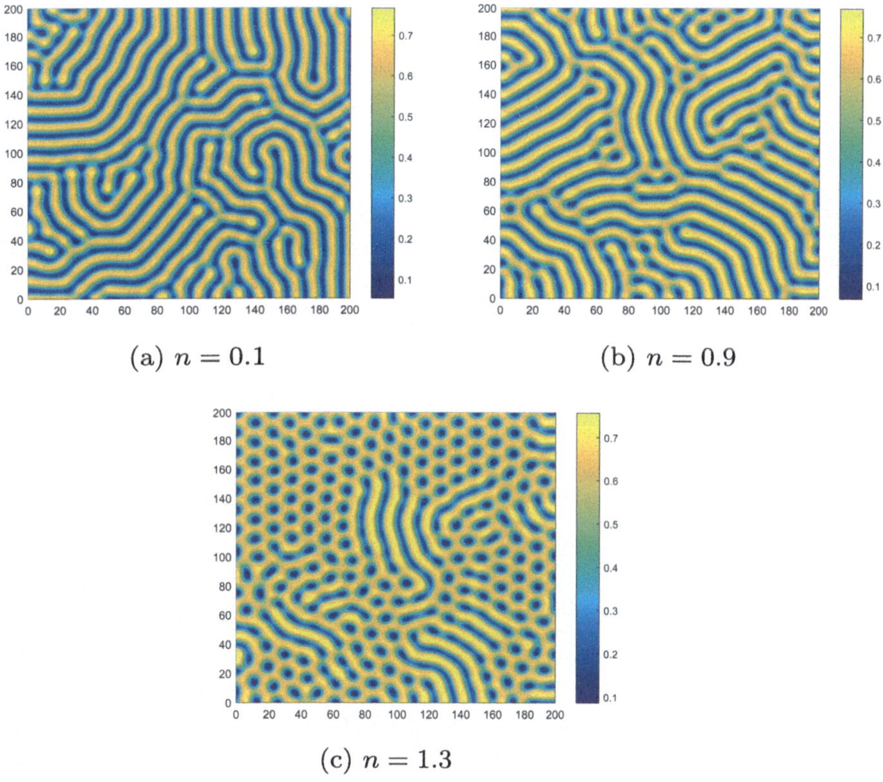

Fig. 10.10 Turing patterns of phytoplankton of model (10.3), where $d_1 = 0.1,\ \ d_2 = 29.7,\ d = 12.6,\ \alpha = 0.5,\ \delta = 4.2,\ \gamma = 2,\ m = 2.6,\ h = 9.3$ and $\xi = 0.73$

Chapter 11
The Effect of Advection

A spatial plankton community which comprises phytoplankton (prey) and zooplankton (predators) with advection has been studied. It is well known that when the advection velocities of all plankton remain the same, the plankton community is relatively stationary. And the speed difference will make the system unstable because that the convection velocities become different. As the original system without advection becomes Turing stable, there should exist a wave bifurcation due to the advection speed difference, driving the system with advection wave instability. As the original system without convection remains Turing unstable, there will always exist wave instability for the system with advection. Numerical simulations not only validated our analysis of stability but also showed more phenomena. If the original system without advection remains Turing unstable, and when we take the coefficient of one advection term as zero and the other as nonzero, there will emerge traveling patterns (patterns on the traveling waves) produced by advection speed difference. When the speed difference becomes small enough, the patterns become stationary. And as the speed difference increases, there appear the traveling patterns. Thus, a traveling bifurcation should be generated which separates stationary patterns and the traveling ones. The results about analysis and simulations enrich the dynamics study of the plankton models and have contribution for a better understanding of the planktonic ecosystem in the real environment.

Even planktons are tiny in size, they play vital role in the development of the marine biosphere [241]. For example, we can see them floating in surface water and drifting with ocean currents [236, 256]. Importantly, phytoplankton as a type of plankton can produce carbohydrates and the primary energy to support the aquatic food webs [266, 268]. Subsequently, when there exist enough sunlight and nutrients, they can grow rapidly and even cause blooms [233, 263]. Oppositely, zooplankton, another type of plankton, can feed on phytoplankton and then suppress the increase of phytoplankton [245, 249]. However, fishes from higher layers of food webs catch them as food at the same time [235, 265]. Then the above phenomena mean that zooplankton can be a bridge connecting "plants" and "animals" in water

S. T. Liu and L. Zhang, *Pattern Dynamics of Marine Plankton Behavior*,
https://doi.org/10.1007/978-981-97-5369-7_11

[264]. So that, the interaction between zooplankton and phytoplankton has vital effect on food webs in the ocean [243, 244].

For this interaction, there are many interesting and important studies such as plankton communities as one of the focuses from different perspectives, especially the nature of phytoplankton. Nowadays, the distribution and evolution of phytoplankton can be monitored and investigated through artificial satellites and buoys [231, 255]. Besides, according to rich data sets, deep learning technology has been used to predict changes in plankton communities [238, 269]. However, there still exists black boxes and few approaches have considered the interaction between phytoplankton and other factors. For example, the concrete influences of factors such as sunlight and nutrients on the growth of phytoplankton have been obtained by cultivating in the laboratories [239, 240]. Moreover, appropriate predation functions—the functional responses—between zooplankton and phytoplankton have been verified and selected [257]. In addition to experiments, functional responses through data or inferences have been also obtained [232, 242]. Mathematical models based on the predation functions have helped scholars to describe the relationship between phytoplankton and zooplankton [196]. So the dynamic behaviors of plankton ecosystems can be explained or identified the key factors to affect plankton communities by these models. Moreover, some biomathematicians take phytoplankton/zooplankton as a whole and utilize ordinary differential equations to model its behavior [267]. For example, the conditions for coexistence, upper bounds, extinction, and other properties have been got [271, 272]. And the models have been expanded by adding nutrients, fishes, or other factors to obtain more fruitful results [253, 254, 261]. However, the influence of the space in three-dimensional space is ignored in these models so that there exist further spatiotemporal models [248, 259]. Mostly, reaction–diffusion equations are the chosen models to investigate the interactions within plankton communities. The reason is that they can generate features such as the self-organization case and bifurcations for the growth of plankton and also help researchers study the effects of the sink, diffusion, or sunlight. For convenience, there is only one control variable for experimentation and other factors like the distribution of nutrients, the intensity of sunlight, and other factors taken as uniform and consistent [270].

In this following, we will consider the effect of advection on the self-organization of plankton. Importantly, planktons drift with ocean currents so that the horizontal advection velocity of each plankton may vary with their size, making a difference in the stability of the plankton community [247, 252, 260]. After comparison from many different aspects, the Arditi–Ginzburg type II functional response is chosen as the predation function [257] and allelopathic effect and the self-defense of phytoplankton are considered. In addition, because that the phytoplanktons emerge toxins [237] and store them in vivo, poisoning the zooplankton which feed on them, the poison function can share the same form with the predation function. When there exists no advection and the system remains Hopf stable, plankton community can be driven to form patterns, a kind of self-organization graphs by natural self-diffusion [246, 250]. If ocean current carries algal ecosystem, the advection coefficient can also represent the speed of advection; if we take the coefficients of two advection terms as equal meaning that the advection speeds in both plankton are the same, the plankton in the current remains relatively stationary. If the advection velocities of phytoplankton and zooplankton become different, there emerge intricate patterns [234, 262]. For convenience, because that time-invariant coefficients are assigned as the advection terms which will generate different patterns on the move (patterns on the traveling wave), we name this phenomenon as traveling patterns.

This chapter is organized as follows. At first, the reaction–diffusion equations are introduced with advection to model the plankton community composed of phytoplankton and zooplankton in Sect. 11.1. Then, in Sects. 11.2 and 11.3, the stability of the system is analyzed at the same and different convection velocities and the wave bifurcation (the smallest advection velocity difference that leads to wave instability) is also obtained. Next, the wave bifurcation is verified through numeral simulation in Sect. 11.4. And the new effect of advection on the traveling patterns is found and more simulations are illustrated in the same section. Finally, we give the conclusion.

11.1 Spatiotemporal Plankton Model with Advection

In this following, a general spatiotemporal plankton model by introducing advection is considered as given below:

$$\begin{cases} \frac{\partial P}{\partial t} &= rP\left(1-\frac{P}{K}\right)-\frac{aPZ}{Z+ahP}-v_P\cdot\nabla P+D_1\nabla^2 P, \\ \frac{\partial Z}{\partial t} &= e\frac{aPZ}{Z+ahP}-\mu Z-\eta\frac{aPZ}{Z+ahP}-v_Z\cdot\nabla Z+D_2\nabla^2 Z, \end{cases} \tag{11.1}$$

where P and Z mean the phytoplankton and zooplankton, respectively. And the detailed interpretations about the equations are listed as follows:

(1) Assume $(P, Z) \in [-L, L] \times [-L, L]$, the initial conditions remain positive, and the boundary conditions remain periodic.
(2) r means the growth rate of phytoplankton with logistic growth, and K stands for the natural carrying capacity.
(3) Based on the fact that zooplankton preys phytoplankton, many aspects of comparison [257] and considering the saturation of phytoplankton and zooplankton, functional response of Arditi–Ginzburg type II is chosen as the predation function, defining it as $f(P, Z) = \frac{aP}{Z+ahP}$.
(4) a means the capture rate of zooplankton, e stands for the absorption rate, and h represents the capture time.
(5) μ means the natural death rate of zooplankton.
(6) Because that phytoplankton produces toxins and stores them to defend against predators, we define the poison mortality rate as η.
(7) As phytoplankton and zooplankton diffuse naturally, we define the respective diffusion coefficients as D_1 and D_2.

(8) Assume that there exists a horizontal ocean current carrying phytoplankton and zooplankton with the speed of v_P and v_Z, in which the coefficients of the advection terms also mean their convective velocities and the operator "·" stand for inner product.

For simplicity, the model will be transferred as follows:

$$\hat{t} = Kt, \qquad \hat{P} = \frac{P}{K}, \qquad \hat{Z} = \frac{Z}{ahK}, \qquad \hat{r} = \frac{r}{K},$$
$$\hat{a} = \frac{a}{K}, \qquad \hat{e} = \frac{e-\eta}{ah}, \qquad \hat{\mu} = \frac{\mu}{K}, \qquad \hat{v}_P = \frac{v_P}{K},$$
$$\hat{v}_Z = \frac{v_Z}{K}, \qquad \hat{D}_1 = \frac{D_1}{K}, \qquad \hat{D}_2 = \frac{D_2}{K}.$$

For convenience, we delete the hat "$\hat{\cdot}$" for the rearranged model in the rest of the part.

$$\begin{cases} \frac{\partial P}{\partial t} &= rP(1-P) - \frac{aPZ}{Z+P} - v_P \cdot \nabla P + D_1 \nabla^2 P, \\ \frac{\partial Z}{\partial t} &= e\frac{aPZ}{Z+P} - \mu Z - v_Z \cdot \nabla Z + D_2 \nabla^2 Z. \end{cases} \tag{11.2}$$

Next, the stability of the planktonic ecosystem will be analyzed under the reformed model.

11.2 Stability Analysis

11.2.1 Equilibriums

In the following section, we will ignore the diffusion and convection terms of the following equations and then get the corresponding functions only considering the left ODE equations:

$$\begin{cases} \frac{\partial P}{\partial t} &= rP(1-P) - \frac{aPZ}{Z+P}, \\ \frac{\partial Z}{\partial t} &= e\frac{aPZ}{Z+P} - \mu Z. \end{cases} \tag{11.3}$$

When we define

$$f(P,Z) = r(1-P) - \frac{aZ}{Z+P}, \quad g(P,Z) = e\frac{aP}{Z+P} - \mu,$$

we get

$$\begin{cases} \frac{\partial P}{\partial t} &= Pf(P,Z), \\ \frac{\partial Z}{\partial t} &= Zg(P,Z). \end{cases}$$

Obviously, it is easy to prove that the system is nonnegative [246]. In addition, there exist most three equilibrium points for this system: two trivial steady states (0, 0) standing for the extinction of the ecosystem and $(1, 0)$ meaning the extinction of zooplankton, and one nontrivial steady state (P^*, Z^*) (coexistence). We can obtain (P^*, Z^*) by $f(P^*, Z^*) = 0$ and $g(P^*, Z^*) = 0$:

$$P^* = \frac{er - ea + \mu}{er}, \qquad Z^* = \frac{ea - \mu}{\mu} P^*.$$

In order to ensure the positive feature of the state, we can choose $ea > \mu$ and $er > (ea - \mu)$.

11.2.2 Hopf Stability

In order to ensure the Hopf stability of the system, the system (11.3) will be linearized at the coexistence point. We then obtain the Jacobian matrix of the system (11.3) at (P^*, Z^*) as follows:

$$J_0 = \begin{pmatrix} a_{11} & a_{12} \\ a_{21} & a_{22} \end{pmatrix} = \begin{pmatrix} Pf_P + f & Pf_Z + f \\ Zg_P + g & Zg_Z + g \end{pmatrix}\Bigg|_{(P^*,Z^*)} = \begin{pmatrix} P^* f_{P^*} & P^* f_{Z^*} \\ Z^* g_{P^*} & Z^* g_{Z^*} \end{pmatrix},$$

and

$$f_{P^*} = -r + \frac{aZ^*}{(P^* + Z^*)^2}, \qquad f_{Z^*} = -\frac{aP^*}{(P^* + Z^*)^2} < 0,$$

$$f_{P^*} = \frac{eaZ^*}{(P^* + Z^*)^2} < 0, \qquad g_{Z^*} = -\frac{eaP^*}{(p^* + Z^*)^2} < 0.$$

Then after linearizing the ODE system around the equilibrium point (P^*, Z^*), we have the first-order approximation:

$$\begin{cases} \frac{\partial P}{\partial t} &= a_{11} P + a_{12} Z, \\ \frac{\partial Z}{\partial t} &= a_{21} P + a_{22} Z. \end{cases} \tag{11.4}$$

We easily get the characteristic equation of the linear ODE system (11.4) as follows:

$$\lambda^2 - Tr_0 \lambda + \Delta_0 = 0, \tag{11.5}$$

where

$$Tr_0 = a_{11} + a_{22} = -rP^* + (1 - e)\frac{aP^* Z^*}{(P^* + Z^*)^2},$$

$$\Delta_0 = a_{11}a_{22} - a_{12}a_{21} = reP^* Z^* \frac{aP^* Z^*}{(P^* + Z^*)^2}.$$

When we ensure $\Delta_0 > 0$ and $Tr_0 < 0$, we get the Hopf stability of the ODE system around the coexistence point (P^*, Z^*). The stability of the ecosystem was studied at the extinction points (0, 0) and (1, 0), especially near the point (0, 0), where the system behavior is abundant [251]. Because the Hopf bifurcation is not our current topics, we will not give further details. When we assume that the plankton system remains Hopf stable at the coexistence point, the algae community can be driven to instability according to diffusion and convection. Thus, we introduce diffusion and advection terms into this model.

11.2.3 Stability Analysis with the Same Advection Speed

From a physical point of view, with the condition that convection terms are taken as the same coefficients, $v_P = v_Z = v$, meaning that plankton community is carried away by water flow at the same speed, we get the following model rewritten as

$$\begin{cases} \frac{\partial P}{\partial t} &= rP(1-P) - \frac{aPZ}{Z+P} - v \cdot \nabla P + D_1 \nabla^2 P, \\ \frac{\partial Z}{\partial t} &= e\frac{aPZ}{Z+P} - \mu Z - v \cdot \nabla Z + D_2 \nabla^2 Z. \end{cases} \tag{11.6}$$

Let $(P(x,t), Z(x,t))$ be the solution of model (11.6), and define $x = \hat{x} + vt$ so that $P(x,t) = \hat{P}(\hat{x},t) = \hat{P}(x - vt, t)$, and the same with $Z(x,t)$. Hence, we can transform the model as follows:

$$\frac{\partial P(x,t)}{\partial t} = \frac{\partial \hat{P}(\hat{x},t)}{\partial \hat{x}} \cdot \frac{\partial \hat{x}}{\partial t} + \frac{\partial \hat{P}(\hat{x},t)}{\partial t} = -v \cdot \nabla \hat{P} + \frac{\partial \hat{P}}{\partial t},$$

and

$$\nabla P = \nabla \hat{P} \frac{\partial \hat{x}}{\partial x} = \nabla \hat{P}.$$

After excluding the hat "$\hat{\cdot}$" in order to simplify the expression, we have the new equations as follows:

$$\begin{cases} \frac{\partial P}{\partial t} &= rP(1-P) - \frac{aPZ}{Z+P} + D_1 \nabla^2 P, \\ \frac{\partial Z}{\partial t} &= e\frac{aPZ}{Z+P} - \mu Z + D_2 \nabla^2 Z, \end{cases} \tag{11.7}$$

where the transformed equations are ordinary reaction–diffusion equations. Through diffusion, the Hopf stability of the system can be driven into instability. Next the system is linearized near the homogeneous stable equilibrium (P^*, Z^*) to obtain its bifurcation.

$$\begin{cases} \frac{\partial P}{\partial t} &= a_{11} P + a_{21} Z + D_1 \nabla^2 P, \\ \frac{\partial Z}{\partial t} &= a_{12} P + a_{22} Z + D_2 \nabla^2 Z. \end{cases} \tag{11.8}$$

As we take the general solution of Eq. (11.8) as the form of exponential function:

$$\begin{pmatrix} P \\ Z \end{pmatrix} = \begin{pmatrix} c_k^1 \\ c_k^2 \end{pmatrix} \exp(\lambda t + ik \cdot k), \tag{11.9}$$

we can have the characteristic equation of the linearized Eq. (11.8):

$$\lambda \begin{pmatrix} c_k^1 \\ c_k^2 \end{pmatrix} = \begin{pmatrix} a_{11} - D_1 k^2 & a_{12} \\ a_{21} & a_{22} - D_2 k^2 \end{pmatrix} \begin{pmatrix} c_k^1 \\ c_k^2 \end{pmatrix}, \tag{11.10}$$

where $k = \|k\|$. Then we can obtain the following relationship of dispersion from Eq. (11.10):

$$\lambda^2 - Tr_k \lambda + \Delta_k = 0, \tag{11.11}$$

where

$$\begin{aligned} Tr_k &= a_{11} + a_{22} - (D_1 + D_2)k^2 = Tr_0 - (D_1 + D_2)k^2, \\ \Delta_k &= a_{11}a_{22} - a_{12}a_{21} - (a_{11}D_2 + a_{22}D_2)k^2 + D_1 D_2 (k^2)^2 \\ &= \Delta_0 - (a_{11}D_2 + a_{22}D_1)k^2 + D_1 D_2 (k^2)^2. \end{aligned}$$

Eigenvalues can be obtained:

$$\lambda_k = \frac{Tr_k \pm \sqrt{Tr_k^2 - 4\Delta_k}}{2}, \tag{11.13}$$

where reaction–diffusion equations remain stable when $Re(\lambda_k) < 0$ for any $k > 0$. As we know $Tr_k < Tr_0 < 0$, a Turing bifurcation in the dynamic system indicates that there exists at least one k and $\Delta_k \leq 0$. Then, we can transform the Turing instability condition as follows:

$$k_T^2 = \arg\min_{k^2} \Delta_k = \frac{a_{11}D_2 + a_{22}D_1}{2D_1 D_2} > 0, \tag{11.14}$$

$$\Delta_{k_T} = \Delta_0 - \frac{(a_{11}D_1 + a_{22}D_1)^2}{4D_1 D_2} \leq 0. \tag{11.15}$$

Based on the above conclusion (11.14), we can easily obtain $a_{11} > 0$ and

$$k_T^2 = \frac{a_{11}D_2 + a_{22}D_1}{2D_1 D_2} < \frac{a_{11}}{2D_1}.$$

Thus the unique condition diffusion makes the system generate Hopf stable and loses stability named as Turing unstable, where

$$Tr_0 < 0, \quad \Delta_0 > 0, \quad Tr_{k_T} < 0, \quad \Delta_{k_T} < 0. \tag{11.16}$$

Besides, if the model remains Hopf unstable and Turing unstable, it means that the system gets a new state named as Turing–Hopf unstable, where

$$Tr_0 > 0, \quad \Delta_0 > 0, \quad Tr_{k_T} < 0, \quad \Delta_{k_T} < 0. \tag{11.17}$$

Hence besides the Turing instability and Hopf–Turing instability, we can also obtain the "Turing pattern domain" for self-organization. In addition, the weakly nonlinear analysis is too cumbersome so that the following part ignores it. In short, ordinary reaction–diffusion equations can be transformed by the reaction–diffusion equations with the convection terms of the same velocity. So, we obtain the solution of the dynamic system (11.6) as

$$(P(x,t), Z(x,t)) = \left(\hat{P}(x - vt, t), Z(x - vt, t)\right),$$

where the results stand for traveling patterns with velocity v, in which the traveling waves mean one form of moving patterns, and the patterns replicate another form of standing waves. The results indicate that if a planktonic ecosystem can generate these self-organization patterns in standing water and with both planktons drifting at the same speed, it can also generate that pattern in flowing water. Then, we can name the patterns on the traveling waves as traveling patterns.

11.3 Advection Instability and Waving Bifurcation

From the results in Sect. 11.2.3, it is known that the stability of the system will not be changed by the advection terms of the same speed but the position of the plankton community will be shifted by it. Simultaneously, the plankton system will be caused by advection terms of different speeds falling into an unstable state. Assuming $v_P \neq v_Z$, dividing v_Z into v_P and $v_Z - v_P$, and according to the fact that the stability of this system will not be changed by the same advection velocity, we can convert the model of the planktonic ecosystem into the following way:

$$\begin{cases} \frac{\partial P}{\partial t} &= rP(1-P) - \frac{aPZ}{Z+P} + D_1 \nabla^2 P, \\ \frac{\partial Z}{\partial t} &= e\frac{aPZ}{Z+P} - \mu Z - v \cdot \nabla Z + D_2 \nabla^2 Z, \end{cases} \tag{11.18}$$

where $v = v_Z - v_P$.

Next we get the linearized equations around the coexistence equilibrium as follows:

$$\begin{cases} \frac{\partial P}{\partial t} &= a_{11}P + a_{21}Z + D_1 \nabla^2 P, \\ \frac{\partial Z}{\partial t} &= a_{12}P + a_{22}Z - v \cdot \nabla Z + D_2 \nabla^2 Z. \end{cases} \tag{11.19}$$

Then, dividing v into two parts: v_k, the part parallel to k, and $v_{\perp}$, the part vertical to k, we rewrite the equations as follows:

$$\begin{cases} \frac{\partial P}{\partial t} &= a_{11}P + a_{21}Z + D_1\nabla^2 P, \\ \frac{\partial Z}{\partial t} &= a_{12}P + a_{22}Z - v_k \cdot \nabla Z + D_2\nabla^2 Z. \end{cases} \tag{11.20}$$

Or we can rewrite it as the form of the solution of Eq. 11.20:

$$\begin{pmatrix} P \\ Z \end{pmatrix} = \begin{pmatrix} c_k^1 \\ c_k^2 \end{pmatrix} \exp\left(\lambda t + ik \cdot x\right), \tag{11.21}$$

and therefore obtain the characteristic equation as

$$\lambda \begin{pmatrix} c_k^1 \\ c_k^2 \end{pmatrix} = \begin{pmatrix} a_{11} - D_1k^2 & a_{12} \\ a_{21} & a_{22} - i v_k \cdot k - D_2k^2 \end{pmatrix} \begin{pmatrix} c_k^1 \\ c_k^2 \end{pmatrix}. \tag{11.22}$$

Easily the stability of the system can only be affected by the tuning parameter, v_k so that we can obtain the dispersion equation by defining $v_k = \|v_k\|$ and $v_k \cdot k = v_k k$:

$$\lambda^2 - Tr_{v_k}\lambda + \Delta_{v_k} = 0, \tag{11.23}$$

where

$$\begin{aligned} Tr_{v_k} &= Tr_k - i v_k k = Tr_0 - (D_1 + D_2)k^2 - i v_k k, \\ \Delta_{v_k} &= \Delta_k - i v_k k(a_{11} - D_1k^2) \\ &= \Delta_0 - (a_{11}D_2 + a_{22}D_1)k^2 + D_1D_2(k^2)^2 - i v_k k(a_{11} + D_1k^2). \end{aligned}$$

The root $s_0 < 0$ because that the eigenvalues are $\lambda_k = s_k \pm i\omega_k$ and the system remains Hopf stable. And for some k, $s_k \geq 0$ meaning that the system runs unstably. Therefore when the root locus crosses the imaginary axis, we get that there exists a pair of pure imaginary roots $\lambda_k = 0 \pm i\omega_k$ for a k. Hence we obtain the corresponding characteristic equation as follows:

$$-\omega^2 - \omega v_k k + \Delta_k - i\omega Tr_k - i v_k k(a_{11} - D_1k^2) = 0, \tag{11.25}$$

where the real and imaginary parts of Eq. 11.25 are

$$\begin{aligned} -\omega^2 - \omega v_k k + \Delta_k = 0, \\ \omega Tr_k + v_k k(a_{11} - D_1k^2) = 0. \end{aligned}$$

And we can also obtain the relationship equation:

$$v_k^2(k^2) = -\frac{Tr_k^2 \Delta_k}{k^2(a_{11} - D_1 k^2)(a_{22} - D_2 k^2)} \geq 0. \tag{11.27}$$

Obviously, all v_k obeying the following conditions will drive the system into an unstable state:

(1) If $\Delta_k > 0$, the system without advection becomes stable, where it is easy that the function v_k^2 has a minimum value for $k^2 < \frac{a_{11}}{D_1}$. Besides, when $k^2 > \frac{a_{11}}{D_1}$, $v_k^2 < 0$, it has a contradiction result for the positive definiteness. Thus, the system becomes wave stable only when $\|v_k\|^2 \in [0, v_w^2)$ for all $k > 0$, where v_w means the wave bifurcation.
(2) If $\Delta_k \leq 0$, the original system without convection is Turing unstable and the system with advection terms becomes also wave unstable where for $k_T^2 < \frac{a_{11}}{D_1}$, the parameter v_k^2 belongs to $(0, +\infty)$ for any v_k.

Therefore, the wave bifurcation under the Turing stability, $\Delta_k > 0$, is the only case we considered. The derivation of v_k^2 to k^2 is given as

$$\frac{dv_k^2}{d(k^2)} = \frac{Tr_k S(k^2)}{k^4(a_{11} - D_1 k^2)^2(a_{22} - D_2 k^2)^2} = 0, \tag{11.28}$$

where

$$\begin{aligned} S(k^2) =& (2Tr_0 - Tr_k)\Delta_k(a_{11} - D_1 k^2)(a_{22} - D_2 k^2) \\ &+ a_{12}a_{21}Tr_k(D_2 a_{11} + D_1 a_{22} - 2D_1 D_2 k^2)k^2. \end{aligned}$$

As the function obtains at least one zero point, we get

$$k^2\big|_{S=0} = \left\{ k^2 \middle| \, S(k^2) = 0, 0 < k^2 < \frac{a_{11}}{D_1} \right\}. \tag{11.29}$$

Simultaneously, the minimum velocity difference of advection, v_w, becomes the wave bifurcation:

$$v_w^2 = v_k^2(k_w^2),$$

where

$$k_w^2 = \arg \min_{k^2|_{S=0}} v_k^2(k^2).$$

In short, when we get the corresponding reaction–diffusion equations as Turing stable, we can obtain the Wave stability conditions of the convective system (11.2) as $v_P // v_Z$, $\|v_P - v_Z\|^2 < v_w^2$.

11.4 Numerical Simulation

Through the following numerical simulations, we not only verify above analysis but also obtain more effects of advection on the plankton communities.
The simulated advection–reaction–diffusion equations are listed as follows:

$$\begin{cases} \frac{\partial P}{\partial t} &= rP(1-P) - \frac{aPZ}{Z+P} - v_P \cdot \nabla P + D_1 \nabla^2 P, \\ \frac{\partial Z}{\partial t} &= e\frac{aPZ}{Z+P} - \mu Z - v_Z \cdot \nabla Z + D_2 \nabla^2 Z. \end{cases}$$

In the simulations, we take a grid of 100×100, the spatial grid size as $\Delta_x = \Delta_y = 1$, and the time interval as $\Delta_t = 0.01$. Some parameters are assumed as: $r = 5.90$, $e = 0.64$, $a = 15.60$, $\eta = 7.00$, $D_1 = 1.00$, and the boundaries remain periodic. We also assume the initial conditions as $P = (1 + \xi_{i,j})P^*$ and $Z = (1 + \zeta_{i,j})Z^*$, where $\xi_{i,j}$ and $\zeta_{i,j}$ remain noise with a uniform distribution. Hopf stable is shown clearly in Fig. 11.1.

As calculated shown that the Turing bifurcation is $D_{2T} = 4.3238$, and the critical parameters for separating different types of patterns are taken as $D_2 =$ 3.6207, 4.3238, 5.8509, 16.2779, $D_2 = 3.00, 4.3238, 5.00, 6.50$ and 16.50, we can plot the corresponding eigenvalue curves illustrated in Fig. 11.2.

From the above analysis, it is easy to know that when the two plankton species have the same velocity for advection terms, the systems have the same stability with and without convection. When the reaction–diffusion equations remain Turing unstable, equations with advection terms of different velocities must also become wave unstable. When the model remains Turing stable, there exists a wave bifurcation $v_Z - v_P$, driving the advection–reaction–diffusion equations to be wave instable. For convenience, we set the advection velocity v as parallel to the x-axis, that is, $v = v_x$. Let $V = \|v\|$ and $v_P \cdot \nabla P = V_P \nabla_x P$, $v_Z \cdot \nabla Z = V_Z \nabla_x Z$, then the relative advection velocity of phytoplankton and zooplankton is obtained as $V = V_Z - V_P$. Figure 11.3 illustrates the stability of the plankton model in the $D_2 - V_{min}^2$ space.

Next we list a series of numerical simulations to illustrate the effect of advection for various parameters in the (D_2, V) space.

Case One: $r = 5.90$, $e = 0.64$, $a = 15.60$, $\eta = 7.00$, $D_1 = 1.00$, $D_2 = 3$.

Figure 11.4 shows that when there exists no advection, the ecosystem will be a unified steady state. The corresponding wave bifurcation means $V_{min} = 0.7541$ so that the system remains located in the Turing wave-stable domain, and both planktons enter a uniform steady state as $V = 0.2$ $((V_P, V_Z) = (0, 0.2))$ (Fig. 11.5). As shown in Fig. 11.6, the system loses stability as we adjust $V_Z = 1.0$ (beyond the wave bifurcation).

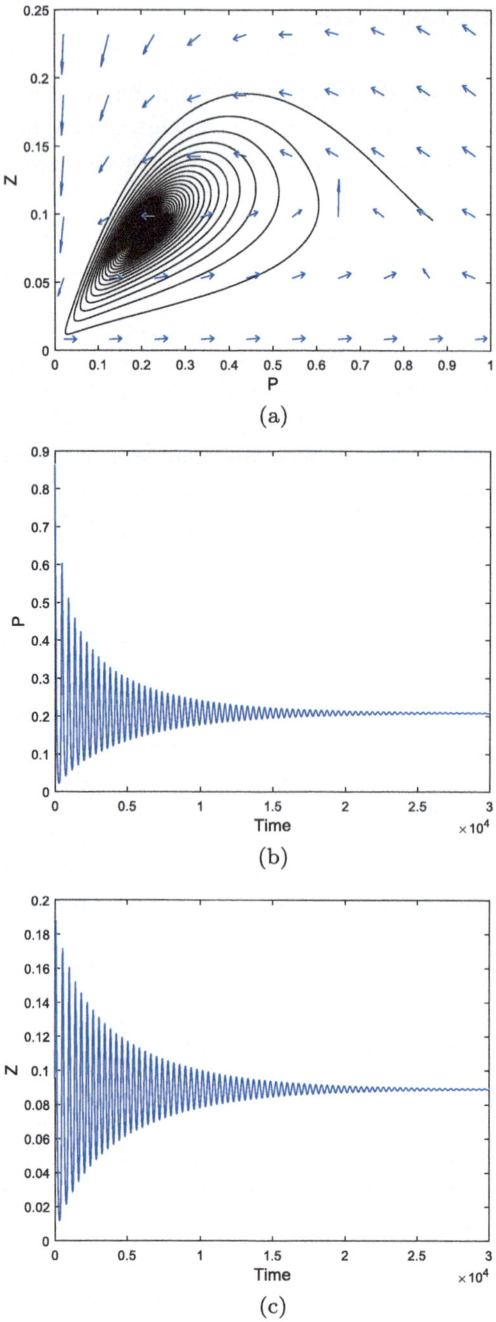

Fig. 11.1 Evolution of plankton community at different instants with parameters $r = 5.90$, $e = 0.64$, $a = 15.60$, $\eta = 7.00$

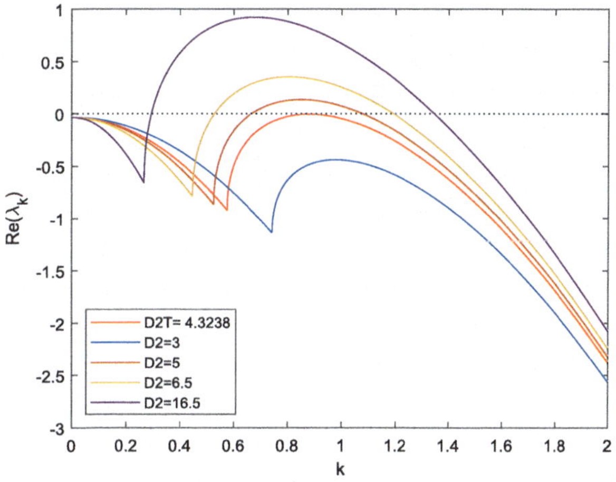

Fig. 11.2 The eigenvalues of the planktonic ecosystem model with different diffusions, where $r = 5.90$, $e = 0.64$, $a = 15.60$, $\eta = 7.00$, $D_1 = 1.00$, $D_2 = 16.50$, $D_{2T} = 4.3238$, $D_2 = 3.00, 5.00, 6.50, 16.50$

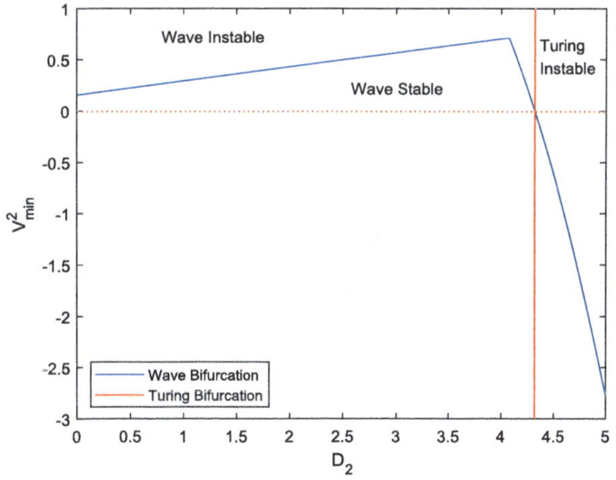

Fig. 11.3 Turing bifurcation and wave bifurcation in the (D_2, V_{min}^2) space for $r = 5.90$, $e = 0.64$, $a = 15.60$, $\eta = 7.00$, and $D_1 = 1.00$

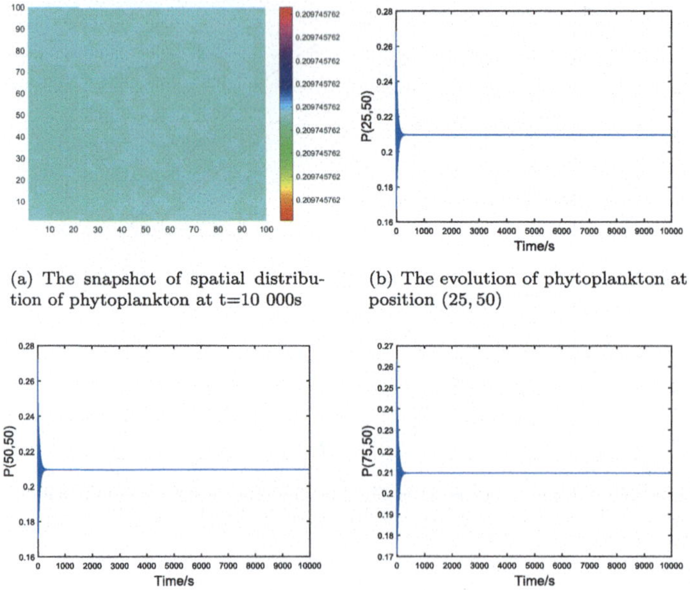

(a) The snapshot of spatial distribution of phytoplankton at t=10 000s
(b) The evolution of phytoplankton at position (25, 50)
(c) The evolution of phytoplankton at position (50, 50)
(d) The evolution of phytoplankton at position (75, 50)

Fig. 11.4 Evolution of phytoplankton at different instants and different grids, where $r = 5.90$, $e = 0.64$, $a = 15.60$, $\eta = 7.00$, $D_1 = 1.00$, $D_2 = 3.00$, $(V_Z, V_P) = (0.00, 0.00)$ $(V = 0.00)$

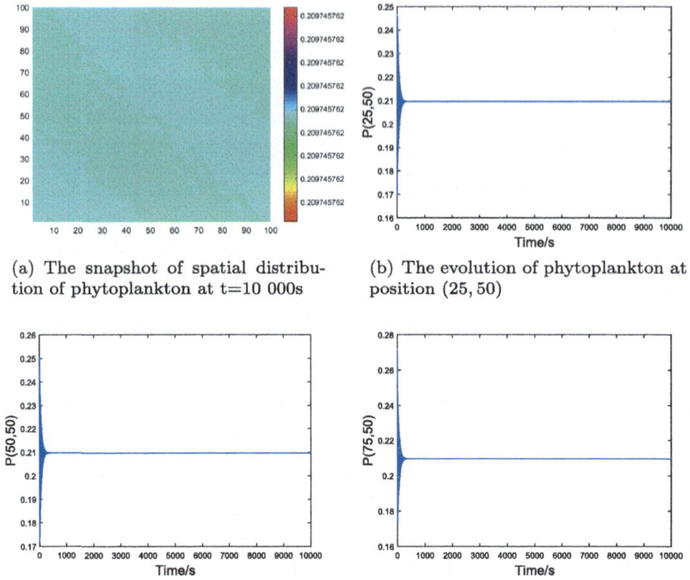

(a) The snapshot of spatial distribution of phytoplankton at t=10 000s
(b) The evolution of phytoplankton at position (25, 50)
(c) The evolution of phytoplankton at position (50, 50)
(d) The evolution of phytoplankton at position (75, 50)

Fig. 11.5 Evolution of phytoplankton at different instants and different grids, where $r = 5.90$, $e = 0.64$, $a = 15.60$, $\eta = 7.00$, $D_1 = 1.00$, $D_2 = 3.00$, $(V_P, V_Z) = (0.00, 0.20)$ $(V = 0.20)$

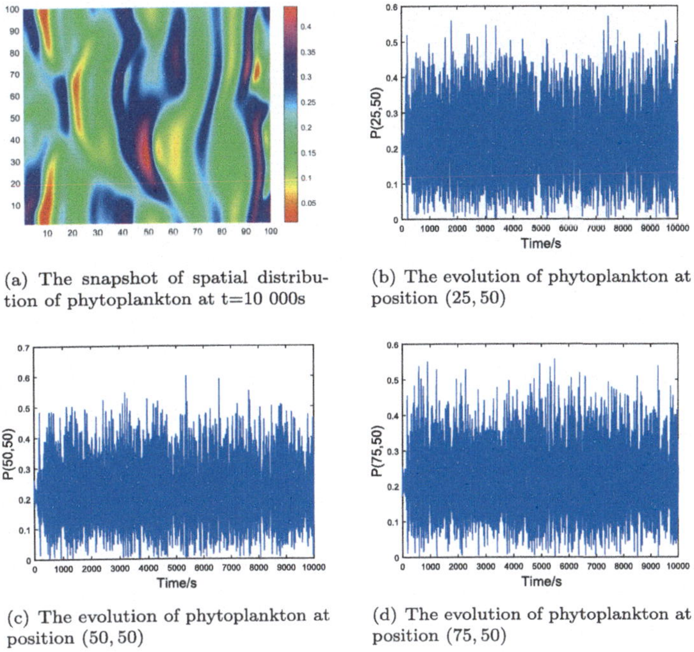

(a) The snapshot of spatial distribution of phytoplankton at t=10 000s

(b) The evolution of phytoplankton at position (25, 50)

(c) The evolution of phytoplankton at position (50, 50)

(d) The evolution of phytoplankton at position (75, 50)

Fig. 11.6 Evolution of phytoplankton at different instants and different grids, where $r = 5.90$, $e = 0.64$, $a = 15.60$, $\eta = 7.00$, $D_1 = 1.00$, $D_2 = 3.00$, $(V_P, V_Z) = (0.00, 1.00)$ $(V = 1.00)$

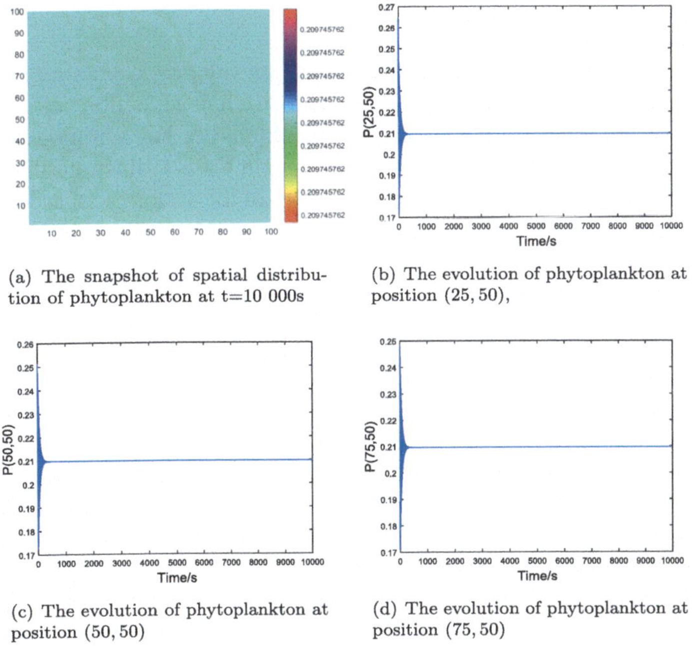

(a) The snapshot of spatial distribution of phytoplankton at t=10 000s

(b) The evolution of phytoplankton at position (25, 50),

(c) The evolution of phytoplankton at position (50, 50)

(d) The evolution of phytoplankton at position (75, 50)

Fig. 11.7 Evolution of phytoplankton at different instants and different grids, where $r = 5.90$, $e = 0.64$, $a = 15.60$, $\eta = 7.00$, $D_1 = 1.00$, $D_2 = 3.00$, $(V_P, V_Z) = (1.00, 1.00)$ $(V = 0.00)$

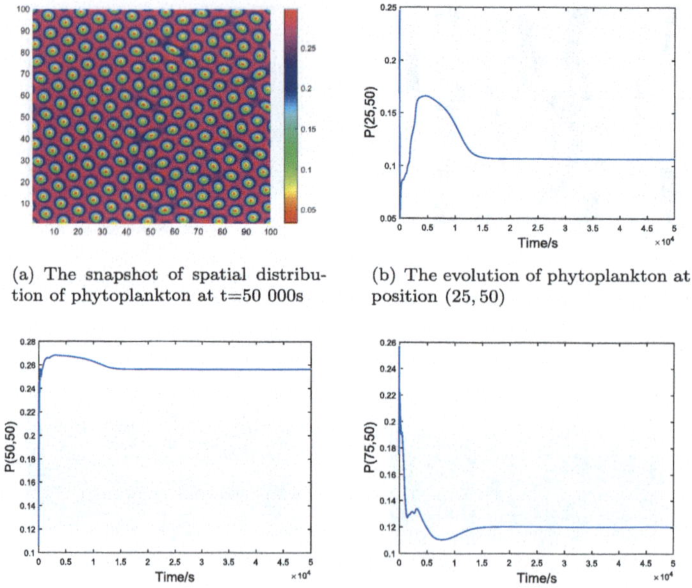

(a) The snapshot of spatial distribution of phytoplankton at t=50 000s

(b) The evolution of phytoplankton at position (25, 50)

(c) The evolution of phytoplankton at position (50, 50)

(d) The evolution of phytoplankton at position (75, 50)

Fig. 11.8 Evolution of phytoplankton at different instants and different grids, where $r = 5.90$, $e = 0.64$, $a = 15.60$, $\eta = 7.00$, $D_1 = 1.00$, $D_2 = 5.00$, $(V_P, V_Z) = (0.00, 0.00)$ $(V = 0.00)$

Because both planktons have the same advection velocity, we assign $(V_P, V_Z) = (1.0, 1.0)$ in order to compare with systems with different advection velocities. Then the system remains relatively static, and both planktons become a homogenous stable state implied in Fig. 11.7. The above simulations in Case 1 illustrate the effects of advection terms at the same speed on the stability of the system so that we will next study the effect of convection on the self-organization.

Case Two: $r = 5.90$, $e = 0.64$, $a = 15.60$, $\eta = 7.00$, $D_1 = 1.00$, $D_2 = 4.5 \in (4.3238, 5.8509)$.

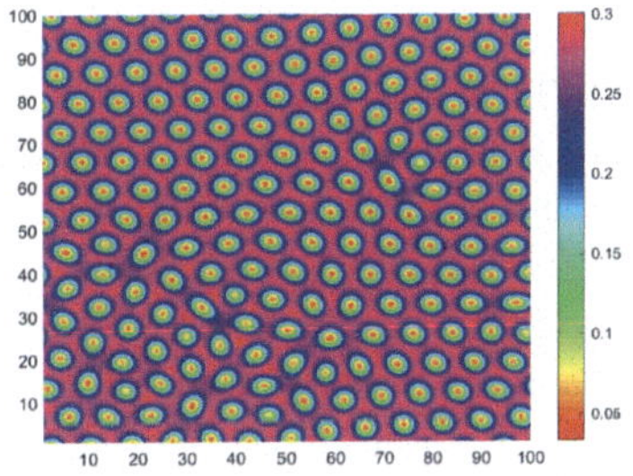

(a) the snapshot of spatial distribution of phytoplankton at t=200 000s

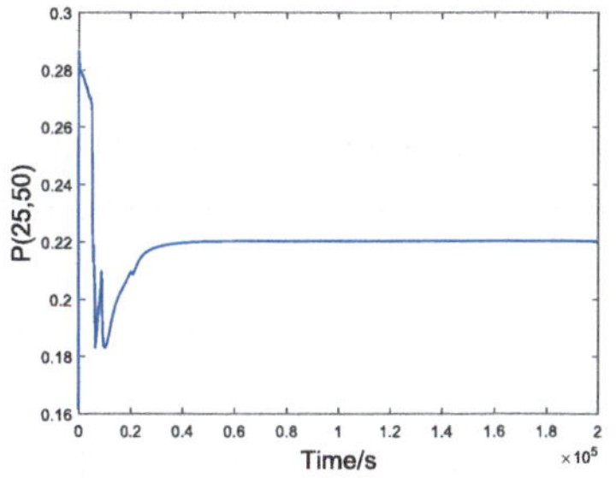

(b) The evolution of phytoplankton at position (25, 50)

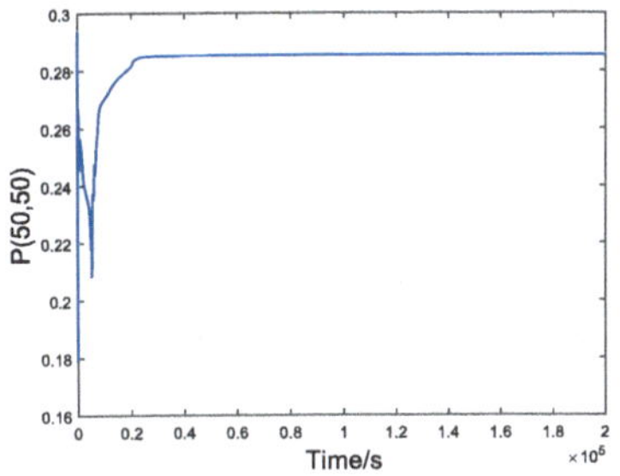

(c) The evolution of phytoplankton at position (50, 50)

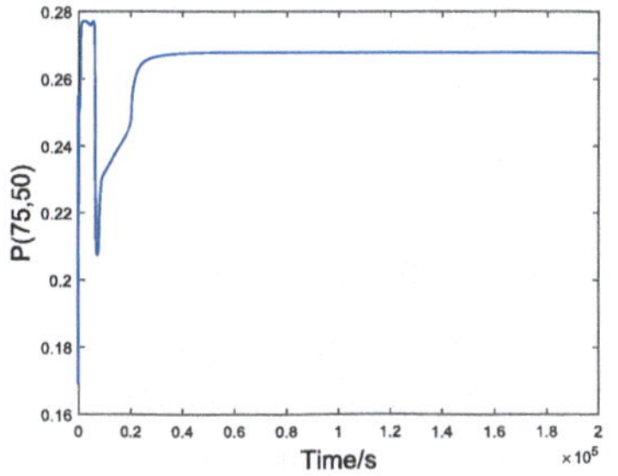

(d) The evolution of phytoplankton at position (75, 50)

Fig. 11.9 Evolution of phytoplankton at different instants and different grids, where $r = 5.90$, $e = 0.64$, $a = 15.60$, $\eta = 7.00$, $D_1 = 1.00$, $D_2 = 5.00$, $(V_P, V_Z) = (0.000, 0.0004)$ $(V = 0.0004)$

The formation of spot patterns without convection is shown in Fig. 11.8.

When we take $(V_Z, V_P) = (0.0004, 0.0000)$, meaning $V = 0.0004$, the system still remains the static spot patterns (Fig. 11.9).

Figure 11.10 shows the traveling spot patterns compared with this phenomenon, where $(V_P, V_Z) = (0.0004, 0.0004)$ $(V = 0.00)$.

Then, adjust the parameters to $(V_P, V_Z) = (0.00, 0.02)$ $(V = 0.02)$, the stationary spot patterns will be traveling ones shown in Fig. 11.11.

Figure 11.12 illustrates that the system generates traveling spot patterns at a higher speed than Fig. 11.11 with $(V_P, V_Z) = (0.02, 0.02)$ $(V = 0.00)$.

Case Three: $r = 5.90$, $e = 0.64$, $a = 15.60$, $\eta = 7.00$, $D_1 = 1.00$, $D_2 = 6.5$.

Figure 11.13 shows these parameters located in the "mixed pattern domain".

Similar to Case Two, the simulations are compared under two sets of parameters: $(V_P, V_Z) = (0.000, 0.002)$ $(V = 0.002)$ and $(V_P, V_Z) = (0.002, 0.002)$ $(V = 0.000)$ (Fig. 11.14).

Figure 11.14a–d shows stationary mixed patterns as the systems with different advection velocities $(V_P, V_Z) = (0.000, 0.002)$ $(V = 0.002)$. For the same advection velocity $(V_P, V_Z) = (0.002, 0.002)$ $(V = 0.000)$, Fig. 11.14e–h illustrates traveling mixed patterns consistent with the above theory. In addition, enlarging the advection speed to $(V_P, V_Z) = (0.00, 0.02)$ $(V = 0.02)$ and $(V_P, V_Z) = (0.02, 0.02)$ $(V = 0.00)$ we can find the traveling mixed patterns arise in both systems (Fig. 11.15), while we also find that the system with the same advection speed has much higher speed of traveling patterns (Fig. 11.15e, f).

Case Four: $r = 5.90$, $e = 0.64$, $a = 15.60$, $\eta = 7.00$, $D_1 = 1.00$, $D_2 = 16.5$. With these parameters, we will get the system to be stripe patterns (Fig. 11.16).

Similar to Cases Two and Three, the stripe patterns keep stationary (Fig. 11.17a–d) as different advection speeds are assigned $(V_P, V_Z) = (0.00, 0.02)$ $(V = 0.02)$, while traveling stripe patterns are produced for the system with the same advection coefficient, $V_P = V_Z = 0.02$ (Fig. 11.17e–h).

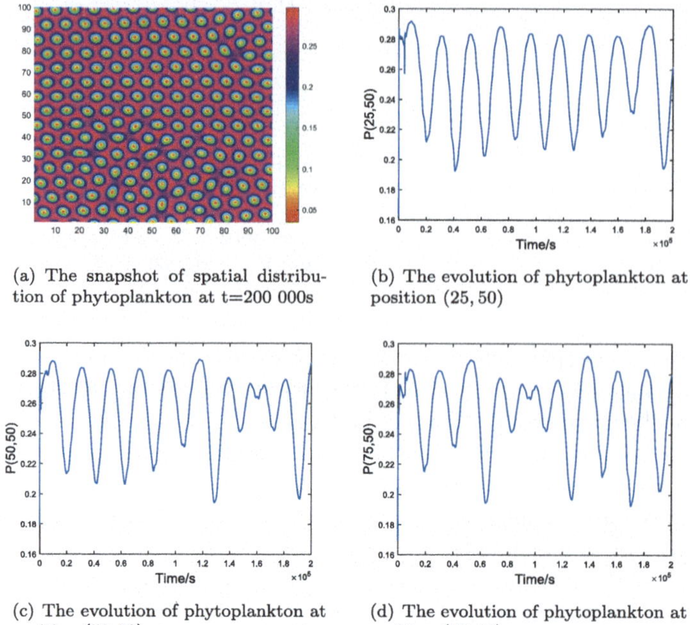

(a) The snapshot of spatial distribution of phytoplankton at t=200 000s

(b) The evolution of phytoplankton at position (25, 50)

(c) The evolution of phytoplankton at position (50, 50)

(d) The evolution of phytoplankton at position (75, 50)

Fig. 11.10 Evolution of phytoplankton at different instants and different grids, where $r = 5.90$, $e = 0.64$, $a = 15.60$, $\eta = 7.00$, $D_1 = 1.00$, $D_2 = 5.00$, $(V_P, V_Z) = (0.0004, 0.0004)$ $(V = 0.00)$

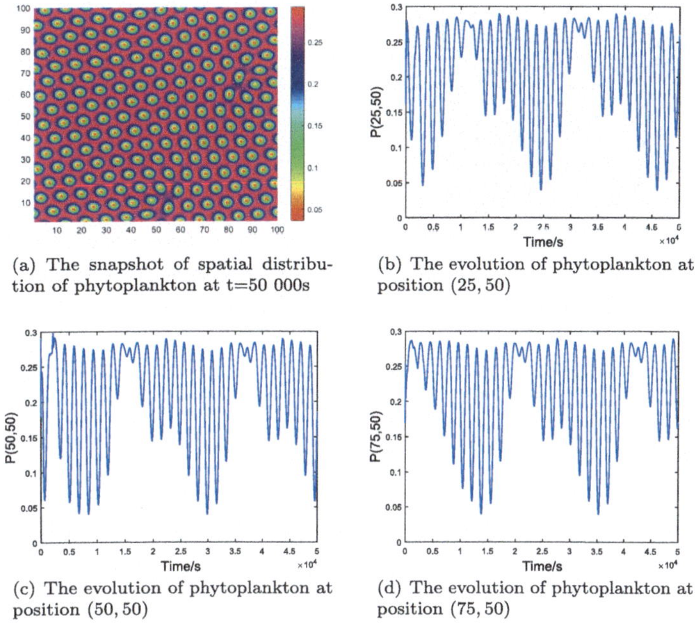

(a) The snapshot of spatial distribution of phytoplankton at t=50 000s

(b) The evolution of phytoplankton at position (25, 50)

(c) The evolution of phytoplankton at position (50, 50)

(d) The evolution of phytoplankton at position (75, 50)

Fig. 11.11 Evolution of phytoplankton at different instants and different grids, where $r = 5.90$, $e = 0.64$, $a = 15.60$, $\eta = 7.00$, $D_1 = 1.00$, $D_2 = 5.00$, $(V_P, V_Z) = (0.00, 0.02)$ $(V = 0.02)$

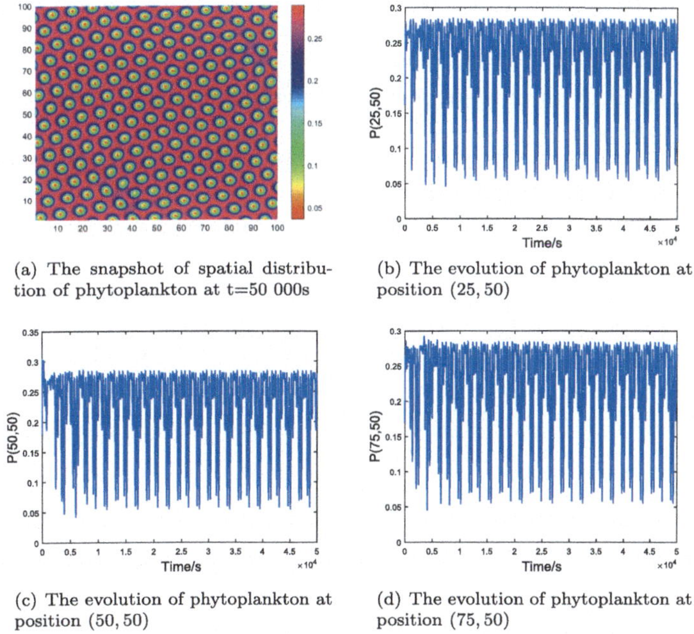

(a) The snapshot of spatial distribution of phytoplankton at t=50 000s

(b) The evolution of phytoplankton at position (25, 50)

(c) The evolution of phytoplankton at position (50, 50)

(d) The evolution of phytoplankton at position (75, 50)

Fig. 11.12 Evolution of phytoplankton at different instants and different grids, where $r = 5.90$, $e = 0.64$, $a = 15.60$, $\eta = 7.00$, $D_1 = 1.00$, $D_2 = 5.00$, $(V_P, V_Z) = (0.02, 0.02)$ $(V = 0.00)$

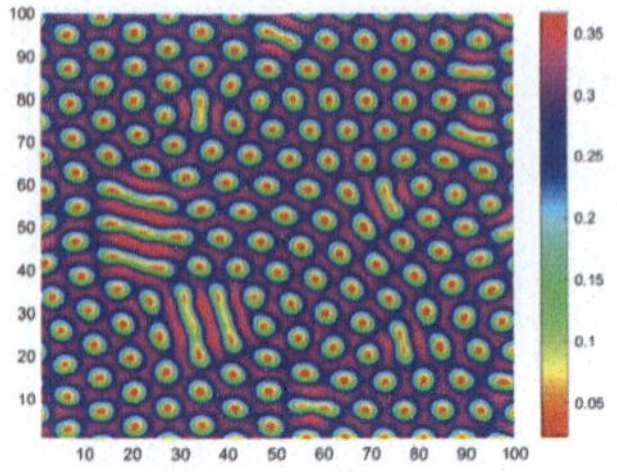

(a) The snapshot of spatial distribution of phytoplankton at t=50 000s

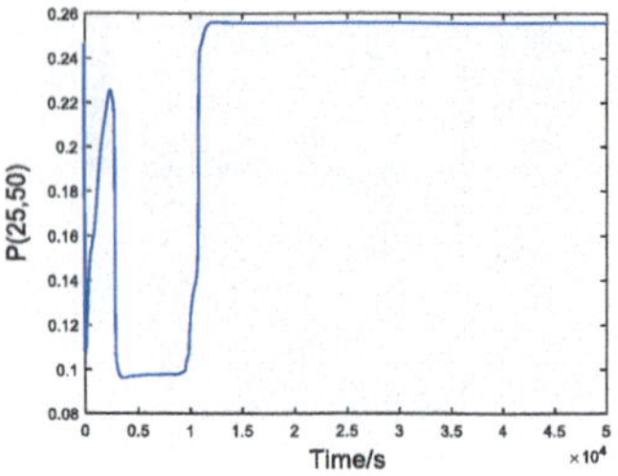

(b) The evolution of phytoplankton at position (25, 50)

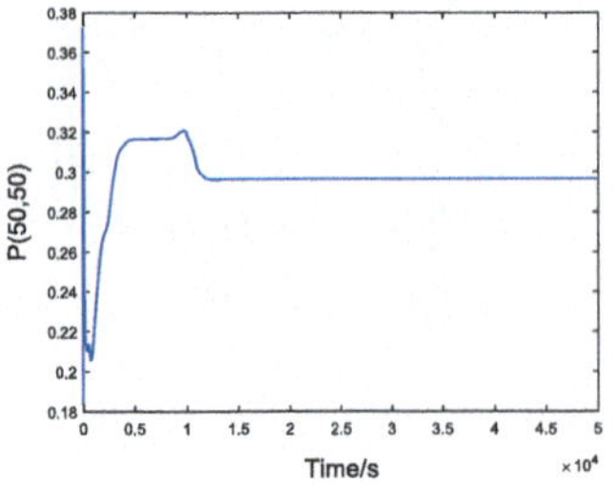

(c) The evolution of phytoplankton at position (50, 50)

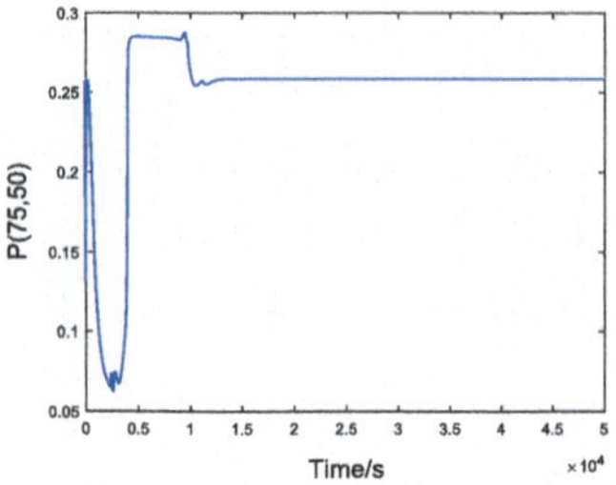

(d) The evolution of phytoplankton at position (75, 50)

Fig. 11.13 Evolution of phytoplankton at different instants and different grids, where $r = 5.90$, $e = 0.64$, $a = 15.60$, $\eta = 7.00$, $D_1 = 1.00$, $D_2 = 6.50$, $(V_P, V_Z) = (0.00, 0.00)$ $(V = 0.00)$

At the same time, traveling stripe patterns will be produced for a system with advection speeds of $(V_P, V_Z) = (0.00, 0.20)$ $(V = 0.20)$ or $(V_P, V_Z) = (0.02, 0.20)$ $(V = 0.00)$ (Fig. 11.18), while the traveling patterns with different convection speeds remain at a lower traveling speed (Fig. 11.18a–d).

From the simulation results, it is known that when one advection coefficient is kept as zero and the other is nonzero, and the advection velocity difference becomes insignificant, the patterns remain still; when the difference becomes large enough, the patterns will move. Thus, there exists a traveling bifurcation meaning that the variation of the smallest convective velocity will drive the pattern to move.

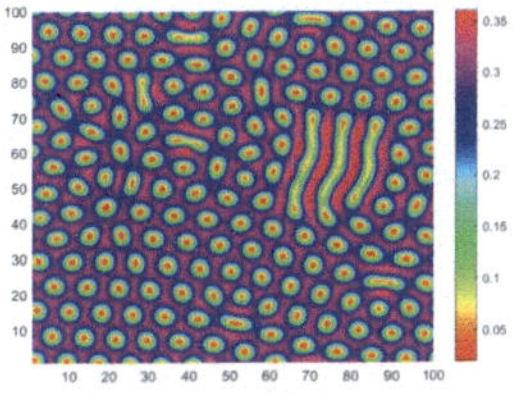

(a) $(V_P, V_Z) = (0.000, 0.002)$ $(V = 0.002)$, the snapshot of spatial distribution of phytoplankton at t=50,000s

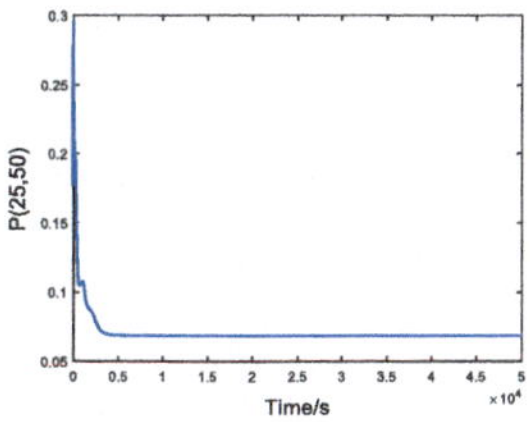

(b) $(V_P, V_Z) = (0.000, 0.002)$ $(V = 0.002)$, the evolution of phytoplankton at position $(25, 50)$

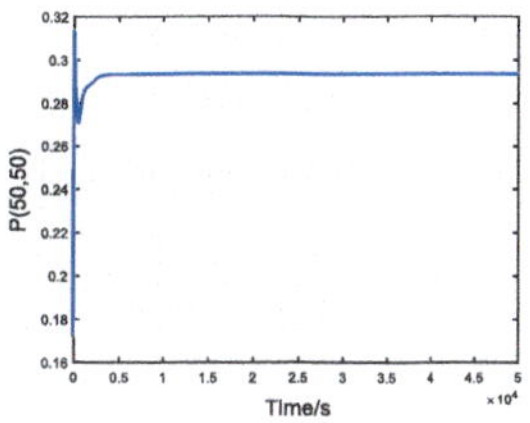

(c) $(V_P, V_Z) = (0.000, 0.002)$ $(V = 0.002)$, the evolution of phytoplankton at position $(50, 50)$

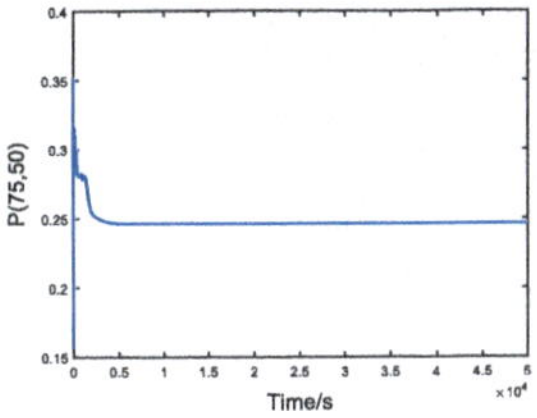

(d) $(V_P, V_Z) = (0.000, 0.002)$ $(V = 0.002)$, the evolution of phytoplankton at position $(75, 50)$

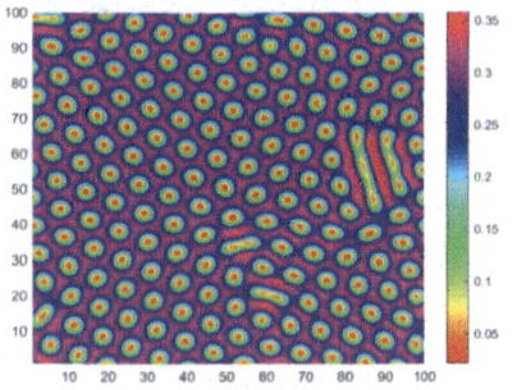

(e) $(V_P, V_Z) = (0.002, 0.002)$ $(V = 0.000)$, the snapshot of spatial distribution of phytoplankton at t=50 000s

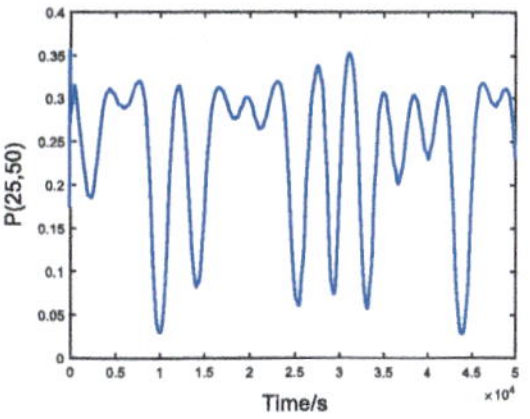

(f) $(V_P, V_Z) = (0.002, 0.002)$ $(V = 0.000)$, the evolution of phytoplankton at position $(25, 50)$

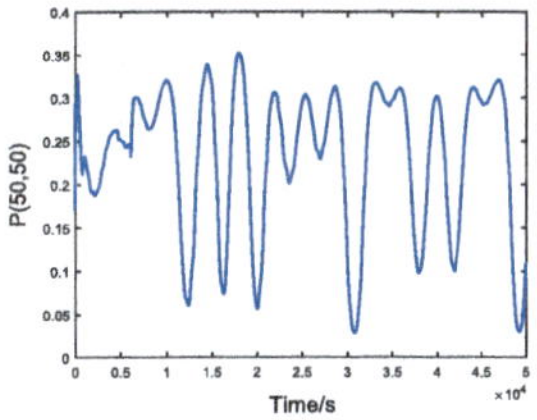

(g) $(V_P, V_Z) = (0.002, 0.002)$ $(V = 0.000)$, the evolution of phytoplankton at position $(50, 50)$

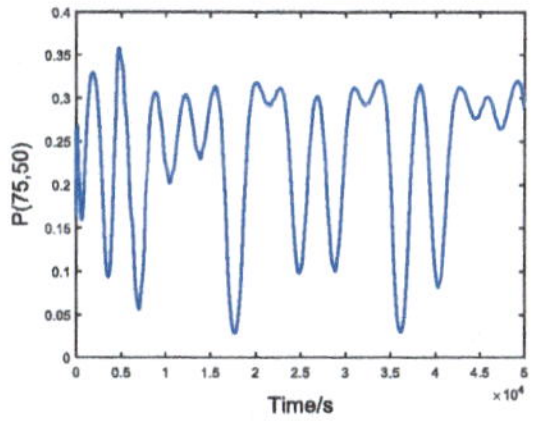

(h) $(V_P, V_Z) = (0.002, 0.002)$ $(V = 0.000)$, the evolution of phytoplankton at position $(75, 50)$

Fig. 11.14 Evolution of phytoplankton at different instants and different grids, where $r = 5.90$, $e = 0.64$, $a = 15.60$, $\eta = 7.00$, $D_1 = 1.00$, $D_2 = 6.50$

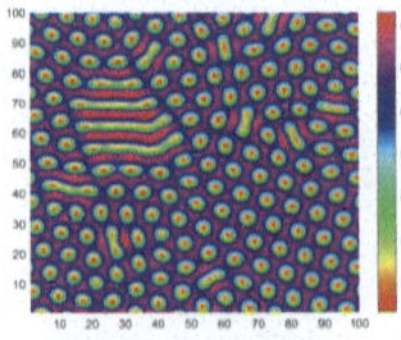

(a) $(V_P, V_Z) = (0.00, 0.02)$ $(V = 0.02)$, the snapshot of spatial distribution of phytoplankton at t=50 000s

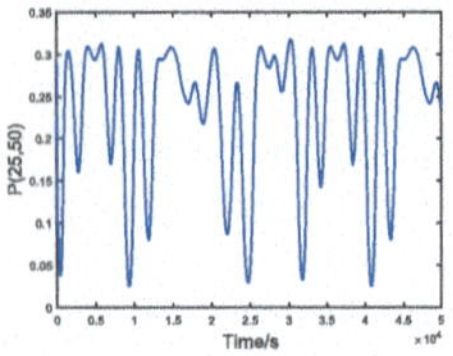

(b) $(V_P, V_Z) = (0.00, 0.02)$ $(V = 0.02)$, the evolution of phytoplankton at position$(25, 50)$

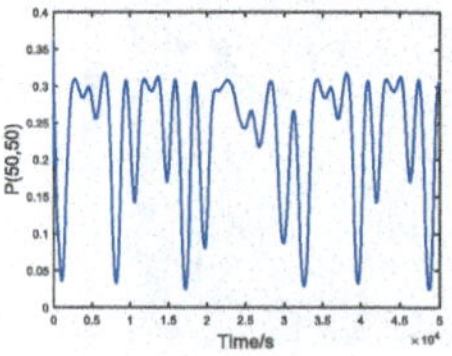

(c) $(V_P, V_Z) = (0.00, 0.02)$ $(V = 0.02)$, the evolution of phytoplankton at position $(50, 50)$

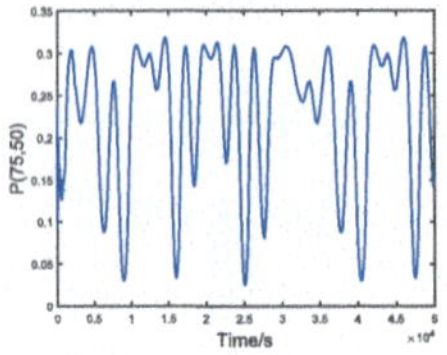

(d) $(V_P, V_Z) = (0.00, 0.02)$ $(V = 0.02)$, the evolution of phytoplankton at position $(75, 50)$

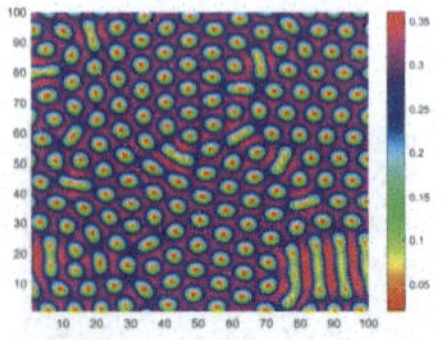

(e) $(V_P, V_Z) = (0.02, 0.02)$ $(V = 0.00)$, the snapshot of spatial distribution of phytoplankton at t=50 000s

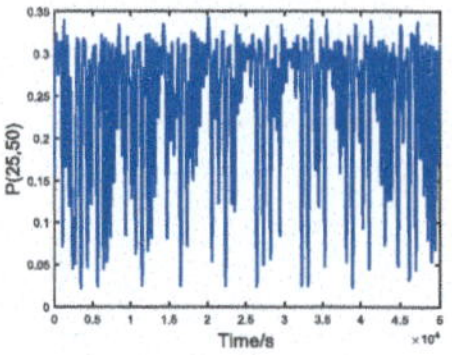

(f) $(V_P, V_Z) = (0.02, 0.02)$ $(V = 0.00)$, the evolution of phytoplankton at position$(25, 50)$

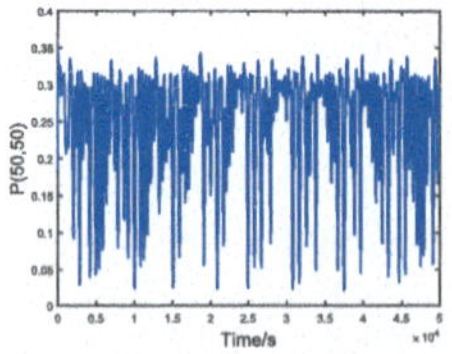

(g) $(V_P, V_Z) = (0.02, 0.02)$ $(V = 0.00)$, the evolution of phytoplankton at position $(50, 50)$

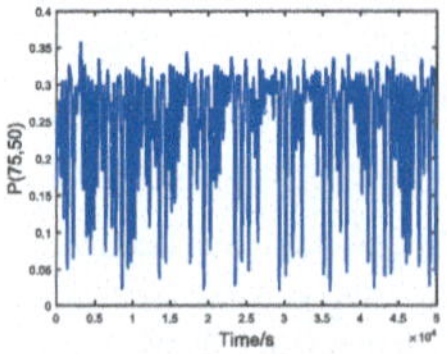

(h) $(V_P, V_Z) = (0.02, 0.02)$ $(V = 0.00)$, the evolution of phytoplankton at position $(75, 50)$

Fig. 11.15 Evolution of phytoplankton at different instants and different grids, where $r = 5.90$, $e = 0.64$, $a = 15.60$, $\eta = 7.00$, $D_1 = 1.00$, $D_2 = 6.50$

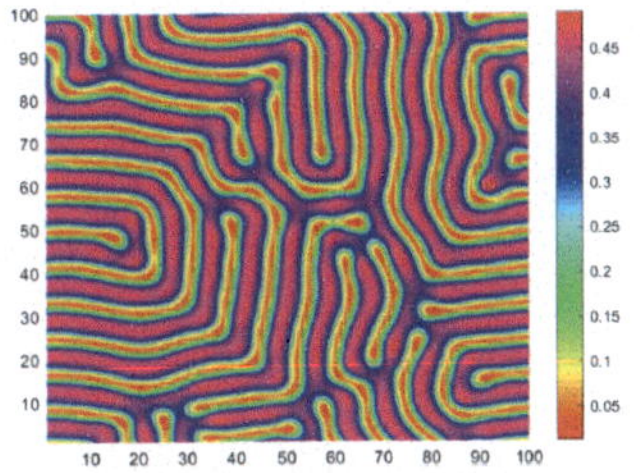

(a) The snapshot of spatial distribution of phytoplankton at t=100 000s

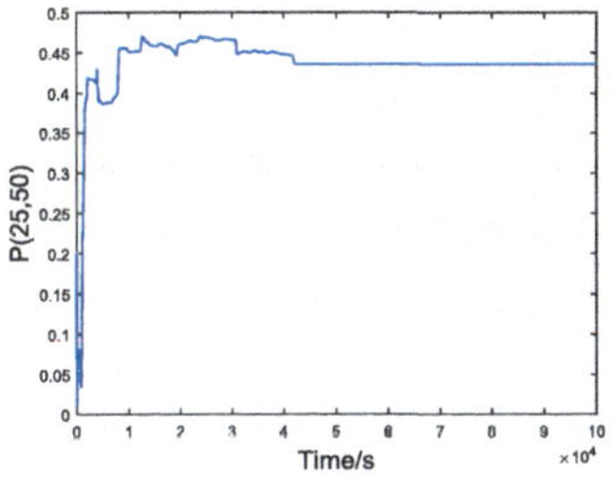

(b) The evolution of phytoplankton at position (25, 50)

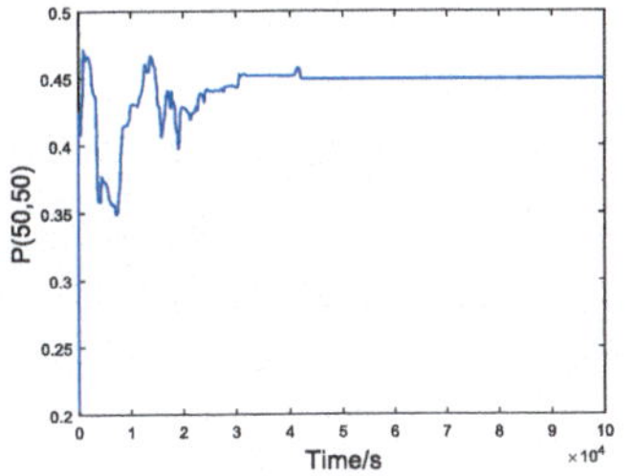

(c) The evolution of phytoplankton at position (50, 50)

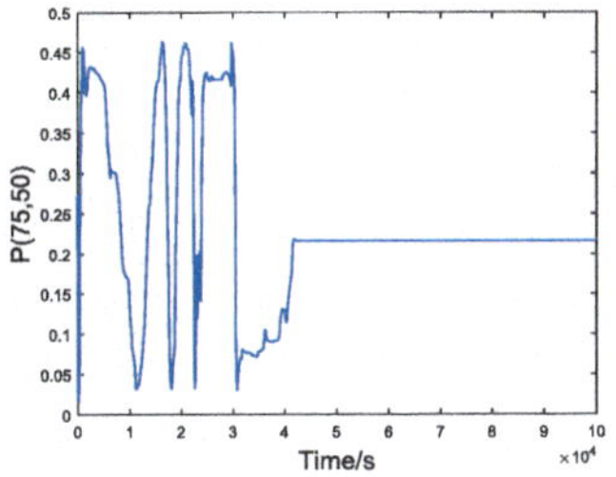

(d) The evolution of phytoplankton at position (75, 50)

Fig. 11.16 Evolution of phytoplankton at different instants and different grids, where $r = 5.90$, $e = 0.64$, $a = 15.60$, $\eta = 7.00$, $D_1 = 1.00$, $D_2 = 16.50$, $(V_P, V_Z) = (0.00, 0.00)$

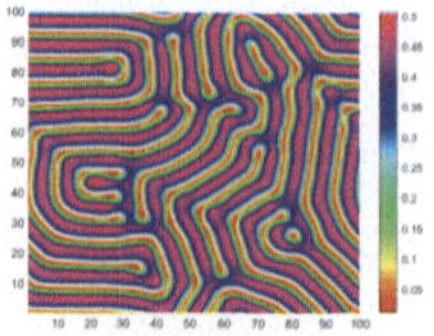

(a) $(V_P, V_Z) = (0.00, 0.02)$ $(V = 0.02)$, the snapshot of spatial distribution of phytoplankton at t=100 000s

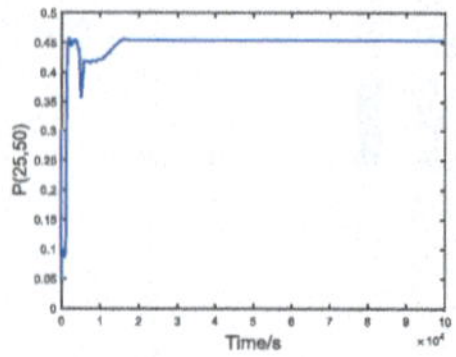

(b) $(V_P, V_Z) = (0.00, 0.02)$ $(V = 0.02)$, the evolution of phytoplankton at position$(25, 50)$

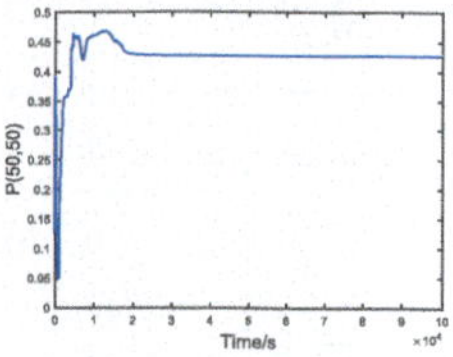

(c) $(V_P, V_Z) = (0.00, 0.02)$ $(V = 0.02)$, the evolution of phytoplankton at position $(50, 50)$

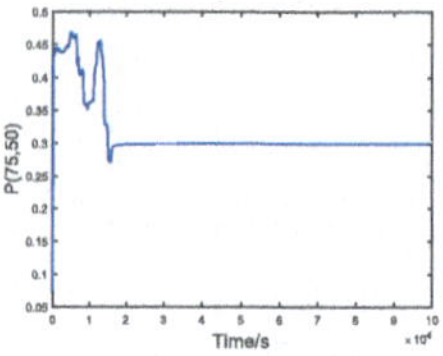

(d) $(V_P, V_Z) = (0.00, 0.02)$ $(V = 0.02)$, the evolution of phytoplankton at position $(75, 50)$

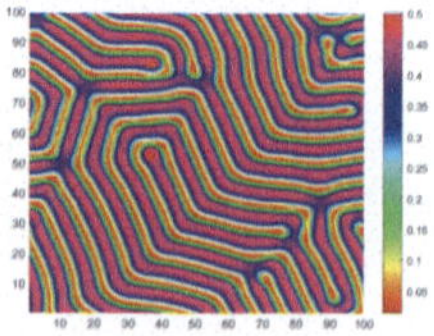

(e) $(V_P, V_Z) = (0.02, 0.02)$ $(V = 0.00)$, the snapshot of spatial distribution of phytoplankton at t=100 000s

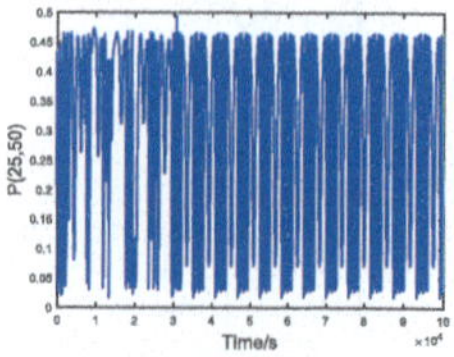

(f) $(V_P, V_Z) = (0.02, 0.02)$ $(V = 0.00)$, the evolution of phytoplankton at position$(25, 50)$

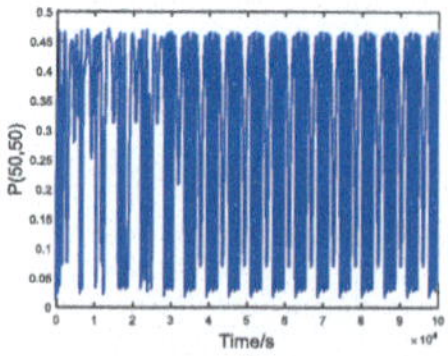

(g) $(V_P, V_Z) = (0.02, 0.02)$ $(V = 0.00)$, the evolution of phytoplankton at position $(50, 50)$

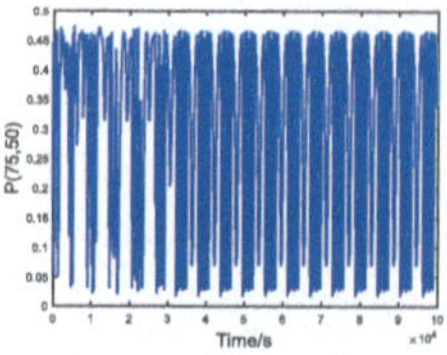

(h) $(V_P, V_Z) = (0.02, 0.02)$ $(V = 0.00)$, the evolution of phytoplankton at position $(75, 50)$

Fig. 11.17 Evolution of phytoplankton at different instants and different grids, where $r = 5.90$, $e = 0.64$, $a = 15.60$, $\eta = 7.00$, $D_1 = 1.00$, $D_2 = 16.50$

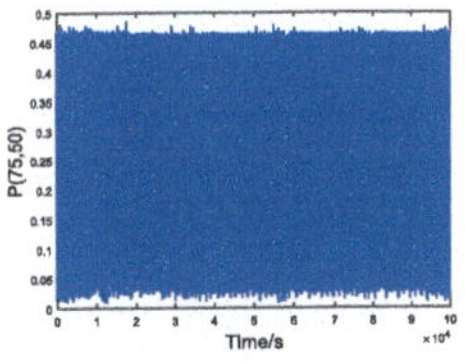

(a) $(V_P, V_Z) = (0.00, 0.2)$ $(V = 0.2)$, the snapshot of spatial distribution of phytoplankton at t=100 000s

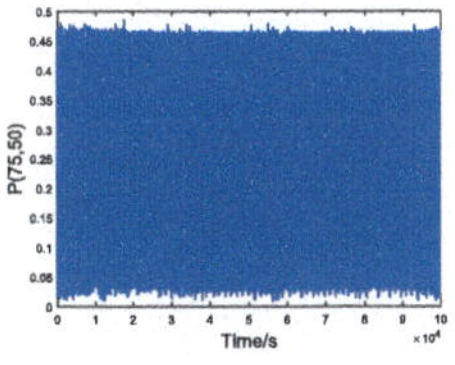

(b) $(V_P, V_Z) = (0.00, 0.2)$ $(V = 0.2)$, the evolution of phytoplankton at position$(25, 50)$

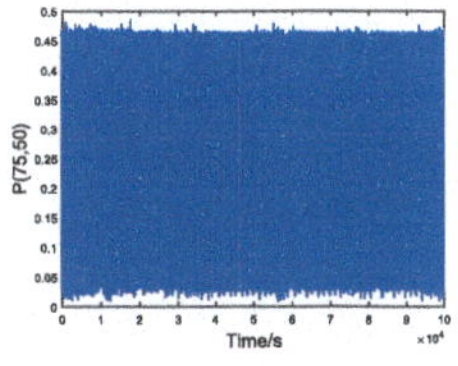

(c) $(V_P, V_Z) = (0.00, 0.2)$ $(V = 0.2)$, the evolution of phytoplankton at position $(50, 50)$

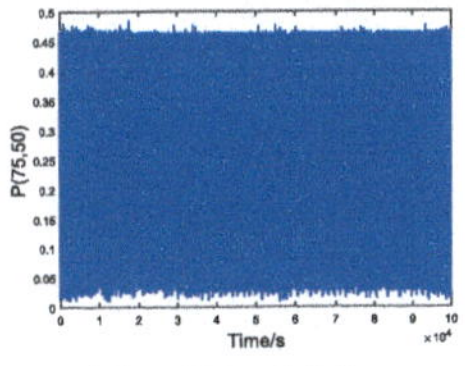

(d) $(V_P, V_Z) = (0.00, 0.2)$ $(V = 0.2)$, the evolution of phytoplankton at position $(75, 50)$

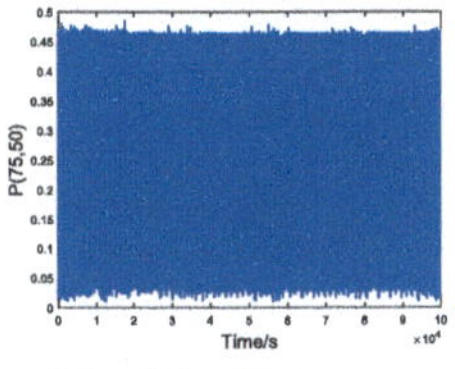

(e) $(V_P, V_Z) = (0.2, 0.2)$ $(V = 0.0)$, the snapshot of spatial distribution of phytoplankton at t=100 000s

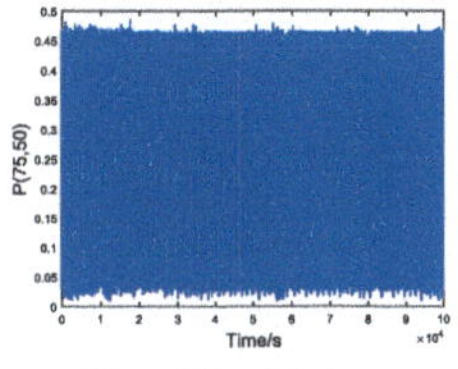

(f) $(V_P, V_Z) = (0.2, 0.2)$ $(V = 0.0)$, the evolution of phytoplankton at position$(25, 50)$

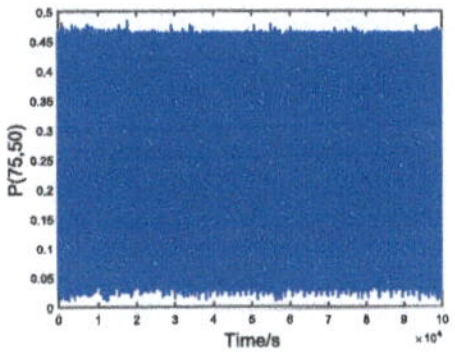

(g) $(V_P, V_Z) = (0.2, 0.2)$ $(V = 0.0)$, the evolution of phytoplankton at position $(50, 50)$

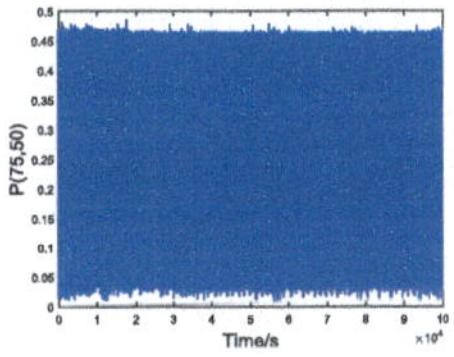

(h) $(V_P, V_Z) = (0.2, 0.2)$ $(V = 0.0)$, the evolution of phytoplankton at position $(75, 50)$

Fig. 11.18 Evolution of phytoplankton at different instants and different grids, where $r = 5.90$, $e = 0.64$, $a = 15.60$, $\eta = 7.00$, $D_1 = 1.00$, $D_2 = 16.50$

Chapter 12
The Effect of Herd-Taxis

In this chapter, the herd-taxi is considered for a plankton community composed of phytoplankton (prey) and zooplankton (predators) because that the herd-taxis indicates the effect of the zooplankton on the higher density zone of phytoplankton and generation of the conformity behavior leading to the nonlinear cross-diffusion phenomena. We use the stability analysis to get the conditions of the Hopf stability and the Turing bifurcation. Moreover, the active foraging behavior of zooplankton plays important role in spatially stabilizing the algal system and smoothing the inhomogeneous distribution of plankton. Significantly, it is also in line with the economic basic idea that there exists an invisible hand regulating and balancing the distribution of plankton. In addition, the weakly nonlinear analysis is employed to establish amplitude equations in order to explore the effect of herd-taxis on the self-organization. Successfully, three types of patterns are found out such as spot patterns, stripe patterns, and mixed patterns of the spots and stripes. There are also present some numerical simulations to validate the self-organization behaviors and help us further understand the interactions in plankton communities in the real world.

It is well known that plankton as the basis of food webs in the marine ecosystem can make a huge difference in the aquatic biosphere [241]. And there are two kinds of plankton: phytoplankton and zooplankton, in which phytoplankton absorbs light and nutrients to support photosynthesis and is fed on by zooplankton, but the zooplanktons are gazed at by the fishes [264]. Both kinds of plankton are two important parts for forming microscopic communities, because of their floating in the surface water and supporting the marine food webs. For example, when there support enough nutrients and sunlight, we can see the increasing growth of phytoplankton quickly and finally leading to algal blooms [274]. At the same time, zooplankton can subsequently grow due to enough food so that it can eventually suppress such a disaster [249]. Based on the fact that those planktons form a primary oceanic micro-ecosystem [287], it obviously makes sense for us to explore the plankton communities.

Nowadays, there are many works from different perspectives for plankton communities, especially for phytoplankton. For instance, the distribution of phytoplankton

S. T. Liu and L. Zhang, *Pattern Dynamics of Marine Plankton Behavior*,
https://doi.org/10.1007/978-981-97-5369-7_12

throughout the global oceans was monitored [280, 284]. The effects of sunlight and nutrients on the variations of phytoplankton were investigated [281, 295]. The phytoplankton growth in different seasons was recorded and analyzed [279]. Some gave the flows of energy and materials [265]. Some other scholars' attention was also apart from biologists and focuses on the plankton ecosystems by other experts like mathematicians. Some literature's aim is using the latest AI techniques to directly predict the evolution of phytoplankton [282, 283, 293]. While, predation functions (functional response) were still investigated through data records [278, 288], logical analysis traditionally [277], and experiments [275, 276]. The plankton community models consisting of phytoplankton and zooplankton have been studied based on the functional responses.

Two mathematical models: ordinary differential equations (ODEs) and partial differential equations (PDEs) have been modified to study and analyze the dynamics of plankton communities. Most works have studied the following items: the spatial effect as a secondary factor is regarded for taking the phytoplankton/zooplankton as a whole, and ODEs are used to model the planktonic system. For instance, the dynamics of plankton ecosystems was shown by these models: the conditions were given for the survival and extinction of the plankton ecosystem with impulse disturbance and a scheme of external was provided to help for the survival of plankton [271]. The role was discussed which is played by the toxin-producing phytoplankton in an impulsive environment and the conditions were obtained for the survival and bloom of phytoplankton [290, 292]. Then the effects of stochasticity and seasonality on a nutrient-phytoplankton–zooplankton ecosystem were also studied and the results show that periodic algal blooms could be resulted in by stochasticity and seasonality [291]. The influence of infochemicals was investigated from phytoplankton on the feeding behavior and selectivity of zooplanktonic predators was also investigated. And infochemicals can make a great difference for the dynamics of marine food webs by the above results [253]. Obviously, with the condition that there are many important and interesting results about the ODEs model clearly and relatively maturely, there exists an increasing number of fruits about the feature of the plankton ecosystem. However, there are not only one or two variables in the evolution for the plankton ecosystem so that plankton can behave with complexity features like moving within different grids actively or passively. In order to solve these problems, PDEs have been employed for better modeling the spatiotemporal plankton community. For example, the dynamics of oxygen in a plankton system were discussed with and without diffusion terms [261]. Wang [289] and Zhang [294], respectively, indicated the linear and nonlinear cross-diffusion effect on self-organization and Turing–Hopf bifurcation for the planktonic system. The effect of toxins produced by algae on the Turing pattern formation of phytoplankton–zooplankton model was investigated in which the results indicate that the toxicity plays a vital role for phytoplankton and the death rate of zooplankton plays an important role in the spatial distribution of population density [250]. Importantly, the horizontal distribution of plankton has been paid more attention. At the same time, the vertical case is also important and it is significant to study its effect on the dynamics of the algae ecosystems. Fortunately, Mellard investigated the vertical case in stratified water columns [285]. Moreover, there needs more related works about the vertical dynamics of algae ecosystems.

Motivated by these works, the plankton community was depicted by the reaction–diffusion predation equations such as the classical Holling I, II, III types and the ratio-dependent functional responses considering the saturation of predator/prey. Naturally, based on the scope of these equations and after comparing these predation functions, the Arditi–Ginzburg type II functional response as the most suitable predation function was chosen for the plankton community [273]. Moreover, toxins would be produced by phytoplankton and adhere to their surface. So that phytoplankton would kill zooplankton touched at a certain toxins' extent [237], where the kill rate remains in direct proportion to the densities of both. Subsequently, zooplanktons have some toxins fedback to phytoplankton and follow the complexity stream, which results in complex nonlinear cross-diffusion [286]. Furthermore, when we assume that the primary movements of zooplankton display as random walks, the predominant diffusion remains also linear. However, the nonlinear feature of the algae growth is well known. So in this chapter, we focus on the self-organization feature of the planktonic ecosystem and mainly study the effect of the herd-taxis on it.

This chapter is arranged in the order: at first, the ratio-dependent plankton model is illustrated and explained when there exists the nonlinear cross-diffusion term. And then the conditions for Hopf stability are ensured around the coexistence equilibrium and the Turing bifurcation is also calculated. Close to this Turing bifurcation, the dynamics of such a planktonic ecosystem with herd-taxis is analyzed by weakly nonlinear analysis and the self-organization behaviors are got in the "Turing pattern domain". Finally, we give some numeral simulations partly to verify our approach and results.

12.1 Mathematic Plankton Community Model with Herd-Taxis

We first introduce the planktonic community consisting of phytoplankton and zooplankton into a spatiotemporal model and present the following nonlinear partial differential equations:

$$\begin{cases} \frac{\partial P}{\partial t} &= rP\left(1-\frac{P}{K}\right)-\frac{aPZ}{Z+ahP}+D_1\nabla^2 P, \\ \frac{\partial Z}{\partial t} &= e\frac{aPZ}{Z+ahP}-\xi Z-\eta PZ-\nabla(D\Psi(Z)\nabla P)+D_2\nabla^2 Z, \end{cases} \tag{12.1}$$

where P, Z mean phytoplankton and zooplankton, respectively. The other conditions and more explanations are listed below:

[(1)] The boundary conditions are assumed by periodic. r represents the growth rate where the phytoplankton grows logistically with it and K means the natural carrying capability; As the fact that the zooplankton preys phytoplankton, and based on comparison of many aspects, the functional response of Arditi–Ginzburg type II is chosen as the predation function [257]; e is the consumption rate of zooplankton; ξ means the natural death rate of zooplankton; Toxins are released and adhere to phytoplankton so that phytoplankton can defend against predation and kill the predators by touching them with rate direct proportion to PZ. D_1 and D_2 stand for the respective diffusion coefficients in which phytoplankton and zooplankton naturally disperse with them. Based on the fact that different with natural diffusion, the zooplankton follows the trend of zooplankton by moving with taxis toward higher density phytoplankton, the movement is modeled by the nonlinear cross-diffusion, $\nabla(D\Psi(Z)\nabla P)$, where D is the taxis strength, and $\Psi(Z)$ named as the trend function indicates the conformity behavior of the zooplankton and satisfies the third-order differentiable.

Let's get a simple bicontinuous map:

$$\begin{array}{llll} \hat{t} = Kt, & \hat{P} = \dfrac{P}{K}, & \hat{Z} = \dfrac{Z}{ahK}, & \hat{r} = \dfrac{r}{K}, \\ \hat{a} = \dfrac{a}{K}, & \hat{e} = \dfrac{e}{ah}, & \hat{\xi} = \dfrac{\xi}{K}, & \hat{D} = \dfrac{D}{ahK}, \\ \hat{\Psi}(\hat{Z}) = \Psi(ahK\hat{Z}), & \hat{D}_1 = \dfrac{D_1}{K}, & \hat{D}_2 = \dfrac{D_2}{K}, & \end{array}$$

and also obtain a couple of brief equations as follows where we leave out the signal "$\hat{\cdot}$" for convenience:

$$\begin{cases} \frac{\partial P}{\partial t} &= rP(1-P) - \frac{aPZ}{Z+P} + D_1\nabla^2 P, \\ \frac{\partial Z}{\partial t} &= e\frac{aPZ}{Z+P} - \xi Z - \eta PZ - \nabla(D\Psi(Z)\nabla P) + D_2\nabla^2 Z. \end{cases} \tag{12.2}$$

In the following, the stability of the simplified model is our focus relievedly.

12.2 Stability Analysis

12.2.1 Hopf Stability

First of all, the model 12.2 degenerates into an ODEs model by leaving out the diffusion terms:

$$\begin{cases} \frac{dP}{dt} = rP(1-P) - \frac{aPZ}{Z+P}, \\ \frac{dZ}{dt} = e\frac{aPZ}{Z+P} - \xi Z - \eta PZ. \end{cases} \tag{12.3}$$

After defining

$$f(P,Z)=r(1-P)-\frac{aZ}{Z+P},\qquad g(P,Z)=e\frac{aP}{Z+P}-\xi-\eta P,$$

the ODEs model is rewritten as

$$\begin{cases}\frac{dP}{dt}=Pf(P,Z),\\ \frac{dZ}{dt}=Zg(P,Z),\end{cases}\tag{12.4}$$

where P and Z are assumed as nonnegative-definite with positive initial conditions [246], $(0,0)$ (not any plankton) and $(1,0)$ (only phytoplankton) are the unstable equilibrium points, and (P^*,Z^*) (the coexistence of both kinds of plankton) means the interior equilibrium point, where

$$P^*=\frac{ea-er-\xi}{\eta-er}\qquad Z^*=\frac{eaP^*}{\xi+\eta P^*}-P^*.$$

When we assign $\delta P=P-P^*$ and $\delta Z=Z-Z^*$, Eq. (12.4) can be linearized around the inter-equilibrium:

$$\frac{d}{dt}\begin{pmatrix}\delta P\\ \delta Z\end{pmatrix}=\begin{pmatrix}a_{11}&a_{12}\\ a_{21}&a_{22}\end{pmatrix}\begin{pmatrix}\delta P\\ \delta Z\end{pmatrix}=\begin{pmatrix}P^*f_P&P^*f_Z^*\\ Z^*g_P^*&Z^*g_Z^*\end{pmatrix}\begin{pmatrix}\delta P\\ \delta Z\end{pmatrix},\tag{12.5}$$

where

$$\begin{aligned}f_P^*&=-r+\frac{aZ^*}{(Z^*+P^*)^2}, & f_Z^*&=-\frac{aP^*}{(Z^*+P^*)^2},\\ g_P^*&=e\frac{aZ^*}{(Z^*+P^*)^2}-\eta, & g_Z^*&=-e\frac{aP^*}{(Z^*+P^*)^2}.\end{aligned}$$

Then we obtain the respective characteristic equation as

$$\lambda^2-Tr_0\lambda+\Delta_0=0,\tag{12.6}$$

where

$$Tr_0=a_{11}+a_{22},\qquad \Delta_0=a_{11}a_{22}-a_{12}a_{21},$$

and the roots of the characteristic Eq. (12.6) are got by the following:

$$\lambda=\frac{Tr_0\pm\sqrt{Tr_0^2-4\Delta_0}}{2},\tag{12.7}$$

Hence we get that the ecosystem remains Hopf stable, when

$$Tr_0 < 0, \qquad \Delta_0 > 0.$$

Remark 5 The stability of system (12.3) was studied at the extinction points (0, 0) and (1, 0), especially near the point (0, 0), because of its abundant dynamic behaviors [251]. As we would not study the Hopf bifurcation theoretically, we would not present too many details about it.

With the conditions that the plankton system becomes Hopf stable around the coexistence point and with diffusion terms, the algae community can be driven to instability. Thus, we will study its dynamical behavior with diffusion terms.

12.2.2 Turing Bifurcation

Based on the fact that we can see the ecosystem as unstable driven by diffusion even being in Hopf stable, the model with dispersion must be analyzed. Firstly, we linearize the reaction nonlinear reaction-diffusion equations as

$$\frac{d}{dt}\begin{pmatrix}\delta P\\ \delta Z\end{pmatrix} = \begin{pmatrix}a_{11} & a_{12}\\ a_{21} & a_{22}\end{pmatrix}\begin{pmatrix}\delta P\\ \delta Z\end{pmatrix} + \begin{pmatrix}D_1\nabla^2\delta P\\ -D\Psi(Z^*)\nabla^2\delta P + D_2\nabla^2\delta Z\end{pmatrix}, \tag{12.8}$$

where the form of the solution of (12.8) is taken as

$$\begin{pmatrix}\delta P\\ \delta Z\end{pmatrix} = \sum\begin{pmatrix}A_k^P\\ A_k^Z\end{pmatrix} e^{\lambda t + ik\cdot r}, \quad k = \|k\|. \tag{12.9}$$

And we also obtain its corresponding characteristic equation as

$$\lambda\begin{pmatrix}\delta P\\ \delta Z\end{pmatrix} = \begin{pmatrix}a_{11} - D_1k^2 & a_{12}\\ a_{21} + D\Psi(Z^*)k^2 & a_{22} - D_2k^2\end{pmatrix}\begin{pmatrix}\delta P\\ \delta Z\end{pmatrix}, \tag{12.10}$$

and its dispersion equation as

$$\lambda^2 - Tr_k\lambda + \Delta_k = 0, \tag{12.11}$$

where

$$Tr_k = Tr_0 - (D_1 + D_2)k^2 \le Tr_0 < 0, \tag{12.12}$$

$$\Delta_k = D_1D_2(k^2)^2 - (a_{11}D_2 + a_{22}D_1 + a_{12}D\Psi(Z^*))k^2 + \Delta_0. \tag{12.13}$$

In addition, we have the roots of the dispersion Eq. (12.11) as

$$\lambda = \frac{Tr_k \pm \sqrt{Tr_k^2 - 4\Delta_k}}{2}. \tag{12.14}$$

Importantly, the system loses its Turing stability when $\Delta_k \leq 0$, which means

$$k_T^2 = \frac{a_{11}D_2 + a_{22}D_1 + a_{12}D\Psi(Z^*)}{2D_1D_2} \geq 0, \tag{12.15}$$

and

$$\Delta_{k_T} = -\frac{(a_{11}D_2 + a_{22}D_1 + a_{12}D\Psi(Z^*))^2}{4D_1D_2} + \Delta_0 \leq 0. \tag{12.16}$$

And D_T becomes the Turing bifurcation when

$$\Delta_{k_T}(D_T) = -\frac{(a_{11}D_2 + a_{22}D_1 + a_{12}D_T\Psi(Z^*))^2}{4D_1D_2} + \Delta_0 = 0, \tag{12.17}$$

and

$$D_T = \frac{2\sqrt{D_1D_2\Delta_0} - (a_{11}D_2 + a_{22}D_1)}{a_{12}\Psi(Z^*)}. \tag{12.18}$$

From Eq. (12.15), it is easy to know $a_{12} < 0$ and $\Psi(Z^*) > 0$. So k_T^2 will vanish when D is big enough, such as

$$D > -\frac{a_{11}D_2 + a_{22}D_1}{a_{12}\Psi(Z^*)}.$$

Therefore, the cross-diffusion plays an important role in destroying the self-organizing behaviors caused by Turing instability.

Remark 6 The above result shows that the plankton community will become uniformly stable, and the distribution of zoo/phytoplankton will remain homogeneous when the zooplanktons have a stronger capability of actively foraging. It is common that the high-density phytoplankton community has important effect on economic development. So when it attracts the zooplankton naturally and the conformity effect intensifies this trend at the same time, we can see the excessive investment of the zooplankton, and the global plankton back to the homogeneous equilibrium. However, the real world remains heterogeneous indicating that there is limited capability for the zooplankton of active forage and then it can't smooth the spatial heterogeneous distribution actually. Hence, we will keep the conformity coefficient D small enough to make it be the self-organization patterns.

12.3 Weakly Nonlinear Analysis

When the conformity coefficient D changes close to the Turing bifurcation, there would exist many kinds of dynamic behaviors like the self-organization phenomenon around the coexistence steady state. With the help of weakly nonlinear analysis, we have conditions about pattern formation. Let P, Z replace δP, δZ and define $F = Pf$, $G = Zg$ for convenience. Firstly, the third-order approximation of the trend function $\Psi(Z)$ is calculated around the coexistence equilibrium:

$$\Psi(Z) = \Psi_Z Z + \frac{1}{2}\Psi_{ZZ} Z^2 + \frac{1}{6}\Psi_{ZZZ} Z^3. \tag{12.19}$$

Then, after taking Eq. (12.19) into Eq. (12.2) and extending the reaction–diffusion equations, we obtain the following third-order Taylor series expansion approximation:

$$\begin{aligned}\frac{\partial}{\partial t}\begin{pmatrix}P\\Z\end{pmatrix} \approx & L_T\begin{pmatrix}P\\Z\end{pmatrix} + (D_T - D)\Psi(Z^*)\nabla^2\begin{pmatrix}0\\P\end{pmatrix}\\ & + (D_T - D - D_T)\begin{pmatrix}0\\(\Psi_Z + \Psi_{ZZ}Z)\nabla Z\cdot\nabla P + (\Psi_Z Z + \frac{1}{2}\Psi_{ZZ}Z^2)\nabla^2 P\end{pmatrix}\\ & + \frac{1}{2}\begin{pmatrix}F_{PP}P^2 + 2F_{PZ}PZ + F_{ZZ}Z^2\\ G_{PP}P^2 + 2G_{PZ}PZ + G_{ZZ}Z^2\end{pmatrix}\\ & + \frac{1}{6}\begin{pmatrix}F_{PPP}P^3 + 3F_{PPZ}P^2Z + 3F_{PZZ}PZ^2 + F_{ZZZ}Z^3\\ G_{PPP}P^3 + 3G_{PPZ}P^2Z + 3G_{PZZ}PZ^2 + G_{ZZZ}Z^3\end{pmatrix} + \dots,\end{aligned} \tag{12.20}$$

where

$$L_T = \begin{pmatrix} a_{11} + D_1\nabla^2 & a_{12}\\ a_{21} - D_T\Psi(Z^*)\nabla^2 & a_{22} + D_2\nabla^2\end{pmatrix},$$

and the coefficients of (12.20) are in **??**. Obviously, as the tuning parameter D_2 becomes near the Turing bifurcation, we can get the solution of (12.20) expressed in the following form:

$$U = \begin{pmatrix}P\\Z\end{pmatrix} = \sum_{j=1}^{3}\begin{pmatrix}P_0\\Z_0\end{pmatrix}\left[A_j e^{ik_j\cdot r} + \overline{A}_j e^{-ik_j\cdot r}\right], \tag{12.21}$$

where $\begin{pmatrix}P_0 & Z_0\end{pmatrix}^T$ means the direction of the eigenmodes, and A_j, $\overline{A_j}$ stand for the amplitudes of the mode k_j, $-k_j$, respectively. In the following, the perturbations of the tuning parameter D_2 as well as P, Z, t are extended as follows:

$$P = \epsilon P_1 + \epsilon^2 P_2 + \epsilon^3 P_3 + \cdots, \tag{12.22a}$$

$$Z = \epsilon Z_1 + \epsilon^2 Z_2 + \epsilon^3 Z_3 + \cdots, \tag{12.22b}$$

$$\frac{\partial}{\partial t} = \epsilon\frac{\partial}{\partial T_1} + \epsilon^2\frac{\partial}{\partial T_2} + \epsilon^3\frac{\partial}{\partial T_3} + \cdots, \tag{12.22c}$$

$$D_T - D = \epsilon d_1 + \epsilon^2 d_2 + \cdots. \tag{12.22d}$$

Then we can extend Eq. (12.20) into perturbation series and divide them into different equations with respect to ϵ:
The first order of ϵ:

$$L_T\begin{pmatrix}P_1\\Z_1\end{pmatrix} = 0. \tag{12.23}$$

The second order of ϵ:

$$\begin{aligned}L_T\begin{pmatrix}P_2\\Z_2\end{pmatrix} =& \frac{\partial}{\partial T_1}\begin{pmatrix}P_1\\Z_1\end{pmatrix} - d_1\Psi(Z^*)\nabla^2\begin{pmatrix}0\\P_1\end{pmatrix}\\&+ D_T\begin{pmatrix}0\\\Psi_Z\left(\nabla P_1\cdot\nabla Z_1 + Z_1\nabla^2 P_1\right)\end{pmatrix}\\&- \frac{1}{2}\begin{pmatrix}F_{PP}P_1^2 + 2F_{PZ}P_1Z_1 + F_{ZZ}Z_1^2\\G_{PP}P_1^2 + 2G_{PZ}P_1Z_1 + G_{ZZ}Z_1^2\end{pmatrix}\\=&\begin{pmatrix}R_2^P\\R_2^Z\end{pmatrix}.\end{aligned} \tag{12.24}$$

The third order of ϵ:

$$\begin{aligned}L_T\begin{pmatrix}P_3\\Z_3\end{pmatrix} =& \frac{\partial}{\partial T_1}\begin{pmatrix}P_2\\Z_2\end{pmatrix} + \frac{\partial}{\partial T_2}\begin{pmatrix}P_1\\Z_1\end{pmatrix} - d_1\Psi(Z^*)\nabla^2\begin{pmatrix}0\\P_2\end{pmatrix} - d_2\Psi(Z^*)\nabla^2\begin{pmatrix}0\\P_1\end{pmatrix}\\&+ D_T\begin{pmatrix}0\\\Psi_Z\left(\nabla P_1\cdot\nabla Z_2 + \nabla P_2\cdot\nabla Z_1 + Z_1\nabla^2P_2 + Z_2\nabla^2P_1\right)\end{pmatrix}\\&+ D_T\begin{pmatrix}0\\\Psi_{ZZ}\left(Z_1\nabla Z_1\nabla P_1 + \frac{1}{2}Z_1^2\nabla^2P_1\right)\end{pmatrix}\\&- d_1\begin{pmatrix}0\\\Psi_Z\left(\nabla Z_1\cdot\nabla P_1 + Z_1\nabla^2P_1\right)\end{pmatrix}\\&- \begin{pmatrix}F_{PP}P_1P_2 + F_{PZ}(P_1Z_2 + P_2Z_1) + F_{ZZ}Z_1Z_2\\G_{PP}P_1P_2 + G_{PZ}(P_1Z_2 + P_2Z_1) + G_{ZZ}Z_1Z_2\end{pmatrix}\\&- \frac{1}{6}\begin{pmatrix}F_{PPP}P_1^3 + 3F_{PPZ}P_1^2Z_1 + 3F_{PZZ}P_1Z_1^2 + F_{ZZZ}Z_1^3\\G_{PPP}P_1^3 + 3G_{PPZ}P_1^2Z_1 + 3G_{PZZ}P_1Z_1^2 + G_{ZZZ}Z_1^3\end{pmatrix}\\=&\begin{pmatrix}R_3^P\\R_3^Z\end{pmatrix}.\end{aligned} \tag{12.25}$$

By solving (12.23), we have

$$\begin{pmatrix} P_1 \\ Z_1 \end{pmatrix} = \begin{pmatrix} l \\ 1 \end{pmatrix} \left(\sum_{j=1}^{3} W_j e^{ik_j \cdot r} + c.c. \right), \tag{12.26}$$

where $l = -\frac{a_{12}}{a_{11} - D_1 k_T^2}$, W_j represents the amplitude of the mode $e^{ik_j \cdot r} (j = 1, 2, 3)$. Based on Eq. (12.24) and the Fredholm solvability condition, it is easy to know that the right-hand side of the equations must be orthogonal to the eigenvector of zero eigenvalue of L_T^* (the adjoint operator of L_T), where the eigenvector of the operator L_T^* is defined as $\begin{pmatrix} 1 \\ l^* \end{pmatrix} e^{-ik_j \cdot r} + c.c. (j = 1, 2, 3)$ with $l^* = -\frac{a_{12}}{a_{22} - D_2 k_T^2}$, and the orthogonality condition is obtained by

$$\begin{pmatrix} 1 & l^* \end{pmatrix} \begin{pmatrix} R_2^P \\ R_2^Z \end{pmatrix} = 0. \tag{12.27}$$

Thus we get the solution of Eq. (12.24) as follows:

$$\begin{aligned} \begin{pmatrix} P_2 \\ Z_2 \end{pmatrix} = & \begin{pmatrix} X_0 \\ Y_0 \end{pmatrix} + \sum_{j=1}^{3} \begin{pmatrix} X_j \\ Y_j \end{pmatrix} e^{ik_j \cdot r} + \sum_{j=1}^{3} \begin{pmatrix} X_{jj} \\ Y_{jj} \end{pmatrix} e^{i2k_j \cdot r} + \begin{pmatrix} X_{12} \\ Y_{12} \end{pmatrix} e^{i(k_1 - k_2) \cdot r} \\ & + \begin{pmatrix} X_{23} \\ Y_{23} \end{pmatrix} e^{i(k_2 - k_3) \cdot r} + \begin{pmatrix} X_{31} \\ Y_{31} \end{pmatrix} e^{i(k_3 - k_1) \cdot r} + c.c, \end{aligned} \tag{12.28}$$

where we get the coefficients in (12.28) by substituting (12.28) into 12.24) and dividing according to the modes e^0, $e^{ik_j \cdot r}$, $e^{i2k_j \cdot r}$, $e^{i(k_j - k_s) \cdot r} (j, s = 1, 2, 3, j \neq s)$. Then we assume

$$L_{k_T^2} = \begin{pmatrix} a_{11} - D_1 k_T^2 & a_{12} \\ a_{21} + D_T \Psi(Z^*) k_T^2 & a_{22} - D_2 k_T^2 \end{pmatrix}, \quad k_T = \|k_T\|_2, \tag{12.29}$$

and obtain the coefficients:

$$\begin{aligned} & X_j = l Y_j \quad j = 1, 2, 3, \\ & \begin{pmatrix} X_0 \\ Y_0 \end{pmatrix} = -L_0^{-1} \begin{pmatrix} H_2^f \\ H_2^g \end{pmatrix} \left(\sum_{j=1}^{3} |W_j|^2 \right) = \begin{pmatrix} U_0 \\ V_0 \end{pmatrix} \left(\sum_{j=1}^{3} |W_j|^2 \right), \\ & \begin{pmatrix} X_{js} \\ Y_{js} \end{pmatrix} = -L_{3k_T^2}^{-1} \begin{pmatrix} H_2^f \\ H_2^g + 3 D_T \Psi_Z l k_T^2 \end{pmatrix} W_j \overline{W}_s = \begin{pmatrix} U_{js} \\ V_{js} \end{pmatrix} W_j \overline{W}_s, \\ & \begin{pmatrix} X_{jj} \\ Y_{jj} \end{pmatrix} = -L_{4k_T^2}^{-1} \begin{pmatrix} \frac{1}{2} H_2^f \\ \frac{1}{2} H_2^g + 2 D_T \Psi_Z l k_T^2 \end{pmatrix} W_j^2 = \begin{pmatrix} U_{jj} \\ V_{jj} \end{pmatrix} W_j^2, \end{aligned}$$

where $H_2^f = F_{PP}l^2 + 2F_{PZ}l + F_{ZZ}$, $H_2^g = G_{PP}l^2 + 2G_{PZ}l + G_{ZZ}$.
At ϵ^3, the solvability condition indicates that

$$\begin{pmatrix} 1 & l^* \end{pmatrix} \begin{pmatrix} R_3^P \\ R_3^Z \end{pmatrix} = 0. \tag{12.30}$$

When we collect the coefficients of $e^{ik_j \cdot r}(j = 1, 2, 3)$, we obtain

$$\begin{cases} (l + l^*)(\frac{\partial W_1}{\partial T_2} + \frac{\partial Y_1}{\partial T_1}) = & H(d_1 Y_1 + d_2 W_1) \\ & +d_1 H_1 \overline{W}_2 \overline{W}_3 + H_2(\overline{W}_2 \overline{Y}_3 + \overline{W}_3 \overline{Y}_2) \\ & +[G_1 |W_1|^2 + G_2(|W_2|^2 + |W_3|^2)] W_1, \\ (l + l^*)(\frac{\partial W_2}{\partial T_2} + \frac{\partial Y_2}{\partial T_1}) = & H(d_1 Y_2 + d_2 W_2) \\ & +d_1 H_1 \overline{W}_1 \overline{W}_3 + H_2(\overline{W}_1 \overline{Y}_3 + \overline{W}_3 \overline{Y}_1) \\ & +[G_1 |W_2|^2 + G_2(|W_1|^2 + |W_3|^2)] W_2, \\ (l + l^*)(\frac{\partial W_3}{\partial T_2} + \frac{\partial Y_3}{\partial T_1}) = & H(d_1 Y_3 + d_2 W_3) \\ & +d_1 H_1 \overline{W}_1 \overline{W}_2 + H_2(\overline{W}_1 \overline{Y}_2 + \overline{W}_2 \overline{Y}_1) \\ & +[G_1 |W_3|^2 + G_2(|W_1|^2 + |W_2|^2)] W_3, \end{cases} \tag{12.31}$$

where

$$\begin{aligned} H =& -\Psi(Z^*) l^* l k_T^2, \\ H_1 =& -\Psi_Z l^* l k_T^2, \\ H_2 =& D_T \Psi_Z l^* l k_T^2 + [F_{PP} l + F_{PZ} + l^*(G_{PP} l + G_{PZ})] l \\ & + [F_{PZ} l + F_{ZZ} + l^*(G_{PZ} l + G_{ZZ})], \\ G_1 =& D_T \Psi_Z l^* k_T^2 \left(l V_0 - l V_{jj} + 2 U_{jj} \right) + \frac{1}{2} D_T \Psi_{ZZ} l^* l k_T^2 \\ & + [F_{PP} l + F_{PZ} + l^*(G_{PP} l + G_{PZ})](U_0 + U_{jj}) \\ & + [F_{PZ} l + F_{ZZ} + l^*(G_{PZ} l + G_{ZZ})](V_0 + V_{jj}) + \frac{1}{2}(H_3^f + l^* H_3^g), \\ G_2 =& D_T \Psi_Z l^* k_T^2 \left(l V_0 - \frac{1}{2} l V_{js} + \frac{3}{2} U_{js} \right) + D_T \Psi_{ZZ} l^* l k_T^2 \\ & + [F_{PP} l + F_{PZ} + l^*(G_{PP} l + G_{PZ})](U_0 + U_{js}) \\ & + [F_{PZ} l + F_{ZZ} + l^*(G_{PZ} l + G_{ZZ})](V_0 + V_{js}) + (H_3^f + l^* H_3^g), \end{aligned}$$

$$\begin{aligned} H_3^f =& FPPPl^3 + 3F_{PPZ} l^2 + 3F_{PZZ} l + F_{ZZZ}, \\ H_3^g =& G_{PPP} l^3 + 3G_{PPZ} l^2 + 3G_{PZZ} l + G_{ZZZ}. \end{aligned}$$

As we expand the amplitude $A_j (j = 1, 2, 3)$ as

$$A_j = \epsilon W_j + \epsilon^2 Y_j + o(\epsilon^2), \tag{12.32}$$

and use the multi-scale extension (12.22), we can obtain the amplitude equations:

$$\begin{cases} \tau_0 \frac{\partial A_1}{\partial t} = \mu A_1 + (\mu h_1 + h_2)\overline{A}_2\overline{A}_3 - [g_1|A_1|^2 + g_2(|A_2|^2 + |A_3|^2)]A_1, \\ \tau_0 \frac{\partial A_2}{\partial t} = \mu A_2 + (\mu h_1 + h_2)\overline{A}_1\overline{A}_3 - [g_1|A_2|^2 + g_2(|A_1|^2 + |A_3|^2)]A_2, \\ \tau_0 \frac{\partial A_3}{\partial t} = \mu A_3 + (\mu h_1 + h_2)\overline{A}_1\overline{A}_2 - [g_1|A_3|^2 + g_2(|A_1|^2 + |A_2|^2)]A_3, \end{cases} \tag{12.33}$$

where

$$\tau_0 = \frac{l + l^*}{D_T H}, \qquad \mu = \frac{D_T - D}{D_T}, \qquad h_1 = \frac{H_1}{H},$$
$$h_2 = \frac{H_2}{D_T H}, \qquad g_1 = -\frac{G_1}{D_T H}, \qquad g_2 = -\frac{G_2}{D_T H}.$$

12.3.1 Amplitude Stability Analysis

Noticeably, we obtain the form of the stable solutions of (12.33) as $A_j = \rho_j e^{i\theta_j}$ $(j = 1, 2, 3)$, where $\rho_j = |A_j|$ and θ_j mean the corresponding norm and phase angle [151]. Thus we have the following four equations of the polar angle and polar radius:

$$\begin{cases} \tau_0 \frac{\partial \theta}{\partial t} = -(\mu h_1 + h_2)\frac{\rho_1^2\rho_2^2 + \rho_2^2\rho_3^2 + \rho_3^2\rho_1^2}{\rho_1\rho_2\rho_3} \sin\theta, \\ \tau_0 \frac{\partial \rho_1}{\partial t} = \mu\rho_1 + (\mu h_1 + h_2)\rho_2\rho_3 \cos\theta - [g_1\rho_1^2 + g_2(\rho_2^2 + \rho_3^2)]\rho_1, \\ \tau_0 \frac{\partial \rho_2}{\partial t} = \mu\rho_2 + (\mu h_1 + h_2)\rho_1\rho_3 \cos\theta - [g_1\rho_2^2 + g_2(\rho_1^2 + \rho_3^2)]\rho_2, \\ \tau_0 \frac{\partial \rho_3}{\partial t} = \mu\rho_3 + (\mu h_1 + h_2)\rho_1\rho_2 \cos\theta - [g_1\rho_3^2 + g_2(\rho_1^2 + \rho_2^2)]\rho_3, \end{cases} \tag{12.34}$$

where $\theta = \theta_1 + \theta_2 + \theta_3$.
When we assume $s = \frac{\mu}{(\mu h_1 + h_2)^2}$, and have the conditions that $\mu < 1$ and $s < \frac{1}{(h_1 + h_2)^2}$ because of surely positive of the coefficient of active predation D, we can get the solutions for the amplitude Eq. (12.34) as follows:

1. For the spatial homogeneous state:

$$\rho_1 = \rho_2 = \rho_3 = 0,$$

   the state becomes stable if $s < s_2 = 0$ and unstable if $s > s_2$.
2. For stripe patterns:

$$\rho_1 = \sqrt{\frac{\mu}{g_1}} \neq 0, \quad \rho_2 = \rho_3 = 0, \quad g_1 > 0,$$

the system becomes stable if $s > s_3 = \frac{g_1}{(g_1-g_2)^2}$ and unstable if $s < s_3$.

3. For spot patterns:

$$\rho_1 = \rho_2 = \rho_3 = \rho^{\pm} = \frac{|\mu h_1 + h_2| \pm \sqrt{(\mu h_1 + h_2)^2 + 4(g_1 + 2g_2)\mu}}{2(g_1 + 2g_2)},$$

they arise if $s > s_1 = -\frac{1}{4(g_1+2g_2)}$. And only the patterns corresponding to the solution ρ^+ become stable if $s < s_4 = \frac{2g_1+g_2}{(g_2-g_1)^2}$ and unstable if $s > s_4 = \frac{2g_1+g_2}{(g_2-g_1)^2}$. Besides, the solution ρ^- is always unstable.

4. For mixed patterns:

$$\rho_1 = \frac{|h|}{g_2 - g_1}, \quad \rho_2 = \rho_3 = \sqrt{\frac{\mu - g_1\rho_1^2}{g_1 + g_2}},$$

the patterns remains stable if $s > s_3$ and $g_2 > g_1 > 0$.

Remark 7 The above result indicates that the algal ecosystem can be self-organized and form at most three types of patterns when the active herd-taxis of zooplankton becomes weak enough. These patterns display that algae can naturally aggregate and form stable patches.

12.4 Numerical Experiment

In the following, we would validate our results about the self-organization phenomenon through numerical simulation, where we would take some specific parameter settings to help us ensure our analysis of the planktonic ecosystem with the weakly nonlinear cross-diffusion term. In the simulation, the diffusion terms are discretized by a five-point stencil finite difference scheme, and the left part of the system (12.2) is discretized by Eulerian difference method. For the boundaries taken as periodic, the domain is divided into 256×256 with $\Delta x = \Delta y = 1$, the temporal interval is taken as $\Delta t = 0.01$. And the initial conditions are assigned as follows: $P(i, j) = P^* + 0.5P^*\zeta^P_{i,j}$ and $Z(i, j) = Z^* + 0.5Z^*\zeta^Z_{i,j}$, where $\zeta^P_{i,j}$ and $\zeta^Z_{i,j}$ means white noise. At the same time, we keep $r = 4.90$, $e = 0.90$, $a = 12.00$, $\xi = 7.10$, $\eta = 0.01$, $D_1 = 1.00$, $D_2 = 5.10$, and assign the trend function $\Psi(Z) = Z$ as a linear function, for simplicity, where $\Psi_Z = 1$, $\Psi_{ZZ} = 0$. The following simulations consider the effect of the strength of active foraging D on the coexistence equilibrium point $(P^*, Z^*) = (0.1614, 0.0840)$, where the system remains Hopf stable, just as shown in Fig. 12.1.

Hence we get the corresponding Turing bifurcation as $D_T = 2.4320$, and four critical parameters, $s_1 = -31.7435 \times 10^{-6}$, $s_2 = 0 \times 10^{-6}$, $s_3 = 14.1922 \times 10^{-6}$, $s_4 = 316.2769 \times 10^{-6}$, which divide the "Turing pattern domain" into four parts.

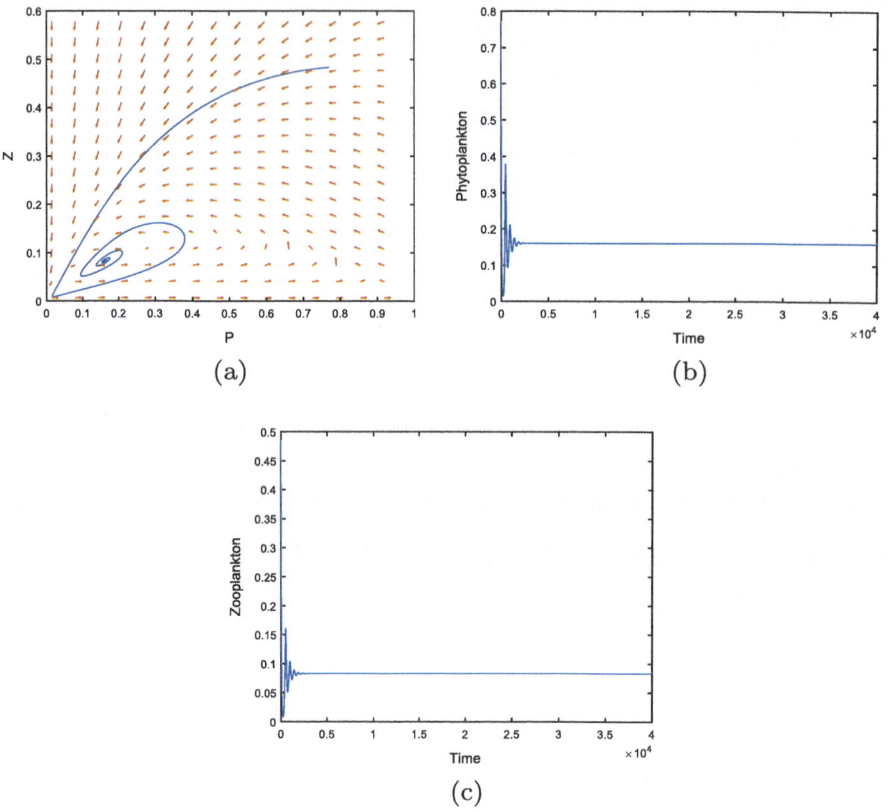

Fig. 12.1 Evolution of plankton ecosystem without any diffusion at different time, where $r = 4.90$, $e = 0.90$, $a = 12.00$, $\xi = 7.10$, $\eta = 0.01$

Theoretically, because D is defined positive, we have "Turing pattern domain" starting from point $D = 4.4538$ ($s_1 = -31.7435 \times 10^{-6}$) and ending at point $D = 0$ ($s = 50.9916 \times 10^{-6}$). So the intervals (s_1, s_2), (s_2, s_3), $(s_3, s_{D=0})$ separately display three types of dynamic behaviors: the homogeneous steady state, spot patterns, and mixed patterns. The eigenvalue curves are illustrated in Fig. 12.2 as the tuning parameters are taken as in the "Turing pattern domain". And those curves indicate the sets of Turing instability. Next the evolution of phytoplankton similar to corresponding patterns generated by the zooplankton is used to verify results by the weakly nonlinear analysis. When we take $D = 0.1$ covered by the domain $(0, 1.6736)$, $s \in (s_3, s_{D=0})$, Fig. 12.3 indicates the mixed patterns produced by phytoplankton on spatial concentration graph. When the parameter D increases to a value in $(1.6736, 2.4320)$, $S \in (s_2, s_3)$, such as $D = 2.00$, Fig. 12.4 shows the spot patterns produced by the phytoplankton on the 2-D concentration graph. Figure 12.5 shows the homogeneous stable, and the coexistence equilibrium for both kinds of plankton when $D = 3$ ($s = -9.7570 \times 10^{-6} \in (s_1, s_2)$). The above simulations illustrate that when the

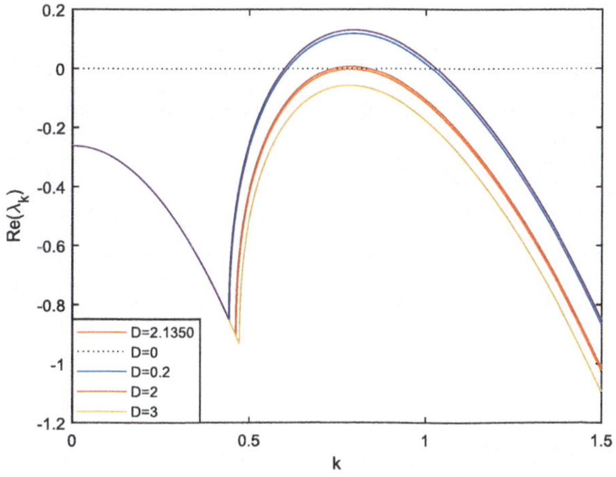

Fig. 12.2 Eigenvalue curves of the planktonic ecosystem models, where $r = 4.90$, $e = 0.90$, $a = 12.00$, $\xi = 7.10$, $\eta = 0.01$, $D_1 = 1.00$, $D_2 = 5.10$, $D_T = 2.43$, $D = 0.00, 0.10, 2.00, 3.00$

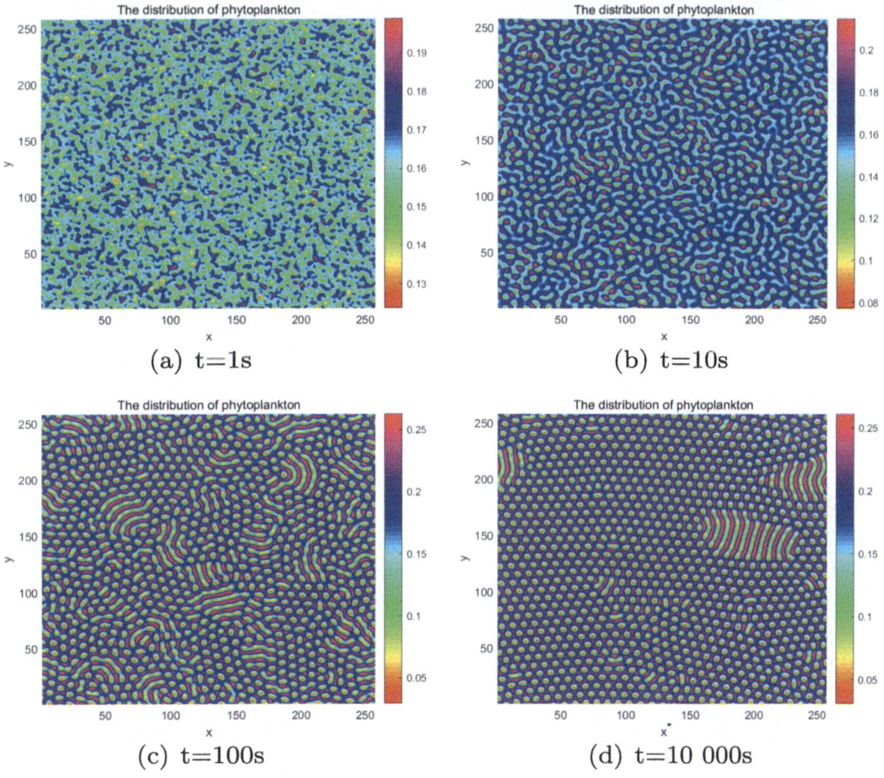

Fig. 12.3 Evolution of phytoplankton at different instants, where $r = 4.90$, $e = 0.90$, $a = 12.00$, $\xi = 7.10$, $\eta = 0.01$, $D_1 = 1.00$, $D_2 = 5.10$, $D = 0.10$ ($s = 48.5550 \times 10^{-6} \in (s_3, s_{D=0})$)

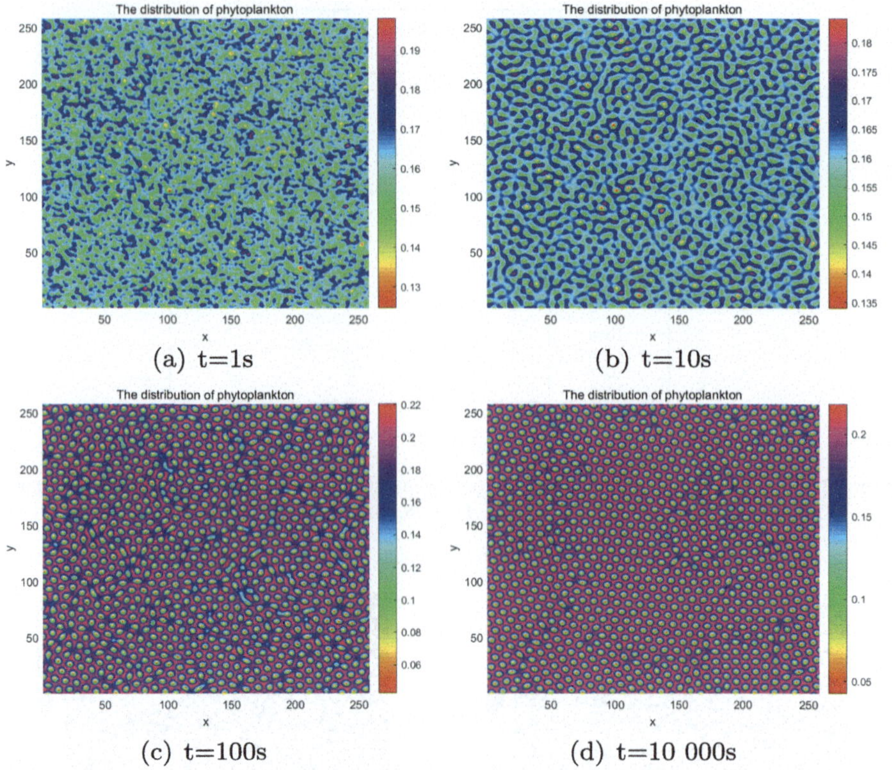

Fig. 12.4 Evolution of phytoplankton at different instants, where $r = 4.90$, $e = 0.90$, $a = 12.00$, $\xi = 7.10$, $\eta = 0.01$, $D_1 = 1.00$, $D_2 = 5.10$, $D = 2.00$ ($s = 7.9129 \times 10^{-6} \in (s_2, s_3)$)

active predation of zooplankton becomes weak enough, the algal ecosystem can form patterns by self-organization. When value D increases from zero, across and far away from the Turing bifurcation, the patterns vary: mixed patterns $\rightarrow$ spot patterns $\rightarrow$ homogeneous steady state.

Remark 8 When the zooplankton has weak active predation, we can see the algae naturally living together and forming stable high-density zones (regular patterns). And obviously because the reason that the zooplankton in the patches can survive more easily than the others do, both planktons will form stable algae communities. Moreover, when zooplankton forage more actively, the patches will vary from being smaller, unconnected, and even vanishing.

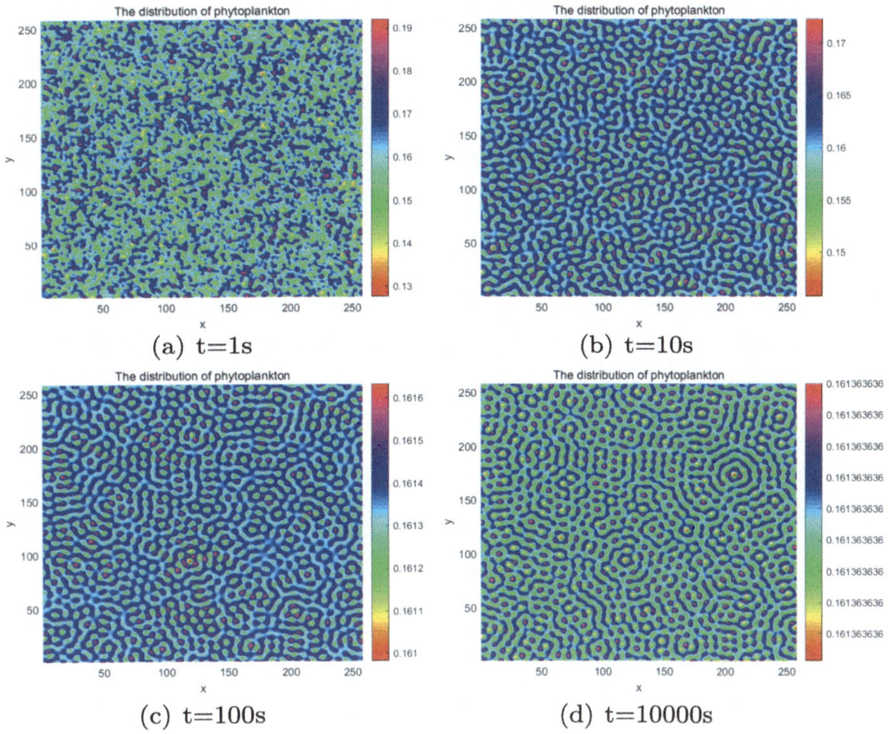

Fig. 12.5 Evolution of phytoplankton at different instants, where $r = 4.90$, $e = 0.90$, $a = 12.00$, $\xi = 7.10$, $\eta = 0.01$, $D_1 = 1.00$, $D_2 = 5.10$, $D = 3.00(s = -9.7570 \times 10^{-6} \in (s_1, s_2))$

Through our simulations, we can indicate the self-organization pattern partly and there also exist other more focus on it based on the following three possible reasons:

- the high-order part effect on the simulation;
- the enlarged effect of the nonlinear-diffusion term on the nonlinearity;
- the lower feasibility of the discretization effect on simulation.

Chapter 13
The Effect of Weakly Nonlinear Diffusion

Plankton communities play a significant role in the global biosphere, especially in the marine ecosystem. They, floated in the surface water, are composed of phytoplankton (prey) and zooplankton (predators) in which the zooplankton aggregates to be small groups, cooperatively forages phytoplankton, and grows nonlinearly and adaptively to form predation strategies and to avoid fierce interspecific competition. It is well known that plankton communities constitute the beginning role in food chains and support the property of marine lives [318]. Importantly, phytoplankton and zooplankton are two kinds of the most important planktons in the plankton communities. For instance, phytoplankton drifts on the surface water to obtain sunlight for photosynthesis, while zooplankton has ability to swim and feeds on phytoplankton [297]. Hence, phytoplankton (prey) and zooplankton (predators) compose a micro algae ecosystem by their interaction effects on the survival of both. Especially, phytoplankton in plankton communities can grow increasingly or decreasingly, giving rise to be a slow-fast processes rapidly [307, 311]. Sometimes, researchers call the high concentration of phytoplankton in the surface water by growth rapidly as "bloom" from the perspective of anthropocentrism [263]. To solve the "bloom" harm, fortunately, researchers found out that zooplankton can prey on phytoplankton [264] and contribute to preventing the rapid growth of phytoplankton [249]. Most of the time, plankton (including zooplankton and phytoplankton) aggregates into stable patches and scatters throughout the ocean. These patches formed naturally can be assumed as the self-organization pattern of plankton because of the turbulence of ocean currents.

Nowadays, we have obtained the conditions of Hopf stability and the Turing bifurcation by linear stability analysis. And noticeably, we found out that weakly nonlinear analysis can help us establish the amplitude equations determined different types of patterns such as spot patterns, stripe patterns, and mixed patterns. Through the numerical simulations, we can see the illustrated stability of the system and the patterns like self-organizing behaviors of planktons. The results play a validated role partly in helping us better understand the dynamics of the algae ecosystem.

S. T. Liu and L. Zhang, *Pattern Dynamics of Marine Plankton Behavior*,
https://doi.org/10.1007/978-981-97-5369-7_13

What's more, the weakly nonlinear analysis can be used to model related phenomena without linearizing the weakly nonlinear diffusion term.

In order to explore the nature of planktonic communities, especially the nature of phytoplankton, there have many different works and there exist two main roads. One is to investigate the ecological or physiological effect on algae in nature and the other is to examine the dynamic behaviors of planktonic communities by mathematical tools. For instance, in the guidance of the first road, the flow of matter and energy [235, 265], the global distribution of phytoplankton [280, 316], the growth law of phytoplankton by cultivation [324, 326] were investigated, respectively. The latest data-driven technology was used to predict the algal blooms by the field sampling data [269, 283]. Under the guidance of the second road, the dynamics of planktonic ecosystems were studied from the perspectives of mathematicians. Generally, researchers build the mathematical models firstly according to the primary interaction between phytoplankton and zooplankton driven by the data and the special predator–predator relationship. Furthermore, in these phytoplankton–zooplankton models, the choose of the predation function plays a key role in these related studies. Therefore, to gain the functional response from experiments was one of the focuses [320], or results driven by data [312], and inferences were also the aim [300, 310]. Based on the models and functions, the dynamics of plankton communities can be studied and can also explain and infer the evolution of phytoplankton [227].

Importantly, there exist usually two kinds of prey–predator models depicting the interaction of phytoplankton-zooplankton. The first case is that phytoplankton/zooplankton was treated as a whole and the heterogeneity was eliminated based on ordinary differential equations. Conditions were gained for the steady coexistence, extinction, periodical behaviors, and some other results [296, 309]. Moreover, more comprehensive solutions were modeled by introducing some other factors, such as nutrients and fishes [322].

The second kind of model is related about partial differential equations. Because those neighbors affect plankton in three-dimensional space, there were many interesting works to present some spatial factors into the models for studying the evolution of the plankton communities, in which the spatial factors include diffusion [270], nonlocal predation [308], sinking [313], depth [303], advection [260].

Furthermore, based on the fact that plankton lives near the water surface, there were related works about the interaction of plankton ecosystems between two adjacent layers for divided layers of the surface water [315, 317]. However, spatiotemporal models have been considered directly or limitedly in two dimensions and also did not consider the influence of sunlight, depth, sinking and some other secondary factors.

Reaction–diffusion equations can be regarded as one of the useful models for predation and the diffusion. As the corresponding reaction equations remain homogeneous Hopf stable, the ecosystem spatial Turing unstable can be driven and self-organization patterns can be obtained by the diffusion terms represented by different coefficients in these equations. Additionally, the effects of different factors like cross-diffusion [289], diffusive toxin [250], and delay [319] on the pattern formation of plankton were studied. What's more, because predation is the primary relationship between phytoplankton and zooplankton, some works were presented by prey–predation relationship [314, 321]. In addition, there were also many related researches about the pattern formation like the Josephson junction [323], economic [325] excitatory media [298, 299], or protein models [306]. To be honest, although more and more works focus on the pattern formation of planktons like self-organization pattern, we cannot obtain this ideal pattern in the real environment [301, 305]. But based on the following three reasons, we insist on the assumption that the planktonic patches, large and small, behave as the self-organization pattern of planktons in the ocean [302]:

(i) Self-organization pattern is one outstanding feature of the complexity system like the marine environments remain too heterogeneous and complexity.

(ii) The self-organization will be driven by ocean currents to some extent where the oceans are full of turbulence with different scales ranging from a few centimeters to thousands of kilometers.

(iii) Plankton competes in the same or different layers of the complex food webs composed of algae communities with more than one kind of phytoplankton (prey) or zooplankton (predators). In addition, larger carnivores in the higher layers also play important role in the dynamics of the plankton ecosystems.

Shortly, most of the mathematical models about algae are too simple and have too idealistic assumptions to uncover the complex dynamics of the marine environments.

In this chapter, we present a two-dimensional reaction–diffusion model introducing weakly nonlinear diffusion term for the plankton community. Comparing various predation functions, functional responses for the suitability of these predation behaviors [304], we choose the functional response for herd predation as our candidate [277]. Based on the fact that zooplankton aggregates together as small groups looking for and besieging phytoplankton to improve their predation efficiency, its density cannot be too low, meaning that the zooplankton spreads too slow whereas when the density becomes more higher, it diffuses faster to avoid fierce intraspecific competition. Such phenomena can be regarded as weakly nonlinear diffusion by the weakly nonlinear analysis. Hence, to present the effect of the weakly nonlinear diffusion term on the self-organization pattern in the planktonic ecosystem is our main purpose.

Our works are composed of four parts. At first, the Hopf stable near the coexistence equilibria of the ecosystem is ensured. Secondly, we obtain the Turing bifurcation by the characteristic equation. Thirdly, the critical values are obtained by weakly nonlinear analysis for the type of the patterns generated in the "Turing pattern domain". Finally, according to the theoretical analysis, we perform some numeral simulations to illustrate the evolution of the self-organization which verifies our results.

13.1 Model with Weakly Nonlinear Diffusion

The mathematical model of the algal ecosystem is presented as below:

$$\begin{cases} \frac{\partial P}{\partial t} = rP\left(1 - \frac{P}{K}\right) - \frac{Ce_2PZ}{\sqrt[3]{Z}+hCe_2P} + D_1\nabla^2 P, \\ \frac{\partial Z}{\partial t} = e\frac{Ce_2PZ}{\sqrt[3]{Z}+hCe_2P} - \eta Z + D_2\left(\frac{Z}{Z_c}\right)\nabla^2 Z, \end{cases} \tag{13.1}$$

where P (prey) is phytoplankton and Z (predators) means zooplankton as the factors of the plankton community. More explanations and assumption about the model are listed as following:

[(1)] There is no flux on the boundaries; r is the growth rate of phytoplankton logistically, and K is the natural carrying capacity; A new kind of functional response – $\frac{Ce_2P}{\sqrt[3]{Z}+hCe_2P}$ is chosen to present the fact that the zooplankton aggregates into small groups and forages phytoplankton within the ocean tridimensionally [151]; D_1 stands for the diffusion coefficient of phytoplankton naturally; $D_2\left(\frac{Z}{Z_c}\right)$ represents the generalized diffusion coefficient for the nonlinearity of the zooplankton growth, where when the density remains lower than the critical parameter Z_c, the coefficient means the slower diffusion rate than natural diffusion rate; otherwise, it stands for the faster case to avoid fierce intraspecific competition.

After making a one-one mapping to simplify the ecosystem

$$\frac{P}{K} \mapsto P, \qquad \frac{Z}{(hCe_2K)^3} \mapsto Z, \qquad Ce_2(hCe_2K)^2 \mapsto a,$$

$$\frac{e}{hCe_2(hCe_2K)^2} \mapsto e, \qquad \frac{D_2(hCe_2K)^3}{Z_c} \mapsto D_2,$$

we have the following rearranged equations:

$$\begin{cases} \frac{\partial P}{\partial t} &= rP(1-P) - \frac{aPZ}{\sqrt[3]{Z}+P} + D_1\nabla^2 P, \\ \frac{\partial Z}{\partial t} &= e\frac{aPZ}{\sqrt[3]{Z}+P} - \eta Z + D_2 Z\nabla^2 Z. \end{cases} \tag{13.2}$$

In the following, the dynamical behaviors of this model are our focus.

13.2 Stability Analysis

13.2.1 Hopf Stability

Firstly, the following homogeneous solution of the model without the diffusion terms will be considered:

$$\begin{cases} \frac{dP}{dt} &= rP(1-P) - \frac{aPZ}{\sqrt[3]{Z}+P}, \\ \frac{dZ}{dt} &= e\frac{aPZ}{\sqrt[3]{Z}+P} - \eta Z. \end{cases} \tag{13.3}$$

Let

$$F = Pf, \qquad f = r(1-P) - \frac{aZ}{\sqrt[3]{Z}+P};$$

$$G = Zg, \qquad g = e\frac{aP}{\sqrt[3]{Z}+P} - \eta.$$

Then, we can express the above ODE system as below:

$$\begin{cases} \frac{dP}{dt} = Pf(P, Z), \\ \frac{dZ}{dt} = Zg(P, Z). \end{cases} \tag{13.4}$$

We can easily prove the nonnegative definiteness for this planktonic ecosystem [246], where the system has one positive coexistence equilibrium point (P^*, Z^*) and two unstable equilibrium points: $(0, 0)$ meaning the extinction of two plankton and $(1, 0)$ meaning the extinction of zooplankton.

Then we obtain the coexistence equilibrium as following:

$$P^* = \frac{-er\eta^2 + \sqrt{(er\eta^2)^2 + 4er\eta^2(ea-\eta)^3}}{2(ea-\eta)^3}, \qquad Z^* = \frac{erP^*(1-P^*)}{\eta}.$$

A linearized model can be obtained by the Hopf stability conditions by linearizing the equations around the coexistence equilibrium point:

$$\frac{d}{dt}\begin{pmatrix}\delta P\\ \delta Z\end{pmatrix} = \begin{pmatrix}a_{11} & a_{12}\\ a_{21} & a_{22}\end{pmatrix}\begin{pmatrix}\delta P\\ \delta Z\end{pmatrix} = \begin{pmatrix}P^*f_P^* & P^*f_Z^*\\ Z^*g_P^* & Z^*g_Z^*\end{pmatrix}\begin{pmatrix}\delta P\\ \delta Z\end{pmatrix}, \tag{13.5}$$

where $\delta P = P - P^*$ and $\delta Z = Z - Z^*$,

$$f_P^* = -r + \frac{aZ^*}{(\sqrt[3]{Z^*}+P^*)^2}, \qquad f_Z^* = -\frac{a}{\sqrt[3]{Z^*}+P^*} + \frac{a\sqrt[3]{Z^*}}{3(\sqrt[3]{Z^*}+P^*)^2},$$

$$g_P^* = e\frac{a\sqrt{Z^*}}{(\sqrt[3]{Z^*}+P^*)^2}, \qquad g_Z^* = -e\frac{aP^*}{3\sqrt[3]{Z^*}^2(\sqrt[3]{Z^*}+P^*)^2}.$$

Hence, we get the Hopf stability for The ODE system 13.3 when

$$Tr_0 = a_{11} + a_{22} > 0, \qquad \Delta_0 = a_{11}a_{22} - a_{12}a_{21} < 0.$$

Obviously, an ecosystem as Hopf stable can be driven into instability by diffusion term.

13.2.2 Turing Bifurcation

In the following, we assume the Hopf stability of the system considering the effect of diffusion. As we obtain the linearized equations of the model (13.2) as following:

$$\frac{d}{dt}\begin{pmatrix}\delta P\\ \delta Z\end{pmatrix} = \begin{pmatrix}a_{11} & a_{12}\\ a_{21} & a_{22}\end{pmatrix}\begin{pmatrix}\delta P\\ \delta Z\end{pmatrix} + \begin{pmatrix}D_1\nabla^2\delta P\\ D_2 Z^*\nabla^2\delta Z\end{pmatrix}, \tag{13.6}$$

we obtain the form of the solution of (13.6) as below:

$$\begin{pmatrix}\delta P\\ \delta Z\end{pmatrix} = \sum\begin{pmatrix}A_k^P\\ A_k^Z\end{pmatrix} e^{\lambda t+ik\cdot r}, \quad k = \|k\|.$$

Therefore, we have the characteristic equation as following:

$$\lambda\begin{pmatrix}A_k^P\\ A_k^Z\end{pmatrix} = \begin{pmatrix}a_{11}-D_1k^2 & a_{12}\\ a_{21} & a_{22}-D_2Z^*k^2\end{pmatrix}\begin{pmatrix}A_k^P\\ A_k^Z\end{pmatrix}, \tag{13.7}$$

and the corresponding dispersion equation as:

$$\lambda^2 + Tr_k\lambda + \Delta_0 = 0, \tag{13.8}$$

where

$$\begin{aligned} Tr_k &= a_{11} - D_1k^2 + a_{22} - D_2Z^*k^2 = Tr_0 - (D_1 + D_2Z^*)k^2,\\ \Delta_k &= (a_{11} - D_1k^2)(a_{22} - D_2Z^*k^2) - a_{12}a_{21}\\ &= D_1D_2Z^*(k^2)^2 - (a_{11}D_2Z^* + a_{22}D_1)k^2 + \Delta_0. \end{aligned}$$

And we get the results as below:

$$\lambda_k = \frac{Tr_k \pm \sqrt{Tr_k^2 - 4\Delta_k}}{2}. \tag{13.9}$$

Then when $Re(\lambda_k) < 0$, the reaction–diffusion system remains Turing stable meaning:

$$Tr_k < 0, \qquad \Delta_k > 0.$$

The term Δ_k is the only factor causing Turing instability because $Tr_k < Tr_0 < 0$. Hence, the system becomes Turing unstable when there exists a k satisfying $\Delta_k \le 0$ like the following case:

$$\Delta_{k_T} = \Delta_0 - \frac{(a_{11}D_2Z^* + a_{22}D_1)^2}{4D_1D_2Z^*} \le 0, \tag{13.10}$$

$$k_T^2 = \frac{a_{11}D_2Z^* + a_{22}D_1}{2D_1D_2Z^*} > 0. \tag{13.11}$$

When we assume D_2 as the tuning parameter, the Turing bifurcation can be obtained as:

$$D_{2T} = \{D_2 | \Delta_{k_T} = 0, Tr_k < 0, \Delta_k \geq 0\}. \tag{13.12}$$

In addition to Turing bifurcation, the self-organization behaviors of the planktonic ecosystem can also be obtained through weakly nonlinear analysis.

13.3 Weakly Nonlinear Analysis

In order to study the dynamical behaviors such as self-organization of the ecosystem around the Turing bifurcation, perturbation analysis would be employed around the homogeneous steady-state point and the third-order Taylor series expansion approximation would be calculated in the following:

$$\begin{aligned}\frac{\partial}{\partial t}\begin{pmatrix}P\\Z\end{pmatrix} \approx & L_T\begin{pmatrix}P\\Z\end{pmatrix} \\ & + D_{2T}\begin{pmatrix}0\\Z\nabla^2 Z\end{pmatrix} + (D_2 - D_{2T})\begin{pmatrix}0\\Z^*\nabla^2 Z + Z\nabla^2 Z\end{pmatrix} \\ & + \frac{1}{2}\begin{pmatrix}F_{PP}P^2 + 2F_{PZ}PZ + F_{ZZ}Z^2\\G_{PP}P^2 + 2G_{PZ}PZ + G_{ZZ}Z^2\end{pmatrix} \\ & + \frac{1}{6}\begin{pmatrix}F_{PPP}P^3 + 3F_{PPZ}P^2Z + 3F_{PZZ}PZ^2 + F_{ZZZ}Z^3\\G_{PPP}P^3 + 3G_{PPZ}P^2Z + 3G_{PZZ}PZ^2 + G_{ZZZ}Z^3\end{pmatrix} \\ & + \ldots,\end{aligned} \tag{13.13}$$

where

$$L_T = \begin{pmatrix}a_{11} + D_1\nabla^2 & a_{12}\\ a_{21} & a_{22} + D_{2T}Z^*\nabla^2\end{pmatrix}, \tag{13.14}$$

and all the coefficients of (13.13) are listed in the Appendices.

When the adjustment parameter D_2 changes near the Turing bifurcation, we get the form of the solution of (13.13) as following:

$$\begin{pmatrix}P\\Z\end{pmatrix} = \sum_{j=1}^{3} U_0\left(A_j e^{ik_j\cdot r} + \overline{A}_j e^{-ik_j\cdot r}\right), \ \|k_j\| = k_T, \tag{13.15}$$

where U_0 means the direction vector and A_j, $\overline{A_j}$ stand for the amplitudes of mode k_j, $-k_j$, respectively. The perturbations of the adjustment parameter D_2, time t and plankton P, Z are extended to different scales:

$$P = \epsilon P_1 + \epsilon^2 P_2 + \epsilon^3 P_3 + \cdots , \tag{13.16a}$$

$$Z = \epsilon Z_1 + \epsilon^2 Z_2 + \epsilon^3 Z_3 + \cdots , \tag{13.16b}$$

$$\frac{\partial}{\partial t} = \epsilon \frac{\partial}{\partial T_1} + \epsilon^2 \frac{\partial}{\partial T_2} + \epsilon^3 \frac{\partial}{\partial T_3} + \cdots , \tag{13.16c}$$

$$D_2 - D_{2T} = \epsilon D_{21} + \epsilon^2 D_{22} + \cdots . \tag{13.16d}$$

According to the different order of ϵ, the following equations for different scales would be obtained:
the first order of ϵ:

$$L_T \begin{pmatrix} P_1 \\ Z_1 \end{pmatrix} = 0, \tag{13.17}$$

the second order of ϵ:

$$\begin{aligned} L_T \begin{pmatrix} P_2 \\ Z_2 \end{pmatrix} =& \frac{\partial}{\partial T_1} \begin{pmatrix} P_1 \\ Z_1 \end{pmatrix} \\ &- D_{2T} \begin{pmatrix} 0 \\ Z_1 \nabla^2 Z_1 \end{pmatrix} - D_{21} \begin{pmatrix} 0 \\ Z^* \nabla^2 Z_1 \end{pmatrix} \\ &- \frac{1}{2} \begin{pmatrix} F_{PP} P_1^2 + 2F_{PZ} P_1 Z_1 + F_{ZZ} Z_1^2 \\ G_{PP} P_1^2 + 2G_{PZ} P_1 Z_1 + G_{ZZ} Z_1^2 \end{pmatrix}, \end{aligned} \tag{13.18}$$

and the third order of ϵ:

$$\begin{aligned} L_T \begin{pmatrix} P_3 \\ Z_3 \end{pmatrix} =& \frac{\partial}{\partial T_1} \begin{pmatrix} P_2 \\ Z_2 \end{pmatrix} + \frac{\partial}{\partial T_2} \begin{pmatrix} P_1 \\ Z_1 \end{pmatrix} \\ &- D_{2T} \begin{pmatrix} 0 \\ Z_1 \nabla^2 Z_2 + Z_2 \nabla^2 Z_1 \end{pmatrix} \\ &- \begin{pmatrix} 0 \\ D_{21} Z^* \nabla^2 Z_2 + D_{22} Z^* \nabla^2 Z_1 \end{pmatrix} - D_{21} \begin{pmatrix} 0 \\ Z_1 \nabla^2 Z_1 \end{pmatrix} \\ &- \begin{pmatrix} F_{PP} P_1 P_2 + F_{PZ}(P_1 Z_2 + P_2 Z_1) + F_{ZZ} Z_1 Z_2 \\ G_{PP} P_1 P_2 + G_{PZ}(P_1 Z_2 + P_2 Z_1) + G_{ZZ} Z_1 Z_2 \end{pmatrix} \\ &- \frac{1}{6} \begin{pmatrix} F_{PPP} P_1^3 + 3F_{PPZ} P_1^2 Z_1 + 3F_{PZZ} P_1 Z_1^2 + F_{ZZZ} Z_1^3 \\ G_{PPP} P_1^3 + 3G_{PPZ} P_1^2 Z_1 + 3G_{PZZ} P_1 Z_1^2 + G_{ZZZ} Z_1^3 \end{pmatrix}, \end{aligned} \tag{13.19}$$

for the ϵ:

$$L_T \begin{pmatrix} P_1 \\ Z_1 \end{pmatrix} = 0, \tag{13.20}$$

the solution is taken as the linear combination of the eigenvectors of the eigenvalue 0 for the linear operator L_T,

$$\begin{pmatrix} P_1 \\ Z_1 \end{pmatrix} = \begin{pmatrix} l \\ 1 \end{pmatrix} \left(\sum_{j=1}^{3} W_j e^{ik_j \cdot r} + c.c. \right), \quad k = \|k_j\|, \tag{13.21}$$

where $l = -\frac{a_{12}}{a_{11} - D_1 k_T^2}$, and W_j means the corresponding amplitude of the mode $e^{ik_j \cdot r}$. We then get the following solution by the higher-order ϵ, such as ϵ^2 and ϵ^3.

For the ϵ^2:

$$\begin{aligned} L_T \begin{pmatrix} P_2 \\ Z_2 \end{pmatrix} &= \frac{\partial}{\partial T_1} \begin{pmatrix} l \\ 1 \end{pmatrix} Z_1 \\ &\quad - D_{2T}(-k_T^2) \begin{pmatrix} 0 \\ Z_1^2 \end{pmatrix} - D_{21}(-k_T^2) \begin{pmatrix} 0 \\ Z^* Z_1 \end{pmatrix} \\ &\quad - \frac{1}{2} \begin{pmatrix} F_{PP} l^2 + 2F_{PZ} l + F_{ZZ} \\ G_{PP} l^2 + 2G_{PZ} l + G_{ZZ} \end{pmatrix} Z_1^2 \\ &= \begin{pmatrix} R_2^P \\ R_2^Z \end{pmatrix}. \end{aligned} \tag{13.22}$$

The above is named as the Fredholm Solvability Condition, which determines the existence of the solution of (13.22), meaning that the right-hand side of Eq. (13.22) becomes orthogonal with the eigenvectors of the eigenvalue, zero, of L_T^* (the adjoint operator of L_T).

Naturally, we get the eigenvectors of zero as:

$$\begin{pmatrix} 1 \\ l^* \end{pmatrix} e^{ik_j \cdot r} + c.c. \quad (j = 1, 2, 3),$$

where $l^* = -\frac{a_{11} - D_1 k_T^2}{a_{21}}$.

So we obtain the following condition by the solvability condition:

$$\begin{pmatrix} 1 & l^* \end{pmatrix} \begin{pmatrix} R_2^P \\ R_2^Z \end{pmatrix} = 0. \tag{13.23}$$

Because of the form of the solution of (13.22) as:

$$\begin{aligned} \begin{pmatrix} P_2 \\ Z_2 \end{pmatrix} &= \begin{pmatrix} X_0 \\ Y_0 \end{pmatrix} + \sum_{j=1}^{3} \begin{pmatrix} X_j \\ Y_j \end{pmatrix} e^{ik_j \cdot r} + \sum_{j=1}^{3} \begin{pmatrix} X_{jj} \\ Y_{jj} \end{pmatrix} e^{i2k_j \cdot r} + \begin{pmatrix} X_{12} \\ Y_{12} \end{pmatrix} e^{i(k_1 - k_2) \cdot r} \\ &\quad + \begin{pmatrix} X_{23} \\ Y_{23} \end{pmatrix} e^{i(k_2 - k_3) \cdot r} + \begin{pmatrix} X_{31} \\ Y_{31} \end{pmatrix} e^{i(k_3 - k_1) \cdot r} + c.c. \end{aligned} \tag{13.24}$$

we obtain the solution of (13.22) by substituting (13.24) into the Eq. (13.22), and recollecting the coefficients of different modes:

$$X_j = lY_j$$

$$\begin{pmatrix} X_0 \\ Y_0 \end{pmatrix} = -L_0^{-1} \begin{pmatrix} H_2^f \\ H_2^g + D_{2T}(-2k_T^2) \end{pmatrix} \left(\sum_{j=1}^{3} |W_j|^2 \right) = \begin{pmatrix} U_0 \\ V_0 \end{pmatrix} \left(\sum_{j=1}^{3} |W_j|^2 \right),$$

$$\begin{pmatrix} X_{jj} \\ Y_{jj} \end{pmatrix} = -L_{4k_T^2}^{-1} \begin{pmatrix} \frac{1}{2} H_2^f \\ \frac{1}{2} H_2^g + D_{2T}(-k_T^2) \end{pmatrix} W_j^2 = \begin{pmatrix} U_{jj} \\ V_{jj} \end{pmatrix} W_j^2,$$

$$\begin{pmatrix} X_{js} \\ Y_{js} \end{pmatrix} = -L_{3k_T^2}^{-1} \begin{pmatrix} H_2^f \\ H_2^g + D_{2T}(-2k_T^2) \end{pmatrix} W_j \overline{W}_s = \begin{pmatrix} U_{js} \\ V_{js} \end{pmatrix} W_j \overline{W}_s,$$

where $j, s = 1, 2, 3$, $j \neq s$ and

$$L_{k_T^2} = \begin{pmatrix} a_{11} - D_1 k_T^2 & a_{12} \\ a_{21} & a_{22} - D_{2T} Z^* k_T^2 \end{pmatrix},$$

$$H_2^f = F_{PP} l^2 + 2F_{PZ} l + F_{ZZ}, \quad H_2^g = G_{PP} l^2 + 2G_{PZ} l + G_{ZZ}.$$

For the ϵ^3

$$\begin{aligned} L_T \begin{pmatrix} P_3 \\ Z_3 \end{pmatrix} =& \frac{\partial}{\partial T_1} \begin{pmatrix} P_2 \\ Z_2 \end{pmatrix} + \frac{\partial}{\partial T_2} \begin{pmatrix} l \\ 1 \end{pmatrix} Z_1 \\ & - D_{2T} \begin{pmatrix} 0 \\ Z_1 \nabla^2 Z_2 + (-k_T^2) Z_2 Z_1 \end{pmatrix} \\ & - \begin{pmatrix} 0 \\ D_{21} Z^* \nabla^2 Z_2 + D_{22}(-k_T^2) Z^* Z_1 \end{pmatrix} - D_{21} \begin{pmatrix} 0 \\ (-k_T^2) Z_1^2 \end{pmatrix} \\ & - \begin{pmatrix} F_{PP} l Z_1 P_2 + F_{PZ} l Z_1 Z_2 + F_{PZ} Z_1 P_2 + F_{ZZ} Z_1 Z_2 \\ G_{PP} l Z_1 P_2 + G_{PZ} l Z_1 Z_2 + G_{PZ} Z_1 P_2 + G_{ZZ} Z_1 Z_2 \end{pmatrix} \\ & - \frac{1}{6} \begin{pmatrix} F_{PPP} l^3 + 3F_{PPZ} l^2 + 3F_{PZZ} l + F_{ZZZ} \\ G_{PPP} l^3 + 3G_{PPZ} l^2 + 3G_{PZZ} l + G_{ZZZ} \end{pmatrix} Z_1^3 \\ =& \begin{pmatrix} R_3^P \\ R_3^Z \end{pmatrix}, \end{aligned} \tag{13.25}$$

and the solvability condition, we can also obtain:

$$(1 \;\; l^*) \begin{pmatrix} R_3^P \\ R_3^Z \end{pmatrix} = 0.$$

Then we have the amplitude equations of mode $e^{ik_j \cdot r}$ $(j = 1, 2, 3)$ as:

$$\begin{cases} (l + l^*)(\frac{\partial W_1}{\partial T_2} + \frac{\partial Y_1}{\partial T_1}) = & \Gamma(D_{21}Y_1 + D_{22}W_1) \\ & +\Omega D_{21}\overline{W}_2\overline{W}_3 + H_0(\overline{W}_2\overline{Y}_3 + \overline{W}_3\overline{Y}_2) \\ & +[G_1|W_1|^2 + G_2(|W_2|^2 + |W_3|^2)]W_1, \\ (l + l^*)(\frac{\partial W_2}{\partial T_2} + \frac{\partial Y_2}{\partial T_1}) = & \Gamma(D_{21}Y_2 + D_{22}W_2) \\ & +\Omega D_{21}\overline{W}_1\overline{W}_3 + H_0(\overline{W}_1\overline{Y}_3 + \overline{W}_3\overline{Y}_1) \\ & +[G_1|W_2|^2 + G_2(|W_1|^2 + |W_3|^2)]W_2, \\ (l + l^*)(\frac{\partial W_3}{\partial T_2} + \frac{\partial Y_3}{\partial T_1}) = & \Gamma(D_{21}Y_3 + D_{22}W_3) \\ & +\Omega D_{21}\overline{W}_1\overline{W}_2 + H_0(\overline{W}_1\overline{Y}_2 + \overline{W}_2\overline{Y}_1) \\ & +[G_1|W_3|^2 + G_2(|W_1|^2 + |W_2|^2)]W_3, \end{cases} \tag{13.26}$$

where

$$\begin{aligned} \Gamma =& - k_T^2 l^* Z^*, \\ \Omega =& - 2k_T^2 l^*, \\ H_0 =& - 2k_T^2 l^* D_{2T} + [F_{PP}l + F_{PZ} + l^*(G_{PP}l + G_{PZ})]l \\ &+ [F_{PZ}l + F_{ZZ} + l^*(G_{PZ}l + G_{ZZ})], \\ G_1 =& - k_T^2 l^* D_{2T}(V_0 + V_{jj}) - 4k_T^2 l^* D_{2T} V_{jj} \\ &+ [F_{PP}l + F_{PZ} + l^*(G_{PP}l + G_{PZ})](U_0 + U_{jj}) \\ &+ [F_{PZ}l + F_{ZZ} + l^*(G_{PZ}l + G_{ZZ})](V_0 + V_{jj}) + \frac{1}{2}(H_3^f + l^* H_3^g), \\ G_2 =& - k_T^2 l^* D_{2T}(V_0 + V_{js}) - 3k_T^2 l^* D_{2T} V_{js} \\ &+ [F_{PP}l + F_{PZ} + l^*(G_{PP}l + G_{PZ})](U_0 + U_{js}) \\ &+ [F_{PZ}l + F_{ZZ} + l^*(G_{PZ}l + G_{ZZ})](V_0 + V_{js}) + (H_3^f + l^* H_3^g), \end{aligned}$$

and

$$\begin{aligned} H_3^f =& FPPPl^3 + 3F_{PPZ}l^2 + 3F_{PZZ}l + F_{ZZZ}, \\ H_3^g =& G_{PPP}l^3 + 3G_{PPZ}l^2 + 3G_{PZZ}l + G_{ZZZ}. \end{aligned}$$

We then expand the amplitude A_j $(j = 1, 2, 3)$ as following:

$$A_j = \epsilon W_j + \epsilon^2 Y_j + o(\epsilon^2). \tag{13.27}$$

The amplitude equations can be obtained by the multi-scale extension of time 13.16:

$$\begin{cases} \tau_0 \frac{\partial A_1}{\partial t} = \mu A_1 + \omega\mu \overline{A}_2\overline{A}_3 + h_0\overline{A}_2\overline{A}_3 - \left[g_1|A_1|^2 + g_2\left(|A_2|^2 + |A_3|^2\right)\right] A_1, \\ \tau_0 \frac{\partial A_2}{\partial t} = \mu A_2 + \omega\mu \overline{A}_1\overline{A}_3 + h_0\overline{A}_2\overline{A}_3 - \left[g_1|A_2|^2 + g_2\left(|A_1|^2 + |A_3|^2\right)\right] A_2, \\ \tau_0 \frac{\partial A_3}{\partial t} = \mu A_3 + \omega\mu \overline{A}_1\overline{A}_2 + h_0\overline{A}_2\overline{A}_3 - \left[g_1|A_3|^2 + g_2\left(|A_1|^2 + |A_2|^2\right)\right] A_3. \end{cases} \tag{13.28}$$

where:

$$\tau_0 = \frac{l + l^*}{D_{2T}\Gamma}, \mu = \frac{D_2 - D_{2T}}{D_{2T}}, \omega = \frac{\Omega}{D_{2T}\Gamma}, h_0 = \frac{H_0}{D_{2T}\Gamma}, g_1 = -\frac{G_1}{D_{2T}\Gamma}, g_2 = -\frac{G_2}{D_{2T}\Gamma}.$$

13.3.1 Analysis on the Amplitude Equations

In the following, the stable solutions of the amplitude equations and the corresponding patterns will be presented, respectively.

After defining $A_j = \rho_j e^{i\theta_j}$ $(j = 1, 2, 3)$ and substituting the solution into (13.28), we can get the following equations by separating the real and imaginary parts:

$$\begin{cases} \tau_0 \frac{\partial \theta}{\partial t} &= -h\frac{\rho_1^2\rho_2^2 + \rho_2^2\rho_3^2 + \rho_3^2\rho_1^2}{\rho_1\rho_2\rho_3} \sin\theta, \\ \tau_0 \frac{\partial \rho_1}{\partial t} &= \mu\rho_1 + h\rho_2\rho_3\cos\theta - [g_1\rho_1^2 + g_2(\rho_2^2 + \rho_3^2)]\rho_1, \\ \tau_0 \frac{\partial \rho_2}{\partial t} &= \mu\rho_2 + h\rho_1\rho_3\cos\theta - [g_1\rho_2^2 + g_2(\rho_1^2 + \rho_3^2)]\rho_2, \\ \tau_0 \frac{\partial \rho_3}{\partial t} &= \mu\rho_3 + h\rho_1\rho_2\cos\theta - [g_1\rho_3^2 + g_2(\rho_1^2 + \rho_2^2)]\rho_3, \end{cases} \tag{13.29}$$

where $\theta = \theta_1 + \theta_2 + \theta_3$ and $h = \mu\omega + h_0$.

When we define $s = \frac{\mu}{h^2} = \frac{\mu}{(\mu\omega + h_1)^2}$, we can obtain four kinds of solutions for the system:

1. For the spatial homogeneous state:

$$\rho_1 = \rho_2 = \rho_3 = 0,$$

the system becomes stable when $s < s_2 = 0$; the system remains unstable when $s > s_2$.

2. For the Stripe patterns:

$$\rho_1 = \sqrt{\frac{\mu}{g_1}} \neq 0, \quad \rho_2 = \rho_3 = 0, \quad g_1 > 0,$$

the state becomes stable when $s > s_3 = \frac{g_1}{(g_1 - g_2)^2}$; the state remains unstable when $s < s_3$.

3. For the Spot patterns:

$$\rho_1 = \rho_2 = \rho_3 = \rho^{\pm} = \frac{|h| \pm \sqrt{h^2 + 4(g_1 + 2g_2)\mu}}{2(g_1 + 2g_2)},$$

there are spot patterns when $s > s_1 = \frac{-1}{4(g_1+2g_2)}$; solution ρ^+ remains stable when $s < s_4 = \frac{2g_1+g_2}{(g_2-g_1)^2}$; solution ρ^+ becomes unstable when $\mu > \mu_4 = \frac{2g_1+g_2}{(g_2-g_1)^2}$; in addition, solution ρ^- remains always unstable.

4. For the mixed patterns:

$$\rho_1 = \frac{|h|}{g_2 - g_1}, \quad \rho_2 = \rho_3 = \sqrt{\frac{\mu - g_1\rho_1^2}{g_1 + g_2}},$$

The state remains stable when $s > s_3$ and $g_2 > g_1 > 0$.

13.4 Numerical Experiment

We next present numerical simulations to validate our results and help us better understand the role of the weakly nonlinear diffusion in the self-diffusion process. In the simulations, the five-point stencil finite difference scheme is used to discretize the diffusion term and the Euler difference method is applied to discretize the others. The simulations run on a 256×256 grid with no flux boundaries, a spatial step size is taken as $\Delta x = \Delta y = 1$, and a time step size is taken as $\Delta t = 0.01$. Assume five parameters, r, e, a, η, and D_1, as:

$$r = 1.00, e = 0.12, a = 29.50, \eta = 1.70, D_1 = 0.10.$$

Figure 13.1 illustrates the Hopf stability for the system at the coexistence equilibrium point $(P^*, Z^*) = (0.2097, 0.0117)$.

We take the initial population distribution of phytoplankton and zooplankton as $P(i, j) = P^* + 0.5P^*\zeta_{i,j}$ and $Z(i, j) = Z^* + 0.5Z^*\xi_{i,j}$, where $\zeta_{i,j}$ and $\xi_{i,j}$ mean the white noise. Without loss of generality, the distribution of phytoplankton, and zooplankton are assumed to share similar spatial distribution. The Turing bifurcation is taken as $D_{2T} = 654.9$, and the corresponding critical parameters are listed as $s_i (i = 1, 2, 3, 4)$, and separating the patterns of different kinds, are -3.65×10^{-8}, 0×10^{-8}, 92.13×10^{-8}, 356.18×10^{-8}. The eigenvalue curves of the reaction–diffusion equations are illustrated in Fig. 13.2 under different tuning parameters, $D_{2T} = 654.9$, $D_2 = 653.8, 683.7, 769.2, 1025.6$.

According to the eigenvalue curves, the four D_2s are chosen to illustrate three kinds of Turing patterns. Figure 13.3 shows that as we take the parameter $D_2 = 653.8$

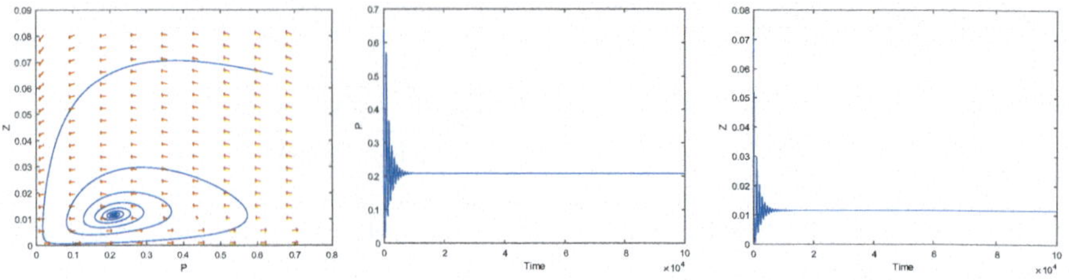

(a) (b) (c)

Fig. 13.1 Evolution of planktonic ecosystem at different times, where $r = 1.00$, $e = 0.12$, $a = 29.50$, $\eta = 1.70$

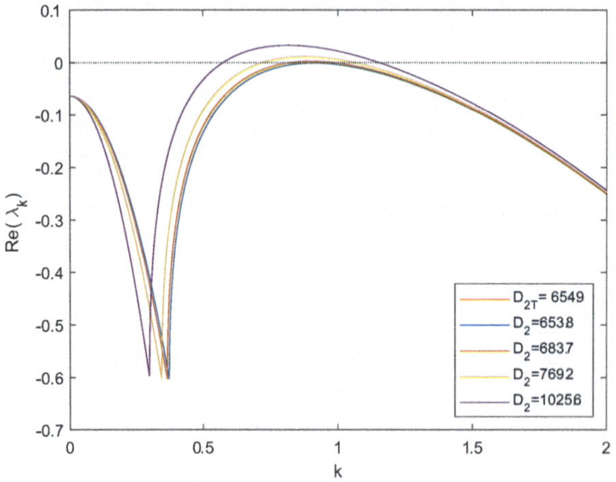

Fig. 13.2 The eigenvalue curves of the planktonic ecosystem model under different diffusion, where $r = 1.00$, $e = 0.12$, $a = 29.50$, $\eta = 1.70$, $D_1 = 0.10$, $D_{2T} = 654.9$, $D_2 = 653.8, 683.7, 769.2, 1025.6$

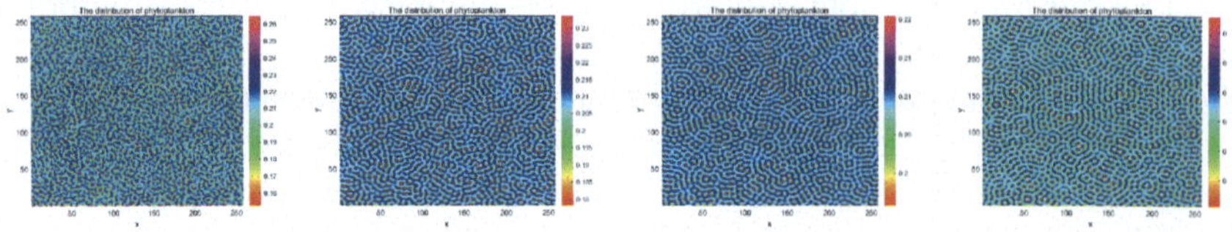

(a) t=10s (b) t=100s (c) t=1 000s (d) t=100 000s

Fig. 13.3 Evolution of phytoplankton at different instants, where $r = 1.00$, $e = 0.12$, $a = 29.50$, $\eta = 1.70$, $D_1 = 0.10$, $D_2 = 653.8$

$(s = -1.23 \times 10^{-8} \in (s_1, s_2))$, we can see that there exists no pattern, in which the phytoplankton remains a uniform steady state.

When the tuning parameter is taken as $D_2 = 683.7$ $(s = 32.93 \times 10^{-8} \in (s_2, s_3))$, we can see spot patterns in Fig. 13.4.

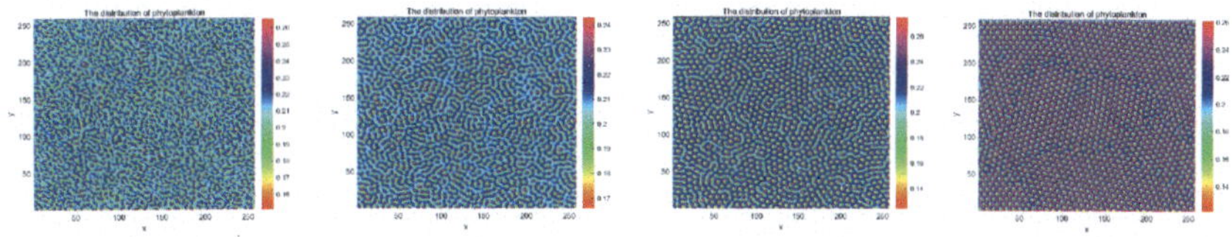

(a) t=10s (b) t=100s (c) t=1000s (d) t=100 000s

Fig. 13.4 Evolution of phytoplankton at different instants, where $r = 1.00$, $e = 0.12$, $a = 29.50$, $\eta = 1.70$, $D_1 = 0.10$, $D_2 = 683.7$

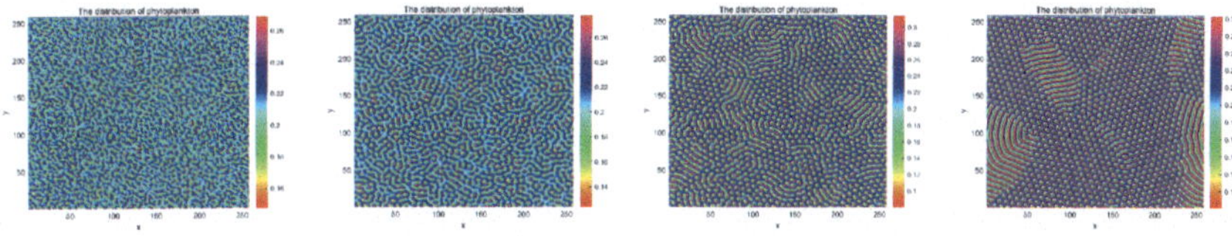

(a) t=10s (b) t=100s (c) t=1000s (d) t=100 000s

Fig. 13.5 Evolution of phytoplankton at different instants, where $r = 1.00$, $e = 0.12$, $a = 29.50$, $\eta = 1.70$, $D_1 = 0.10$, $D_2 = 769.2$

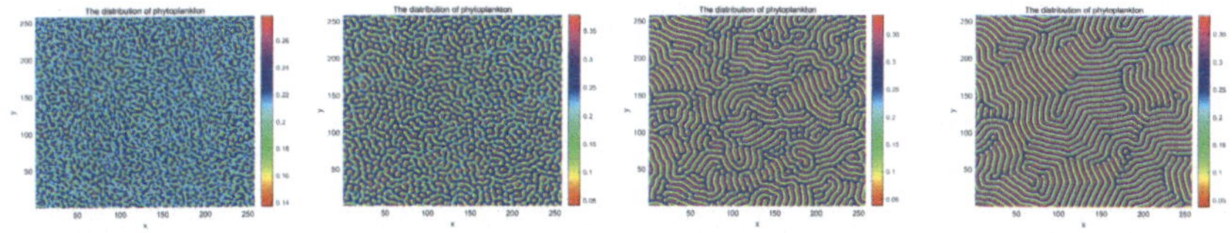

(a) t=10s (b) t=100s (c) t=1 000s (d) t=100 000s

Fig. 13.6 Evolution of phytoplankton at different instants, where $r = 1.00$, $e = 0.12$, $a = 29.50$, $\eta = 1.70$, $D_1 = 0.10$, $D_2 = 1025.6$

When s is adjusted in the domain $[s_3, s_4]$, for instance, $s = 130.57 \times 10^{-8}$ ($D_2 = 769.2$), Fig. 13.5 illustrates the mixed patterns composing of spots and stripes.

Because that $D_2 = 1025.6$ ($s = 423.70 \times 10^{-8} > s_4$) meaning that D_2 increases beyond 966.6 ($s_4 = 356.18 \times 10^{-8}$), Fig. 13.6 illustrates only the striped pattern.

The simulations above indicate that the reaction–diffusion equations with a nonlinear diffusion term $Z\nabla^2 Z$ can be analyzed by our approach. However, based on more simulations, the three sets of patterns cannot be perfectly predicted. The three factors may be the reasons:

1. the high-order extensions play an important role in the pattern formation but we only include the third-order approximation;
2. the nonlinear diffusion term also takes the larger nonlinearity of the model;
3. the discretization method will reduce the calculation accuracy.

Chapter 14
Temporal Forcing Induced Pattern Transitions

Fleming was the first researcher to demonstrate the control of phytoplankton blooming by zooplankton through mathematical modeling [328]. Afterwards, there exist various functional responses such as Ivlev-type [333, 335] and Holling-type [329, 330] to describe the relation between phytoplankton and zooplankton. Based on these works, there are greatly and deeply understanding of dynamical properties for plankton system, which provides insight into solving some ecological problems [331, 340].

According to the fact that periodic forcing due to the cycle of many environmental factors including season, nutrient availability, temperature, etc., inevitably affects the real plankton ecosystem, dynamics of externally forced system has been one of the most focuses in the theory of plankton model. And considerable literatures have found that forcing term plays significant role in affecting the plankton community functioning [327, 338, 339, 342, 350]. On this issue, there generates one of the most important questions that how will the forcing term affect the evolution of the dynamics of system [332, 344]? Moreover, besides the external forcing effect, cross-diffusion term between phytoplankton and zooplankton also plays an important role in deciding the population dynamics [192, 345]. In fact, authors deduced that the cross-diffusion term can account for the population fluxes of one species owing to the presence of another species [190, 191]. For example, we can observe a widely scenario that predator hunts for prey and prey prevents from predator. Subsequently, the diffusion tendency of two plankton populations relies partly on the action of each other. The effect of the model with cross-diffusion terms adds some degree of realism compared with those cases with self-diffusion.

In the following, we will reveal the destabilizing effect of the weak temporal forcing on spatially homogeneous stationary state, in which triggering a switch of states can force the dynamics of system evolving into temporally periodic oscillations. Hence, it is obvious that the loss of stable steady states of plankton or oscillating with the population densities can be obtained by cycling fluctuations in living environment. We use the amplitude equations to obtain the results in the presence of forcing term. Here, we consider the following complex phenomenon by the inter-

S. T. Liu and L. Zhang, *Pattern Dynamics of Marine Plankton Behavior*,
https://doi.org/10.1007/978-981-97-5369-7_14

play between Turing and Hopf instabilities with forcing's effect on the vicinity of Turing-Hopf bifurcation. The main methods used in the following include the weakly nonlinear analysis and a linear stability analysis so that we can track the behavior of the most unstable modes near bifurcation onset varying the states induced by forcing.

14.1 Reaction–Diffusion Model with Temporal Forcing

The original model named as a reaction–diffusion plankton model was studied and listed as below: [336]

$$\begin{aligned}\frac{\partial u}{\partial t} &= d\Delta u + \alpha u(1-\frac{u}{b}) - \gamma\frac{u}{u+H}v,\\ \frac{\partial v}{\partial t} &= d\Delta v + \kappa\gamma\frac{u}{u+H}v - \mu v,\end{aligned} \tag{14.1}$$

where $u(\boldsymbol{x},t)$ and $v(\boldsymbol{x},t)$ mean the abundance of prey and predator at the spatial position $\boldsymbol{x}$ and time t respectively. Parameters α, b, γ, H, κ, and μ stand for the positive constants, the prey intrinsic growth rate, the environmental carrying capacity for prey, the attack rate for predator, the half-saturation abundance for prey, the food utilization rate and the natural mortality for predator, respectively. Parameter d indicates the self-diffusion rate of prey and predator. After introducing suitable dimensionless variables [336], we get the following model (14.1)

$$\begin{aligned}\frac{\partial u}{\partial t} &= \Delta u + u(1-u) - \frac{u}{u+h}v,\\ \frac{\partial v}{\partial t} &= \Delta v + k\frac{u}{u+h}v - mv,\end{aligned} \tag{14.2}$$

where $h = H/b$, $m = \mu/\alpha$ and $k = \kappa\gamma/\alpha$, in which this model has also been applied to study the predator–prey interactions between phyto- and zooplankton (more details as well as spatiotemporal dynamics can be found in [336]. In order to consider the important effect of external forcing and cross-diffusion between different species on their dynamical behavior, model (14.2) can be refined as follows

$$\begin{aligned}\frac{\partial u}{\partial t} &= d_{11}\Delta u + d_{12}\Delta v + u(1-u) - \frac{u}{u+h}v + \delta f(t),\\ \frac{\partial v}{\partial t} &= d_{21}\Delta u + d_{22}\Delta v + k\frac{u}{u+h}v - mv + \delta f(t),\end{aligned} \tag{14.3}$$

where $\delta f(t)$ stands for the forcing with the intensity δ; d_{11} and d_{22} represent the differing self-diffusion coefficients of two species; d_{12} and d_{21} mean cross-diffusion coefficients. Without loss of generality, we assume the forcing being time-periodic type with weak intensity and having zero mean. Based on the strongly effect of the environmental periodic forcing on the growth of phytoplankton and its further influence on the survival of zooplankton which feeds on it, it is natural to perceive the effect of the natural forcing decreasingly on the abundance of these two species. In truth, phytoplankton, as a primary producer in marine food web, plays a great role in material cycle and energy transportation by considering the influence of the states of phytoplankton on populations of zooplankton. Moreover, the intrinsic characteristics of zooplankton are not unchanging, for instance, its natural mortality often varies along with the external forcing. Under these circumstances, we therefore consider the temporal forcing effect on the variation of the phytoplankton and zooplankton populations. Based on the fact that prey and predator face the same fluctuating environment, the same periodic forcing term has been considered [341, 344]. Hence, we study the role of temporal forcing in switching between different dynamical behavior for model (14.3) in this chapter.

14.2 Linear Stability Analysis

In the following, we only derive the conditions for the occurrence of Turing-Hopf bifurcation by linear stability analysis for model (14.3) without forcing term.

In absence of forcing term, we obtain three equilibria for model (14.3): $E_0 = (0, 0)$ (total extinction), $E_b = (1, 0)$ (extinction of the predator) and $E^* = (u^*, v^*)$ (coexistence of prey and predator), where

$$u^* = \frac{rh}{1-r}, \quad v^* = (1-u)(u+h), \quad \text{and } r = \frac{m}{k}.$$

From a biological point of view, the study of the positive equilibrium E^* is our focus. Easily, equilibria E_0 and E_b become both saddle points when E^* is feasible in the parameter region. At the equilibrium E^*, the linearized model can be written as following

$$\begin{aligned} \frac{\partial \tilde{u}}{\partial t} &= d_{11}\Delta\tilde{u} + d_{12}\Delta\tilde{u} + J_{11}\tilde{u} + J_{12}\tilde{v}, \\ \frac{\partial \tilde{v}}{\partial t} &= d_{21}\Delta\tilde{u} + d_{22}\Delta\tilde{v} + J_{21}\tilde{u} + J_{22}\tilde{v}, \end{aligned} \tag{14.4}$$

where $(\tilde{u}, \tilde{v})$ means small deviations from E^* and

$$J_{11} = (-1 + \frac{v^*}{(u^*+h)^2})u^*, \quad J_{12} = -\frac{u^*}{u^*+h}, \quad J_{21} = \frac{khv^*}{(u^*+h)^2}, \quad J_{22} = 0.$$

After omitting the tildes as below, the spatiotemporal stability analysis of E^* can be addressed to the usual Fourier-type perturbation

$$\begin{pmatrix} u \\ v \end{pmatrix} = \sum_q \begin{pmatrix} c_q^1 \\ c_q^2 \end{pmatrix} e^{\lambda t + i\boldsymbol{q}\cdot\boldsymbol{r}}, \tag{14.5}$$

where λ represents the growth rate of perturbation; i denotes the imaginary unit; $\boldsymbol{r}$ means the spatial vector in two-dimensional space; q stands for the wave number satisfying $\boldsymbol{q}\cdot\boldsymbol{q} = q^2$. Substituting Eq. (14.5) into model (14.4), we find out that parameters, perturbation growth rate λ and wave number q, determined the stability of E^*, have the dispersion relation between them, where

$$\lambda^2 - \mathrm{Tr}(q)\,\lambda + D(q) = 0 \tag{14.6}$$

$$\mathrm{Tr}(q) = J_{11} + J_{22} - (d_{11} + d_{22})q^2,$$

and

$$D(q) = (d_{11}d_{22} - d_{12}d_{21})q^4 - (J_{11}d_{22} + J_{22}d_{11} - J_{12}d_{21} - J_{21}d_{12})q^2 + J_{11}J_{22} - J_{12}J_{21}.$$

Because that there emerges Hopf bifurcation as $\mathrm{Im}\,\lambda_1 = -\mathrm{Im}\,\lambda_2 \neq 0,\ \mathrm{Re}\,\lambda_{1,2} = 0,$ at $q = 0$, we have $h - \frac{1-r}{1+r} = 0$ as the conditions listed below are satisfied: $\mathrm{Tr}(0) = J_{11} + J_{22} = 0$ and $D(0) = J_{11}J_{22} - J_{12}J_{21} > 0$. When the frequency $\omega_H = \sqrt{-J_{12}J_{21}}$, we can generate a spatially homogeneous oscillation by the bifurcation. Hence, we obtain the critical eigenvector as $R_H = (J_{12}, i\omega_H - J_{11})^T$ and its complex conjugate.

Obviously, when $\mathrm{Re}\,\lambda_1 < \mathrm{Re}\,\lambda_2 = 0,$ at $q = q_T \neq 0$ where q_T represents the critical wave number, Turing bifurcation emerges. Naturally, this condition is equal to $D(q_T) = 0$. In addition, all modes except 0 and q_T are assumed to be damped at the critical point, which guarantees the minimum of $D(q_T)$ obtained at wave number q_T. For the case of Turing instability, there will emerge a time-independent spatially periodic pattern, where the critical eigenvector reads $R_T = (-d_{12}q_T^2 + J_{12}, d_{11}q_T^2 - J_{11})^T$.

Hopf instability thresholds and Turing coincide at the codimension-2 Turing-Hopf bifurcation. Nearby, the different unstable modes generate interaction and competition so that there exists rich dynamical behavior. In order to locate this bifurcation point, parameters r and h are taken as control parameters. Based on the above conditions, we have the conditions for the critical values r_c and h_c for Turing-Hopf bifurcation:

$$h_c - \frac{1 - r_c}{1 + r_c} = 0$$

and

$$(r_c(1 + r_c)d_{21})^2 + (k(1 - r_c)d_{12})^2 + 2kr_c(1 - r_c^2)(d_{12}d_{21} - 2d_{11}d_{22}) = 0,$$

for the frequency, the wave number, and the critical eigenvectors as following

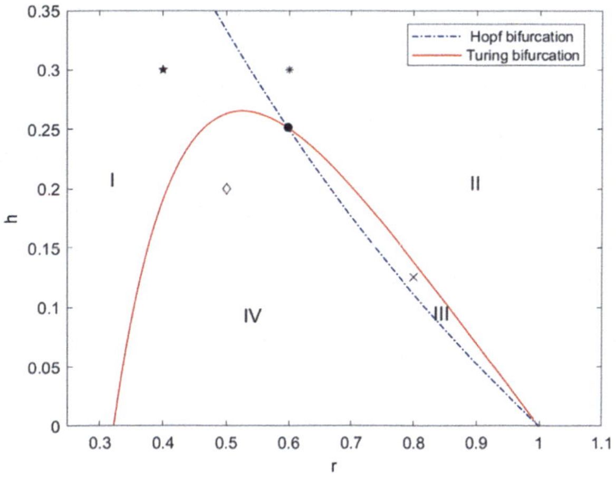

Fig. 14.1 Bifurcation diagram in $r-h$ parameter plane for $d_{11}=0.25$, $d_{12}=0.5$, $d_{21}=0.25$, $d_{22}=0.6$, $k=0.5$, where Regions I-IV means Hopf instability, asymptotic stability of equilibrium E^*, Turing instability and Turing-Hopf instability respectively. The full circle represents the Turing-Hopf bifurcation point. $\star$, $*$, $\times$ and $\diamond$ mark the parameter settings used in Fig. 14.2b–e

$$\omega_c=\sqrt{kh_cr_c},\ q_c^2=\sqrt{\frac{kh_cr_c}{d_{11}d_{22}-d_{12}d_{21}}},\ R_H=\begin{pmatrix} s-r_c \\ i\omega_c \end{pmatrix},\ R_T=\begin{pmatrix} -d_{12}q_c^2-r_c \\ d_{11}q_c^2 \end{pmatrix}.$$

Note that the effect of self-diffusion of populations is our focus which is assumed to be stronger than that of cross-diffusion.

Figure 14.1 illustrates the Turing and Hopf bifurcation in the parameter plane spanned by r and h, where other parameters are taken as $d_{11}=0.25$, $d_{12}=0.5$, $d_{21}=0.25$, $d_{22}=0.6$, $k=0.5$. Significantly, the figure shows an intersection point between bifurcation lines clearly like Turing-Hopf bifurcation, and four stability regions for the equilibrium E^*. We calculate straightforwardly the critical values for Turing-Hopf bifurcation as $r_c=0.59823$ and $h_c=0.25138$. Obviously, Regions I–IV indicate, the occurrence of Hopf instability, Turing instability and Turing-Hopf instability, asymptotic stability of equilibrium E^*, respectively.

Figure 14.2 shows the dispersion relation more intuitively, such as the largest real part and the corresponding imaginary part for growth rate λ against wave number q for different values of r and h. Figure 14.2b–e illustrate the parameter settings used in sub-figures corresponded to regions I-V in Fig. 14.1 respectively.

14.3 Amplitude Equation

In the following, the shift of stationary states would be revealed which is generated by the addition of temporal forcing near Turing-Hopf bifurcation. Since amplitude

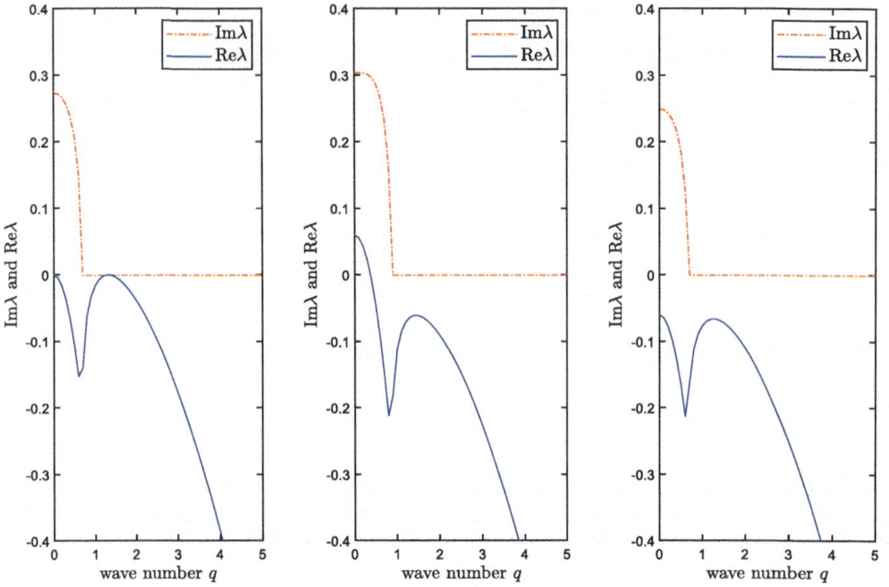

(a) $(0.59823, 0.25138)$ (b) $(0.4, 0.3)$ (c) $(0.6, 0.3)$

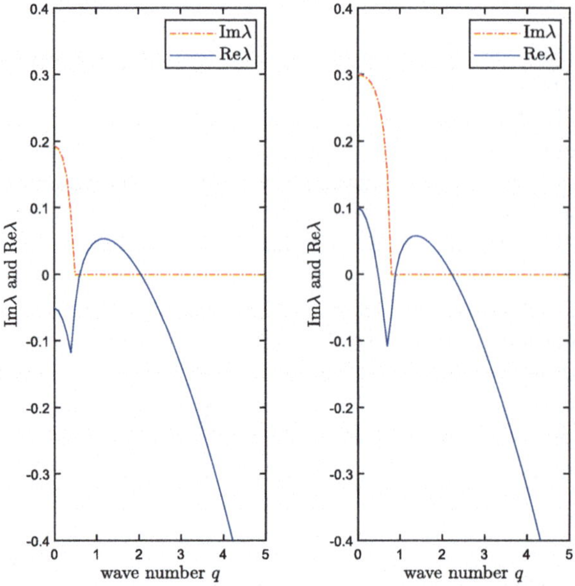

(d) $(0.8, 0.125)$ (e) $(0.5, 0.2)$

Fig. 14.2 Dispersion relation diagram for $d_{11} = 0.25$, $d_{12} = 0.5$, $d_{21} = 0.25$, $d_{22} = 0.6$ and $k = 0.5$

equation can uncover the structure of the solutions of model (14.3) on spatial domain, we use it to study the shift. Therefore, a weakly nonlinear analysis is employed to obtain the amplitude equation for the effect of the forcing term.

At first, the linearized form of model (14.3) is written at the equilibrium E^* by Taylor expansion (up to cubic order) as follows

$$\begin{aligned}\frac{\partial u}{\partial t} &= d_{11}\Delta u + d_{12}\Delta v + J_{11}u + J_{12}v + a_{11}u^2 + a_{12}uv + b_{11}u^3 + b_{12}u^2v + \delta f(t),\\ \frac{\partial v}{\partial t} &= d_{21}\Delta u + d_{22}\Delta v + J_{21}u + J_{22}v + a_{21}u^2 + a_{22}uv + b_{21}u^3 + b_{22}u^2v + \delta f(t),\end{aligned} \tag{14.7}$$

where

$$\begin{aligned} a_{11} &= -1 + \frac{hv^*}{(u^*+h)^3},\ a_{12} = -\frac{h}{(u^*+h)^2},\ a_{21} = -\frac{khv^*}{(u^*+h)^3},\ a_{22} = \frac{kh}{(u^*+h)^2},\\ b_{11} &= -\frac{hv^*}{(u^*+h)^4},\ b_{12} = \frac{h}{(u^*+h)^3},\ b_{21} = \frac{khv^*}{(u^*+h)^4},\ b_{22} = \frac{-kh}{(u^*+h)^3}.\end{aligned}$$

Then we can see the bifurcation of Turing modes and Hopf modes simultaneously closing to the Turing-Hopf bifurcation point, and it results in

$$\begin{pmatrix} u \\ v \end{pmatrix} = \begin{pmatrix} A_1^u \\ A_1^v \end{pmatrix} e^{iq_1\cdot r} + \begin{pmatrix} A_2^u \\ A_2^v \end{pmatrix} e^{iq_2\cdot r} + \begin{pmatrix} B^u \\ B^v \end{pmatrix} e^{i\omega_c t} + c.c. + \cdots \tag{14.8}$$

where $\begin{pmatrix} A_1^u \\ A_1^v \end{pmatrix}$, $\begin{pmatrix} A_2^u \\ A_2^v \end{pmatrix}$, and $\begin{pmatrix} B^u \\ B^v \end{pmatrix}$ stand for the slowly varying amplitudes corresponding to Turing and Hopf modes, respectively; the dots represent the damped modes. Next, our aim is to describe the evolution of these critical amplitudes by applying original parameters in model (14.3), using the derived so-called amplitude equation.

Assume

$$U = \begin{pmatrix} u \\ v \end{pmatrix},\ L = \begin{pmatrix} d_{11}\Delta + J_{11} & d_{12}\Delta + J_{12} \\ d_{21}\Delta + J_{21} & d_{22}\Delta + J_{22} \end{pmatrix},$$

$$N = \begin{pmatrix} a_{11}u^2 + a_{12}uv + b_{11}u^3 + b_{12}uv^2 \\ a_{21}u^2 + a_{22}uv + b_{21}u^3 + b_{22}uv^2 \end{pmatrix},\ F = \begin{pmatrix} \delta f(t) \\ \delta f(t) \end{pmatrix},$$

and we can get the Eq. (14.7) as

$$\frac{\partial U}{\partial t} = LU + N + F. \tag{14.9}$$

Obviously, as the reason is that the slow time variables can determine the dynamics in the end, our key focus is to separate the dynamical behavior of system by using the weakly nonlinear analysis in terms of different time scales. In the following, the fast time scale t_0 is introduced with the same order as variable t and slow time scale

$t_2 = \epsilon^2 t$, where the derivation of time satisfies the following conditions:

$$\frac{\partial}{\partial t} = \frac{\partial}{\partial t_0} + \epsilon^2 \frac{\partial}{\partial t_2} + \cdots, \tag{14.10}$$

ϵ stands for infinitesimal. Then it yields the following as expanding control parameters r and h upon ϵ referring to the dimensionless distance from the Turing-Hopf bifurcation threshold

$$r = r_c + \epsilon^2 r_2 + \cdots, \ h = h_c + \epsilon^2 h_2 + \cdots. \tag{14.11}$$

On the other hand, let the intensity of forcing be sufficiently weak and small comparing with ϵ^2, where $\delta = \epsilon^3 \delta_3 + \cdots$. In this case, we can set the remained elements in system (14.9) in the weakly nonlinear form as below

$$U = \epsilon \begin{pmatrix} u_1 \\ v_1 \end{pmatrix} + \epsilon^2 \begin{pmatrix} u_2 \\ v_2 \end{pmatrix} + \epsilon^3 \begin{pmatrix} u_3 \\ v_3 \end{pmatrix} + \cdots, \tag{14.12}$$

$$L = L_c + \epsilon^2 L_2 + \cdots, \ N = \epsilon^2 N_2 + \epsilon^3 N_3 + \cdots, \ F = \epsilon^3 F_3 + \cdots \tag{14.13}$$

where

$$L_c = \begin{pmatrix} d_{11}\Delta & d_{12}\Delta - r_c \\ d_{21}\Delta + kh_c & d_{22}\Delta \end{pmatrix}, L_2 = \begin{pmatrix} \frac{-2r_c r_2}{1-r_c^2} - \frac{r_c h_2}{h_c} & -r_2 \\ -k(r_2 + h_c r_2 + r_c h_2) & 0 \end{pmatrix},$$

$$F_3 = \begin{pmatrix} \delta_3 f(t) \\ \delta_3 f(t) \end{pmatrix},$$

and

$$N_2 \triangleq \begin{pmatrix} a_{11}u_1^2 + a_{12}u_1 v_1 \\ a_{21}u_1^2 + a_{22}u_1 v_1 \end{pmatrix},$$

$$N_3 \triangleq \begin{pmatrix} 2a_{11}u_1u_2 + a_{12}u_1v_2 + a_{12}u_2v_1 \\ 2a_{21}u_1u_2 + a_{22}u_1v_2 + a_{22}u_2v_1 \end{pmatrix} + \begin{pmatrix} b_{11}u_1^3 + b_{12}u_1^2 v_1 \\ b_{21}u_1^3 + b_{22}u_1^2 v_1 \end{pmatrix},$$

where $(u_j, v_j)^T$ $(j = 1, 2, 3)$ stands for the solution behavior under different order of perturbation. Hence, a sequence of problems can be obtained by substituting expansions (14.10)–(14.13) into system (14.9) and collecting the same order of ϵ:

$$O(\epsilon): \ \frac{\partial}{\partial t_0}\begin{pmatrix} u_1 \\ v_1 \end{pmatrix} - L_c \begin{pmatrix} u_1 \\ v_1 \end{pmatrix} = 0, \tag{14.14}$$

$$O(\epsilon^2): \ \frac{\partial}{\partial t_0}\begin{pmatrix} u_2 \\ v_2 \end{pmatrix} - L_c \begin{pmatrix} u_2 \\ v_2 \end{pmatrix} = N_2, \tag{14.15}$$

$$O(\epsilon^3): \frac{\partial}{\partial t_0}\begin{pmatrix}u_3\\v_3\end{pmatrix} - L_c\begin{pmatrix}u_3\\v_3\end{pmatrix} = -\frac{\partial}{\partial t_2}\begin{pmatrix}u_1\\v_1\end{pmatrix} + L_2\begin{pmatrix}u_1\\v_1\end{pmatrix} + N_3 + F_3. \quad (14.16)$$

We get the following by solving linear problem Eq. (14.14):

$$\begin{pmatrix}u_1\\v_1\end{pmatrix} = R_T(W_1 e^{i\boldsymbol{q}_1\cdot\boldsymbol{r}} + W_2 e^{i\boldsymbol{q}_2\cdot\boldsymbol{r}}) + R_H W_H e^{i\omega_c t_0} + c.c., \quad (14.17)$$

where $W_{1,2}$ and W_H represent amplitudes under the first-order perturbation, and their forms are controlled by higher-order perturbation terms. Then, after plugging solution $(u_1, v_1)^T$ into Eq. (14.15), we can obtain the solution to $O(\epsilon^2)$ problem as

$$\begin{pmatrix}u_2\\v_2\end{pmatrix} = \begin{pmatrix}u_{21}\\v_{21}\end{pmatrix}\sum_{j=1}^{2}|W_j|^2 + \begin{pmatrix}u_{22}\\v_{22}\end{pmatrix}|W_H|^2 + \begin{pmatrix}u_{23}\\v_{23}\end{pmatrix} W_H e^{i\omega_c t_0}\sum_{j=1}^{2} W_j e^{i\boldsymbol{q}_j\cdot\boldsymbol{r}} + \quad (14.18)$$

$$\begin{pmatrix}u_{24}\\v_{24}\end{pmatrix}\overline{W}_H e^{-i\omega_c t_0}\sum_{j=1}^{2} W_j e^{i\boldsymbol{q}_j\cdot\boldsymbol{r}} + \begin{pmatrix}u_{25}\\v_{25}\end{pmatrix}\sum_{j=1}^{2} W_j^2 e^{2i\boldsymbol{q}_j\cdot\boldsymbol{r}} + \begin{pmatrix}u_{26}\\v_{26}\end{pmatrix}$$

$$W_H^2 e^{2i\omega_c t_0} + \begin{pmatrix}u_{27}\\v_{27}\end{pmatrix} W_1 W_2 e^{i(\boldsymbol{q}_1+\boldsymbol{q}_2)\cdot\boldsymbol{r}} + \begin{pmatrix}u_{28}\\v_{28}\end{pmatrix} W_1 \overline{W}_2 e^{i(\boldsymbol{q}_1-\boldsymbol{q}_2)\cdot\boldsymbol{r}} + c.c.,$$

where the coefficients can be calculated by solving corresponding linear problems for different types of terms. Assume

$$u_T = -d_{12}q_c^2 - r_c,\ v_T = d_{11}q_c^2,\ u_H = -r_c \text{ and } v_H = i\omega_c,$$

and we can write the coefficients of Eq. (14.18) as

$$\begin{pmatrix}u_{21}\\v_{21}\end{pmatrix} = \frac{1}{\omega_c^2}\begin{pmatrix}2J_{12}l_2\\2J_{21}l_1\end{pmatrix},\ \begin{pmatrix}u_{22}\\v_{22}\end{pmatrix} = \frac{1}{\omega_c^2}\begin{pmatrix}2J_{12}l_4\\2J_{21}l_3\end{pmatrix},$$

$$\begin{pmatrix}u_{23}\\v_{23}\end{pmatrix} = \frac{2}{i\omega_c(d_{11}+d_{22})q_c^2 - \omega_c^2}\begin{pmatrix}(i\omega_c + d_{22}q_c^2)l_5 + (J_{12} - d_{12}q_c^2)l_6\\(J_{21} - d_{21}q_c^2)l_5 + (i\omega_c + d_{11}q_c^2)l_6\end{pmatrix},$$

$$\begin{pmatrix}u_{24}\\v_{24}\end{pmatrix} = \begin{pmatrix}\bar{u}_{23}\\\bar{v}_{23}\end{pmatrix},\ \begin{pmatrix}u_{25}\\v_{25}\end{pmatrix} = \frac{1}{9\omega_c^2}\begin{pmatrix}4d_{22}q_c^2 l_1 + (J_{12} - 4d_{12}q_c^2)l_2\\(J_{21} - 4d_{21}q_c^2)l_1 + 4d_{11}q_c^2 l_2\end{pmatrix},$$

$$\begin{pmatrix}u_{26}\\v_{26}\end{pmatrix} = \frac{1}{-3\omega_c^2}\begin{pmatrix}2i\omega_c l_7 + J_{12}l_8\\J_{21}l_7 + 2i\omega_c l_8\end{pmatrix},$$

$$\begin{pmatrix}u_{27}\\v_{27}\end{pmatrix} = \begin{pmatrix}u_{28}\\v_{28}\end{pmatrix} = \frac{2}{\omega_c^2}\begin{pmatrix}2d_{22}q_c^2 l_1 + (J_{12} - 2d_{12}q_c^2)l_2\\(J_{21} - 2d_{21}q_c^2)l_1 + 2d_{11}q_c^2 l_2\end{pmatrix},$$

where

$$l_1 = a_{11}u_T^2 + a_{12}u_T v_T,\ l_2 = a_{21}u_T^2 + a_{22}u_T v_T,$$

$$l_3 = a_{11}|u_H|^2 + \frac{1}{2}a_{12}(\bar{u}_H v_H + u_H \bar{v}_H),\ l_4 = a_{21}|u_H|^2 + \frac{1}{2}a_{22}(\bar{u}_H v_H + u_H \bar{v}_H),$$

$$l_5 = a_{11}u_H u_T + \frac{1}{2}a_{12}(u_H v_T + u_T v_H),\ l_6 = a_{21}u_H u_T + \frac{1}{2}a_{22}(u_H v_T + u_T v_H),$$

$$l_7 = a_{11}u_H^2 + a_{12}u_H v_H,\ l_8 = a_{21}u_H^2 + a_{22}u_H v_H.$$

Next, turn to the case as $O(\epsilon^3)$. We apply the Fredholm alternative conditions to ensure the existence of nontrivial solutions of $O(\epsilon^3)$ problem. To be specific, we assume that the vector functions of the right side of Eq. (14.16) should remain orthogonal with the zero eigenvectors of the adjoint operator of L_c, i.e., $L_H e^{-i\omega_c t_0}$, $L_T e^{-i\mathbf{q}_1\cdot\mathbf{r}}$ and $L_T e^{-i\mathbf{q}_2\cdot\mathbf{r}}$ with $L_H = (J_{21}, i\omega_c)$ and $L_T = (-d_{21}q_c^2 + J_{21}, d_{11}q_c^2)$. Making use of this approach yields in

$$\begin{aligned}
\tau_T \frac{dW_1}{dt_2} &= \mu_T W_1 + g_{1T}|W_1|^2 W_1 + g_{2T}|W_2|^2 W_1 + g_{3T}|W_H|^2 W_1,\\
\tau_T \frac{dW_2}{dt_2} &= \mu_T W_2 + g_{1T}|W_2|^2 W_2 + g_{2T}|W_1|^2 W_2 + g_{3T}|W_H|^2 W_2,\\
\tau_H \frac{dW_H}{dt_2} &= \mu_H W_H + g_{1H}|W_H|^2 W_H + g_{2H}(|W_1|^2 + |W_2|^2) W_H + \xi.
\end{aligned} \tag{14.19}$$

The coefficients of Eq. (14.19) are listed as below

$$\tau_T = L_T R_T,\ \mu_T = L_T L_2 R_T,\ g_{1T} = L_T G_{1T},\ g_{2T} = L_T G_{2T},\ g_{3T} = L_T G_{3T},$$

$$\tau_H = L_H R_H,\ \mu_H = L_H L_2 R_H,\ g_{1H} = L_H G_{1H},\ g_{2H} = L_H G_{2H},$$

$$\xi = \frac{\omega_c}{2\pi}\int_0^{2\pi/\omega_c} L_H e^{-i\omega_c t_0} F_3 dt,$$

where

$$G_{1T} = \begin{pmatrix} (2a_{11}u_T + a_{12}v_T)(u_{21} + u_{25}) + a_{12}u_T(v_{21} + v_{25}) + 3b_{11}u_T^3 + 3b_{12}u_T^2 v_T \\ (2a_{21}u_T + a_{22}v_T)(u_{21} + u_{25}) + a_{22}u_T(v_{21} + v_{25}) + 3b_{21}u_T^3 + 3b_{22}u_T^2 v_T \end{pmatrix},$$

$$G_{2T} = \begin{pmatrix} (2a_{11}u_T + a_{12}v_T)(u_{21} + 2u_{27}) + a_{12}u_T(v_{21} + 2v_{27}) + 6b_{11}u_T^3 + 6b_{12}u_T^2 v_T \\ (2a_{21}u_T + a_{22}v_T)(u_{21} + 2u_{27}) + a_{22}u_T(v_{21} + 2v_{27}) + 6b_{21}u_T^3 + 6b_{22}u_T^2 v_T \end{pmatrix},$$

$$G_{3T} = \begin{pmatrix} (2a_{11}u_T + a_{12}v_T)u_{22} + a_{12}u_T v_{22} + 2\mathrm{Re}\,((2a_{11}u_H + a_{12}v_H)u_{24} + a_{12}u_H v_{24}) \\ (2a_{21}u_T + a_{22}v_T)u_{22} + a_{22}u_T v_{22} + 2\mathrm{Re}\,((2a_{21}u_H + a_{22}v_H)u_{24} + a_{22}u_H v_{24}) \end{pmatrix}$$

$$+\begin{pmatrix} 6b_{11}|u_H|^2u_T + b_{12}(2v_T|u_H|^2 + 2v_Hu_T\bar{u}_H + 2\bar{v}_Hu_Tu_H) \\ 6b_{21}|u_H|^2u_T + b_{22}(2v_T|u_H|^2 + 2v_Hu_T\bar{u}_H + 2\bar{v}_Hu_Tu_H) \end{pmatrix},$$

$$G_{1H} = \begin{pmatrix} (2a_{11}u_H + a_{12}v_H)u_{22} + a_{12}u_Hv_{22} + (2a_{11}\bar{u}_H + a_{12}\bar{v}_H)u_{26} + a_{12}\bar{u}_Hv_{26} \\ (2a_{21}u_H + a_{22}v_H)u_{22} + a_{22}u_Hv_{22} + (2a_{21}\bar{u}_H + a_{22}\bar{v}_H)u_{26} + a_{22}\bar{u}_Hv_{26} \end{pmatrix}$$

$$+\begin{pmatrix} 3b_{11}|u_H|^2u_H + b_{12}(\bar{v}_Hu_H^2 + 2v_H|u_H|^2) \\ 3b_{21}|u_H|^2u_H + b_{22}(\bar{v}_Hu_H^2 + 2v_H|u_H|^2) \end{pmatrix},$$

$$G_{2H} = \begin{pmatrix} (2a_{11}u_H + a_{12}v_H)u_{21} + a_{12}u_Hv_{21} + 2(2a_{11}u_T + a_{12}v_T)u_{23} + 2a_{12}u_Tv_{23} \\ (2a_{21}u_H + a_{22}v_H)u_{21} + a_{22}u_Hv_{21} + 2(2a_{21}u_T + a_{22}v_T)u_{23} + 2a_{22}u_Tv_{23} \end{pmatrix}$$

$$+\begin{pmatrix} 6b_{11}u_Hu_T^2 + b_{12}(2v_Hu_T^2 + 4u_Tv_Tu_H) \\ 6b_{21}u_Hu_T^2 + b_{22}(2v_Hu_T^2 + 4u_Tv_Tu_H) \end{pmatrix}.$$

Based on the formation for the amplitudes of Turing modes and Hopf modes as

$$A_1 = A_1^u = \epsilon u_T W_1 + \cdots, \quad A_2 = A_2^v = \epsilon u_T W_2 + \cdots, \quad B = B^u = \epsilon u_H W_H + \cdots,$$

the coupled amplitude equations are written finally as follows

$$\begin{aligned} \frac{dA_1}{dt} &= \tilde{\mu}_T A_1 + \tilde{g}_{1T}|A_1|^2A_1 + \tilde{g}_{2T}|A_2|^2A_1 + \tilde{g}_{3T}|B|^2A_1, \\ \frac{dA_2}{dt} &= \tilde{\mu}_T A_2 + \tilde{g}_{1T}|A_2|^2A_2 + \tilde{g}_{2T}|A_1|^2A_2 + \tilde{g}_{3T}|B|^2A_2, \\ \frac{dB}{dt} &= \tilde{\mu}_H B + \tilde{g}_{1H}|B|^2B + \tilde{g}_{2H}(|A_1|^2 + |A_2|^2)B + \tilde{\xi}, \end{aligned} \tag{14.20}$$

where

$$\tilde{\mu}_T = \frac{L_T(L - L_c)R_T}{\tau_T}, \quad \tilde{g}_{1T} = \frac{g_{1T}}{|u_T|^2\tau_T}, \quad \tilde{g}_{2T} = \frac{g_{2T}}{|u_T|^2\tau_T}, \quad \tilde{g}_{3T} = \frac{g_{3T}}{|u_H|^2\tau_T},$$

$$\tilde{\mu}_H = \frac{L_H(L - L_c)R_H}{\tau_H}, \quad \tilde{g}_{1H} = \frac{g_{1H}}{|u_H|^2\tau_H}, \quad \tilde{g}_{2H} = \frac{g_{2H}}{|u_T|^2\tau_H}, \quad \tilde{\xi} = \frac{u_H}{\tau_H}\xi.$$

For simplicity, we omit the tildes. Importantly, a kind of stable pattern formation corresponds to a stable steady state of Eq. (14.20) so that the changes of states named as the response of system to external forcing can be studied by the stability analysis of amplitude equations. As a weakly nonlinear analysis can be valid in the vicinity of supercritical bifurcation, the case as $g_{1T} < 0$ and Re $g_{1H} < 0$ is our focus because of the generation of the supercritical behavior for both Turing and Hopf bifurcations. For this aim, we use the rescaling

$$\sqrt{-g_{1T}}A_1 \to A_1,\ \sqrt{-g_{1T}}A_2 \to A_2,\ \varsigma B \to B,$$

and have the amplitude Eq. (14.20) as the scaled form

$$\begin{aligned}
\frac{dA_1}{dt} &= \mu_T A_1 - |A_1|^2 A_1 + \sigma |A_2|^2 A_1 + \eta |B|^2 A_1,\\
\frac{dA_2}{dt} &= \mu_T A_2 - |A_2|^2 A_2 + \sigma |A_1|^2 A_2 + \eta |B|^2 A_2,\\
\frac{dB}{dt} &= \mu_H B + \phi |B|^2 B + \varphi(|A_1|^2 + |A_2|^2) B + \delta,
\end{aligned} \tag{14.21}$$

where

$$\sigma = -\frac{g_{2T}}{g_{1T}},\ \eta = \frac{g_{3T}}{|\varsigma|^2},\ \phi = \frac{g_{1H}}{|\varsigma|^2},\ \varphi = -\frac{g_{2H}}{g_{1T}},\ \varsigma = \frac{\delta}{\xi},$$

amplitudes $A_{1,2}$ and B depend on the intensity of forcing. Actually, without the forcing term, amplitudes will decay to 0 when $\mu_T < 0$ and Re $\mu_H < 0$ meaning that the steady state (u^*, v^*) becomes stable without forcing. In order to see effect of the weak forcing on the perturbation of this state, we apply a regular perturbation expansion. Assume δ in Eq. (14.21) is $O(\zeta)$ and ζ is infinitesimal. Let

$$A_1 = A_1^{(0)} + \zeta A_1^{(1)} + \cdots,\ A_2 = A_2^{(0)} + \zeta A_2^{(1)} + \cdots,\ B = B^{(0)} + \zeta B^{(1)} + \cdots. \tag{14.22}$$

Substituting Eq. (14.22) into Eq. (14.21) and gathering terms of the same order of ζ, we calculate and get the perturbed solution at each power

$$A_1 = A_2 = 0,\ B = -\frac{\delta}{\phi}.$$

Thus, after a long time, perturbed amplitudes converge to a positive stationary state corresponding to a $O(\delta)$ spatially homogeneous oscillation with frequency ω_c. It can be observed that the existence of weak forcing responses by driving Hopf modes whereas Turing modes are same as it. Furthermore, it deduces that forcing can play an important role in the evolution of system in time but not in the spatial structure. As the above shift generated by forcing, we infer that weak forcing has destabilization effect on the spatially homogeneous stationary state of system and produce a pattern transition for time-periodic oscillations.

Biologically, based on our results, external periodic forcing has a constructive effect on the formation of spatial patterns of species. For the ocean system, it indicates that abiotic environmental factors instead of pure ecological interactions may generate oceanic plankton fluctuations partially. In fact, many works have distinguished many mechanisms closely relating to the pattern formation of populations. For example, the Hopf bifurcation can excite the nonlinear prey–predator interactions and then generate spatial patterns [336]. Besides that, diffusion can destabilize a stable spatially homogeneous state to further result in the emergence of Turing patterns [346–348]. In particular, the plankton models with cross-diffusion can generate complex structures by varying corresponding parameter such as spots, stripes and their mixture [192, 337, 345]. Our study enriches the way to generate pattern transitions, for example, even forcing of weak intensity can make the stable states of plankton system be oscillatory. In addition, we only present the resulting oscillation in time without affecting the homogeneity of plankton distribution. The above results help us further understand the role of environmental events in population dynamics in real ecosystem.

14.4 Example

In the following, simulation examples are given to confirm the above theoretical results. Generally, we take the control parameters in model (14.3) as $r = 0.6$, $h = 0.255$, the forcing as $0.01\cos(\omega_c t)$ with $\omega_c = 0.2742$ and other parameter values as the same in Fig. 14.1. Then the stability of the equilibrium E^* is simulated taking numerical values as $\mu_T = \operatorname{Re}\mu_H = -0.06$ without forcing. In order to illustrate the pattern formation of model (14.3) in two-dimensional space, we subdivide the discrete domain uniformly into 200×200 lattice sites, use five-point difference method for diffusion terms and apply Euler forward finite difference method for reaction parts. Hence, we solve the model on this spatial grid in which the space step $\Delta h = 1$ and the time step $\Delta t = 0.1$. The initial conditions are assumed that the solution E^* plus random disturbances and produce some inhomogeneities in distribution. We take the Neumann boundary conditions as on the boundary. Subsequently, the time evolution and spatial structure of model (14.3) are illustrated with and without forcing, respectively. Importantly, the time stopping simulations are sufficient to ensure the unchange of states of system dynamically.

In Fig. 14.3, the time evolution and spatial structure of system (14.3) are demonstrated without forcing. Firstly, Fig. 14.3a shows the mean densities $< u >$ and $< v >$ with respect to time averaging over the whole domain. It is readily seen that the system remains asymptotically stable. The temporal dynamics of prey and predator is illustrated at a fixed space point (100, 100) in Fig. 14.3b. Figure 14.3a also indicates the same feature for the depicted time series. In fact, without forcing, the decay of the exact solutions of system is deduced by the stability of steady state E^*. Furthermore, the pattern formation of u and v is presented by Fig. 14.3d, e, respectively. Subsequently, the distribution of populations in spatial domain becomes being homogeneous after the initial irregular transient period.

Numerical results of system (14.3) are given with the addition of weak forcing shown in Fig. 14.4. Obviously, Figs. 14.3 and 14.4 illustrate the averaging effect of the time-series of spatially on population densities, and the oscillatory behavior of the system is dynamically revealed by temporal dynamics of two species located in fixed spatial location (100, 100). However, Figs. 14.3 and 14.4 also present the homogeneous feature of the spatial patterns of u and v. The above simulations consistent with theoretical results demonstrate the significant role of the forcing term

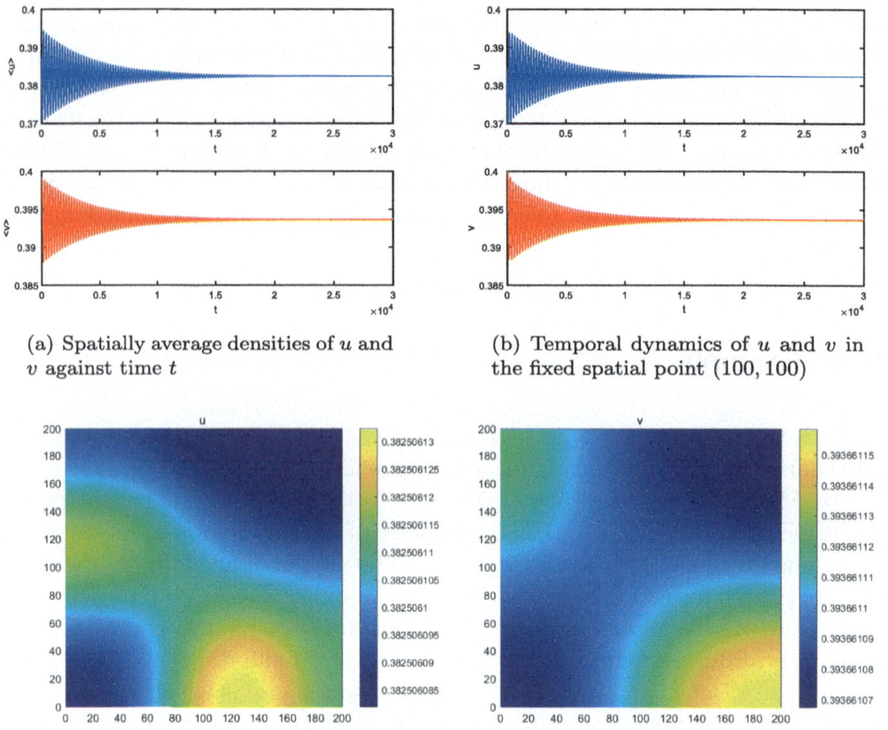

(a) Spatially average densities of u and v against time t

(b) Temporal dynamics of u and v in the fixed spatial point $(100, 100)$

(c) Spatial patterns developed by u

(d) Spatial patterns developed by v

Fig. 14.3 Time evolution and spatial distribution of prey and predator in system (14.3) in the absence of forcing

in the stability of system, in which the forcing term can induce switch of patterns between homogeneous stationary state spatially and time-periodic oscillations.

In this chapter, the effect of weak time-periodic forcing is investigated near Turing-Hopf bifurcation with cross-diffusion in a plankton system. We obtain the conditions for the occurrence of Turing-Hopf bifurcation applying the linear stability analysis. Together using the weakly nonlinear analysis, we derive the expressions of amplitude equations and consider the effect of external forcing on the change of states. The results demonstrate that weak forcing can drive a transition from spatially homogeneous stationary state to temporally periodic state. We also show the influence in the time evolution of system but not in the spatial structure of solutions. However, we must point out that we can only obtain good results when the parameters of amplitude equations are taken in a sufficiently small neighborhood of Turing-Hopf bifurcation point. As parameters are in far away from the bifurcation point, we can not get valid theoretical predictions about the time and the position of the patterns.

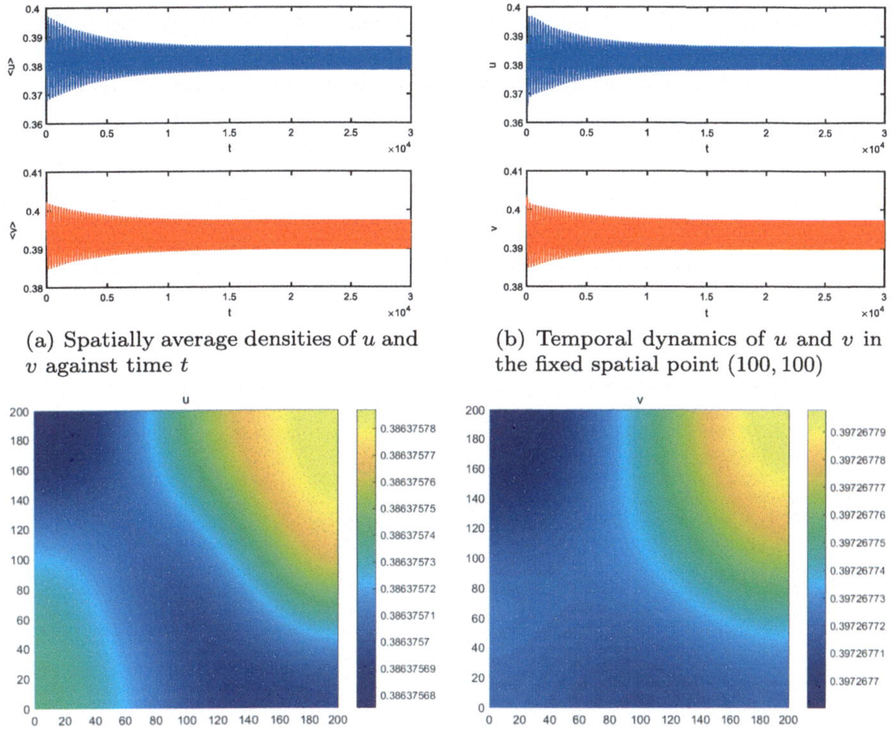

(a) Spatially average densities of u and v against time t

(b) Temporal dynamics of u and v in the fixed spatial point $(100, 100)$

(c) Spatial patterns developed by u

(d) Spatial patterns developed by v

Fig. 14.4 Time evolution and spatial distribution of prey and predator in system (14.3) in the presence of forcing

On the other hand, it is well known that environmental periodic forcing is an important driver of chaotic dynamics. It is widely believed that because of the less restriction of the chaotic patterns than other patterns like Turing patterns, the chaotic spatial patterns have high ecological relevance [347, 349]. Therefore, there is another meaningful direction that considers the effect of the forcing on the transitions between regular and chaotic phases. In addition, there exists an interesting challenge for the effect of the forcing on Turing patterns. Actually, new complex and localized patterns can be generated by the temporal, spatial, and spatiotemporal forcing. Furthermore, the forcing term can also simply enhance and suppress these patterns. Through deep research on this subject, one can gain insight into the random fluctuations in oceanic plankton species abundance.

There are many extensions about these results in order to enhance understanding about the ecological and the forcing's effect in advance. For instance, even though the general expressions are used for amplitude equations, we can use the analytical framework established here to other spatial models with other more complex functional response. In addition, further considering the instabilities of increasing codimension may shed more light on the basic features of the model. We believe that our research enriches the theoretical theory about the related ecological systems and provide inspiration for pattern formation in other application fields such as in epidemic [347] and neural systems [334, 351].

Chapter 15
Spatiotemporal Dynamics Near the Turing-Hopf Bifurcation

During the last few decades, there develop greatly many theoretical results to describe the dynamics of pattern formation [149, 354]. Noticeably, the type of instability remains the main reason for the variation of the spatiotemporal patterns. But when two or more instabilities coincide, system will behave more complicated patterns than patterns generated by sole instability. A prominent example is the Turing-Hopf bifurcation in which two instabilities break the spatial or temporal symmetry, compete and interact with each other, then lead to various patterns like mixed modes, bistability, and spatiotemporal chaos [208]. Such pattern features were experimentally observed in chlorite-iodide-malonic acid (CIMA) reaction [355]. To address this effect of the instability, here we consider the model with phytoplankton-zooplankton interactions. In fact, there are attempts to investigate pattern dynamics near the Turing-Hopf bifurcation for plankton system [356, 357]. However, most studies just focus on the control effect of the model parameters on the bifurcation variation or present some numerical results of patterns. There are few research about conditions that how different types of spatiotemporal patterns emerge.

On the other hand, spatial patterns generated by the distribution of plankton organisms have been investigated in many ecological issues, such as the outbreak of harmful algal blooms (HABs). With the global increase of HABs, many researches focus on the role of toxin-producing phytoplankton (TPP) in planktonic blooms and succession. Moreover, there exists another diffusion in plankton populations, named as cross-diffusion. It can enunciate the population fluxes of one species influenced by the density gradient of another species. Kerner first proposed the cross-diffusion problem [190] and Shigesada et al. applied it in the competitive population systems [191]. Then, the important role of cross-diffusion was explored to the generation of the spatiotemporal patterns for planktonic systems, and the introduction of cross-diffusion was recognized by providing some additional degree of realism [360, 361]. Unfortunately, little attention has been paid on the effect of cross-diffusion on the spatiotemporal dynamics near Turing-Hopf bifurcation whereas most works mainly

S. T. Liu and L. Zhang, *Pattern Dynamics of Marine Plankton Behavior*,
https://doi.org/10.1007/978-981-97-5369-7_15

focus on pattern formation generated by the instability of Turing [362] or Hopf [188] form.

15.1 Spatial Model with TPP

The following framework is presented by Chattopadhayay [227] to reveal that TPP may play a biocontrol role in terminating the planktonic blooms, and describing the functioning of community consisting of TPP

$$\begin{aligned} \frac{dP}{dt} &= rP(1-\frac{P}{K}) - \beta f(P)Z, \\ \frac{dZ}{dt} &= \alpha f(P)Z - mZ - \delta g(P)Z, \end{aligned} \tag{15.1}$$

where P and Z represent the abundance of TPP and zooplankton, respectively, $f(P)$ means the predational response function, $g(P)$ stands for the distribution of toxic substances which ultimately generate the death of zooplankton populations, and Table 15.1 lists the description of the parameters. Afterwards, based on the fact that the interactions between plankton populations generally emerge in a wide range of spatial and temporal scales, we pay attention to the following shifted spatial version of model (15.1)

$$\begin{aligned} \frac{\partial P}{\partial t} &= d_{11}\nabla^2 P + rP(1-\frac{P}{K}) - \frac{\beta PZ}{h+P}, \\ \frac{\partial Z}{\partial t} &= d_{22}\nabla^2 Z + \frac{\alpha PZ}{h+P} - mZ - \frac{\delta PZ}{h+P}, \end{aligned} \tag{15.2}$$

where ∇^2 denotes the Laplacian operator in two-dimensional space, h means the half-saturation constant, d_{11}, d_{22} stands for the self-diffusion coefficients, $f(P)$ and $g(P)$ both represent Holling type II functional response as a common and intuitively obvious choice. For example, a complete and rigorous analysis was given for the global dynamics of model (15.2) [358]. The evolutionary processes of pattern formation were presented in model (15.2) to indicate the driving effect of the forces of diffusion on the interactions between TPP and zooplankton in marine environments partly [258]. Motivated by the above-mentioned work, we mainly focus on the model as below

$$\begin{aligned} \frac{\partial P}{\partial t} &= d_{11}\nabla^2 P + d_{12}\nabla^2 Z + rP(1-\frac{P}{K}) - \frac{\beta PZ}{h+P}, \\ \frac{\partial Z}{\partial t} &= d_{21}\nabla^2 P + d_{22}\nabla^2 Z + \frac{\alpha PZ}{h+P} - mZ - \frac{\delta PZ}{h+P}, \end{aligned} \tag{15.3}$$

with positive initial conditions and periodic boundary conditions, where Table 15.1 lists the description of the parameters of model (15.3).

Table 15.1 The description and values of the parameters in the model (15.3)

Parameter	Description	Value
r	Intrinsic growth rate of TPP	0.39
K	Environmental carrying capacity of TPP	1.00
β	Predation rate of zooplankton	0.47
α	Ratio of biomass conversion	0.39
m	Natural mortality rate of zooplankton	0.11
δ	Rate of toxin liberation by TPP	–
h	Half saturation constant	0.32
d_{11}	Self-diffusion coefficient of TPP	0.47
d_{12}	Cross-diffusion coefficient of TPP	0.66
d_{21}	Cross-diffusion coefficient of zooplankton	–
d_{22}	Self-diffusion coefficient of zooplankton	0.52

15.2 Linear Stability Analysis

At first, we present the uniform steady states of the model (15.3), give the conditions for the Turing and Hopf instability, and further locate the point for the codimension-two Turing-Hopf bifurcation.

Easily, we give the equilibria of the non-spatial version for the model (15.3) as: $E_0 = (0, 0)$, $E_b = (K, 0)$ and $E_* = (P_*, Z_*)$ with $P_* = \frac{mh}{\alpha-\delta-m}$, $Z_* = \frac{r}{\beta}(1 - \frac{P_*}{K})(h + P_*)$. And we get that the coexistence equilibrium E_* becomes feasible as $\delta < \alpha - m - \frac{mh}{K}$, where E_0 and E_b are saddle points. Next, we only concentrate on the stability of E_* considering the biological significance.

Rescaled the model (15.3) by $\hat{P} = P - P_*$ and $\hat{Z} = Z - Z_*$, the linearized form can be obtained

$$
\begin{aligned}
\frac{\partial \hat{P}}{\partial t} &= d_{11}\nabla^2 \hat{P} + d_{12}\nabla^2 \hat{Z} + J_{11}\hat{P} + J_{12}\hat{Z}, \\
\frac{\partial \hat{Z}}{\partial t} &= d_{21}\nabla^2 \hat{P} + d_{22}\nabla^2 \hat{Z} + J_{21}\hat{P} + J_{22}\hat{Z},
\end{aligned}
\tag{15.4}
$$

where $J_{11} = r - \frac{2r}{K}P_* - \frac{\beta h Z_*}{(h+P_*)^2}$, $J_{12} = -\frac{\beta P_*}{h+P_*}$, $J_{21} = \frac{(\alpha-\delta)hZ_*}{(h+P_*)^2}$ and $J_{22} = 0$. Then after omitting the hats for convenience and considering the Fourier expansion for the solutions of Eq. (15.4), we have

$$
\begin{pmatrix} P \\ Z \end{pmatrix} = \sum_q \begin{pmatrix} c_q^1 \\ c_q^2 \end{pmatrix} e^{\lambda t + i\boldsymbol{q}\cdot\boldsymbol{r}}
\tag{15.5}
$$

where λ denotes the growth rate and $|\boldsymbol{q}| = q$ means the wave number. The stability is determined by the eigenvalues $\lambda_{1,2}(q)$ of the matrix

$$L(q) = \begin{pmatrix} J_{11} - d_{11}q^2 & J_{12} - d_{12}q^2 \\ J_{21} - d_{21}q^2 & J_{22} - d_{22}q^2 \end{pmatrix}. \tag{15.6}$$

When we take the conditions for the Hopf bifurcation as $\mathrm{Re}\,\lambda_{1,2}(0) = 0$ and $\mathrm{Im}\,\lambda_1(0) = -\mathrm{Im}\,\lambda_2(0) \neq 0$, it yields $\mathrm{tr}\,L(0) = 0$. So we get the critical condition as $J_{11} = 0$. Under the unstable case, spatially homogeneous oscillatory behavior can be shown with Hopf frequency $\omega_H = \sqrt{-J_{12}J_{21}}$. We calculate the critical left and right eigenvectors as $L_H = \left(J_{21}, i\omega_H - J_{11}\right)$ and $R_H = \left(J_{12}, i\omega_H - J_{11}\right)^T$, respectively.

The occurrence of Turing instability needs the generation of the critical wave number with $\mathrm{Re}\,\lambda_1(q_T) = \mathrm{Im}\,\lambda_1(q_T) = 0$, where$q_T \neq 0$, implying $\det L(q_T) = 0$. Besides that, bifurcation point can be regarded as the most important point for all other modes so that the minimum of $\det L(q_T)$ reaches at q_T where $q_T^2 = \sqrt{\frac{J_{11}J_{22} - J_{12}J_{21}}{d_{11}d_{22} - d_{12}d_{21}}}$. The critical left and right eigenvectors are calculated as $L_T = \left(-d_{21}q_T^2 + J_{21}, -d_{11}q_T^2 - J_{11}\right)$ and $R_T = \left(-d_{12}q_T^2 + J_{12}, d_{11}q_T^2 - J_{11}\right)^T$, respectively. Furthermore, Turing instability will produce stationary spatially periodic patterns.

Therefore, choosing control parameters like d_{21} as cross-diffusion coefficient and δ as toxin liberation rate, we can simulate the bifurcation via Turing and Hopf modes by the d_c as critical values and δ_c for Turing-Hopf bifurcation where

$$J_{11}(\delta_c) = 0,$$

$$2\sqrt{(d_{11}d_{22} - d_{12}d_c)(-J_{12}(\delta_c)J_{21}(\delta_c))} = d_{22}J_{11}(\delta_c) - d_c J_{12}(\delta_c) - d_{12}J_{21}(\delta_c),$$

and

$$\omega_c = \sqrt{-J_{12}(\delta_c)J_{21}(\delta_c)}, \ q_c^2 = \sqrt{\frac{-J_{12}(\delta_c)J_{21}(\delta_c)}{d_{11}d_{22} - d_{12}d_c}},$$

denote the critical frequency and wave number.

15.3 Weakly Nonlinear Analysis

In order to reveal the evolution of Turing and Hopf modes, we apply weakly nonlinear analysis as an effective motion to give further understanding for the effect of slow degrees of freedom near the bifurcation point.

At first, we expand the model (15.3) by Taylor expansion near the equilibrium E_* and get

$$\frac{\partial p}{\partial t} = d_{11}\nabla^2 p + d_{12}\nabla^2 z + J_{11}p + J_{12}z + a_{11}p^2 + a_{12}pz + b_{11}p^3 + b_{12}p^2 z,$$
$$\frac{\partial z}{\partial t} = d_{21}\nabla^2 p + d_{22}\nabla^2 z + J_{21}p + J_{22}z + a_{21}p^2 + a_{22}pz + b_{21}p^3 + b_{22}p^2 z, \tag{15.7}$$

where $(p, z)^T$ represents the perturbations around E_* and

$$\begin{aligned} a_{11} &= -\frac{r}{K} + \frac{\beta h Z_*}{(h+P_*)^3}, \ a_{12} = -\frac{\beta h}{(h+P_*)^2}, \ b_{11} = -\frac{\beta h Z_*}{(h+P_*)^4}, \\ b_{12} &= \frac{\beta h}{(h+P_*)^3}, a_{21} = -\frac{(\alpha-\delta)hZ_*}{(h+P_*)^3}, \ a_{22} = \frac{(\alpha-\delta)h}{(h+P_*)^2}, \\ b_{21} &= \frac{(\alpha-\delta)hZ_*}{(h+P_*)^4}, \ b_{22} = -\frac{(\alpha-\delta)h}{(h+P_*)^3}. \end{aligned}$$

We then give the following compact form as a representation for Eq. (15.7)

$$\frac{\partial X}{\partial t} = LX + N, \tag{15.8}$$

where

$$X = \begin{pmatrix} p \\ z \end{pmatrix}, \ L = \begin{pmatrix} d_{11}\nabla^2 + J_{11} & d_{12}\nabla^2 + J_{12} \\ d_{21}\nabla^2 + J_{21} & d_{22}\nabla^2 + J_{22} \end{pmatrix},$$

$$N = \begin{pmatrix} a_{11}p^2 + a_{12}pz + b_{11}p^3 + b_{12}p^2 z \\ a_{21}p^2 + a_{22}pz + b_{21}p^3 + b_{22}p^2 z \end{pmatrix}.$$

In order to study the effect of the multiple time scales on dynamical behavior, perturbation analysis is proceeded with by introducing different but mutually independent time scales. The perturbation analysis includes, the fast time scale t, appearance of the oscillation frequency ω_c, and slow time scale $t_2 = \epsilon^2 t$ $(0 < \epsilon \ll 1)$ accounted for slow time evolution for critical modes in the neighborhood of Turing-Hopf bifurcation. Applied weakly nonlinear expansion to the solutions to Eq. (15.8), the following is obtained

$$X = \begin{pmatrix} p \\ z \end{pmatrix} = \epsilon \begin{pmatrix} p_1 \\ z_1 \end{pmatrix} + \epsilon^2 \begin{pmatrix} p_2 \\ z_2 \end{pmatrix} + \epsilon^3 \begin{pmatrix} p_3 \\ z_3 \end{pmatrix} + \cdots,$$

where $(p_n, z_n)^T$ $(n = 1, 2, 3)$ stand for functions of t_2 describing the slow time scale behavior after all decay of the transients. Similarly, we set other elements of Eq. (15.8) as

$$d_{21} = d_c + \epsilon^2 d_2 + \cdots, \ \delta = \delta_c + \epsilon^2 \delta_2 + \cdots,$$

$$L = L_c + \epsilon^2 L_2 + \cdots, \ N = \epsilon^2 N_2 + \epsilon^3 N_3 + \cdots,$$

where

$$L_c = \begin{pmatrix} d_{11}\nabla^2 & d_{12}\nabla^2 - \frac{\beta(K-h)}{K+h} \\ d_c\nabla^2 + \frac{hr(\alpha-\delta_c)}{\beta K} & d_{22}\nabla^2 \end{pmatrix},$$

$$L_2 = \begin{pmatrix} (\frac{K+h}{(\alpha-\delta_c)^2} - \frac{2h}{(\alpha-\delta_c-m)^2})\frac{mr\delta_2}{K} & -\frac{m\beta\delta_2}{(\alpha-\delta_c)^2} \\ d_2\nabla^2 - \frac{r\delta_2}{\beta} & 0 \end{pmatrix},$$

and

$$N_2 = \begin{pmatrix} a_{11}p_1^2 + a_{12}p_1z_1 \\ a_{21}p_1^2 + a_{22}p_1z_1 \end{pmatrix},\ N_3 = \begin{pmatrix} 2a_{11}p_1p_2 + a_{12}(p_1z_2 + p_2z_1) + b_{11}p_1^3 + b_{12}p_1^2z_1 \\ 2a_{21}p_1p_2 + a_{22}(p_1z_2 + p_2z_1) + b_{21}p_1^3 + b_{22}p_1^2z_1 \end{pmatrix}.$$

Substituting the above decomposition into Eq. (15.8) and gathering the same order of ϵ, we calculate the solution $(p_n, z_n)^T$ that satisfied the problems as below

$$O(\epsilon):\ \frac{\partial}{\partial t}\begin{pmatrix} p_1 \\ z_1 \end{pmatrix} - L_c\begin{pmatrix} p_1 \\ z_1 \end{pmatrix} = 0, \tag{15.9}$$

$$O(\epsilon^2):\ \frac{\partial}{\partial t}\begin{pmatrix} p_2 \\ z_2 \end{pmatrix} - L_c\begin{pmatrix} p_2 \\ z_2 \end{pmatrix} = N_2, \tag{15.10}$$

$$O(\epsilon^3):\ \frac{\partial}{\partial t}\begin{pmatrix} p_3 \\ z_3 \end{pmatrix} - L_c\begin{pmatrix} p_3 \\ z_3 \end{pmatrix} = -\frac{\partial}{\partial t_2}\begin{pmatrix} p_1 \\ z_1 \end{pmatrix} + L_2\begin{pmatrix} p_1 \\ z_1 \end{pmatrix} + N_3. \tag{15.11}$$

Then, the solution $(p_1, z_1)^T$ is given by solving Eq. (15.9) as following

$$\begin{pmatrix} p_1 \\ z_1 \end{pmatrix} = R_H B e^{i\omega_c t} + R_T(A_1 e^{i\boldsymbol{q}_1\cdot\boldsymbol{r}} + A_2 e^{i\boldsymbol{q}_2\cdot\boldsymbol{r}} + A_3 e^{i\boldsymbol{q}_2\cdot\boldsymbol{r}}) + c.c. \tag{15.12}$$

where B and A_n $(n = 1, 2, 3)$ stand for, respectively, amplitudes of Hopf mode and Turing modes with the perturbations of first order, in which the form of them are controlled by the perturbational terms of higher order. In the case of $O(\epsilon^2)$ problem, we plug Eq. (15.12) into it and obtain its solutions as the form listed below

$$\begin{aligned} \begin{pmatrix} p_2 \\ z_2 \end{pmatrix} = &\begin{pmatrix} p_{21} \\ z_{21} \end{pmatrix}\sum_{n=1}^{3}|A_n|^2 + \begin{pmatrix} p_{22} \\ z_{22} \end{pmatrix}|B|^2 + 2\mathrm{Re}\begin{pmatrix} p_{23} \\ z_{23} \end{pmatrix}Be^{i\omega_c t}\cdot\sum_{n=1}^{3}A_n e^{i\boldsymbol{q}_n\cdot\boldsymbol{r}} \\ &+ \begin{pmatrix} p_{25} \\ z_{25} \end{pmatrix}\sum_{n=1}^{3}A_n^2 e^{2i\boldsymbol{q}_n\cdot\boldsymbol{r}} + \begin{pmatrix} p_{26} \\ p_{26} \end{pmatrix}B^2 e^{2i\omega_c t} + \sum_{n=1}^{3}R_T W_n e^{i\boldsymbol{q}_n\cdot\boldsymbol{r}} \\ &+ \begin{pmatrix} p_{28} \\ z_{28} \end{pmatrix}\sum_{n_1\neq n_2}A_{n_1}\bar{A}_{n_2}e^{i(\boldsymbol{q}_{n_1}-\boldsymbol{q}_{n_2})\cdot\boldsymbol{r}} + c.c. \end{aligned} \tag{15.13}$$

where the coefficients are determined by solving corresponding equations to different terms, and Appendix E lists all the detailed presentations. Next, we turn to the case of $O(\epsilon^3)$. In order to eliminate secular terms, we apply the Fredholm alternative theorem as below. We first write the critical left eigenvectors as $L_H e^{-i\omega_c t}$ and $L_T e^{-i\mathbf{q}_n \cdot \mathbf{r}}$ $(n = 1, 2, 3)$. Then, each left eigenvector derives the second-order perturbations for the following amplitude equations by taking an inner product for the right-hand side of Eq. (15.11):

$$\begin{aligned}
\frac{\partial B}{\partial t} &= \eta B + \tilde{\gamma}|B|^2 B + \tilde{\sigma}(|A_1|^2 + |A_2|^2 + |A_3|^2)B, \\
\frac{\partial A_1}{\partial t} &= \xi A_1 + \tilde{\mu}|A_1|^2 A_1 + \tilde{g}(|A_2|^2 + |A_3|^2)A_1 + \tilde{b}\bar{A}_2\bar{A}_3 + \tilde{s}|B|^2 A_1,
\end{aligned} \tag{15.14}$$

where cyclic permutation means the other two Turing mode equations by $(1 \to 2 \to 3 \to 1)$, and the coefficients are explicitly calculated based on the parameters of the full system (15.3). As the amplitude of Turing and Hopf modes under different orders and coefficients $e^{i\mathbf{q}_n \cdot \mathbf{r}}$ and $e^{i\omega_c t}$ is listed as below [194]

$$T_n = \epsilon\varphi A_n + \epsilon^2 \varphi W_n + \cdots, \quad H = \epsilon\psi B + \cdots, \quad \text{with } \varphi = J_{12} - d_{12}q_c^2 \text{ and } \psi = J_{12},$$

we get the final amplitude equations by merging different variables based on Eq. (15.14)

$$\begin{aligned}
\frac{\partial H}{\partial t} &= \eta H + \gamma|H|^2 H + \sigma(|T_1|^2 + |T_2|^2 + |T_3|^2)H, \\
\frac{\partial T_1}{\partial t} &= \xi T_1 + \mu|T_1|^2 T_1 + g(|T_2|^2 + |T_3|^2)T_1 + b\bar{T}_2\bar{T}_3 + s|H|^2 T_1,
\end{aligned} \tag{15.15}$$

where we can get the remaining Turing amplitude equations by the change of the subscript of T_n. Here, the calculation is given in detail in the Appendix, and readers can also refer more information in [363].

15.4 Amplitude Stability Analysis

As we all know, the coefficients of amplitude Eq. (15.15) play an important role in determining the type of patterns. Notably, as the feature of the weakly nonlinear analysis that it is only suitable for the vicinity of supercritical bifurcation, we study the cases generating supercritical behavior of both Hopf and Turing lines: $\mathrm{Re}\,\gamma < 0$ and $\mu < 0$. Hence, we employ the rescaling method to normalize $\mathrm{Re}\,\gamma$ and μ to be unity,

$$\sqrt{-\mu}\, T_1 \to T_1, \ \sqrt{-\mathrm{Re}\,\gamma}\, \exp(-i\, \mathrm{Im}\,\eta t) H \to H,$$

and get the results as

$$1+ic=\frac{\gamma}{\mathrm{Re}\,\gamma},\ \hat{\sigma}=\frac{\sigma}{-\mu},\ \hat{g}=\frac{g}{-\mu},\ \hat{b}=\frac{b}{\sqrt{-\mu}},\ \hat{s}=\frac{s}{-\mathrm{Re}\,\gamma}.$$

In the following, we omit the hats below for brevity. After we decompose Turing amplitudes into moduli ρ_n ($n = 1, 2, 3$) and total complex phase ϕ and separate the real and imaginary parts, we write the amplitude equations as

$$\begin{aligned}
\frac{\partial\phi}{\partial t} &= -b(\frac{\rho_2\rho_3}{\rho_1}+\frac{\rho_1\rho_3}{\rho_2}+\frac{\rho_1\rho_2}{\rho_3})\sin\phi,\\
\frac{\partial H}{\partial t} &= \mathrm{Re}\,\eta H-(1+ic)|H|^2H+\sigma(\rho_1^2+\rho_2^2+\rho_3^2)H,\\
\frac{\partial\rho_1}{\partial t} &= \xi\rho_1-\rho_1^3+g(\rho_2^2+\rho_3^2)\rho_1+b\rho_2\rho_3\cos\phi+s|B|^2\rho_1.\\
&(1\to2\to3\to1)
\end{aligned}\tag{15.16}$$

Table 15.2 summarizes the similar existence and stability conditions for different solutions. Thus, we can predict the emergence of spatiotemporal patterns based on the results.

Table 15.2 Different solutions for amplitude Eq. (15.16) and their stability

Solution	Expression formula	Existence and stability conditions
Hopf mode	$\rho_1=\rho_2=\rho_3\equiv0$ $H=\sqrt{\mathrm{Re}\,\eta}\exp(-ic\,\mathrm{Re}\,\eta t)$	$\mathrm{Re}\,\eta\geq0$ $\xi+s\,\mathrm{Re}\,\eta<0$
Truing mode (plane wave)	$\rho_1=\sqrt{\xi},\ \rho_2=\rho_3\equiv0$ $H\equiv0$	$\xi\geq0$ $g<-1$ $\mathrm{Re}\,\eta+\xi\,\mathrm{Re}\,\sigma<0$ $\xi(1+g)^2>b^2$
Turing mode (hexagonal patterns)	$\rho_n=\frac{\lvert b\rvert+\sqrt{b^2+4\xi(1-2g)}}{2(1-2g)}$ $H\equiv0$	$1-2g>0$ $4(1-2g)\xi>-b^2$ $\lvert b\rvert+(1+g)\rho_1>0$ $\mathrm{Re}\,\eta+3\,\mathrm{Re}\,\sigma\rho_1^2<0$
Mixed mode (plane wave)	$\rho_1=\sqrt{\frac{\xi+s\,\mathrm{Re}\,\eta}{1-s\,\mathrm{Re}\,\sigma}},\ \rho_2=\rho_3\equiv0$ $H=\sqrt{\frac{\mathrm{Re}\,\eta+\xi\,\mathrm{Re}\,\sigma}{1-s\,\mathrm{Re}\,\sigma}}\exp(i\Delta\omega t)$	$g<-1$ $1-s\,\mathrm{Re}\,\sigma>0$ $b^2<(1+g)^2\frac{\xi+s\,\mathrm{Re}\,\eta}{1-s\,\mathrm{Re}\,\sigma}$
Mixed mode (hexagonal patterns)	$\rho_n=\frac{\lvert b\rvert+\sqrt{b^2+4(1-2g-3s\,\mathrm{Re}\,\sigma)(\xi+s\,\mathrm{Re}\,\sigma)}}{2(1-2g-3s\,\mathrm{Re}\,\sigma)}$ $H=\theta\exp(i\Delta\omega t)$	$1-2g-3s\,\mathrm{Re}\,\sigma>0$ $\lvert b\rvert\rho_1-2(1-2g)\rho_1^2<2\theta^2,$ $\lvert b\rvert+(1+g)\rho_1>0,$ $\mathrm{Re}\,\eta+3\,\mathrm{Re}\,\sigma\rho_1^2>0,$ $4(\xi+s\,\mathrm{Re}\,\eta)>\frac{-b^2}{(1-2g-3s\,\mathrm{Re}\,\sigma)}.$

Notes (i) $\Delta\omega$ in plane wave $\Delta\omega=-c|H|^2+\mathrm{Im}\,\sigma\rho_1^2$.
(ii) In hexagonal patterns, $\theta^2=\mathrm{Re}\,\eta+3\,\mathrm{Re}\,\sigma\rho_1^2$, $\Delta\omega=3\,\mathrm{Im}\,\sigma\rho_1^2-\theta^2c$

15.5 Numerical Simulations

Next, we simulate particular examples for model (15.3) to confirm the aforementioned results. All parameter settings are shown in Table 15.1. The critical parameter values are straightforward calculated for the occurrence of the supercritical Turing-Hopf bifurcation for model (15.3) as $\delta_c = 0.18$, $d_c = 0.32$. Without special explanations, we assume these parameter values remained unchanged below. Hence, we calculate the coefficients of amplitude Eq. (15.16) as $\mathrm{Re}\,\sigma = -23.2085$, $\mathrm{Im}\,\sigma = -2.2477$, $g = -0.4823$, $b = 0.8006$, $s = -1.6668$. It is already known that Turing patterns not only formed from plane waves or squares but also all the mixed modes remain unstable so that we cannot observe them under the present parameter values. For visualization, the results of pattern formations are summarized in a parameter plane spanned by parameters δ and d_{21} shown in Fig. 15.1. From the visualization, we get the information clearly that, for the vicinity of Turing-Hopf bifurcation point, there exist three regions: Region I (green) refers to the bistability between trivial solution and hexagonal patterns, Region II (yellow) refers to the coexistence of Hopf state and hexagonal patterns, and Region III (blue) represents the stability of hexagonal patterns. Significantly, even in Region I, the stable range of the homogeneous state, there still emerge the Turing patterns. This is special compared with the behavior for the Turing instability. Besides, as the stable region for the hexagons becomes relatively large, the competition between different modes favoring Turing modes can be inferred compared to Hopf mode. Nevertheless, because of a larger distance from the Turing-Hopf bifurcation which cannot satisfy the condition of the sufficiently small

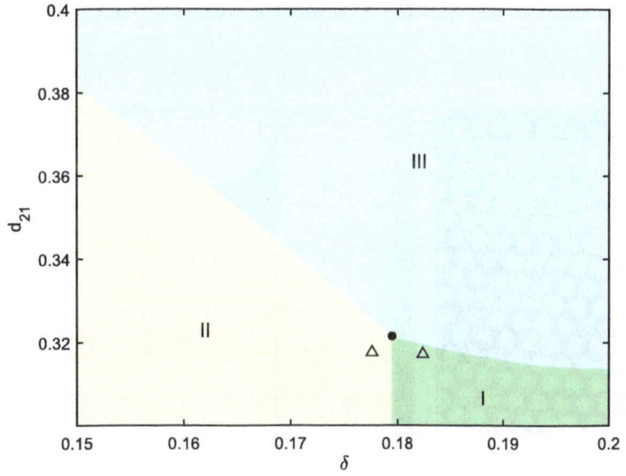

Fig. 15.1 Stability regions near the Turing-Hopf bifurcation (filled circle) in model (15.3) in (δ, d_{21}) parameter plane, where Region I (green) means the bistability between trivial solution and hexagonal patterns, Region II (yellow) refers to the coexistence of hexagonal patterns and Hopf state, Region III (blue) represents the stability of hexagonal patterns. The green and yellow triangles mark the parameter settings used in Figs. 15.2 and 15.3

neighborhood of the critical point, we may not see the outcome of weakly nonlinear analysis effectively. Afterwards, these predictions are compared with the numerical simulations of the model (15.3). We use here the finite difference method to simulate for the model (15.3) in a two-dimensional domain [364, 365]. The boundary condition is set as periodic form and the initial condition is taken as the uniform steady state E_* superimposed Gaussian white noise in order to generate some inhomogeneities in the distribution. This choice can guarantee the necessary asymmetry for the growth of modes. Because the case that the population's distribution will become spatially uniform all the time when the initial data of TPP and zooplankton remains homogeneous, it is not our current interest. Besides, it is assumed that the noise amplitude should be small enough to make sure the relaxation time remaining small too. In addition, different initial conditions could derive to different steady states as the pattern variation is partly sensitive to the initial setting (see more related details in Refs. [366, 367]). In order to apply the finite difference method, we divide the spatial domain into 150×150 lattice sites and take the spacing between the lattice points as $\Delta h = 1$. And we use the Euler forward difference method with a time step size $\Delta t = 0.1$ for the time evolution. The space step is defined as Δh for the five-point difference scheme in the diffusion terms. During the simulations, we always take the spatiotemporal patterns of P and Z as the same type so that there exist only patterns of P illustrated for instance. It is to be noted that we present the simulations until they do not change with time. Then, we take parameter values from Region I and II shown in Fig. 15.1 to present numerical results, respectively.

Figure 15.2 shows the the well-distributed hexagonal patterns taking parameters as $\delta = 0.182$ and $d_{21} = 0.317$ in Region I. Moreover, we also see the dependence of the trivial solution on initial condition. The trivial solution is characterized by the spatial average of P against time and the corresponding variance $\mathrm{Var}(P) = <P^2> - <P>^2$.

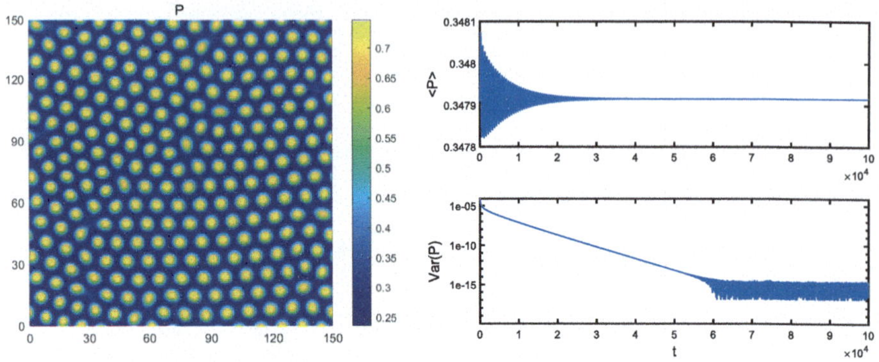

(a) Hexagonal patterns

(b) Trivial solution reflected by the spatial average of P against time and the corresponding variance $\mathrm{Var}(P) = < P^2 > - <P>^2$

Fig. 15.2 Spatiotemporal patterns of P at $\delta = 0.182$ and $d_{21} = 0.317$ depending on initial data

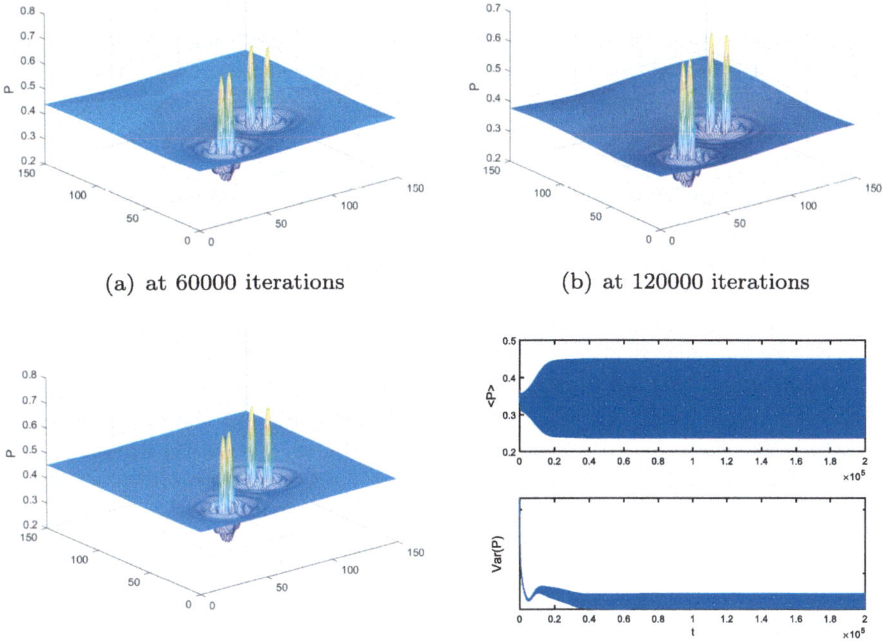

(a) at 60000 iterations

(b) at 120000 iterations

(c) at 200000 iterations

(d) The spatial average of P against time and the corresponding variance $\mathrm{Var}(P) = <P^2> - <P>^2$

Fig. 15.3 Coexistence patterns of P between hexagons and Hopf states for parameters $\delta = 0.178$ and $d_{21} = 0.318$

Figure 15.3 illustrates the evolution of spatiotemporal patterns of P taking as parameters $\delta = 0.178$ and $d_{21} = 0.318$ in Region II, after 60000, 120000, and 200000 Iterations, respectively. In the initial period with the domination of the irregular transient distribution, there remain patterns like Turing structures and Hopf oscillations in the whole domain. The periodic oscillation is exhibited by the spatial average of P. Obviously, our theoretical outcome is proved effectively by the numerical results.

In this chapter, spatiotemporal dynamics are investigated near the Turing-Hopf bifurcation for a diffusive plankton model. We uncover the corresponding pattern structures by linear stability analysis and weakly nonlinear analysis. Specifically, we find out the control parameters spanning a diagram to generate the stability conditions for different patterns. That is the reason that we can predict the variation of the patterns theoretically. Our findings illustrate that there emerges a wide region where there exist spatially homogeneous oscillations and hexagonal patterns together. Significantly, we can see the hexagonal patterns easily even though the homogeneous state remains stable. These works will provide abundant results in dynamical patterns for plankton system and also play important role in giving further insight about interaction between toxin-producing phytoplankton and zooplankton. What's more, the effect of the external action like noise and forcing on the stability and pattern formation for reaction–diffusion systems also attract attention among researchers. It is found that the existence of noise can destroy, enhance or decay the patterns of Turing and Hopf. But the secret behind these phenomena has been remained mysterious yet and needs further study in future (related contents seen from Refs. [368]).

Additionally, pattern formation and spiral wave in neural networks like the case in reaction–diffusion systems are also active research fields [205]. Especially more and more works focus on the stability and breakup mechanism of these patterns. For instance, the pattern selection and stability of a neuronal network about memory were discussed for the effect of electromagnetic induction [369]. Spiral waves were induced in a neuronal network as the condition taken as random values for the boundary [200]. Collapse and instability of spiral wave were studied, states were ordered and the prediction of them was controlled by monitoring and analyzing the time series for some nodes [370]. Certainly, what is worth mentioning is the investigations that the mechanisms of spiral wave breakup also have an important role in further understanding the cardiac electrical activity and in the treatment of cardiac arrhythmias and heart disease [371].

Chapter 16
Fractional Herd Behavior with Cross-Diffusion

16.1 Introduction

Survival behaviors among marine biological populations have been studied by quite many researchers [270, 395]. For example, diffusion as the most significant phenomenon can be regarded as the diffusion of standardized random walks linearly with time. But the life activities remain complicated. Predators in the close-range scale can adjust their foraging with near-end clues by using the information of resource distribution [393]. Furthermore, predators mainly apply their foraging strategies to get the foraging success in a wide range of mesoscale like from several kilometers to hundreds of kilometers [394]. However, within the mesoscale boundary, there exists no uniform foraging activities for some predators such as a big divergent between the typical random walk theory and the statistic feature of their real data. So that to study the spatial scale dynamically becomes important and valuable for understanding the related biological behavior. In addition, some experiment evidence proves that there occurs abnormal diffusion in nature [377, 379, 387]. Lévy flight model, as one of the most widely effective models for these abnormal diffusion phenomena, has been applied for the foraging phenomena by marine predators empirically. For instance, based on the ocean remote sensing data, it is noted that predation process of penguins, the sharks, and other ocean creatures can be modeled effectively by the Lévy flying model [394] not the original random walk model. According to the believe that Lévy flight model can be used commonly for open ocean foragers as diffusion behavior, a fractional-order model is introduced in the diffusion process.

So far, we have presented many integer order dynamical models for marine biological behavior and there are also some extended models like fractional-order dynamical models for the dynamics of many complex systems. However, there exists rare elucidates biophysical evidence and sufficient scientific background. It is well known that introducing fractional operation must have reliable physical evidence and scientific clarification. For this reason, more researchers use the fractional-order calculation in the diffusion process and spatiotemporal phenomena. Furthermore, the fractional-

S. T. Liu and L. Zhang, *Pattern Dynamics of Marine Plankton Behavior*,
https://doi.org/10.1007/978-981-97-5369-7_16

order calculation instead of integer order can also be applied to reproduce the related scientific problems [400].

Over the years, a lot of research work about fractional reaction–diffusion systems have been done for some nonlinear-diffusion process in applied science and engineering [388, 396]. For example, because that the super-diffusion order $\alpha(\nabla^{\alpha})$ as the control term plays important role in the dynamical properties of a predator–prey model, a fractional reaction–diffusion system was introduced to expand the original space-time domain. Then the model explained the process of pattern formation effectively and also regenerate the actual behavior of the marine biological population scientifically [381]. Hence, fractional diffusion has significant effect on the dynamics of the marine organisms.

Generally, there exist a wide variety of biological aggregation in the ocean [390] like capelin (Mallotus villosus), a fish (such as Engraulis mordax), tuna (Thunnus spp.), and sardines. These fish live in groups to reduce the risk of predation, increase the probability of successful mating, breeding, and improve the foraging success. Herd behaviors have been revealed by some studies. For example, experts gave the result that juvenile tuna explore different foraging strategies at different times like having herd behavior during the day [383]. Or adult cod usually lives in herds [185]. Furthermore, many fish undergo complicated relationship with their natural enemies. For instance, tuna (Thunnus spp.) catches fish such as Engraulis mordax and sardines such as Sardinops sagax [373, 382] for food. Arctic cod (Gadus morhua) is the most endanger natural enemy for capelin (Mallotus villosus) or herring (Clupea harengus) is [376]. These attacked fish collect response caused by the predator attack, transfer rapid information between individual members, and finally show respective reactions in the fish herd [378]. For these relationship, although there have been many predator–prey models about fish populations in ecology, we have not fully elucidated the impact of educational behavior [189]. In addition, the related herd population behavior of certain fish as one of the most practical behavior has been ignored and not been studied. Therefore, the herd behavior in fish predator–predator systems becomes our aim as a vital task.

In these population interactions, predators always want to catch the prey but prey don't. This interactions between predator and prey usually have significant effect on maintaining the balance of the ecology in nature. The necessary role of the refuge has been demonstrated in the coexistence of prey and predator by continuous theories and experiments [384–386, 391]. For example, the refuge prevents the extinction of the prey population by providing a certain degree of protection for them so that it has impact on the population of the predator and the prey [372, 384–386, 389, 391, 392]. Nowadays, more and more predator–predator models with refuges have been proposed and there have been obtained some valuable results [375, 397]. For instance, a predator–prey model with prey refuge was discussed by fractional diffusion, and the existence and uniqueness of the solution was also proved by applying the Lyapunov method and graph theory method [398]. Based on the reason that cross-diffusion has crucial effect on pattern formation in the fields of ecology, biology and chemistry, mutual interference between predators was also introduced into a predator-prey model with cross-fractional diffusion [399]. Importantly, the work proves that there exists no pattern when the model remains bistable without cross-fractional diffusion term, but the Turing pattern would appear considering an appropriate cross-fractional diffusion term into this model. Subsequently, adding Michaelis–Menten-type predator harvesting term, the Turing instability and other pattern formation were also studied in a predator–prey system with cross-fractional diffusion term, such as equilibrium instability generated by the cross-fractional diffusion term [380].

According to the above results, we consider a kind of spatial fractional predator–prey model with refuge effect for the prey and the predator living in herds. As a result, we obtain many exciting results regarding stability. The structure of this chapter is listed as follows. At first, a mathematical model with factors like refuge and herd term is introduced. Then, we carry out the stability analysis of the non-spatial diffusion model. Thirdly, the detailed derivation and stability analysis are contained for the amplitude equation of the three-dimensional spatial pattern. Fourthly, we simulate the system model by MATLAB software. Finally, a general discussion and the related biological significance of our findings are presented.

16.2 Fractional Reaction–Diffusion Mathematical Model With Herds and Cross-Diffusion

As predators and preys living in the marine environment tend to live in herds, they exhibit herd behaviors at the same time. The phenomenon that the predators live in groups with prey can be indicated by many economically influential fish species in ecology. Based on the group defense and group predation of these population, prey can be only hunted by predators on the group's boundary. Regarding that the interaction term can be directly indicated by the proportional one of the square root of the density for the two populations [270], we propose the following model:

$$\begin{aligned} \frac{du}{dt} &= au(1-\frac{u}{K}) - b\sqrt{u}\sqrt{v} - h_1 u, \\ \frac{dv}{dt} &= c\sqrt{u}\sqrt{v} - dv - h_2 v, \end{aligned} \tag{16.1}$$

where u and v are the population density of prey and predator, respectively, a refers to the growth rate of the prey, K means the environmental capacity, b represents the capture rate of the predator, c stands for the energy conversion rate of the predator, d means the natural mortality of predators, and $h_1 u$ and $h_2 v$ represent the output of prey and predator separately. Table 16.1 shows all the meanings and values of parameters in the model. Considering the protective role of the refuge in the prey population, we suppose that eu means the protection by the predation wholly for the prey, and $(1-e)u$ stands for the interacting predation activities between the predators. e means the protection parameter for the prey in the habitat, $e \in [0, 1)$. Hence, we can rewrite the system (16.1) as

Table 16.1 Parameters definitions in model (16.3) and their value

Parameter	Parameter definition	Value
a	Intrinsic growth rate of prey	0.919
K	Carrying capacity of environment	7.053
b	Consumption rate of predator	1.081
c	Conversion rate of predator	0.748
d	Death rate of predator	0.392
e	Coefficient protected by ecological refuge	$[0, 1)$
h_1	Catchability coefficient of prey	0.246
h_2	Catchability coefficient of predator	0.185
d_{11}	Self-fractional diffusion coefficient of prey	0.093
d_{22}	Self-fractional diffusion coefficient of predator	3.277
d_{12}	Cross-fractional diffusion coefficient of prey	2.444

$$\begin{cases} \frac{du}{dt} = au\left(1 - \frac{u}{K}\right) - b(1-e)\sqrt{u}\sqrt{v} - h_1 u, \\ \frac{dv}{dt} = c(1-e)\sqrt{u}\sqrt{v} - dv - h_2 v. \end{cases} \tag{16.2}$$

Interestingly, there often indicate Lévy-walk-like search and predation behavior by many predators in the marine environment. Moreover, some common features like the scale invariance and long-range correlation in marine predator movement can be shown in the spatial variation of prey density. And we can also see the speedy diffusion after the predator attacks the prey. Similarly, a fractal pattern like the Lévy's in the prey density distribution can also be shown because of the reaction of the predator with the prey distribution. The behavior like the Lévy-walk-like search and predation like sharks, sea turtles, bony fishes, and penguins has been very close to the theoretical optimal [394]. Significantly, the work also describes the Lévy-like vertical movement reflecting the more complex 3D movement for some marine vertebrate species with different phylogenies, and it also indicates that high sea foragers may generate a common strategy as the Lévy-like walking. This inspire us to introduce a three-dimensional anomalous diffusion term into a reactive diffusion system. Naturally, the predator wants to chase the prey, and the prey tries to escape. And this interaction means the density gradient of one species population influences the diffusive movements of another species. Then a cross-diffusion term can be important in one model to enunciate the driving effect of population fluxes of one species on the presence of the other species. Thus we introduce a inclusion of cross-diffusion to study the formation of spatial patterns in the fractional predator–prey model with herd behaviors. The following fractional reaction–diffusion equation is introduced by introducing the cross-diffusion term into model (16.2):

$$\begin{cases} \frac{\partial u}{\partial t} = au\left(1-\frac{u}{K}\right) - b(1-e)\sqrt{u}\sqrt{v} - h_1 u + d_{11}\nabla^{\alpha}u + d_{12}\nabla^{\alpha}v, \\ (x,y,z)\in\Omega,\ t>0, \\ \frac{\partial v}{\partial t} = c(1-e)\sqrt{u}\sqrt{v} - dv - h_2 v + d_{22}\nabla^{\alpha}v, \\ (x,y,z)\in\Omega,\ t>0, \\ \frac{\partial u}{\partial \eta} = \frac{\partial v}{\partial \eta}, \\ (x,y,z)\in\partial\Omega,\ t>0, \\ u(x,y,z,0) = u_0(x,y,z),\ v(x,y,z,0) = v_0(x,y,z), \\ (x,y,z)\in\Omega, \end{cases} \tag{16.3}$$

where d_{11} and d_{22} mean the self-diffusion coefficients, d_{12} stands for the cross-diffusion coefficient, $\Omega \subset R^3$ represents a smooth bounded domain, and η means the upper outward unit normal vector $\partial\Omega$. $u_0(x,y,z)$ and $v_0(x,y,z)$ stand for nonnegative smooth functions, and both are not less than 0. ∇^{α} $(1<\alpha<2, (x,y,z)\in R^3)$ indicates the *Fractional order Laplacian Operator*. We give the abnormal diffusion operator ∇^{α} as the following Riesz fractional derivative:

$$\begin{aligned} \nabla^{\alpha}u &= \frac{\partial^{\alpha}u}{\partial|x|^{\alpha}} + \frac{\partial^{\alpha}u}{\partial|y|^{\alpha}} + \frac{\partial^{\alpha}u}{\partial|z|^{\alpha}} \\ &= -\frac{1}{2\cos(\pi\alpha/2)}\left({}_{RL}D^{\alpha}_{-\infty,x}u + {}_{RL}D^{\alpha}_{x,+\infty}u\right) - \frac{1}{2\cos(\pi\alpha/2)}\left({}_{RL}D^{\alpha}_{-\infty,y}u + {}_{RL}D^{\alpha}_{y,+\infty}u\right) \\ &\quad - \frac{1}{2\cos(\pi\alpha/2)}\left({}_{RL}D^{\alpha}_{-\infty,z}u + {}_{RL}D^{\alpha}_{z,+\infty}u\right) \end{aligned}$$

${}_{RL}D^{\alpha}_{-\infty,x}u$ and ${}_{RL}D^{\alpha}_{x,+\infty}u$ are indicated as

$$\begin{aligned} {}_{RL}D^{\alpha}_{-\infty,x}u &= \frac{1}{\Gamma(2-\alpha)}\frac{\partial^2}{\partial x^2}\int_{-\infty}^{x}(x-s)^{1-\alpha}u(s,y,t)\,ds, \\ {}_{RL}D^{\alpha}_{x,+\infty}u &= \frac{1}{\Gamma(2-\alpha)}\frac{\partial^2}{\partial x^2}\int_{x}^{+\infty}(s-x)^{1-\alpha}u(s,y,t)\,ds, \end{aligned}$$

where $\Gamma(\cdot)$ means the Gamma function.

16.3 Linear Stability Analysis

In order to calculate the static case of the model (16.3), we have the following conditions:

$$\begin{cases} au\left(1-\frac{u}{K}\right) - b(1-e)\sqrt{u}\sqrt{v} - h_1 u = 0, \\ c(1-e)\sqrt{u}\sqrt{v} - dv - h_2 v = 0. \end{cases} \tag{16.4}$$

Then we obtain three nonnegative equilibrium points from model (16.4), namely, $E_1 = (0,0)$, $E_2 = \left(K\left(1-\frac{h_1}{a}\right), 0\right)$, and

$$E_*\left(u_*, v_*\right) = \left(K\left(1 - \frac{h_1}{a} - \frac{bc(1-e)^2}{a\left(d+h_2\right)}\right), \frac{c^2(1-e)^2 u_*}{\left(d+h_2\right)^2}\right).$$

Besides, E_1 always exists but E_2 only emerges as the condition of $a - h_1 > 0$ is satisfied. Furthermore, $E_*(u_*, v_*)$ only exists as the condition of $bc(1-e)^2 - (a-h_1)(d+h_2) < 0$ is established.

Because we cannot linearize the system (16.4) near the equilibrium points E_1 and E_2, we would not study their stability.

Furthermore, the linear stability of the coexistence equilibrium point $E_*(u_*, v_*)$ is our main focus. When we do not consider the diffusion term, we can get the following linearized form of Eq. (16.4) at the equilibrium point $E_*(u_*, v_*)$:

$$\begin{cases} \frac{du}{dt} = J_{11}u + J_{12}v, \\ \frac{dv}{dt} = J_{21}u + J_{22}v, \end{cases} \tag{16.5}$$

where

$$J_{11} = \frac{3bc(1-e)^2}{2\left(d+h_2\right)} - a + h_1,\ J_{12} = -\frac{b\left(d+h_2\right)}{2c},\ J_{21} = \frac{c^2(1-e)^2}{2\left(d+h_2\right)},\ J_{22} = -\frac{d+h_2}{2}.$$

Then, we obtain the characteristic equation of Eq. (16.4)

$$\lambda_0{}^2 - T\lambda_0 + D = 0, \tag{16.6}$$

where

$$T = J_{11} + J_{22},\ D = J_{11}J_{22} - J_{12}J_{21}.$$

Therefore, the following results are obtained.

(i) The system equilibrium point $E_*(u_*, v_*)$ becomes locally asymptotically stable as $T < 0$ and $D > 0$ based on the Routh–Hurwitz criterion.

(ii) There emerges Hopf bifurcation for the predator–prey system (16.4) with the condition of $e = e_H$, where

$$e_H = 1 - \sqrt{\frac{(h_2+d)^2 - 2\left(h_1-a\right)\left(h_2+d\right)}{3bc}}.$$

Here, the horizontal conditions of the Hopf bifurcation will be only verified. When the parameter e indicates the bifurcation parameter, and the conditions $T = 0$ and $D > 0$ are satisfied, we obtain the Hopf bifurcation immediately. After taking the eigenvalue as $\lambda_0(e) = p(e) + iq(e)$, substituting it into the characteristic Eq. (16.5) and separating the real and imaginary parts, we derive the two sides to obtain

$$\begin{cases} 2p\frac{dp}{de} - 2q\frac{dq}{de} - p\frac{dT}{de} + \frac{dD}{de} - T\frac{dp}{de} = 0, \\ 2q\frac{dp}{de} + 2p\frac{dq}{de} - q\frac{dT}{de} - T\frac{dq}{de} = 0. \end{cases} \quad (16.7)$$

Noticeably, we cannot get the $\frac{d(Re(\lambda_0))}{de}|e_H = \frac{3bc(e-1)}{d+h_2} \neq 0$ at $e = e_H$. Clearly, the system (16.5) emerges Hopf bifurcation at the equilibrium point $E_*(u_*, v_*)$ according to the Poincaré–Hopf bifurcation theorem at $e = e_H$.

In the following, we derive the conditions of Turing instability for the predator–prey system (16.3). For easy writing, we assume that $u = \overline{u} + u_*$, $v = \overline{v} + v_*$, where $(\overline{u}, \overline{v})$ stands for the micro-disturbances near the equilibrium point $E_*(u_*, v_*)$. In the next following part, $\overline{u}$, $\overline{v}$ are also written as u, v. Then we get the linearized form of Eq. (16.3) at the equilibrium point $E_*(u_*, v_*)$ is

$$\begin{cases} \frac{\partial u}{\partial t} = J_{11}u + J_{12}v + d_{11}\nabla^{\alpha}u + d_{12}\nabla^{\alpha}v, \\ \frac{\partial v}{\partial t} = J_{21}u + J_{22}v + d_{22}\nabla^{\alpha}v. \end{cases} \quad (16.8)$$

Suppose the following form expanding the perturbation variable in Fourier space:

$$\begin{pmatrix} u \\ v \end{pmatrix} = \sum \begin{pmatrix} Z^u \\ Z^v \end{pmatrix} e^{\lambda_k t + i\mathbf{k}\cdot\mathbf{r}}, \quad (16.9)$$

where λ_k represents the growth rate of the perturbation, k means the modulus, and r stands for the direction vector. After substituting the perturbation variable (16.9) into Eq. (16.8), we get the characteristic equation as follows:

$$\lambda_k^2 - A_1\lambda_k + A_2 = 0, \quad (16.10)$$

where

$$A_1 = J_{11} + J_{22} - (d_{11} + d_{22})\,k^{\alpha},$$

$$A_2 = d_{11}d_{22}k^{2\alpha} - (J_{11}d_{22} + J_{22}d_{11} - J_{21}d_{12})\,k^{\alpha} + J_{11}J_{22} - J_{12}J_{21}.$$

We calculate the above equation and get its eigenvalue as

$$\lambda_k = \frac{A_1 \pm \sqrt{A_1^2 - 4A_2}}{2}.$$

Comparing with the linear stability analysis between ordinary differential equations and partial differential equations, we obtain the decaying effect to zero for the perturbation dynamically only when the real parts of all eigenvalues λ_k become less than zero. In addition, we can get the Turing instability as $A_2 < 0$ for some k values. Assuming A_2 be the quadratic function of k^{α}, and calculating the minimum value of

k^α, we can gain the most dangerous modulus value k_c^α of the system with perturbation as

$$k_c^\alpha = \frac{J_{11}d_{22} + J_{22}d_{11} - J_{21}d_{12}}{2d_{11}d_{22}}.$$

Therefore, we get the following conditions for the Turing bifurcation of system (16.3)
(i) $J_{11} + J_{22} < 0$,
(ii) $J_{11}d_{22} + J_{22}d_{11} - J_{21}d_{12} > 0$,
(iii) $2\sqrt{d_{11}d_{22}(J_{11}J_{22} - J_{12}J_{21})} < J_{11}d_{22} + J_{22}d_{11} - J_{21}d_{12}$.

After choosing the cross-diffusion coefficient d_{12} as the bifurcation parameter, we obtain the critical value d_T for Turing bifurcation as

$$d_T = \frac{J_{11}d_{22} + J_{22}d_{11} - 2\sqrt{d_{11}d_{22}(J_{11}J_{22} - J_{12}J_{21})}}{J_{21}}.$$

Next, in order to verify the Turing instability existence of system (16.3) at the equilibrium point E_*, we take the parameters as $d_{11} = 0.093$, $d_{22} = 3.277$, $a = 0.92$, $K = 7.053$, $b = 1.081$, $e = 0.35$, $c = 0.748$, $d = 0.392$, $h_1 = 0.246$ and $h_2 = 0.185$ in the following numerical simulation. We assume the function A_2 and the maximum real part for the eigenvalue $Re(\lambda_k)$ as functions of wavenumber k^α shown in Fig. 16.1. Here, the critical value $d_T = 2.444$ can get the Turing bifurcation in the system. Taking $d_{12} = 1.5, 1.8, 2.35, 2.7$ and 3.0 separately on both sides of the critical value, we observe the local stability transition of E_*. Figure 16.1a depicts the relationship between function A_2 and wavenumber k^α. Figure 16.1a also illustrates the negative effect of the cross-diffusion on the wavenumber and A_2: the smaller cross-diffusion coefficient d_{12} means the larger range of the wavenumber k^α and makes a negative A_2. This effect indicates the more likely Turing bifurcation for the system. In Fig. 16.1b, we exhibit the relationship between the real part of the eigenvalue $Re(\lambda_k)$ and the wavenumber k^α. Figure 16.1b shows that the range of k^α at the location of $Re(\lambda_k)$ remaining positive shrinks gradually when parameter d_{12} increases. As d_{12} remains larger than the critical value d_T, the values of $Re(\lambda_k)$ become all negative, which indicates the stability of the equilibrium point E_* and the nonexistence of the Turing bifurcation of the system.

Figure 16.2 indicates the effect of fractional diffusion on Turing instability with variation orders of fractional diffusion, variation curves of the function A_2, change of the real part of the eigenvalue $Re(\lambda_k)$, and the variation for the wavenumber k. Figure 16.2a, b shows the relationship curves that the order of fractional diffusion has no control effect on moving up or down for the relationship curves, but has effect on causing the relationship curves to shrink or expand. It is apparent that the increasing fractional diffusion indicated as the increasing order α gets the shrink relationship between the function A_2 and the wavenumber k illustrated in Fig. 16.2a. The related

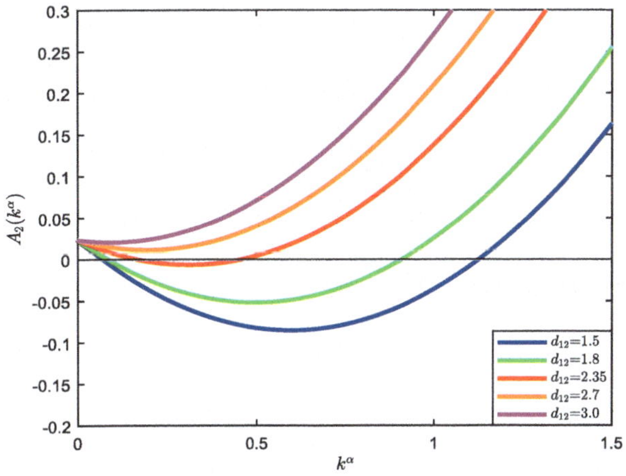

(a)

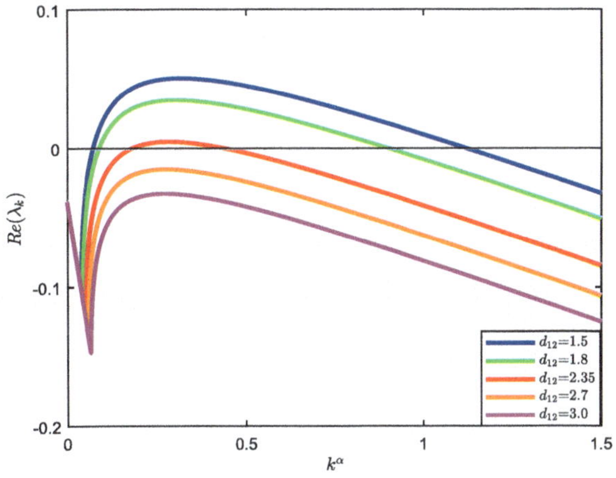

(b)

Fig. 16.1 The graphic of **a** A_2 and **b** $Re(\lambda_k)$ against k^α for $d_{12} = 1.5,\ 1.8,\ 2.35,\ 2.7,\ 3.0$

detail is that when the order α of fractional diffusion becomes larger, the wavenumber range gets more smaller where the function A_2 is less than 0. Figure 16.2b illustrates four different relationships between the real part of the eigenvalue $Re(\lambda_k)$ and the wavenumber k. Based on this relationships, we get the range of the wavenumber k obviously smaller than that of the case at the order of fractional diffusion $\alpha = 1.12$, where the order of fractional diffusion $\alpha = 2$ and the system has Turing instability. Therefore, the wavenumber range of Turing instability has negative relationship with the order α of fractional diffusion, meaning that as the order of fractional diffusion becomes larger, it is less easy to emerge Turing bifurcation for the system.

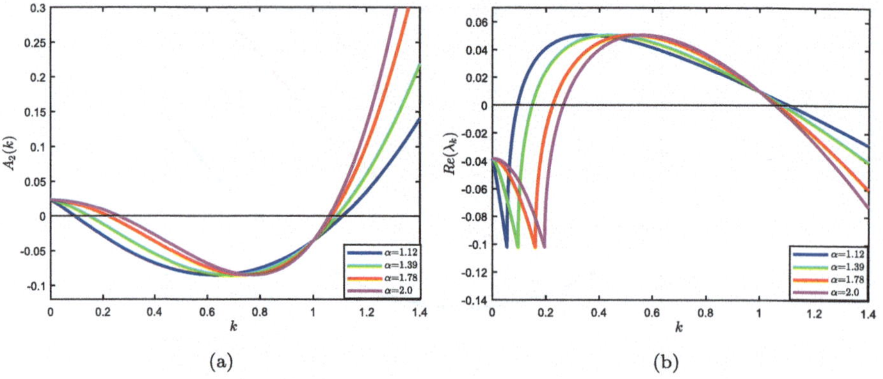

Fig. 16.2 The graphic of **a** A_2 and **b** $Re(\lambda_k)$ against k. The cross-diffusion coefficient $d_{12} = 1.5$, and the parameters $\alpha = 1.12,\ 1.39,\ 1.78,\ 2.0$

16.4 Amplitude Equations of Turing Patterns

In the following, the amplitude equation and its stability are extensively studied in three-dimensional space. We introduce the nonlinear coupling term into the systems corresponding to different wave vectors to understand the pattern selection of the system. For this aim, we use the central manifold reduction method and obtain the critical amplitude evolution equation for the critical value d_T of Turing bifurcation. Next, the general amplitude equation will be derived and discussed for the system as 3D pattern.

First of all, the spatial fractional predator–prey model is written coupling two species as

$$\frac{\partial X}{\partial t} = D\nabla^{\alpha} X + N_i X + N_{ij} X_i X_j + N_{ijk} X_i X_j X_w + \dots, \tag{16.11}$$

where

$$D = \begin{pmatrix} d_{11} & d_{12} \\ d_{21} & d_{22} \end{pmatrix}.$$

After adopting Einstein's summation convention for indicators $i,\ j$ and w in the two species, ignoring the diffusion term, not considering the covariance and contravariance, and arranging The tensors N^i, N^{ij}, and N^{ijw} as symmetrical [374], we get the following equation as writing the number of species s as a superscript, and the number of discrete grid points l as a subscript and substituting it into Eq. (16.11):

$$\frac{dX_l}{dt} = -D_l\left|\overrightarrow{k_l}\right|^{\alpha} X_l + N^i X_l^i + N^{ij} \sum_{l_1+l_2=l} X_{l_1}^i X_{l_2}^j + N^{ijw} \sum_{l_1+l_2+l_3=l} X_{l_1}^i X_{l_2}^j X_{l_3}^w + o\left(X^3\right). \tag{16.12}$$

For the $s-th$ population, we obtain the first term on the right side of Eq. (16.12):

$$A_l^{s,i} = -D_l^s k_l^{\alpha} \delta^{s,i} - D_l^{s+1} k_l^{\alpha} \delta^{s+1,i} + N^{s,i}, \quad s = 1.2.$$

Secondly, xx$X$ by a new representation. Let

$$X_l = P_l W_l, \tag{16.13}$$

where

$$P_l = \begin{pmatrix} p_l^{11} & p_l^{12} \\ p_l^{21} & p_l^{22} \end{pmatrix} = \begin{pmatrix} G_1 P_2 & G_2 P_2 \\ P_2 & P_2 \end{pmatrix}$$

where

$$R = d_{11}k_c^{\alpha},\ T = d_{22}k_c^{\alpha},\ G_1 = \frac{J_{11} - J_{22} - R + T + P_1}{2\left(J_{21} - M\right)},$$
$$G_2 = \frac{J_{11}-J_{22}-R+T-P_1}{2(J_{21}-M)},$$
$$G = d_{12}k_c^{\alpha},\ M = 0,\ P_1 = \sqrt{(J_{11} - J_{22} - R + T)^2 - 4\left(J_{12} - G\right)\left(J_{21} - M\right)},$$
$$P_2 = \sqrt{\frac{J_{21}-M}{P_1}}.$$

Then, we have

$$P_s^{-1} = \begin{pmatrix} q_l^{11} & q_l^{12} \\ q_l^{21} & q_l^{22}, \end{pmatrix} = \begin{pmatrix} P_2 & -G_2 P_2 \\ -P_2 & G_1 P_2 \end{pmatrix}$$

and

$$P_l^{-1} A_l P_l = \begin{pmatrix} \lambda_l^1 & 0 \\ 0 & \lambda_l^2 \end{pmatrix}.$$

These λ_l^i remain different and follow the $\lambda_l^1 > \lambda_l^2$ arrangement.

In the following, we substitute Eq. (16.13) into Eq. (16.12) and simplify it. Thus, we obtain the $s-th$ population model.

$$\begin{aligned} \frac{dW_l^h}{dt} &= \lambda_l^h W_l^h + q_l^{hs} A^{s,ij} \sum_{l_1+l_2=l} p_{l_1}^{ii'} W_{l_1}^{i'} p_{l_2}^{jj'} W_{l_2}^{j'} \\ &+ q_l^{hs} A^{s,ijw} \sum_{l_1+l_2+l_3=l} p_{l_1}^{ii'} W_{l_1}^{i'} p_{l_2}^{jj'} W_{l_2}^{j'} p_{l_3}^{ww'} W_{l_3}^{w'}. \end{aligned} \tag{16.14}$$

We define the coefficient λ_l^h of the linear term as Eq. (16.15).

$$\lambda_c = \left.\frac{d\lambda_k}{dd_{12}}\right|_{d_{12}=d_T} \left(d_{12} - d_T\right). \tag{16.15}$$

Note that we can obtain the coefficients of the amplitude equation as $d_{12} = d_T$. We also utilize this condition to have the amplitude equation for the mode on the central manifold.

Finally, after taking $r = k^\alpha / k_c^\alpha$, further simplifying Eq. (16.14), and expressing the critical wave vector m, we obtain the following equation at the beginning ($\lambda_l^h = 0$):

$$\frac{dW_m^1}{dt} = g_1 \sum_{m_1+m_2=m} W_{l_1}^1 W_{l_2}^1 + \sum_{m_1+m_2+m_3=m} H(m_2+m_3) W_{m_1}^1 W_{m_2}^1 W_{m_3}^1, \quad (16.16)$$

where

$$g_1 = q^{1s} A^{s,ij} p^{i1} p^{j1}, \; H = -2q^{1s} A^{s,ij} p^{i1} \left(J_l^{-1}\right)^{ja} A^{a,bc} p^{b1} p^{c1} + q^{1s} A^{s,ijw} p^{i1} p^{j1} p^{w1}.$$

Noticeably, the α power of the length of the parameter r is the only parameter deciding the coefficient H.

$$H(r) = P_2{}^2 \left(\frac{T_3 + r\left(2P_2{}^2 Q_q (J_{21}G - J_{22}R - J_{11}T + rRT) + T_4\right)}{D_0 + r(J_{21}G - J_{22}R - J_{11}T + rRT)} \right),$$

where

$$\begin{gathered}
q_1 = P_2{}^2 (-b_{11} - 2b_{20}G_1 + G_2(p_{11} + 2G_1 p_{20})), \\
q_2 = (-2b_{02} - b_{11}G_1 + 2G_2 p_{02} + G_1 G_2 p_{11}) P_2{}^2, \\
G_3 = J_{22}q_1 - J_{21}q_2, \; G_4 = -(J_{12}q_1 - J_{11}q_2), \; T_1 = b_{02} + b_{11}G_1 + b_{20}G_1^2, \\
T_2 = p_{02} + p_{11}G_1 + p_{20}G_1^2, \\
T_3 = 2P_2^2 Q_q D_0 + G_3 T_1 + G_4 T_2, \; T_4 = T_1(Mq_2 - q_1 T) + T_2(Gq_1 - q_2 R), \\
D_0 = J_{11}J_{22} - J_{12}J_{21}, \\
T_5 = G_2\left(p_{03} + p_{12}G_1 + p_{21}G_1^2 + p_{30}G_1^3\right), \\
Q_q = b_{03} + \left(b_{12} + b_{21}G_1 + b_{30}G_1^2\right) G_1 - T_5.
\end{gathered}$$

Therefore, the previous derivation is general and does not focus on any specific symmetry structure. Furthermore, there exist various arrangement possibilities in three-dimensional space such as simple cubic lattice (SC), face-centered cubic lattice (FCC), and body-centered cubic lattice (BCC) as well as double gyroid structures, and so on. For example, we can define the BCC lattice structure as the following six vectors:

$$\begin{gathered}
\vec{k_1} = \frac{k_c}{\sqrt{2}}(1,1,0), \; \vec{k_2} = \frac{k_c}{\sqrt{2}}(0,1,1), \vec{k_3} = \frac{k_c}{\sqrt{2}}(1,0,1), \\
\vec{k_4} = \frac{k_c}{\sqrt{2}}(1,-1,0), \; \vec{k_5} = \frac{k_c}{\sqrt{2}}(0,1,-1), \; \vec{k_6} = \frac{k_c}{\sqrt{2}}(1,0,-1),
\end{gathered}$$

where k_c means the critical wave number. Additionally, we can rewrite the solutions of model (16.3) for d_{12} near the critical point d_T as

$$\begin{pmatrix} u \\ v \end{pmatrix} = \begin{pmatrix} u_* \\ v_* \end{pmatrix} + \sum_{j=1}^{n} \begin{pmatrix} g_j^u \\ g_j^v \end{pmatrix} e^{ik_j \cdot r} + c.c..$$

where g_j^u and g_j^v stand for the amplitude related to the wave vector $\mathbf{k}_j$, and $c.c.$ means the conjugate term. Based on the perturbation at equilibrium point $E_*(u_*, v_*)$ and the Taylor series expansion, we can have the following form of model (16.3):

$$\frac{\partial \mathbf{X}}{\partial t} = \mathbf{LX} + \mathbf{N}, \tag{16.17}$$

where

$$\mathbf{X} = \begin{pmatrix} u \\ v \end{pmatrix}, \ \mathbf{L} = \begin{pmatrix} J_{11} + d_{11}\nabla^\alpha & J_{12} + d_{12}\nabla^\alpha \\ J_{21} & J_{22} + d_{22}\nabla^\alpha \end{pmatrix},$$

$$\mathbf{N} = \begin{pmatrix} b_{20}u^2 + b_{02}v^2 + b_{11}uv + b_{30}u^3 + b_{03}v^3 + b_{21}u^2v + b_{12}uv^2 \\ p_{20}u^2 + p_{02}v^2 + p_{11}uv + p_{30}u^3 + p_{03}v^3 + p_{21}u^2v + p_{12}uv^2 \end{pmatrix},$$

and

$$\begin{aligned}
b_{20} &= -\frac{a}{K} + \frac{b(1-e)}{8} u_*^{-\frac{3}{2}} v_*^{\frac{1}{2}}, \ b_{30} = -\frac{b(1-e)}{16} u_*^{-\frac{5}{2}} v_*^{\frac{1}{2}}, \\
b_{11} &= -\frac{b(1-e)}{4} u_*^{-\frac{1}{2}} v_*^{-\frac{1}{2}}, \ b_{21} = \frac{b(1-e)}{16} u_*^{-\frac{3}{2}} v_*^{-\frac{1}{2}}, \\
b_{02} &= \frac{b(1-e)}{8} u_*^{\frac{1}{2}} v_*^{-\frac{3}{2}}, \ b_{03} = \frac{b(e-1)}{16} u_*^{\frac{1}{2}} v_*^{-\frac{5}{2}}, \\
b_{12} &= \frac{b(1-e)}{16} u_*^{-\frac{1}{2}} v_*^{-\frac{3}{2}}, \ p_{30} = \frac{c(1-e)}{16} u_*^{-\frac{5}{2}} v_*^{\frac{1}{2}}, \\
p_{03} &= \frac{c(1-e)}{16} u_*^{\frac{1}{2}} v_*^{-\frac{5}{2}}, \ p_{20} = \frac{c(e-1)}{8} u_*^{-\frac{3}{2}} v_*^{\frac{1}{2}}, \\
p_{02} &= \frac{c(e-1)}{8} u_*^{\frac{1}{2}} v_*^{-\frac{3}{2}}, \ p_{11} = \frac{c(1-e)}{4} u_*^{-\frac{1}{2}} v_*^{-\frac{1}{2}}, \\
p_{12} &= -\frac{c(1-e)}{16} u_*^{-\frac{1}{2}} v_*^{-\frac{3}{2}}, \ p_{21} = -\frac{c(1-e)}{16} u_*^{-\frac{3}{2}} v_*^{-\frac{1}{2}}.
\end{aligned}$$

Using the central manifold reduction, we calculate the coefficients of the amplitude equation and have the amplitude equation of the BCC structure:

$$\begin{aligned}
\frac{dW_1}{dt} &= \lambda_c W_1 + g_1 (W_2 W_6 + W_3 W_5) + g_2 |W_1|^2 W_1 \\
&+ g_3 \left(|W_2|^2 + |W_3|^2 + |W_5|^2 + |W_6|^2\right) W_1 \\
&+ g_4 |W_4|^2 W_1 + g_5 \left(W_2 W_4 W_5 + W_3 \bar{W}_4 W_6\right),
\end{aligned} \tag{16.18}$$

where

$$g_1 = \left(b_{02} + b_{11}G_2 + b_{20}G_2^2 - G_1(p_{02} - p_{11}G_2 - p_{20}G_2^2)\right) P_2^3,$$
$$g_2 = 2Q_2\,(0) + Q_2\,(2^{\alpha})\,,$$
$$g_3 = 2Q_2\,(0) + 2Q_2\,(1) + 2Q_2\left(3^{\frac{\alpha}{2}}\right),\ g_4 = 2Q_2\,(0) + 4Q_2\left(2^{\frac{\alpha}{2}}\right),$$
$$g_5 = 2Q_2\left(2^{\frac{\alpha}{2}}\right) + 4Q_2\,(1)\,.$$

16.5 Stability Analysis of Pattern Formation

It is well known that we can use the related theory of pattern formation to study the universal feature of system dynamics near branch points. Specially, the amplitude equation as such an equation can be used to study the dynamic behavior for any system near the instability point. When we obtain the coupled amplitude equations for the predator–prey system, we can use the linear analysis to describe their stability near the steady state for different symmetries.

At first, the steady-state W_c is determined for the amplitude system, and the right term H_i is also derived for the amplitude equation i at the stationary state W_c. And then we construct the corresponding Jacobian matrix A, the elements of matrix A are described as

$$\{A\}_{ij} = \left.\frac{\partial H_i}{\partial |W_j|}\right|_{W_c} = \begin{pmatrix} \left.\frac{\partial H_1}{\partial |W_1|}\right|_{W_c} & \cdots & \left.\frac{\partial H_1}{\partial |W_6|}\right|_{W_c} \\ \cdots & \cdots & \cdots \\ \left.\frac{\partial H_6}{\partial |W_1|}\right|_{W_c} & \cdots & \left.\frac{\partial H_6}{\partial |W_6|}\right|_{W_c} \end{pmatrix}, \tag{16.19}$$

where $i,\ j = 1, 2, ...6$. According to this basis, we can draw the bifurcation diagram for the linear system $\frac{dW}{dt} = AW$ with parameter d_{12} as the control parameter for the pattern selection. Next, the bifurcation diagram will be given for the three-dimensional BCC lattice structure and the stability results will be also obtained for different symmetry steady state.

Here, only two important steady-state conditions will be considered: the planar lamella and the isotropic BCC droplet. We then define the steady state corresponding to a planar lamella ($|W_2^c| = |W_3^c| = |W_4^c| = |W_5^c| = |W_6^c| = 0$) is as

$$|W_1^c| = \sqrt{-\frac{\lambda_c}{g_2}}. \tag{16.20}$$

As for the steady state of the constrained BCC droplet, it yields $|W_1^c| = |W_2^c| = |W_3^c| = |W_4^c| = |W_5^c| = |W_6^c|$.

$$W_i^c = \frac{-g_1 \pm \sqrt{g_1^2 - \lambda_c(g_2 + 4g_3 + g_4 + 2g_5)}}{g_2 + 4g_3 + g_4 + 2g_5}.$$

In the following, the stability of the planar lamella $\left(W_c = \left(W_1^c, 0, 0, 0, 0, 0\right)^T\right)$ will be analyzed and the spherical droplets $\left(W_c = \left(W_1^c, W_2^c, W_3^c, W_4^c, W_5^c, W_6^c\right)^T\right)$ will be organized in the BCC lattice.

Moreover, because that we assume the real part of the eigenvalue as a function of the parameter d_{12} for the linearized amplitude system of the three-dimensional body-centered cubic lattice. Obviously, the stability of the planar lamella is determined by the real part of the eigenvalue μ^L, and the stability of the spherical droplets is determined by $Re(\mu^{BCC})$ and organized in the BCC lattice. Then when $Re\ (\mu) < 0$, we get the stability of the pattern state. As for planar lamellae, we give the eigenvalues of the linearization matrix by $\mu_1^L = -2\lambda_c$, $\mu_2^L = \lambda_c\left(1 - \frac{g_4}{g_2}\right)$, and $\mu_3^L = \lambda_c\left(1 - \frac{g_3}{g_2}\right)$. Note that $\mu_1^L < 0$ means that the stability of the planar lamella structure is determined by the larger eigenvalue of μ_2^L and μ_3^L.

Figure 16.3 exhibits the relationship between the real part of these eigenvalues and d_{12} as the control parameter. It can be illustrated from Fig. 16.3 that when the real parts of the three eigenvalues remain less than zero, the state becomes stable for a lamella pattern. As for the BCC droplet structure, the state remains stable when the parameter range is taken as $2.37 < d_{12} < 2.47$. That is named as bistable at this time. Moreover, as the parameter $d_{12} < 2.37$, only the planar lamella state becomes stable.

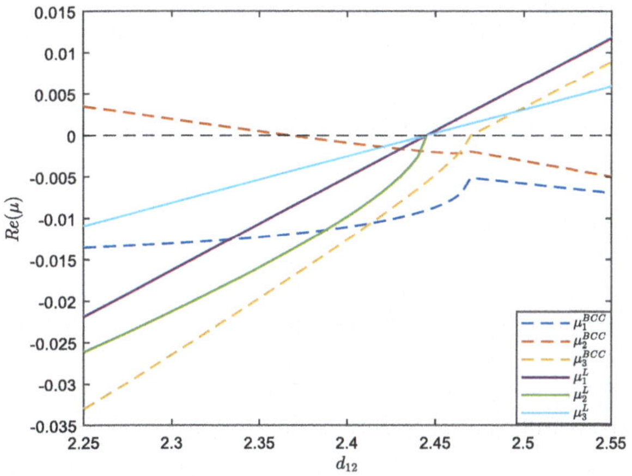

Fig. 16.3 Under different values for parameter d_{12}, the real part $Re(\mu)$ of the eigenvalue for the linearized amplitude system changes

16.6 Numerical Simulations

We next give the numerical simulation for the three-dimensional space-time pattern of the model (16.3) to verify the correctness of the above analysis results. The homogeneous Neumann boundary conditions are assumed and used in all numerical simulations. That is, we take the inflow and outflow of the population on the boundary as 0. And we also carry out all numerical simulations in a discrete grid of $40 \times 40 \times 40$, on the space step as $\Delta h_x = \Delta h_y = \Delta h_z = 1$, and in the time interval step as $\Delta t = 0.1$.

Because of living in the three-dimensional space for both the predator population and the prey population in the actual marine environment, the stability of the Turing pattern is studied in three-dimensional space. Actually, we can consider the pattern selection as the phase space of the corresponding system. So the model parameters can control the dynamics and evolution of a particular form. Then we can get different nonlinear symmetry requirements by many possible three-dimensional structures. When the other parameters remain fixed, the initial population structure becomes crucial to the 3D stability pattern. The closer the initial population structure gets to the symmetry requirement state, the smaller the phase space range for system exploration. If we use the initial conditions of spatial heterogeneity in favor of a particular structure, we may get the system dynamics limited to a small part of the phase space, and we can obtain the expected optimal species density pattern. Similarly, we deduce the initial conditions with the BCC symmetry structure to be a spherical droplet pattern with the same symmetry structure. Numerical simulation illustrates that the final pattern structure does not vary with time, and the pattern types of predator and prey are as the same. Thus, we only give the pattern of prey population as an example.

16.6.1 *Influence of Cross-Fractional Diffusion Coefficient d_{12} on Pattern Formation*

In the following, we observe some interesting phenomena in the spatiotemporal pattern as we take the predator cross-diffusion rate d_{12} changing. Except for the bifurcation parameters, we take the values of system's other parameters as $a = 0.92$, $b = 1.081$, $c = 0.748$, $d = 0.392$, $d_{11} = 0.093$, $d_{22} = 3.277$, $e = 0.35$, $h_1 = 0.246$, $h_2 = 0.185$, $K = 0.705$, and $\alpha = 1.8$. Figures 16.4, 16.5, and 16.6 show the three-dimensional pattern structure for the system (16.3) after 60000 time steps. Of course the initial conditions of these patterns remain different. Among them, the content of Fig. 16.4 displays the iso-density surface shown in Fig. 16.4a, c, e, g and 3D pattern (see Fig. 16.4b, d, f, h) for the prey population as the initial condition of the system (16.3) is adopted as the random distribution near the equilibrium point E_*. The iso-density surface is exhibited in Fig. 16.5 shown left side of Fig. 16.5 and 3D pattern shown right side of Fig. 16.5 for the prey population

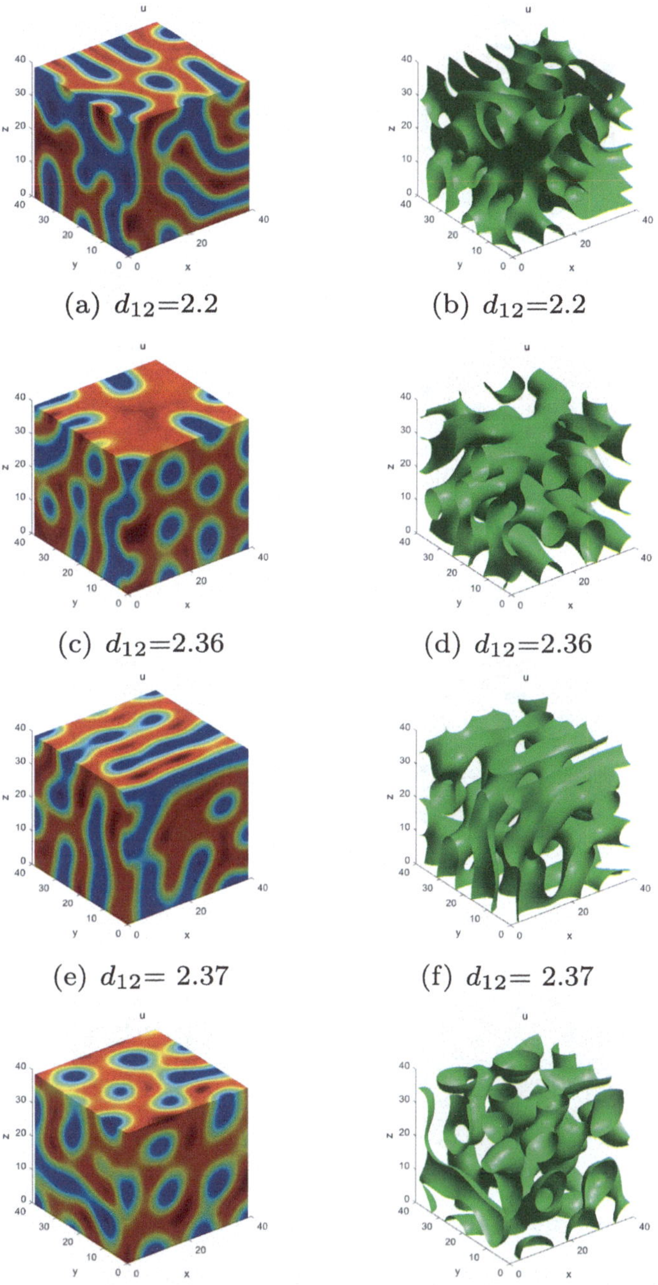

(a) d_{12}=2.2 (b) d_{12}=2.2

(c) d_{12}=2.36 (d) d_{12}=2.36

(e) d_{12}= 2.37 (f) d_{12}= 2.37

(g) d_{12}=2.405 (h) d_{12}=2.405

Fig. 16.4 The Turing variation of the prey population as the initial population is random variables, and other parameters of model (16.3) remain unchanged. **a**, **c**, **e**, and **g** indicate the isosurfaces of the concentration of the prey population; **b**, **d**, **f**, and **h** display the three-dimensional patterns of the prey population

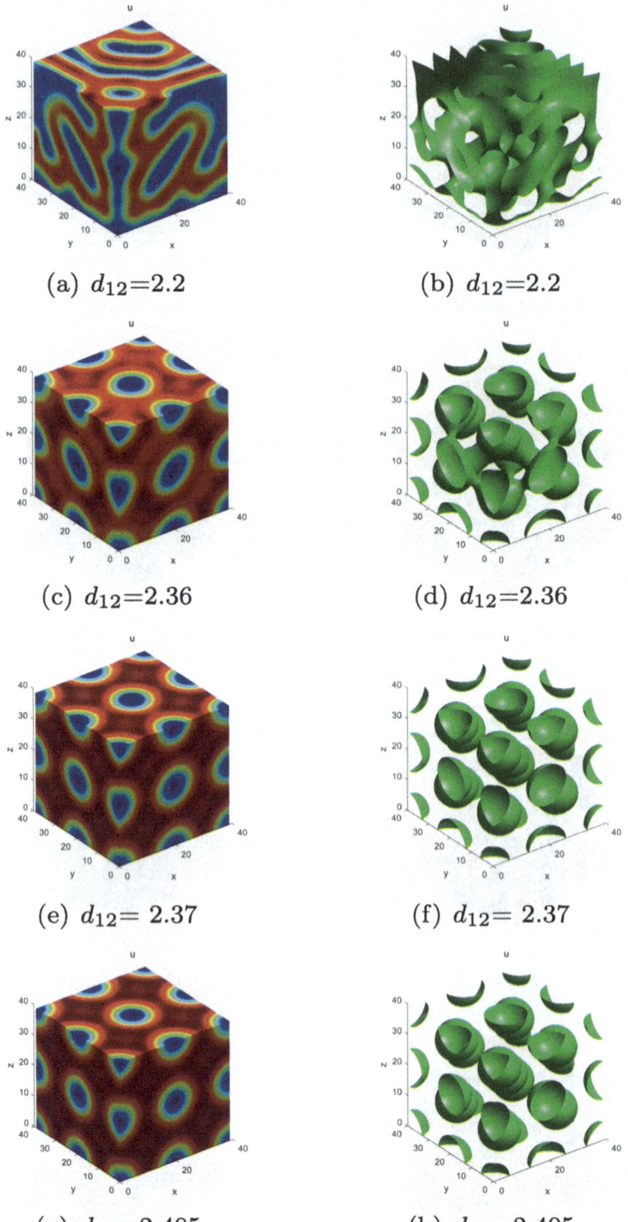

(a) d_{12}=2.2 (b) d_{12}=2.2

(c) d_{12}=2.36 (d) d_{12}=2.36

(e) d_{12}= 2.37 (f) d_{12}= 2.37

(g) d_{12}=2.405 (h) d_{12}=2.405

Fig. 16.5 The Turing variation of the prey population as the initial population is BCC structure, and other parameters of model (16.3) remain unchanged. **a**, **c**, **e**, and **g** stand for the isosurfaces of the concentration of the prey population; **b**, **d**, **f**, and **h** mean the three-dimensional patterns of the prey population had after 60000 time steps, and the initial conditions of the system (16.3) are utilized with the BCC symmetric structure. At first, we apply a single cosine function $cos(\frac{x}{\sqrt{2}})$ disturbance near the equilibrium point E_* to the system (16.3). Then Fig. 16.6 shows the iso-density surface (left side) and 3D pattern shown (right side) of the prey population.

By observing Fig. 16.4, there displays especially the relationship on a certain surface (such as the O-XZ plane) for the iso-density surface of the prey population, apparently. At the same time, the other parameters of the system (16.3) are fixed, the initial condition of the prey population is taken as a random distribution near the equilibrium point E_*, and the control parameter d_{12} increases from 2.2 to 2.405. Then Fig. 16.4a, c, e, and h shows the pattern variation on this plane gradually from a stripe pattern to a spot pattern. It is also shown that the pattern gradually changes from a twisted tube form to a regular tube form in the three-dimensional space, and finally becomes mixture type as spherical droplets and tubes shown in Fig. 16.4b, d, f, and h. Figure 16.4 shows that the system cannot spontaneously emerge a purely lamella or an organized spherical structure with random initial conditions in a three-dimensional space.

Assume the other parameters of the system (16.3) be fixed, the initial population density of the prey population has the unchanged BCC symmetric structure, and the control parameter d_{12} increases from 2.2 to 2.405. Then Fig. 16.5a, c, e, and g illustrates the unchanged transfer patterns from the stripe case to the spot case. In detail, the three-dimensional pattern gradually emerges from a twisted tube case to a deformed spherical droplet-like case, and finally emerges a spherical droplet case with a regular and stable BCC structure shown in Fig. 16.5b, d, f, and h. Moreover, when $d_{12} = 2.36$, the pattern remains unstable, while when $d_{12} = 2.37$, the pattern becomes stable BCC spherical droplet structure. Hence, when $d_{12} < 2.37$, we get the unstable spherical droplet BCC structure for the system (16.3).

For the planar lamella pattern, when other parameters of the system (16.3) remain unchanged, and the initial condition of the prey population is taken as a single cosine disturbance near the equilibrium point ($u_0 = u_* + 0.01cos(\frac{x}{\sqrt{2}})$, $v_0 = v_* + 0.01cos(\frac{x}{\sqrt{2}})$), keeping it immobile, Fig. 16.6 illustrates the stable patch pattern with the control parameter d_{12} increasing from 2.2 to 2.405.

By viewing the contents of Figs. 16.4, 16.5, and 16.6, distinctly, we may not see the alteration of the pattern for the system (16.3) in two dimensions, like spot or stripe, but can see massive changes in three-dimensional space, such as the transformation patterns from twisted tube to deformed droplet case. For example, Fig. 16.4c indicates a spot pattern for the iso-density surface of the prey population , while a tube pattern for the interior in three-dimensional space is shown in Fig. 16.4d. And Fig. 16.5c also illustrates a spot pattern, while a spherical droplet pattern for its interior case is shown in Fig. 16.5d. Thus, there exists more complicated and more sensitive patterns in three-dimensional space than that in two-dimensional plane.

Figure 16.7 illustrates the variations according to the increasing control parameter d_{12}. For instance, when d_{12} increases to 2.37, the system gradually emerges stable spherical droplets, while the planar lamella remains always stable, and the system

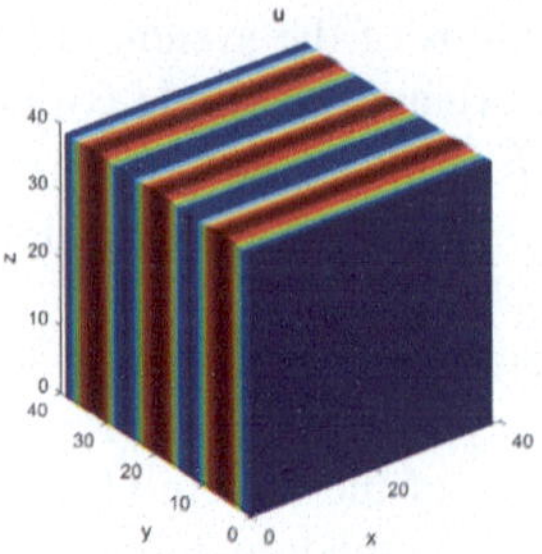

(a) d_{12}=2.2

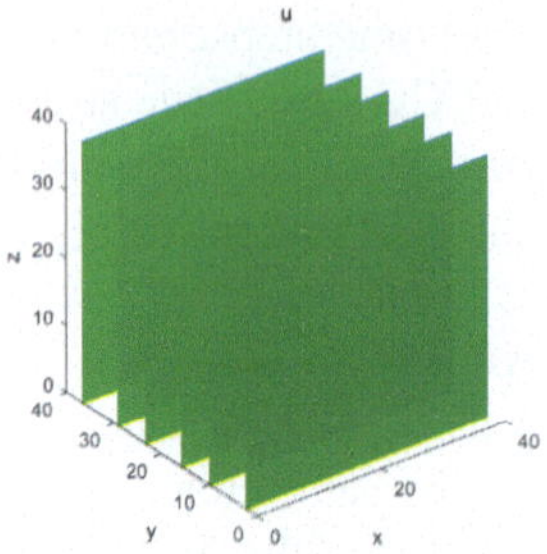

(b) d_{12}=2.2

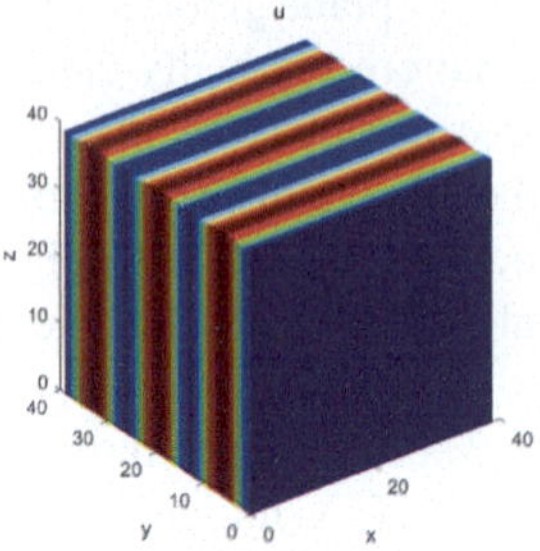

(c) d_{12}=2.36

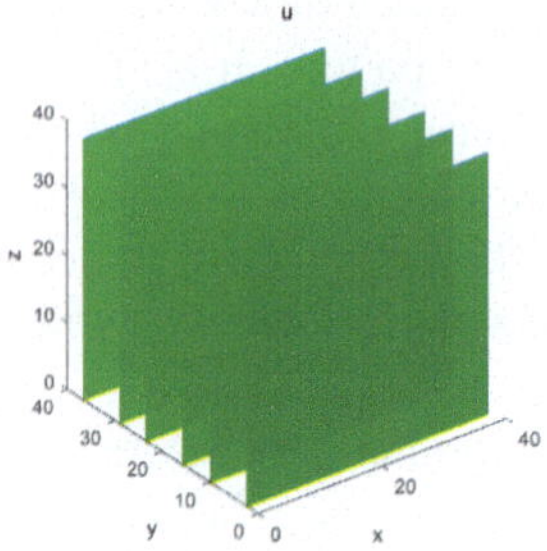

(d) d_{12}=2.36

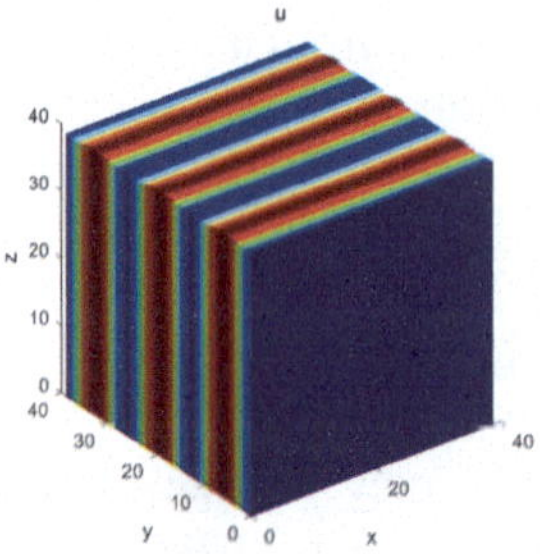

(e) d_{12}= 2.37

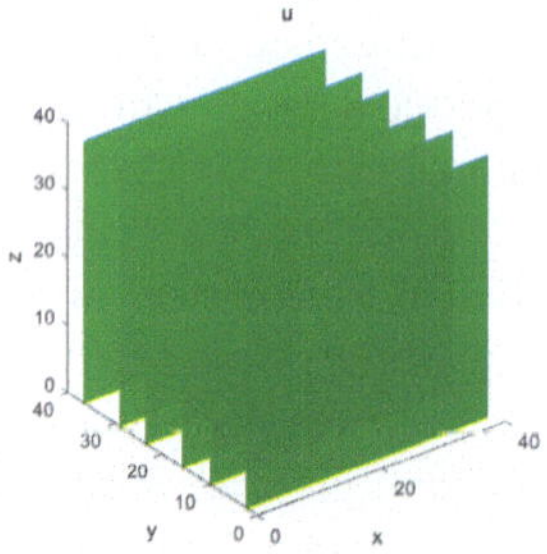

(f) d_{12}= 2.37

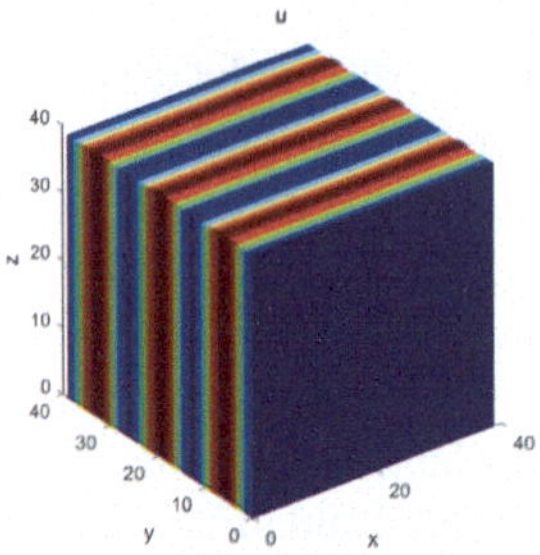

(g) d_{12}=2.405

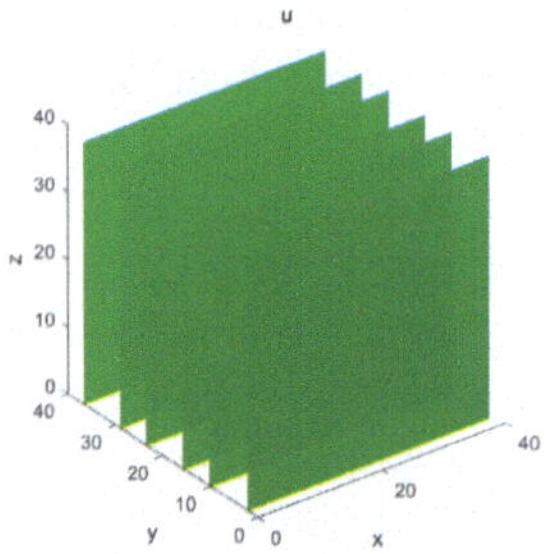

(h) d_{12}=2.405

Fig. 16.6 The Turing variation of the prey population, and other parameters of model (16.3) remain unchanged. **a**, **c**, **e**, and **g** stand for the isosurfaces of the concentration of the prey population; **b**, **d**, **f**, and **h** mean the three-dimensional patterns of the prey population

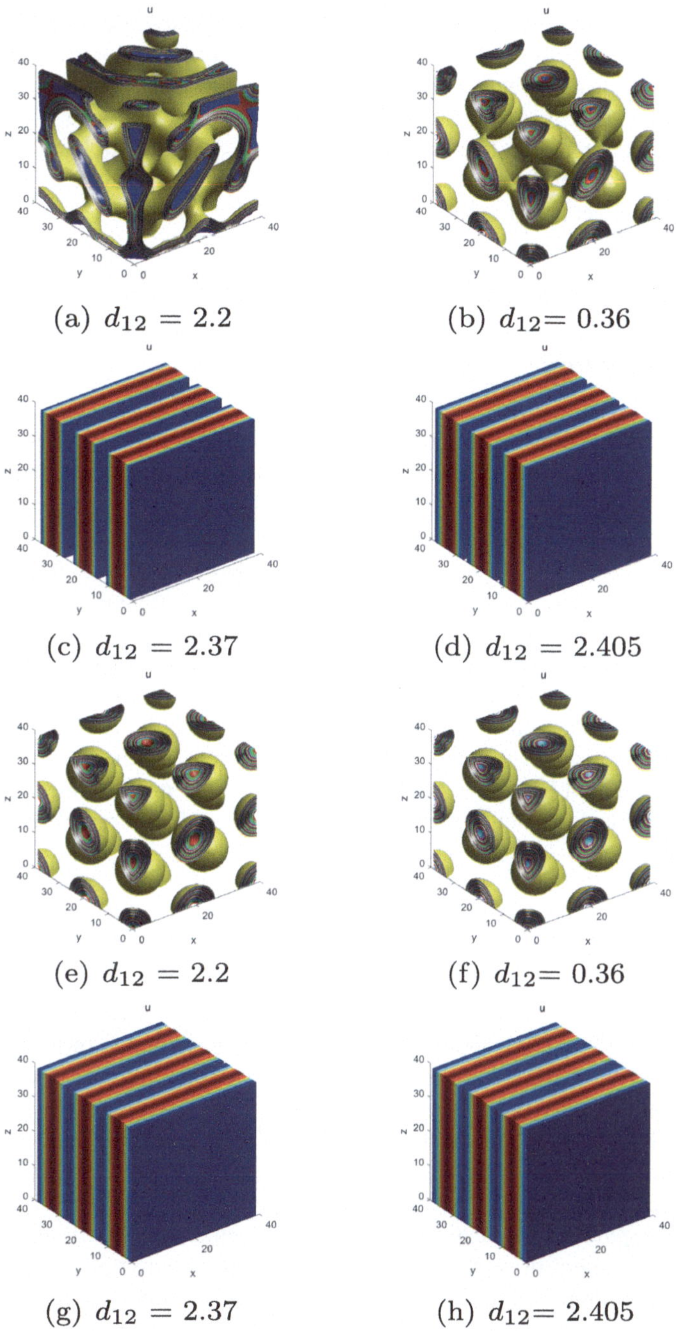

Fig. 16.7 The first row **a–d** means the variation of the BCC spherical droplet; the second row **e–h** stands for the variation of the planar lamella

state varies from monostable to bistable. In addition, when the control parameter $d_{12} < 2.37$, we get the unstable spherical droplet BCC structure for the system (16.3). When $d_{12} > 2.37$, the planar lamella pattern becomes stable all the time. At the same time, the system remains a bistable state. However, the size of the interval between the two lamellae becomes different in the case of the same population density ($u = 0.4$). The larger the value of d_{12}, the smaller the distance between the two planar lamellae, and the denser they become.

16.6.2 Influence of Fractional Diffusion Order α on Pattern Formation

Figure 16.8 demonstrates six different three-dimensional patterns for the prey population at the same time for different diffusion orders α, where the system parameters are assumed as $d_{12} = 2.405$, $d_{11} = 0.093$, $d_{22} = 3.277$, $a = 0.92$, $K = 0.705$, $b = 1.081$, $e = 0.35$, $c = 0.748$, $d = 0.392$, $h_1 = 0.246$ and $h_2 = 0.185$, and the running time is $t = 10000$. The diffusion orders α are taken as 1.2, 1.4, 1.6, 1.823, 1.824, and 2, respectively. The initial condition is selected as the symmetrical BCC structure population distribution. More stable spherical droplets, and the spherical droplets of the BCC structure are shown in Fig. 16.8c. Subsequently, when the diffusion order α increases to 1.823, we can see the stability of the three-dimensional spherical droplet patterns shown in Fig. 16.8c, d. Then, when the diffusion order α increases to 1.824, we can see the variation like the BCC structure pattern, and twisted droplets and tubes patterns shown in Fig. 16.8e. Finally, there emerge more connected tube patterns than spherical droplets patterns as the diffusion order increases to 2 shown in Fig. 16.8f.

16.6.3 Effect of Varied e on the Pattern Formations

Next, the role of the refuge factor e in the formation is studied in three-dimensional space. The value of the parameter e for the prey refuge and the other values of the system are fixed as given below: $d_{12} = 2.405$, $d_{11} = 0.093$, $d_{22} = 3.277$, $a = 0.92$, $K = 0.705$, $b = 1.081$, $\alpha = 1.801$, $c = 0.748$, $d = 0.392$, $h_1 = 0.246$, $h_2 = 0.185$. Then Fig. 16.9 displays spatial patterns of different prey populations. In addition, Fig. 16.9a–e illustrates the spherical patterns for the spatial distribution of the prey population, indicating that the prey population density at the location becomes less than its surroundings. As for the predatory activities of the predators, we obtain a minimum prey population density in some locations, and thus find a lower prey population density than the surroundings. When the refuge $e = 0.35$, the prey population form a stable BCC spherical pattern, and also isolate the low-density area of the prey population. However, taking $e = 0.36$ as a large-scale refuge which

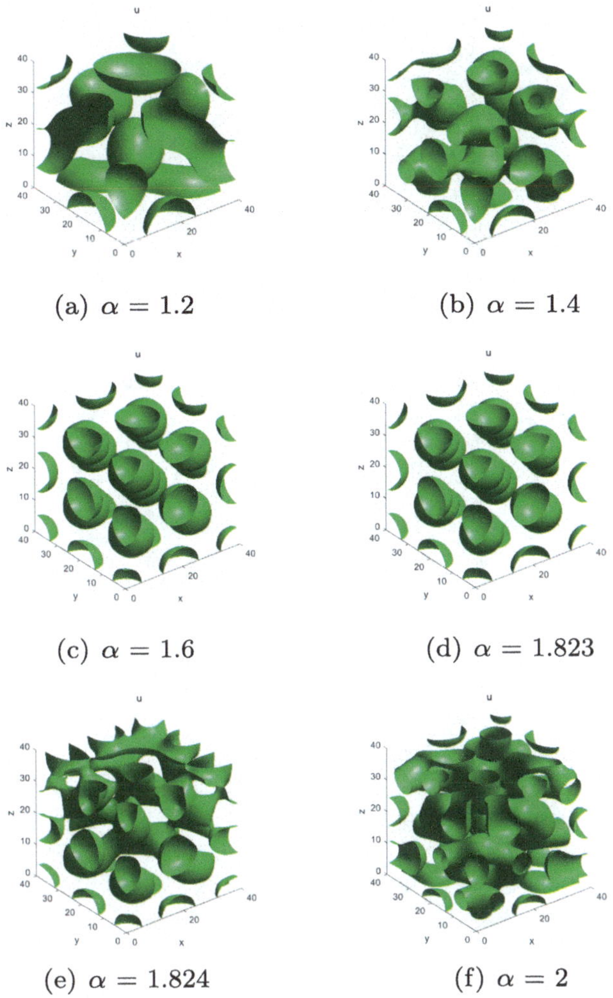

Fig. 16.8 The influence effect of the diffusion order α on the formation of the three-dimensional pattern for BCC spherical droplets at the time $t = 10000$. Other parameters in model (16.3) are fixed at $d_{12} = 2.405$, $d_{11} = 0.093$, $d_{22} = 3.277$, $a = 0.92$, $K = 0.705$, $b = 1.081$, $e = 0.35$, $c = 0.748$, $d = 0.392$, $h_1 = 0.246$ and $h_2 = 0.185$

means a relatively strong protective effect, there exists an area with the density of the prey population exceeding the surroundings and emerging a patch-like pattern similar to "waves" shown in Fig. 16.9f. Thus, increasing the value of the refuge parameter e, we can observe the gradual variation of the distribution pattern for the prey population such as forming a spherical and tubular mixed pattern or a stable spherical BCC structure pattern. Therefore increasing the value of e deduces the unstable structure for the spherical BCC pattern. And finally, there appears a wavy

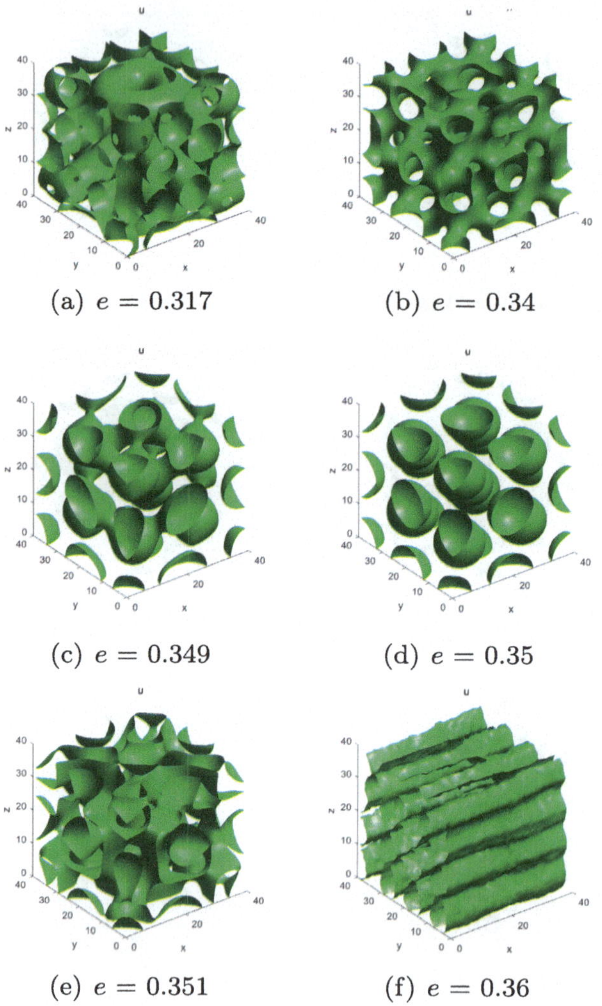

Fig. 16.9 The effect of refuge e on the formation of three-dimensional pattern for BCC spherical droplets at the time $t = 6000$. The system running time is $t = 6000$

flake pattern. The above phenomena display that prey populations can survive longer by living together. In fact, the distribution of species becomes more complicated in the actual natural environment.

16.7 Discussion

In this chapter, the effect of the herd behavior, fractional cross-diffusion, and the refuge protection on pattern dynamics is investigated by a predator–prey model. The herd behavior can make the population model be more social, and the refuge can reflect the complexity of the living environment for the population, such as a very significant role in the spatial distribution of the population. Utilizing the analysis of linear stability theory, we derive the conditions of Hopf bifurcation and Turing instability. Then, we infer the amplitude equation for the three-dimensional pattern of the system by using the center manifold reduction. And we also discuss the stability of the spherical droplet and the lamella in the three-dimensional space. In addition, we apply the numerical simulation to show the wealth of dynamic behaviors for the system, and obtain different spatial patterns, and also verify the correctness of the results. The numerical results illustrate that there exists significant effect of the cross-fractional diffusion and the fractional diffusion on the formation and stability for Turing patterns like spherical droplets, lamellae, and tubes. Thus, fractional diffusion can be regarded as an important factor in deeply understanding the complexity of biological and ecosystem models.

All in all, according to the effect of cross-fractional diffusion on the pattern formation, a spatial predator–prey model is studied by introducing a cross-fractional diffusion theoretically and numerically. Then we draw the following conclusions: (1) we can take the cross-fractional diffusion coefficient as the control parameter, where there occurs Turing bifurcation with the variation of the value of the cross-fractional diffusion coefficient d_{12}, and the population can present rich pattern types like labyrinth tubes, BCC structured spherical droplets, and lamellae. (2) Compared with pattern formation for only one population with herd behaviors, there exist few three-dimensional pattern results for both groups exhibiting herd behaviors. Our model exhibits more complex dynamic behavior for the state of population extinction. In the three-dimensional space, the system also emerges tube, spherical droplet, and lamella patterns. (3) The initial density structure type for the two populations has direct effect on the pattern type of the system performance. For instance, when the initial condition remains a symmetrical BCC structure, the system becomes a spherical droplets BCC structure. On the other hand, the degree of the complicated of the pattern type has positive relationship between the degree of the randomness of the initial population distributed near the equilibrium point. This also enables the pattern of the system to display more complex dynamic behavior in three-dimensional space. (4) Fractional diffusion plays a crucial role in further understanding the complexity of biological and ecosystem models. And the order of fractional diffusion has effect on the stability of the system pattern and on the complexity of the system in three-dimensional space.

In summary, we have investigated the formation of Turing patterns with different control parameters and initial conditions in three-dimensional space. The parameter d_{12} of the predator–prey model can control the pattern formation, and we also explore the generated pattern space. In the simulation, there are very interesting and rich patterns. For instance, we observe a regular BCC spherical droplet structure, a perfect lamella structure, a labyrinth tube structure, and a twisted tube structure in the distribution of prey population density as the control parameter d_{12} changes.

Chapter 17
Spatial Fractional Behavior with Fear Factor and Refuge

17.1 Introduction

There exist many interaction studies between predator and prey [401–408]. For example, a Leslie-Gower predator–prey model has been presented [409], and the Allee effect was combined into growth function of the predator and corresponding predation function [410–412]. Because of the fear of the prey for their predators, they can change the defense ability, significantly reduce their reproduction and finally have different effects on the ecological food chain [404, 413, 415–417]. There are two main kinds of effects in the process of predation. Firstly, the killing behavior of the predators for their prey affects the number of prey; secondly, the intimidation of the predators for their prey, causes their various anti-predator defense capabilities to avoid predation such as altering the living habitat, vigilance, foraging behavior, and physiological changes [418–420]. Among them, refuge as a protection role for many preys plays essential role in protecting the prey, reducing the possibility of extinction, in the coexistence of prey and predator [384, 386, 391]. Hence, there are many studies about the refuge effects for the interaction between predator and prey [421, 422], which display complex feature on population dynamics. First of all, it has an opposite effect for the growth of prey and predators: for the former one the effect is positive and for the latter one the effect is negative. The reason is that less success rate of predation generates a food crisis to predators [392]. Secondly, there is no stable beneficial or harmful effect of the hiding behavior of prey on the interacting population [423–427]. Based on these studies, it is natural and has greater research value to introduce refuges into predator–prey system. Generally, the refuge is usually used as two types of refuges in the predator–prey model: a constant proportion or a constant number [421–424]. In this chapter, we assume the refuge as a proportional form to the overall prey population.

Moreover, we have known that the movement of animals can emerge a new random walk model in a larger spatial range or a longer time range, named as Lévy

S. T. Liu and L. Zhang, *Pattern Dynamics of Marine Plankton Behavior*,
https://doi.org/10.1007/978-981-97-5369-7_17

flight, which represents that animals will grow abnormally in the process of movement [446]. Recently, fractional-order reaction–diffusion systems have been new focus about the dynamical behavior for the predation process. For example, time delay and cross-fraction diffusion terms were introduced into the predation within guilds [381]. The Turing diffusion instability was studied in a cross-fractional diffusion system [399]. In addition, the existence of Turing instability was presented in the fractional reaction–diffusion system about two substances [428]. Hence, fractional diffusion can be a vital term in pattern formation process. According to the above introduction, we analyze a predator–prey system by introducing a fractional diffusion term in the pattern formation process.

Importantly, a "pattern phenomenon" like the patch structure with different densities for uneven distribution of population in the spatial region is essential. Furthermore, the 1D pattern can uniformly extend to 2D as a stripe and uniformly extend to 3D as a lamella. And the 2D hexagonal pattern can extend uniformly to be a tube in the third direction. Obviously, the simplest 3D periodic patterns are cubes or spheres. So far, most studies about the 3D Turing pattern are direct numerical simulation [429, 430]. Furthermore, the Turing pattern was observed experimentally and analyzed simulatively in the three-dimensional system [431–433]. The interconnected periodic structure is also possible in three dimensions, so that there will be more Turing three-dimensional patterns. Although early pattern theory assumed the homogeneous of the systems and researchers obtained the layered, cylindrical and spherical patterns numerically, there is still lack definite conclusions about the formation of interconnect structures [436, 437]. For instance, a homogeneous three-dimensional Turing structure and interconnected structures were obtained like single gyroid structure, single diamond structure, double gyroid structure, and perforated lamellar structures [438]. The periodic structure of interconnection were proved to usually exist in three-dimensional Turing mode [439]. Moreover, Ref. [434] introduced briefly the branches of the 3D pattern, and Ref. [435] explained some results of the 3D pattern. These studies display a connection of the three-dimensional Turing pattern between the reaction–diffusion system and the geometry and topology of the interface. So the three-dimensional predator–prey models can be our important motivation to get deeply the understanding of three-dimensional patterns in the marine ecological environment.

Although predator–prey dynamical systems have been focused, there exist still many challenging open problems. Based on recent studies [412, 414, 417, 440, 441, 447], a spatial fractional predator–prey model is constructed by introducing the terms of fear and refuge. The rest of the chapter is organized as follows. At first, we build a model theoretically. Then we analyze the stability of the model and obtain the necessary conditions for Hopf bifurcation and Turing instability. Thirdly, the scaling method is used to derive the two-dimensional amplitude equation and the stability of the Turing pattern is also analyzed. Fourthly, we derive the three-dimensional amplitude equation by utilizing the central manifold theory, and we discuss the stability of the three-dimensional Turing pattern. Fifthly, we use MATLAB to simulate the rationality of the analysis results. Finally, we give the conclusion.

17.2 Fractional-Order Self-Diffusion Mathematical Model with Fear and Refuge

In this chapter, we focus on the habitat living prey and predator together. Let $X(t)$ and $Y(t)$ stand for the number of prey and predator respectively. Due to fear of predators, prey has anti-predator behavior, which affects the reproduction rate of the prey population. So fear factor $F(f_0, Y)$ was first introduced as below:

$$F(f_0, Y) = \frac{1}{1 + f_0 Y},$$

where parameter f_0 indicates the degree of fear [416]. For the fear factors and the risk of prey, refuge is evolved to generate a strategic method reducing predator predation rate. Here, we assume that part of the prey (eX) remains completely protected from predation. The rest of the prey $(1-e)X$ is allocated to the predator, where $e \in [0, 1)$ is fixed measuring the availability of the refuge. According to the above description and inspired by references [441, 447], we introduce the fear effect, prey refuge, Leslie-Gower functional response, and Allee effect to propose the predator–prey system as below

$$\begin{aligned} \frac{dX}{dT} &= \frac{aX}{1+f_0 Y}\left(1 - \frac{X}{K} - \frac{m_1}{X+b_1}\right) - \frac{c_1(1-e)XY}{(1-e)X+K_1}, \\ \frac{dY}{dT} &= Y\left(d_1 - \frac{s_1 Y}{(1-e)X+K_2}\right), \end{aligned} \tag{17.1}$$

where, a, K, c_1 and K_1 mean the prey's growth rate, the environmental capacity, the maximum predation rate and the protection degree provided by the environment. m_1 and b_1 stand for Allee effect constants. d_1, s_1, and K_2 represent the growth rate, the predators' mortality rate, and the protection degree provided by environment. f_0 means the fear degree of the prey to the predator due to the prey's anti-predator response. Let

$$u = \frac{X}{K},\ v = \frac{Y}{K},\ t = aT,\ m = \frac{m_1}{K},\ b = \frac{b_1}{K},\ c = \frac{c_1}{a}$$

$$k_1 = \frac{K_1}{K},\ K_2 = k_1 K,\ d = \frac{d_1}{a},\ s_1 = ad,\ f = f_0 K.$$

Then after taking dimensionless treatment on Eq. (17.1), we get its transformation as following:

$$\begin{aligned} \frac{du}{dt} &= \frac{u}{1+fv}(1 - u - \frac{m}{u+b}) - \frac{c(1-e)uv}{(1-e)u+k_1}, \\ \frac{dv}{dt} &= dv(1 - \frac{v}{(1-e)u+k_1}). \end{aligned} \tag{17.2}$$

In addition, many animals display optimized behavior as the crucial foraging effect on the survival of many animal populations. For example, many marine organisms obey the Lévy flight mode as their best foraging strategy for the mottled distribution of prey in the marine environment [453–455]. For the results about predation behavior of 14 marine predatory fish species, like sharks, tuna, billfish, and marine sunfish [379], the Lévy search model can be used to strongly support these results. Based on records about more than one million movement displacements by electronic tags of marine animals like sharks, turtles, bony fishes, and penguins [394], the Lévy flight search for different marine predators is also revealed its closeness to the theoretical optimal. Furthermore, stronger helical components and Lévy walking search strategies indicate their improvement for the encounter rate of predators in three-dimensional environment [456]. Additionally, a common feature of marine predator movement is deduced by the long-term correlation between the variation of prey density and scale invariance spatially. Noticeably, predators spread quickly after hitting their prey and then the distribution of prey density displays a fractal pattern similar to Lévy, as reaction to prey distribution. Like in the ocean region of mesoscale eddy current, there exists a long vertical jump for the predator, similar to Lévy. Ref. [394] described the more complex 3D pattern generated by Lévy-like vertical motion reflecting the motion of marine vertebrate species with different phylogeny. It also stands for the adoption of the open-ocean foragers to be a common strategy for Lévy-like walking. However, few works pay more attention on the potential effectiveness of the Lévy flight search in three-dimensional space because of the prohibition of the computational cost for this numerical simulation. Thus, to study a three-dimensional Lévy flight search strategy in marine predator population's behavior is essential and can help obtain an in-depth understanding for it correctly.

Based on experimental data and analysis results in Refs. [453–458], the three-dimensional fractional diffusion is investigated in the reaction–diffusion system. We use this system to study the complexity of the fractional system. In the context of population dynamics, we use the fractional diffusion ∇^α to describe animal movement in some cases. $0 < \alpha < 1$ and $1 < \alpha < 2$ represent sub-diffusion and Lévy super-diffusion, respectively. We consider the Lévy type of population movement to describe the diffusion behavior of prey and predator spatially in the natural ocean. Then the form of reaction–diffusion equation is obtained by introducing the term into the model (17.2),

$$\begin{aligned} \frac{\partial u}{\partial t} &= \frac{u}{1+fv}\left(1-u-\frac{m}{u+b}\right)-\frac{c(1-e)uv}{(1-e)u+k_1}+D_1\nabla^\alpha u,\\ \frac{\partial v}{\partial t} &= dv\left(1-\frac{v}{(1-e)u+k_1}\right)+D_2\nabla^\alpha v, \end{aligned} \tag{17.3}$$

where

$$\begin{aligned} \nabla^\alpha u = \frac{\partial^\alpha u}{\partial|x|^\alpha}+\frac{\partial^\alpha u}{\partial|y|^\alpha}+\frac{\partial^\alpha u}{\partial|z|^\alpha} &= -\frac{1}{2\cos(\pi\alpha/2)}\left({}_{RL}D^\alpha_{-\infty,x}u+{}_{RL}D^\alpha_{x,+\infty}u\right)\\ &-\frac{1}{2\cos(\pi\alpha/2)}\left({}_{RL}D^\alpha_{-\infty,y}u+{}_{RL}D^\alpha_{y,+\infty}u\right)\\ &-\frac{1}{2\cos(\pi\alpha/2)}\left({}_{RL}D^\alpha_{-\infty,z}u+{}_{RL}D^\alpha_{z,+\infty}u\right), \end{aligned}$$

${}_{RL}D^\alpha_{-\infty,x}u$ and ${}_{RL}D^\alpha_{x,+\infty}u$ are listed and defined respectively as below:

$$
\begin{aligned}
{}_{RL}D^{\alpha}_{-\infty,x}u &= \tfrac{1}{\Gamma(2-\alpha)}\tfrac{\partial^2}{\partial x^2}\int_{-\infty}^{x}(x-s)^{1-\alpha}u(s,y,z,t)\,ds,\\
{}_{RL}D^{\alpha}_{x,+\infty}u &= \tfrac{1}{\Gamma(2-\alpha)}\tfrac{\partial^2}{\partial x^2}\int_{x}^{+\infty}(s-x)^{1-\alpha}u(s,y,z,t)\,ds,
\end{aligned}
$$

where $\Gamma(\cdot)$ means the Gamma function. Fractional Laplacian operator stands for that the species obey a continuous-time walk of Lévy distribution and the jump length obeys a large tail variance [442, 443].

In this chapter, we consider the spectral definition for the fractional Laplace operator, where the operator is defined as the following Fourier spectral method:

$$
\nabla^{\alpha}u(x) = -\sum\nolimits_{j=0}^{\infty}\hat{u}_j\lambda_j^{\frac{\alpha}{2}}\varphi_j,\ u\in I_\alpha,
$$

where, $I_\alpha := \left\{u=\sum\limits_{j=0}^{\infty}\hat{u}_j\varphi_j,\ \hat{u}_j=\langle u,\varphi_j\rangle,\ \sum\limits_{j=0}^{\infty}\left|\hat{u}_j\right|^2\left|\lambda_j\right|^{\frac{\alpha}{2}}<\infty\right\}$, λ_j and φ_j represent the eigenvalues and orthogonal eigenfunctions for the standard Laplace operator ∇^2, respectively, and the boundary condition is the standard condition on the bounded region $\Omega\subset\mathbb{R}^n$, that is, $\nabla^2\varphi_j=-\lambda_j\varphi_j$. In one-dimensional range with length L, and for Neumann boundary conditions, we have its eigenpairs $(\varphi_j,\lambda_j)=(\cos(j\pi x/L),(j\pi/L)^{\alpha})$ [449].

17.3 Stability Analysis of the Non-spatial Model

17.3.1 Existence of Nonnegative Equilibria

Consider the following ordinary differential system corresponding to system (17.3)

$$
\begin{cases}
\frac{u}{1+fv}\left(1-u-\frac{m}{u+b}\right)-\frac{c(1-e)uv}{(1-e)u+k_1}=0.\\
dv\left(1-\frac{v}{(1-e)u+k_1}\right)=0.
\end{cases}
\tag{17.4}
$$

In view of the ecological significance, the positive equilibrium of system (17.3) and system (17.4) is our only focus, where there exists a unique positive equilibrium point $E_*=(u_*,v_*)$ for system (17.3), where $v_*=(1-e)u_*+k_1$, u_* is

$$
u_*=\frac{1-b-cg\,(bfg+(fk_1+1))+M}{2\left(cfg^2+1\right)},
$$

$M=\sqrt{(b+cg\,(bfg+fk_1+1)-1)^2-4\left(cfg^2+1\right)(bcg\,(fk_1+1)-b+m)}$, $g=1-e$.

Additionally, we obtain the Jacobian matrix at the positive equilibrium E_* as

$$J = \begin{pmatrix} J_{11} & J_{12} \\ J_{21} & J_{22} \end{pmatrix}_{(u_*, v_*)},$$

where
$J_{11} = -\frac{ck_1(1-e)}{k_1+(1-e)u_*} + \frac{b^2(1-2u_*)+u_*^2-2u_*^3-b(m+2u_*(-1+2u_*))}{(b+u_*)^2(1+f(k_1+(1-e)u_*))}$,
$J_{12} = -\frac{cu_*(1-e)}{k_1+(1-e)u_*} - \frac{fu_*((b+u_*)(1-u_*)-m)}{(b+u_*)(1+f(k_1+(1-e)u_*))^2}$,
$J_{21} = (1-e)d, \ J_{22} = -d.$
Thus, we get the following characteristic equation:

$$\lambda_0{}^2 - M_1\lambda_0 + M_2 = 0, \tag{17.5}$$

where

$$M_1 = J_{11} + J_{22}, \ M_2 = J_{11}J_{22} - J_{12}J_{21}.$$

Obviously, as $M_1 < 0$ and $M_2 > 0$, it yields the locally asymptotically stability for the equilibrium point E_* by the Routh-Hurwitz criterion.

17.3.2 *Stability and Hopf Bifurcation Analysis*

If $M_1 = 0$ and $M_2 > 0$ about the equilibrium point E_* in the characteristic Eq. (17.5), we get the variation for the stability of Hopf bifurcation. When we choose the e as the bifurcation parameter, we can obtain the critical value e_H of Hopf bifurcation. Let $\lambda_0(e) = x(e) + iy(e)$ be the eigenvalue, where $e = e_H$. Then we have

$$\frac{d(Re(\lambda_0))}{de}\Big|_{e_H} = \frac{2\frac{dM_1}{de}y^2 + M_1\frac{dM_2}{de}}{M_1^2+4y^2}\Big|_{e_H} \neq 0.$$

Hence, we obtain the following conditions for Hopf bifurcation at the coexistence equilibrium E_*. If the system satisfies the following conditions at $e = e_H$, Hopf bifurcation will occur at equilibrium point E_*.

$$(i)\ M_1(e_H) = 0, \ M_2(e_H) > 0,$$

$$(ii)\ \frac{d(Re(\lambda_0))}{de}|e_H = \frac{2\frac{dM_1}{de}y^2 + M_1\frac{dM_2}{de}}{M_1^2+4y^2}|e_H \neq 0.$$

It is natural that the sign of the real part $Re(\lambda_0)$ of the eigenvalue varies and deduces Hopf bifurcation as e gets past through the critical point e_H from left to right.

17.4 Spatial Dynamics of the Reaction–Diffusion Model

17.4.1 Turing Instability Analysis

Although Turing instability means asymptotically stability for the corresponding ordinary differential predator–prey system, the uniform steady state of the system becomes unstable for the partial differential system with diffusion term. For this unstability, we firstly add the time stability of the uniform steady state to the non-uniform disturbance

$$\begin{pmatrix} u \\ v \end{pmatrix} = \begin{pmatrix} u_* \\ v_* \end{pmatrix} e^{\lambda_k t + i\boldsymbol{k}\cdot\boldsymbol{r}} + c.c., \tag{17.6}$$

where, λ_k means the growth rate of the disturbance at time t, i stands for the imaginary unit, $\boldsymbol{k}$ represents the wave number, $\boldsymbol{r}$ indicates the space vector, $c.c.$ means the corresponding complex conjugate.

After linearizing the model (17.3) near the coexistence equilibrium point E_*, we obtain the following equation,

$$\begin{cases} \frac{\partial u}{\partial t} = J_{11}u + J_{12}v + D_1\nabla^\alpha u, \\ \frac{\partial v}{\partial t} = J_{21}u + J_{22}v + D_2\nabla^\alpha v. \end{cases} \tag{17.7}$$

By substituting Eq. (17.6) into equation (17.7), we have the characteristic equation for growth rate λ_k, where the determinant $detA$ of matrix A is defined as

$$A \triangleq \begin{pmatrix} J_{11} - D_1k^\alpha - \lambda_k & J_{12} \\ J_{21} & J_{22} - D_2k^\alpha - \lambda_k \end{pmatrix}.$$

And the characteristic equation of λ_k is obtained as follows

$$\lambda_k^2 - M_3\lambda_k + M_4 = 0,$$

where $M_3 = M_1 - k^\alpha\left(D_1 + D_2\right),\ \ M_4 = M_2 - k^\alpha\left(J_{22}D_1 + J_{11}D_2\right) + k^{2\alpha}D_1D_2$. As we all know, the necessary condition for Turing instability is that there is stability for uniform perturbation, but unstability for some modulus perturbation, and exists also saddle-node bifurcation. Therefore, let $M_1 < 0,\ M_2 > 0$, which ensure the locally stability for the equilibrium point of the ordinary differential system (17.4). Because $D_1, D_2 > 0$ and based on the expression of M_3, we have $M_3 < 0$. At the same time, the only way of Turing instability emerges when $M_4 < 0$ for some k values. If we take M_4 as a quadratic function of k^α and obtain the minimum value of M_4 according to k^α, we can get the most dangerous modulus value k_c for the system with perturbation,

$$k_c = \left(\frac{J_{11}D_2 + J_{22}D_1}{2D_1D_2}\right)^{\frac{1}{\alpha}}. \tag{17.8}$$

In this case, we get the necessary condition for Turing instability as

$$M_{4\min} = M_2 - \frac{(J_{11}D_2 + J_{22}D_1)^2}{4D_1D_2} < 0.$$

The other fixed system parameters remain fixed, and e as the critical value at which instability occurs is regarded as e_c when $M_4 = 0$ is solved. Hence, the necessary condition is obtained for Turing instability of system (17.2) at equilibrium E_* based on the above analysis,

$$(i)\ M_1 < 0,\ M_2 > 0,$$

$$(ii)\ 4D_1D_2M_2 - (J_{11}D_2 + J_{22}D_1)^2 < 0.$$

Next, numerical simulations are performed to verify the existence of Turing instability for the system (17.3) at the equilibrium point E_*. Let M_4 and the maximum real part $Re(\lambda_k)$ of the eigenvalue be a function of the wave number k^α. Then Fig. 17.1 illustrates their relationships with k^α respectively. Here, we assume the parameters of the system as $b = 0.4,\ c = 0.5,\ f = 0.4,\ d = 0.25,\ k_1 = 0.12,\ D_1 = 0.1,\ D_2 = 2,\ \alpha = 1.8$, and $m = 0.24$ separately. The critical value of Turing instability is taken as $e_c = 0.3299$. Let parameter e be evaluated on both sides of the critical value: $0.27,\ 0.31,\ 0.3298,\ 0.35$, and 0.37 respectively. We then observe the transformation for the local stability of the equilibrium. Figure 17.1a shows the negative effect of the parameter e on the diffusion range: the increasing of parameter e can get the shrinks for the range of k^α where $Re(\lambda_k)$ remains positive. When e is greater than the critical value e_c, we have all negative values of $Re(\lambda_k)$. This indicates the stability of the uniform steady state for the system in time and space, and also displays the absence of the Turing instability. The same trend for the relationship between the

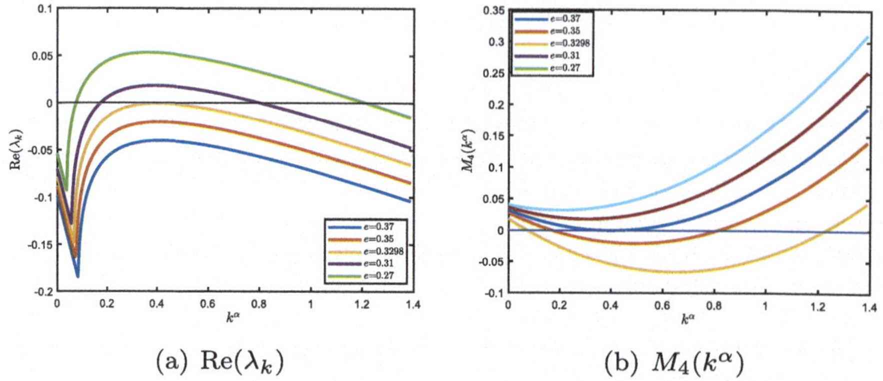

Fig. 17.1 Curves of the functions with respect to k^α at different values of e and other parameters are fixed at $b = 0.4,\ c = 0.5,\ f = 0.4,\ d = 0.25,\ k_1 = 0.12,\ D_1 = 0.1,\ D_2 = 2,\ \alpha = 1.8$, and $m = 0.24$. The horizontal lines display **a** $\text{Re}(\lambda_k) = 0$ and **b** $M_4(k^\alpha) = 0$

functions M_4 and k^α is depicted in Fig. 17.1b. The narrower the range of k^α where M_4 is negative, the lower the possibility of Turing instability. Figure 17.2 exhibits that the higher the fractional diffusion order α, the higher the critical wave number k_c of Turing instability, the narrower the wavelength of the solution, and the narrower the wavenumber range of Turing instability. Thus, the positive effect of the critical wave number k_c of Turing instability on the fractional diffusion order indicates that a smaller fractional diffusion order can more likely cause Turing bifurcation.

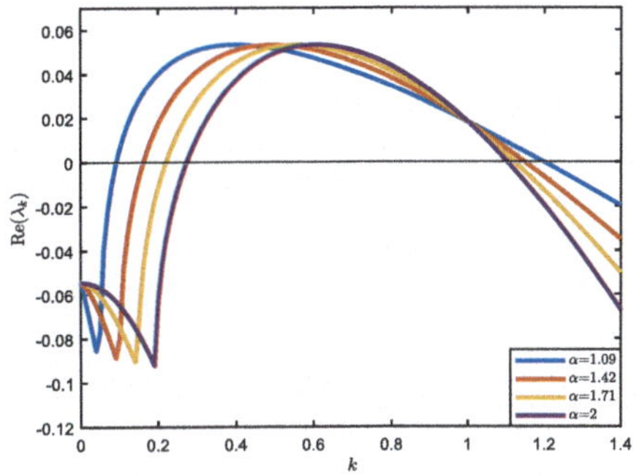

(a) When fractional diffusion order $\alpha = 1.09$, 1.42, 1.71 and 2, function $\text{Re}(\lambda_k)$ changes with wave number k

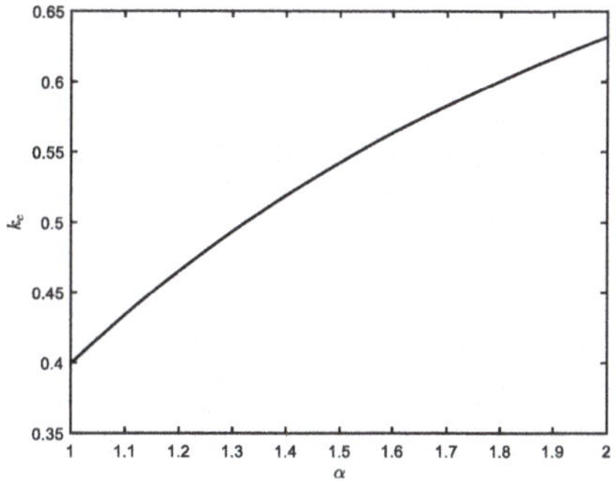

(b) The critical wave number k_c varies with the fractional diffusion order α

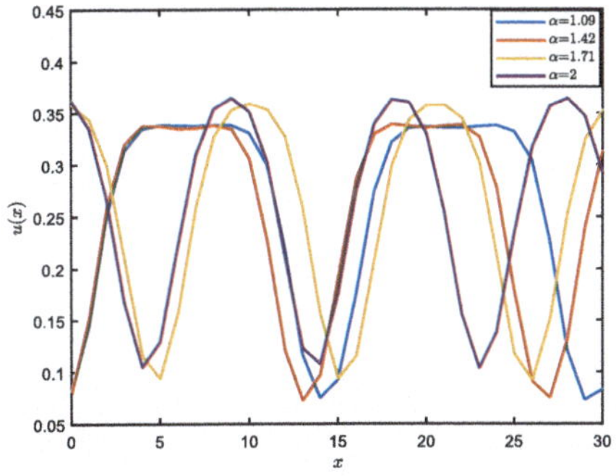

(c) The wavelength of the solution changes with the fractional diffusion order α at $e = e_c$

Fig. 17.2 Fractional diffusion order effect. Other parameters are fixed at $b = 0.4$, $c = 0.5$, $f = 0.4$, $d = 0.25$, $k_1 = 0.12$, $D_1 = 0.1$, $D_2 = 2$, and $m = 0.24$. The horizontal lines indicate $\text{Re}(\lambda_k) = 0$

As shown the behavior far from homogeneous solutions or critical points revealed by the steady-state global bifurcation structure [448], we use the bifurcation diagram measurement to compare the effects of different fractional diffusion on the system independent of the domain size. The normalized L^8-norm is used as $(\|u\|_{L^8} = \left(\frac{1}{|\Omega|}\int_\Omega |u(x)|^8\,\mathrm{d}x\right)^{\frac{1}{8}})$ to measure it [449]. Figure 17.3 shows the bifurcation diagrams for three different fractional orders α for system (17.3) (i.e., α is 1.09, 1.42 and 1.9, respectively). At first, as the fractional index α gradually increases, we can see the gradually moving away from the first bifurcation at $e = e_c$ for bifurcation point on the thin black line. Secondly, the red second bifurcation point moved away from the third green bifurcation point even though they were close together initially. Except for the first bifurcation point, the other bifurcation points also gradually separate as the increase of fractional order index α. The last observation relates about the stability of the solution for the system (17.3). And the first blue bifurcation from the homogeneous state remains entirely stable for fractional diffusion.

Figure 17.4 illustrates that the higher the degree of protection with the refuge deduces the higher the possibility of Turing instability for the system (17.3) as the prey population fears the predator behavior. At the same time, the two populations

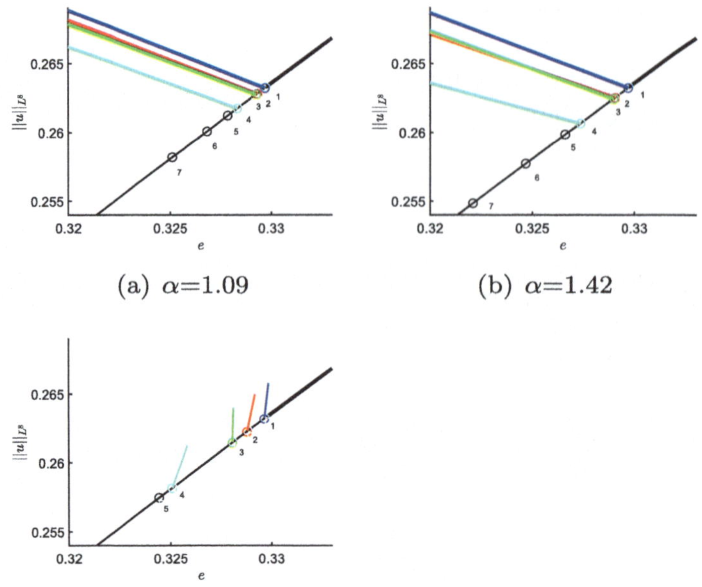

(a) α=1.09 (b) α=1.42 (c) α=1.9

Fig. 17.3 The evolution of the bifurcation graph structure of the system (17.3) decreases with the order of fractional Laplacian. Black: homogeneous state. Thick and thin lines mean stable and unstable solutions, respectively, and circles stand for branch points. Blue, red, green, light blue: branches of non-homogeneous stationary solutions presenting 5, 4.5, 5.5, and 4 bumps, respectively [448]

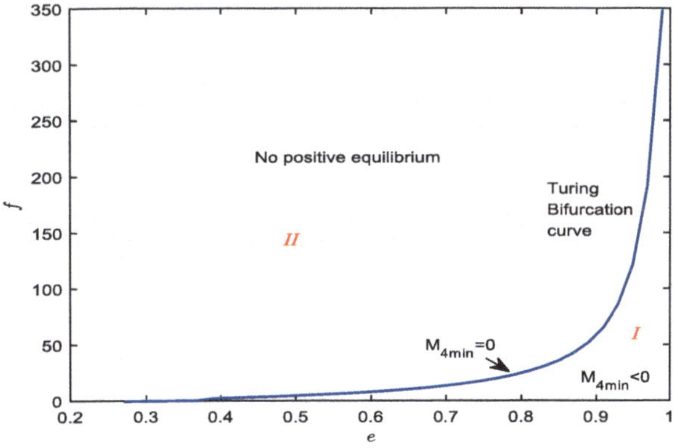

Fig. 17.4 Model (17.3), different stable regions varying with parameters e and f. Other parameters are fixed at $b = 0.4,\ c = 0.5,\ d = 0.25,\ k_1 = 0.12,\ D_1 = 0.1,\ D_2 = 2,\ \alpha = 1.8$, and $m = 0.24$. Region I means Turing unstable region ($M_{4min} < 0$); Region II stands for the extinction parameter region of the prey population

have uneven spatial distribution and then form different pattern types (Region I). If we fix the refuge for the degree of protection of the prey population, and let the fear of predator behavior increase or reach a huge level, there emerges the broken of the balance between the two populations and exists result in the extinction of the prey population (Region II).

17.4.2 2D Amplitude Equations

In this following, we utilize the multi-scale analysis to accurately describe the type of Turing pattern for system (17.3) near the Turing bifurcation point. Let the parameter e as the control parameter of the system, $u_s = u_* + \bar{u}$, and $v_s = v_* + \bar{v}$. For convenience, we still use the original variable. Therefore, at (u_*, v_*), we write the system (17.3) as

$$\begin{cases} \frac{\partial u}{\partial t} = D_1 \nabla^\alpha u + J_{11} u + J_{12} v + N_u, \\ \frac{\partial v}{\partial t} = D_2 \nabla^\alpha v + J_{21} u + J_{22} v + N_v, \end{cases} \tag{17.9}$$

where

$$N_u = b_{20} u^2 + b_{02} v^2 + b_{11} uv + b_{30} u^3 + b_{03} v^3 + b_{21} u^2 v + b_{12} uv^2,$$

$$N_v = p_{20} u^2 + p_{02} v^2 + p_{11} uv + p_{30} u^3 + p_{03} v^3 + p_{21} u^2 v + p_{12} uv^2,$$

and $b_{21} = \frac{cg^2 k_1}{(k_1+gu_*)^3} + \frac{f\left(b^3+3b^2u_*+u_*^2(5u_*-2)+b(m+u_*(11u_*-4))\right)}{(b+u_*)^3(1+f(k_1+gu_*))^2}$,

$$b_{30} = -\frac{cg^3k_1\left((b+u_*)^4-4bu_*^3\right)(1+f(k_1+gu_*))+b\left((k_1+gu_*)^3m+4cg^3k_1u_*^3(1+f(k_1+gu_*))\right)}{(b+u_*)^4(k_1+gu_*)^3(1+f(k_1+gu_*))},$$

$$b_{20} = \frac{cg^2k_1}{(k_1+gu_*)^2} - \frac{(b+u_*)^3-bm}{(b+u_*)^3(1+f(k_1+gu_*))},\ b_{02} = \frac{f^2u_*((b+u_*)(1-u_*)-m)}{(b+u_*)(1+f(k_1+gu_*))^3},$$

$$b_{11}=-\frac{cgk_1}{(k_1+gu_*)^2}+\frac{f\left((b+u_*)^2(2u_*-1)+bm\right)}{(b+u_*)^2(1+f(k_1+gu_*))^2},\ b_{03}=-\frac{f^3u_*((b+u_*)(1-u_*)-m)}{(b+u_*)(1+f(k_1+gu_*))^4},$$
$$b_{12}=-\frac{f^2\left((b+u_*)^2(2u_*-1)+bm\right)}{(b+u_*)^2(1+f(k_1+gu_*))^3},\ p_{20}=-\frac{dg^2}{gu_*+k_1},\ p_{02}=-\frac{d}{gu_*+k_1},$$
$$p_{11}=\frac{2dg}{gu_*+k_1},\ p_{30}=\frac{dg^3}{(gu_*+k_1)^2},\ p_{21}=-\frac{2dg^2}{(gu_*+k_1)^2},\ p_{12}=\frac{dg}{(gu_*+k_1)^2},\ p_{03}=0.$$

Near $e=e_c$, we take the following form for the solution of model (17.9)

$$\begin{pmatrix} u \\ v \end{pmatrix}=\sum_{j=1}^{3}\begin{pmatrix} A_j^u \\ A_j^v \end{pmatrix}e^{i\boldsymbol{k}\cdot\boldsymbol{r}}+c.c.$$

Let $X=\begin{pmatrix} u \\ v \end{pmatrix}$, $N=\begin{pmatrix} N_u \\ N_v \end{pmatrix}$, then we abbreviate model (17.9) as

$$\frac{\partial X}{\partial t}=LX+N, \tag{17.10}$$

where

$$L=\begin{pmatrix} J_{11}+D_u\nabla^\alpha & J_{12} \\ J_{21} & J_{22}+D_v\nabla^\alpha \end{pmatrix},$$

$$N=\begin{pmatrix} b_{20}u^2+b_{02}v^2+b_{11}uv+b_{30}u^3+b_{03}v^3+b_{21}u^2v+b_{12}uv^2 \\ p_{20}u^2+p_{02}v^2+p_{11}uv+p_{30}u^3+p_{21}u^2v+p_{12}uv^2 \end{pmatrix}.$$

The following expansions are listed in the form of a small quantity ε near the critical value of Turing bifurcation,

$$e-e_c=\varepsilon e_1+\varepsilon^2e_2+\varepsilon^3e_3+o\left(\varepsilon^3\right), \tag{17.11}$$

$$X=\varepsilon\begin{pmatrix} u_1 \\ v_1 \end{pmatrix}+\varepsilon^2\begin{pmatrix} u_2 \\ v_2 \end{pmatrix}+\varepsilon^3\begin{pmatrix} u_3 \\ v_3 \end{pmatrix}+o(\varepsilon^3), \tag{17.12}$$

$$N=\varepsilon^2N_2+\varepsilon^3N_3+o\left(\varepsilon^3\right), \tag{17.13}$$

where e is the control parameter, X stands for the variable and N means the nonlinear term,

$$N_2=\begin{pmatrix} b_{20}u_1^2+b_{02}v_1^2+b_{11}u_1v_1 \\ p_{20}u_1^2+p_{02}v_1^2+p_{11}u_1v_1 \end{pmatrix}.$$

$$N_3=\begin{pmatrix} 2b_{20}u_1u_2+2b_{02}v_1v_2+b_{11}u_1v_2+b_{11}u_2v_1+b_{30}u_1^3+b_{03}v_1^3+b_{21}u_1^2v_1+b_{12}u_1v_1^2 \\ 2p_{20}u_1u_2+2p_{02}v_1v_2+p_{11}u_1v_2+p_{11}u_2v_1+p_{30}u_1^3+p_{21}u_1^2v_1+p_{12}u_1v_1^2 \end{pmatrix}.$$

At the same time, we decompose the linear operator L into

$$L=L_c+(e-e_c)\,M, \tag{17.14}$$

where

$$L_c = \begin{pmatrix} J_{11} + D_1\nabla^{\alpha} & J_{12} \\ J_{21} & J_{22} + D_2\nabla^{\alpha} \end{pmatrix},$$

$$M = \begin{pmatrix} m_{11} & m_{12} \\ m_{21} & m_{22} \end{pmatrix},$$

$m_{11} = \frac{dJ_{11}}{de}\Big|_{e=e_c}$, $m_{12} = \frac{dJ_{12}}{de}\Big|_{e=e_c}$, $m_{21} = d$, $m_{22} = 0$. By introducing multiple time scales, we regard the time scales of the system as independent scales. In this way, we then write the derivative of time t as

$$\frac{\partial}{\partial t} = \varepsilon\frac{\partial}{\partial t_1} + \varepsilon^2\frac{\partial}{\partial t_2} + \varepsilon^3\frac{\partial}{\partial t_3} + o\left(\varepsilon^3\right). \tag{17.15}$$

Substituting Eqs. (17.11)–(17.15) into Eq. (17.10), we have different orders of ε.

$$\varepsilon : L_c \begin{pmatrix} u_1 \\ v_1 \end{pmatrix} = 0; \tag{17.16}$$

$$\varepsilon^2 : L_c \begin{pmatrix} u_2 \\ v_2 \end{pmatrix} = \frac{\partial}{\partial t_1}\begin{pmatrix} u_1 \\ v_1 \end{pmatrix} - e_1 M \begin{pmatrix} u_1 \\ v_1 \end{pmatrix} - N_2; \tag{17.17}$$

$$\varepsilon^3 : L_c \begin{pmatrix} u_3 \\ v_3 \end{pmatrix} = \frac{\partial}{\partial t_1}\begin{pmatrix} u_2 \\ v_2 \end{pmatrix} + \frac{\partial}{\partial t_2}\begin{pmatrix} u_1 \\ v_1 \end{pmatrix} - e_2 M \begin{pmatrix} u_1 \\ v_1 \end{pmatrix} - e_1 M \begin{pmatrix} u_2 \\ v_2 \end{pmatrix} - N_3. \tag{17.18}$$

Obviously, for Eq. (17.16), we take L_c as a linear operator for the system at the critical point e_c, and $(u_1, v_1)^T$ as a linear combination of eigenvectors according to the zero eigenvalues of linear operator L_c. Then we expand the general solution of Eq. (17.16) as the form of three corresponding modes for wave vectors.

$$\begin{pmatrix} u_1 \\ v_1 \end{pmatrix} = \begin{pmatrix} \varphi \\ 1 \end{pmatrix} (W_1 e^{i\boldsymbol{k}_1\cdot \boldsymbol{r}} + W_2 e^{i\boldsymbol{k}_2\cdot \boldsymbol{r}} + W_3 e^{i\boldsymbol{k}_3\cdot \boldsymbol{r}} + c.c.), \tag{17.19}$$

where $\varphi = \frac{-J_{12}^c}{J_{11}^c - D_1 k_c^{\alpha}}$, $|\boldsymbol{k_j}| = k_c$, $j = 1, 2, 3$, W_j remains the amplitude of mode $e^{i\boldsymbol{k}_j\cdot \boldsymbol{r}}$ for the system under the first order perturbation.

Based on the Fredholm solvability condition, we make the right-hand term of (17.17) be orthogonal to the zero eigenvectors of operator L_c^+ ensuring the existence of nontrivial solutions of Eq. (17.17), where L_c^+ means the conjugate operator for operator L_c, and its zero eigenvector is defined as

$$\begin{pmatrix} 1 \\ \phi \end{pmatrix} e^{-i\boldsymbol{k}_j\cdot \boldsymbol{r}} + c.c., \qquad j = 1, 2, 3, \tag{17.20}$$

where $\phi = -\frac{J_{11}^c - D_1 k_c^\alpha}{J_{21}^c}$. The L_c^+ and ε^2-order terms meet the orthogonal condition written as

$$(1, \phi)\begin{pmatrix} F_u^j \\ F_v^j \end{pmatrix} = 0, \qquad j = 1, 2, 3, \tag{17.21}$$

where F_u^j and F_v^j stand for the coefficients corresponding to $e^{i\boldsymbol{k}_j\cdot\boldsymbol{r}}$ in F_u and F_v respectively. Making an example of $e^{i\boldsymbol{k}_1\cdot\boldsymbol{r}}$, it has the following form

$$\begin{pmatrix} F_u \\ F_v \end{pmatrix} = \begin{pmatrix} \varphi\frac{\partial W_1}{\partial t_1} \\ \frac{\partial W_1}{\partial t_1} \end{pmatrix} - e_1\begin{pmatrix} \varphi m_{11} W_1 + m_{12} W_1 \\ \varphi m_{21} W_1 + m_{22} W_1 \end{pmatrix} - \begin{pmatrix} G_1 \bar{W}_2 \bar{W}_3 \\ G_2 \bar{W}_2 \bar{W}_3 \end{pmatrix},$$

where $G_1 = 2b_{20}\varphi^2 + 2b_{11}\varphi + 2b_{02},\ G_2 = 2p_{20}\varphi^2 + 2p_{11}\varphi + 2p_{02}$.

According to Fredholm solvability condition (17.21), the amplitude equations are obtained under the first-order perturbation.

$$\begin{cases} (\varphi+\phi)\frac{\partial W_1}{\partial t_1} = e_1[(\varphi m_{11} + m_{12}) + \phi(\varphi m_{21} + m_{22})]W_1 + (G_1 + \phi G_2)\bar{W}_2\bar{W}_3, \\ (\varphi+\phi)\frac{\partial W_2}{\partial t_1} = e_1[(\varphi m_{11} + m_{12}) + \phi(\varphi m_{21} + m_{22})]W_2 + (G_1 + \phi G_2)\bar{W}_1\bar{W}_3, \\ (\varphi+\phi)\frac{\partial W_3}{\partial t_1} = e_1[(\varphi m_{11} + m_{12}) + \phi(\varphi m_{21} + m_{22})]W_3 + (G_1 + \phi G_2)\bar{W}_1\bar{W}_2. \end{cases} \tag{17.22}$$

Because the coefficient of the second-order term remains greater than 0 and the amplitude W_i diverges, we must introduce the higher-order perturbation terms. Therefore, the solution form of Eq. (17.17) is given as follows,

$$\begin{aligned} \begin{pmatrix} u_2 \\ v_2 \end{pmatrix} &= \begin{pmatrix} U_0 \\ V_0 \end{pmatrix} + \sum_{j=1}^{3}\begin{pmatrix} U_j \\ V_j \end{pmatrix} e^{i\boldsymbol{k}_j\cdot\boldsymbol{r}} + \sum_{j=1}^{3}\begin{pmatrix} U_{jj} \\ V_{jj} \end{pmatrix} e^{2i\boldsymbol{k}_j\cdot\boldsymbol{r}} + \begin{pmatrix} U_{12} \\ V_{12} \end{pmatrix} e^{i(\boldsymbol{k}_1-\boldsymbol{k}_2)\cdot\boldsymbol{r}} \\ &+ \begin{pmatrix} U_{23} \\ V_{23} \end{pmatrix} e^{i(\boldsymbol{k}_2-\boldsymbol{k}_3)\cdot\boldsymbol{r}} + \begin{pmatrix} U_{31} \\ V_{31} \end{pmatrix} e^{i(\boldsymbol{k}_3-\boldsymbol{k}_1)\cdot\boldsymbol{r}} + c.c.. \end{aligned} \tag{17.23}$$

Then after obtaining the coefficients of Eq. (17.23) by solving the set of linear equations for e^0, $e^{i\boldsymbol{k}_j\cdot\boldsymbol{r}}$, $e^{i(2\boldsymbol{k}_j)\cdot\boldsymbol{r}}$, and $e^{i(\boldsymbol{k}_j-\boldsymbol{k}_k)\cdot\boldsymbol{r}}$, we get the results as follows,

$$\begin{pmatrix} U_0 \\ V_0 \end{pmatrix} = \begin{pmatrix} u_{00} \\ v_{00} \end{pmatrix}(|W_1|^2 + |W_2|^2 + |W_3|^2),\ U_j = \varphi V_j,$$

$$\begin{pmatrix} U_{jj} \\ V_{jj} \end{pmatrix} = \begin{pmatrix} u_{11} \\ v_{11} \end{pmatrix} W_j^2,\ \begin{pmatrix} U_{jl} \\ V_{jl} \end{pmatrix} = \begin{pmatrix} u^* \\ v^* \end{pmatrix} W_j \bar{W}_l.$$

For the specific expression, we obtain it by the undetermined coefficient method,

$$\begin{pmatrix} u_{00} \\ v_{00} \end{pmatrix} = \frac{-1}{J_{11}^c J_{22}^c - J_{21}^c J_{12}^c} \begin{pmatrix} J_{22}^c G_1 - J_{12}^c G_2 \\ J_{11}^c G_2 - J_{21}^c G_1 \end{pmatrix},$$

$$\begin{pmatrix} u_{11} \\ v_{11} \end{pmatrix} = \frac{-1}{2((J_{11}^c - 2^\alpha D_1 k_c^\alpha)(J_{22}^c - 2^\alpha D_2 k_c^\alpha) - J_{12}^c J_{21}^c)} \begin{pmatrix} (J_{22}^c - 2^\alpha D_2 k_c^\alpha) G_1 - J_{12}^c G_2 \\ (J_{11}^c - 2^\alpha D_1 k_c^\alpha) G_2 - J_{21}^c G_1 \end{pmatrix},$$

$$\begin{pmatrix} u^* \\ v^* \end{pmatrix} = \frac{-1}{(J_{11}^c - \sqrt{3^\alpha} D_1 k_c^\alpha)(J_{22}^c - \sqrt{3^\alpha} D_2 k_c^\alpha) - J_{12}^c J_{21}^c} \begin{pmatrix} (J_{22}^c - \sqrt{3^\alpha} D_2 k_c^\alpha) G_1 - J_{12}^c G_2 \\ (J_{11}^c - \sqrt{3^\alpha} D_1 k_c^\alpha) G_2 - J_{21}^c G_1 \end{pmatrix}.$$

The third-order term (17.18) of ε can be obtained the following results by using the solvability condition,

$$\begin{cases} (\varphi + \phi)(\frac{\partial V_1}{\partial t_1} + \frac{\partial W_1}{\partial t_2}) \\ = [(\varphi m_{11} + m_{12}) + \phi(\varphi m_{21} + m_{22})](e_2 W_1 + e_1 V_1) - [(T_1 + \phi I_1)|W_1|^2 \\ +(T_2 + \phi I_2)(|W_2|^2 + |W_3|^2)]W_1 + 2(b_{20}\varphi^2 + b_{11}\varphi + b_{02} + \phi(p_{20}\varphi^2 \\ +p_{11}\varphi + p_{02}))(\bar{W}_2 \bar{V}_3 + \bar{W}_3 \bar{V}_2), \\ (\varphi + \phi)(\frac{\partial V_2}{\partial t_1} + \frac{\partial W_2}{\partial t_2}) \\ = [(\varphi m_{11} + m_{12}) + \phi(\varphi m_{21} + m_{22})](e_2 W_2 + e_1 V_2) - [(T_1 + \phi I_1)|W_2|^2 \\ +(T_2 + \phi I_2)(|W_1|^2 + |W_3|^2)]W_2 + 2(b_{20}\varphi^2 + 2b_{11}\varphi + b_{02} + \phi(p_{20}\varphi^2 \\ +p_{11}\varphi + p_{02}))(\bar{W}_1 \bar{V}_3 + \bar{W}_3 \bar{V}_1), \\ (\varphi + \phi)(\frac{\partial V_3}{\partial t_1} + \frac{\partial W_3}{\partial t_2}) \\ = [(\varphi m_{11} + m_{12}) + \phi(\varphi m_{21} + m_{22})](e_2 W_3 + e_1 V_3) - [(T_1 + \phi I_1)|W_3|^2 \\ +(T_2 + \phi I_2)(|W_2|^2 + |W_1|^2)]W_3 + 2(b_{20}\varphi^2 + 2b_{11}\varphi + b_{02} + \phi(p_{20}\varphi^2 \\ +p_{11}\varphi + p_{02}))(\bar{W}_2 \bar{V}_1 + \bar{W}_1 \bar{V}_2), \end{cases} \tag{17.24}$$

where

$$T_1 = -(2b_{20}\varphi + b_{11})(u_{11} + u_{00}) - (2b_{02} + b_{11}\varphi)(v_{11} + v_{00}) \\ -3b_{03} - 3b_{12}\varphi - 3b_{21}\varphi^2 - 3b_{30}\varphi^3,$$

$$T_2 = -(2b_{20}\varphi + b_{11})(u_{00} + u^*) - (2b_{02} + b_{11}\varphi)(v_{00} + v^*) \\ -6b_{03} - 6b_{12}\varphi - 6b_{21}\varphi^2 - 6b_{30}\varphi^3,$$

$$I_1 = -(2p_{20}\varphi + p_{11})(u_{11} + u_{00}) - (2p_{02} + p_{11}\varphi)(v_{11} + v_{00}) \\ -3p_{12}\varphi - 3p_{21}\varphi^2 - 3p_{30}\varphi^3,$$

$$I_2 = -(2p_{20}\varphi + p_{11})(u_{00} + u^*) - (2p_{02} + p_{11}\varphi)(v_{00} + v^*)$$
$$-6p_{12}\varphi - 6p_{21}\varphi^2 - 6p_{30}\varphi^3.$$

And we expand the amplitude A_j as follows,

$$A_j = \varepsilon W_j + \varepsilon^2 V_j + O\left(\varepsilon^2\right). \tag{17.25}$$

As we multiply Eqs. (17.22) and (17.24) by ε^2 and ε^3, respectively, and combine these variables by Eq. (17.25), we can obtain the amplitude equation corresponding to A_j as below,

$$\begin{cases} \tau_0 \frac{\partial A_1}{\partial t} = \mu A_1 + h\bar{A}_2\bar{A}_3 - [g_1|A_1|^2 + g_2(|A_2|^2 + |A_3|^2)]A_1, \\ \tau_0 \frac{\partial A_2}{\partial t} = \mu A_2 + h\bar{A}_1\bar{A}_3 - [g_1|A_2|^2 + g_2(|A_1|^2 + |A_3|^2)]A_2, \\ \tau_0 \frac{\partial A_3}{\partial t} = \mu A_3 + h\bar{A}_1\bar{A}_2 - [g_1|A_3|^2 + g_2(|A_1|^2 + |A_2|^2)]A_3, \end{cases} \tag{17.26}$$

where
$\tau_0 = \frac{\varphi+\phi}{e_c[(\varphi m_{11}+m_{12})+\phi(\varphi m_{21}+m_{22})]}$, $h = \frac{G_1+\phi G_2}{e_c[(\varphi m_{11}+m_{12})+\phi(\varphi m_{21}+m_{22})]}$,
$g_1 = \frac{T_1+\phi I_1}{e_c[(\varphi m_{11}+m_{12})+\phi(\varphi m_{21}+m_{22})]}$, $g_2 = \frac{T_2+\phi I_2}{e_c[(\varphi m_{11}+m_{12})+\phi(\varphi m_{21}+m_{22})]}$.
$\mu = \frac{e-e_c}{e_c}$.

17.4.3 2D Amplitude Instability

We next decompose each amplitude in the Eq. (17.26) into the product of the modulus $\omega_j = |A_j|$ and a coincident phase angle ψ_j. In this way, we substitute $A_j = \omega_j e^{i\psi_j}$ into Eq. (17.26) and separate the real and imaginary parts. Then we calculate the four different equations about real variables,

$$\begin{cases} \tau_0 \frac{\partial \psi}{\partial t} = -h\frac{\omega_2^2\omega_3^2+\omega_1^2\omega_3^2+\omega_1^2\omega_2^2}{\omega_1\omega_2\omega_3} sin\psi, \\ \tau_0 \frac{\partial \omega_1}{\partial t} = \mu\omega_1 + h\omega_2\omega_3 cos\psi - [g_1\omega_1^2 + g_2(\omega_2^2 + \omega_3^2)]\omega_1, \\ \tau_0 \frac{\partial \omega_2}{\partial t} = \mu\omega_2 + h\omega_1\omega_3 cos\psi - [g_1\omega_2^2 + g_2(\omega_1^2 + \omega_3^2)]\omega_2, \\ \tau_0 \frac{\partial \omega_3}{\partial t} = \mu\omega_3 + h\omega_1\omega_2 cos\psi - [g_1\omega_3^2 + g_2(\omega_1^2 + \omega_2^2)]\omega_3, \end{cases} \tag{17.27}$$

where $\psi = \psi_1 + \psi_2 + \psi_3$. The four different solutions for dynamical system (17.27) have been listed as below.
(1) The uniform state remains stable when $\mu < \mu_2 = 0$ and unstable when $\mu > \mu_2$ for the following form

$$\omega_1 = \omega_2 = \omega_3 = 0.$$

(2) The stripe pattern state remains stable when$\mu > \mu_3 = \frac{h^2 g_1}{(g_1-g_2)^2}$, otherwise, it is unstable for the following forms

$$\omega_1 = \sqrt{\frac{\mu}{g_1}} \neq 0,\ \omega_2 = \omega_3 = 0.$$

(3) For the hexagonal pattern/spot (I_0, I_π) pattern with the following forms

$$\omega_1 = \omega_2 = \omega_3 = \frac{|h| \pm \sqrt{h^2 + 4\,(g_1 + 2g_2\mu)}}{2\,(g_1 + 2g_2)},$$

where $\psi = 0$ or $\psi = \pi$, the solution exists as the conditions are satisfied as below,

$$\mu > \mu_1 = -\frac{h^2}{4\,(g_1 + 2g_2)}.$$

The solution $\omega^+ = \frac{|h|+\sqrt{h^2+4(g_1+2g_2\mu)}}{2(g_1+2g_2)}$ remains stable and the solution $\omega^- = \frac{|h|-\sqrt{h^2+4(g_1+2g_2\mu)}}{2(g_1+2g_2)}$ remains always unstable,if and only if the following conditions are satisfied,

$$\mu < \mu_4 = \frac{(2g_1 + g_2)\,h^2}{(g_2 - g_1)^2}.$$

(4) When $\mu > \mu_3$, the state remains always unstable for the mixed state:

$$\omega_1 = \frac{|h|}{g_2 - g_1},\ \omega_2 = \omega_3 = \sqrt{\frac{\mu - g_1\omega_1^2}{g_1 + g_2}},$$

where $g_2 > g_1$.

In conclusion, the static state of the system emerges losing stability as the increasing of the control parameter μ to the critical point μ_2, and at the same time, the system first generates the hexagonal pattern by the non-equilibrium phase transition. When the coefficient h of the second term in Eq. (17.27) becomes greater than 0, the hexagon pattern is I_0. Contrarily, the hexagon pattern becomes I_π. In the former case, the pattern for I_π remains always unstable. In the latter case, the pattern for I_0 becomes always unstable. And the subcritical bifurcation causes the appearance of hexagon patterns. In other words, the system gets a bistable region in a range of control parameters $(\mu_1 < \mu < \mu_2)$. Clearly, both hexagon pattern and static state remain stable in this region. Moreover, even though the supercritical bifurcation causes the appearance of stripe pattern, it is unstable for $\mu < \mu_3$ and stable only for $\mu > \mu_3$. Moreover, the hexagonal pattern becomes unstable if $\mu < \mu_4$, so that there exists another bistable area for the system as the control parameter remains in the area $\mu_3 < \mu < \mu_4$. Furthermore, the system pattern varies from hexagonal form to stripe form if the control parameter μ exceeds μ_4. In the other case, the system pattern varies from stripe form to hexagon form when the control parameter μ becomes lower than μ_3.

17.5 3D Pattern Formation

17.5.1 3D Amplitude Equation

It is well known that the predator–prey system in two-dimensional space usually appears stripes or hexagonal spots. Instead, the pattern types show more abundant and more complicated dynamic properties in the three-dimensional space. For instance, there are a variety of arrangement possibilities in three-dimensional space, like simple cubic lattice, face-centered cubic lattice structure, body-centered cubic lattice (BCC), double helix structure, and so on. Callahan and Knobloch first solved the bifurcation problem for 3D Turing system [466, 488]. In the following, the general amplitude equation is derived for the system 3D pattern, and the amplitude equation of the BCC structure is also discussed.

Firstly, a spatial fractional predator–prey model is written for two species as follows

$$\frac{\partial X_s}{\partial t} = D_s \nabla^\alpha X_s + A_{s,i} X_s + A_{s,ij} X_i X_j + A_{s,ijw} X_i X_j X_w + O(X^3) \tag{17.28}$$

where, $s = 1, 2$, X means the population density. In the two species, Einstein summation is used for the indices s, i, j, and w, and covariance or inversion except for the diffusion term is not be considered. Assume recording the number for lattice points in space as l and omitting higher-order terms, we write Eq. (17.28) as:

$$\frac{dX_s^l}{dt} = A_{s,i} X_i^l - D_s |k^l|^\alpha X_s^l + A_{s,ij} \sum_{l_1+l_2=l} X_i^{l_1} X_j^{l_2} + A_{s,ijw} \sum_{l_1+l_2+l_3=l} X_i^{l_1} X_j^{l_2} X_w^{l_3}. \tag{17.29}$$

Diagonalizing the first two terms for the above equation

$$J_{s,i}^l = -D_s k^{l^\alpha} \delta_{s,i} + A_{s,i} \tag{17.30}$$

and secondly, reducing the equation, by

$$X^l = P^l W^l, \tag{17.31}$$

we choose a matrix for each lattice point,

$$P^l = \begin{pmatrix} p_{11}^l & p_{12}^l \\ p_{21}^l & p_{22}^l \end{pmatrix} = \begin{pmatrix} H_2 \sqrt{\frac{dg_c}{P}} & H_1 \sqrt{\frac{dg_c}{P}} \\ \sqrt{\frac{dg_c}{P}} & \sqrt{\frac{dg_c}{P}} \end{pmatrix},$$

with inverse

$$P^{l^{-1}} = \begin{pmatrix} q_{11}^{l} & q_{12}^{l} \\ q_{21}^{l} & q_{22}^{l} \end{pmatrix} = \begin{pmatrix} \sqrt{\frac{dg_c}{P}} & -H_1\sqrt{\frac{dg_c}{P}} \\ -\sqrt{\frac{dg_c}{P}} & H_2\sqrt{\frac{dg_c}{P}} \end{pmatrix}, \tag{17.32}$$

where

$$g_c = 1 - e_c,\ R = D_1 k_c^{\alpha},\ T = -\frac{dJ_{11}^{c} + J_{12}^{c} dg_c}{R},\ H_1 = \frac{J_{11}^{c} + d - R + T - P}{2dg_c},$$

$$H_2 = \frac{J_{11}^{c} + d - R + T + P}{2dg_c},\ P = \sqrt{(J_{11}^{c} + d - R + T)^2 + 4J_{12}^{c} dg_c}.$$

We substitute (17.31) into (17.29), omit the higher order term, and obtain the following results,

$$\begin{aligned} \frac{dW_h^l}{dt} &= \lambda_h^l W_h^l + q_{hs}^l A_{s,ij} \sum_{l_1+l_2=l} p_{ii'}^{l_1} W_{i'}^{l_1} p_{jj'}^{l_2} W_{j'}^{l_2} \\ &+ q_{hs}^l A_{s,ijw} \sum_{l_1+l_2+l_3=l} p_{ii'}^{l_1} W_{i'}^{l_1} p_{jj'}^{l_2} W_{j'}^{l_2} p_{ww'}^{l_3} W_{w'}^{l_3}. \end{aligned} \tag{17.33}$$

Then we define the coefficient λ_l^h of the linear term by the following formula

$$\lambda_c = \left.\frac{d\lambda_k}{de}\right|_{e=e_c} (e - e_c) = \frac{(\varphi m_{11} + m_{12}) + \varphi\phi m_{21}}{\varphi + \phi} (e - e_c). \tag{17.34}$$

Note that when $e = e_c$, we can obtain the coefficients for the central manifold and amplitude equation. Based on this, the amplitude equation is deduced for the mode on the central manifold. Taking $r = k^{\alpha}/k_c^{\alpha}$, we further simply the Eq. (17.31) and express the critical wave vector m for $\lambda_l^h = 0$ as

$$\frac{dW_m^1}{dt} = z_1 \sum_{m_1+m_2=m} W_{l_1}^1 W_{l_2}^1 + \sum_{m_1+m_2+m_3=m} Q\,(m_2 + m_3)\, W_{m_1}^1 W_{m_2}^1 W_{m_3}^1, \tag{17.35}$$

where

$$z_1 = q^{1s} A^{s,ij} p^{i1} p^{j1},$$

$$Q = q^{1s} A^{s,ijk} p^{i1} p^{j1} p^{k1} - 2q^{1s} A^{s,ij} p^{i1} \left(J_l^{-1}\right)^{ja} A^{a,bc} p^{b1} p^{c1}.$$

It should be noted that the coefficient Q is only controlled by the α power for the length of the parameter r.

$$Q\,(r) = \frac{dg_c\left(2dg_c H_3 M_2^c - PH_4 + r\left(2dg_c H_3\left(dR - J_{11}^c T + rM_2^c\right) - PH_5\right)\right)}{P^2\left(M_2^c + r\left(dR - J_{11}^c T + rM_2^c\right)\right)}, \tag{17.36}$$

where

$$q_1 = \frac{dg_c(b_{11}-2b_{20}H_2+(p_{11}+2p_{20}H_2)H_1)}{P},\ q_2 = \frac{dg(2b_{02}-b_{11}H_2+(2p_{02}+p_{11}H_2)H_1)}{P},$$
$$G_1 = d\left(q_1 + g_c q_2\right),\ G_2 = p_{02} + \left(p_{11} + p_{20}H_2\right)H_2,$$
$$H_3 = b_{03} + \left(b_{12} + b_{21}H_2 + b_{30}H_2^2 - p_{12}H_1 - p_{21}H_1H_2 - p_{30}H_1H_2^2\right)H_2,$$
$$H_4 = b_{02} + \left(b_{11} + b_{20}H_2\right)G_1H_2 + \left(J_{12}^c q_1 - J_{11}^c q_2\right)G_2,$$
$$H_5 = p_{02}q_2R + b_{02}q_1T + \left(p_{11}q_2R + b_{11}q_1T\right)H_2 + \left(p_{20}q_2R + b_{20}q_1T\right)H_2^2.$$

Notably, the above processing becomes general and not specific for any particular symmetry. Next, the coefficients of the amplitude equation are calculated for the BCC structure. Firstly, we define the BCC lattice structure as the following six vectors,

$$\vec{k_1} = \frac{k_c}{\sqrt{2}}(1,1,0),\ \vec{k_2} = \frac{k_c}{\sqrt{2}}(0,1,1),\ \vec{k_3} = \frac{k_c}{\sqrt{2}}(1,0,1),$$

$$\vec{k_4} = \frac{k_c}{\sqrt{2}}(1,-1,0),\ \vec{k_5} = \frac{k_c}{\sqrt{2}}(0,1,-1),\ \vec{k_6} = \frac{k_c}{\sqrt{2}}(1,0,-1).$$

And we then obtain the BCC amplitude equation

$$\begin{aligned}\frac{dW_1}{dt} &= \lambda_c W_1 + z_1\left(W_2W_6 + W_3W_5\right) + z_3\left(|W_2|^2 + |W_3|^2 + |W_5|^2 + |W_6|^2\right)W_1 \\ &\quad + z_2|W_1|^2W_1 + z_4|W_4|^2W_1 + z_5\left(W_2W_4W_5 + W_3\bar{W_4}W_6\right),\end{aligned} \tag{17.37}$$

where

$$z_1 = \left(b_{02} + b_{11}H_2 + b_{20}H_2^2 - p_{02}H_1 - p_{11}H_1H_2 - p_{20}H_1H_2^2\right)P^{\frac{3}{2}},$$
$$z_2 = 2Q_2(0) + Q_2(2^{\alpha}),\ z_3 = 2Q_2(0) + 2Q_2(1) + 2Q_2\left(3^{\frac{\alpha}{2}}\right),$$
$$z_4 = 2Q_2(0) + 4Q_2\left(2^{\frac{\alpha}{2}}\right),\ z_5 = 2Q_2\left(2^{\frac{\alpha}{2}}\right) + 4Q_2(1).$$

17.5.2 3D Amplitude Instability

In the following, the linear analysis is used to study the stability for the amplitude equation near the steady-state.

At first, we must determine the steady state for the amplitude system. Here, the stability of lamellas and isotropic BCC droplets are only studied. We get the following for the steady-state for an isotropic BCC droplet $\left|W_1^c\right| = \left|W_2^c\right| = \left|W_3^c\right| = \left|W_4^c\right| = \left|W_5^c\right| = \left|W_6^c\right|$,

$$W_i^c = \frac{-z_1 \pm \sqrt{z_1^2 - \lambda_c\left(z_2 + 4z_3 + z_4 + 2z_5\right)}}{z_2 + 4z_3 + z_4 + 2z_5}. \tag{17.38}$$

The corresponding steady state of lamellas is

$$\left|W_1^c\right| = \sqrt{-\frac{\lambda_c}{z_2}}, \tag{17.39}$$

where $\left|W_2^c\right| = \left|W_3^c\right| = \left|W_4^c\right| = \left|W_5^c\right| = \left|W_6^c\right| = 0$.

Furthermore, we linearize the amplitude system, and construct the corresponding matrix S,

$$S = \left|\frac{\partial F_i}{\partial \left|W_j\right|}\right|_{W^c} = \begin{pmatrix} \left|\frac{\partial F_1}{\partial |W_1|}\right|_{W^c} & \cdots & \left|\frac{\partial F_1}{\partial |W_6|}\right|_{W^c} \\ \vdots & \ddots & \vdots \\ \left|\frac{\partial F_6}{\partial |W_1|}\right|_{W^c} & \cdots & \left|\frac{\partial F_6}{\partial |W_6|}\right|_{W^c} \end{pmatrix}. \tag{17.40}$$

Then we can study the stability of different symmetries by substituting z_1, z_2, z_3, z_4, z_5, and W^c into the above 6×6 matrices. And we regard the real part m of the eigenvalue of S as a function of the parameter e. Here, we observe the stability of the amplitude system as the parameter e changes. It is obvious that the stability of spherical droplets in BCC lattice is determined by the real part of the eigenvalue μ_L, while the stability of planar lamellae is determined by μ_{BCC}. When $\mu(e) < 0$, the amplitude system remains stable. For the planar lamellae, we obtain three eigenvalues for the matrix S: $\mu_L^1 = -2\lambda_c$, $\mu_L^2 = \lambda_c\left(1 - z_4/z_2\right)$ and $\mu_L^3 = \lambda_c\left(1 - z_3/z_2\right)$. Similarly, we use the linearization matrix S to obtain three eigenvalues for BCC structured droplets, which are $\mu_{BCC}^1 = \lambda_c + W^c\left(-2z_1 + 3W^c\left(z_2 + z_4\right)\right)$, $\mu_{BCC}^2 = \lambda_c + W^{c2}\left(3z_2 + 4z_3 - z_4 - 2z_5\right)$, and $\mu_{BCC}^3 = \lambda_c + W^c\left(4z_1 + 3W^c\left(z_2 + 4z_3 + z_4 + 2z_5\right)\right)$, respectively. Figure 17.5 shows the relation between the real part of these eigenvalues and e as the control parameter. It can be seen from Fig. 17.5 thatthe state of the lamella becomes

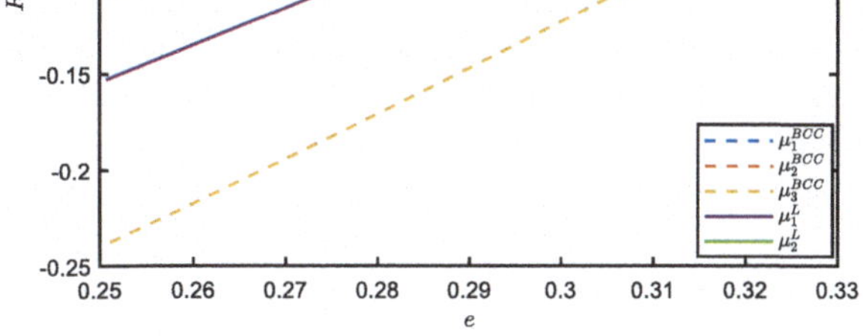

Fig. 17.5 The variation of the real part of the eigenvalue for the linearized amplitude system with the parameter e

stable as the real parts of the three eigenvalues remain less than zero for the lamella. For BCC droplet structure, only as $0.2993 < e < 0.3297$, the state becomes stable. That is to say, it generates bistable. And when the parameter e becomes less than 0.2993, only the lamella becomes stable.

17.6 Numerical Simulation

17.6.1 2D Pattern Formation

In the following, we use MATLAB software to carry out the numerical simulation for the spatial pattern of the system (17.3). First of all, the pattern formation for the system in a two-dimensional space (O-XY) is simulated numerically and gives help for deeply understanding the spatial distribution of the population. All ranges and step are assumed as below: $\Omega = (0, 120) \times (0, 120)$, the space step is $\Delta h = 1$, and the time step is $\Delta t = 0.1$, the boundary condition is the Neumann boundary condition, and the initial condition has a minor random disturbance for spatial uniform steady state. Since the similarity between the spatial distribution of prey and predator, the spatial patterns of prey are obtained in the following numerical simulation.

The system parameter values are fixed as

$$b = 0.4, \ c = 0.5, \ d = 0.25, \ m = 0.24, \ k_1 = 0.1, \ d_1 = 0.1, \ d_2 = 2, \quad (17.41)$$

and the values of e, f, and α are changed. So we can study the effect of the refuge protection, fractional diffusion index, fear effect on pattern formation.

Firstly, the parameter value is taken as $f = 0.4$, where the other system parameter values in Eq. (17.41) are fixed. For example, we take the critical value of control parameter e as 0.3299, the parameter e as 0.322, and the corresponding μ value as 0.0234 locating between μ_2 and μ_3. Figure 17.6a illustrates forming of the regular hexagon patterns gradually and finally stable occupying a two-dimensional space. When the controlled parameter e remains lowered to 0.285, there emerges the hexagon and stripe patterns for the prey population forms (see Fig. 17.6b). As the parameter $e = 0.26$, there appear the stripe patterns gradually after experiencing the initial irregular state. Figure 17.6c shows the smoothly gathered stripes and evenly distribution in the whole area. Simulation results indicate the assumption that the predator–prey system can generate Turing instability and also a self-organizing pattern. Then, the patterns varies as hexagon patterns $\rightarrow$ mixed patterns $\rightarrow$ stripe patterns when the protective effect of the refuge decreases from strong to weak. By observing Fig. 17.6 and c, it is easy to discover that the smaller protective effect as the prey refuge, the more risk to be eaten for the prey. On the contrary, the larger the refuge parameter e, the more easier escape for the prey, presenting spot patterns (hexagon patterns) at the same time.

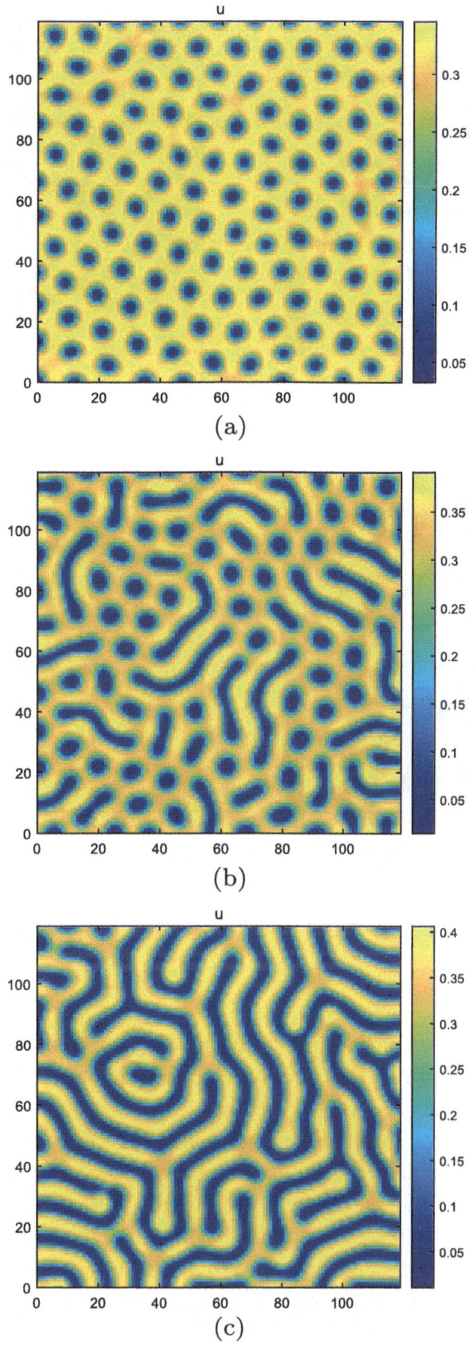

Fig. 17.6 Pattern structures of prey population with different parameters e. **a** $e = 0.322$; **b** $e = 0.285$; **c** $e = 0.26$

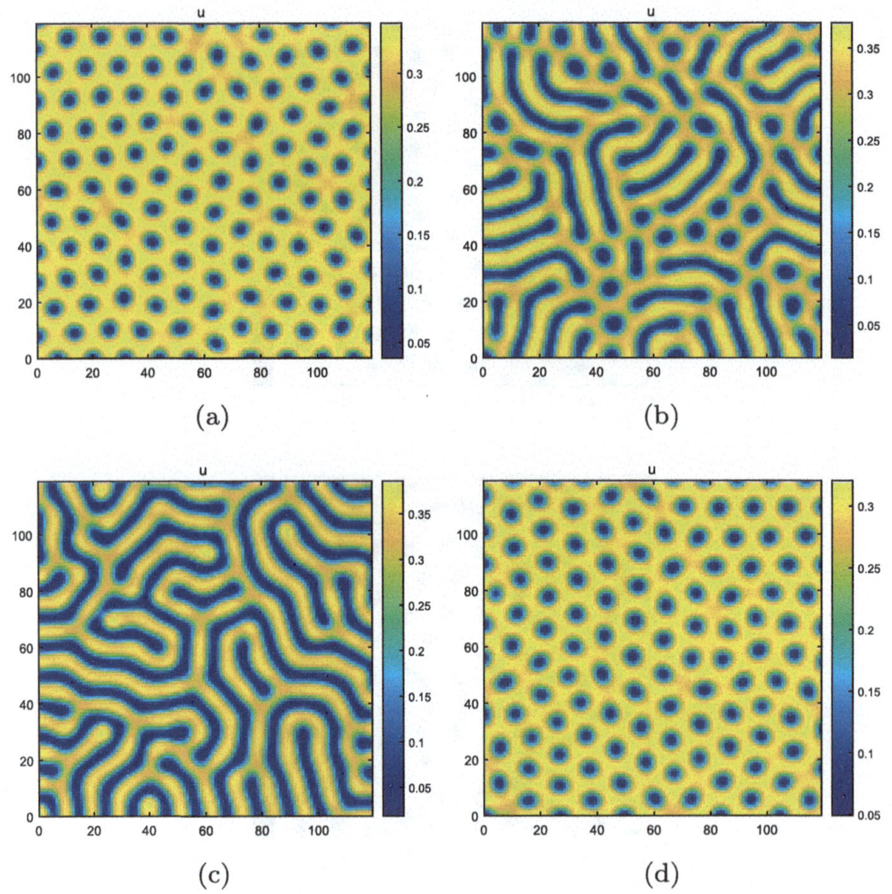

Fig. 17.7 Pattern structures of prey population after 40000 iterations with different f. **a** $f = 0.36,\ e = 0.322$; **b** $f = 0.68,\ e = 0.322$; **c** $f = 0.8,\ e = 0.322$; **d** $f = 0.8,\ e = 0.37$

Then, the effect of the fear term on pattern selection is also considered. And we fix the parameters in Eq. (17.41), and compare the simulation results for three groups of parameters ($f = 0.36$, $f = 0.68$, $f = 0.8$). Significantly, Fig. 17.7a–c shows the pattern structure variation gradually from spot(hexagon) patterns to stripe patterns after the mixing state as increasing the fear factor f. From an ecological point of view, strong fear effect makes the prey individuals gather together, meaning more anti-predator behavior for prey populations with the increase of fear. When we keep the parameter f fixed and take the value of the refuge protection parameter e as 0.37 standing for the enhanced protection effect of the refuge, Fig. 17.7d yields that the system retreats to the spot patterns. It is also shown that as the fear effect remains weak, the system emerges the spot pattern. Figure 17.7c illustrates that the stronger the fear effect, the more tendency of the stripe pattern for the system. These

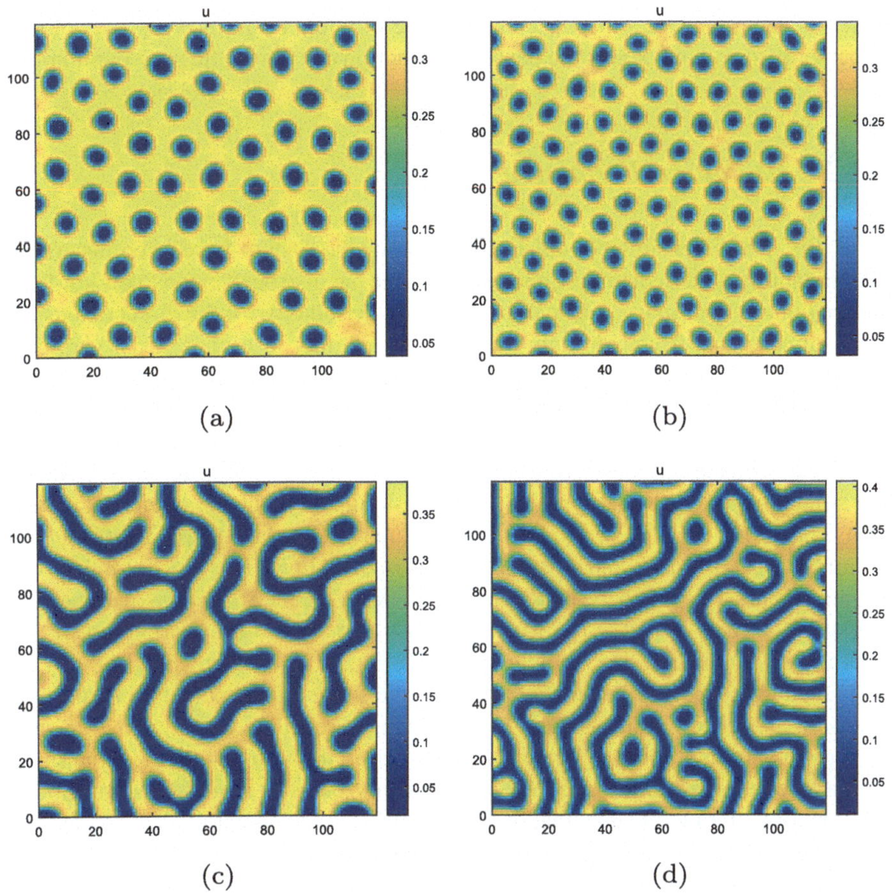

Fig. 17.8 Turing patterns of prey population developed by **a** $\alpha = 0.12$, $e = 0.322$; **b** $\alpha = 0.18$, $e = 0.322$; **c** $\alpha = 0.12$, $e = 0.26$; **d** $\alpha = 0.18$, $e = 0.26$, with other parameters of model (17.3) kept fixed at $b = 0.4$, $c = 0.5$, $d = 0.25$, $f = 0.4$, $k_1 = 0.12$, $D_1 = 0.1$, $D_2 = 2$ and $m = 0.24$

indicate that the fear effect plays a specific role in the population density such as that the greater the fear level in a specific range, the more unfavorable survival for the population.

Finally, the influence of fractional-order α is considered on the pattern formation for the system, where other parameters are equal to that of equation (17.41) and only the fractional-order α changes. It can be seen from Fig. 17.8 that the yellow width area of hexagon and stripe patterns decreases significantly with the increase of α. In addition, there exist the increasing of the critical wave number k_c and the decreasing of the wavelength of the solution significantly. For larger α, more hexagon and stripe patterns can be accommodated in this area because of the long-tailed mechanism for the fractional Laplace operator: the smaller the fractional diffusion index α deduces the larger the wavelength of the solution. From the view of ecology, the larger diffusion index gets the denser distribution density of the prey population.

17.6.2 3D Pattern Formation

So far, the two-dimensional simulation results for the system have been discussed and the stability pattern in three-dimensional space has also been extensively studied. In three dimensional space, we layer a plane with periodicity for the direct extension of two-dimensional fringes in only one direction. In this case, there exist variation from stripes and spots to lamellae and spherical droplets, respectively. In the three-dimensional simulation for the predator–prey model, we set the parameters as the same in two-dimensional simulation shown in Fig. 17.9. Among them, left side of Fig. 17.9 exhibits the concentration isosurface and its right size behaviors 3D pattern after simulating the system. Figure 17.9a, b are simulated as the parameter set for the "spots" pattern in Fig. 17.9a, and Fig. 17.9c, d are simulated as the parameter set for the mixed state with spots and stripes in Fig. 17.9b, and Fig. 17.9e, f are simulated as the parameters set for the stripe pattern in Fig. 17.9c. We reduce the range of numerical simulation to $\Omega = (0, 30) \times (0, 30) \times (0, 30)$ to survey the transformation of 3D pattern more intuitively. The content shown in Fig. 17.10 is corresponding to Fig. 17.9. Obviously, it seems that all patterns of mixed stripes and spots remain transient. That is to say, as the simulation lasts long enough, there will eventually emerge only one type of pattern dominating the system. However, it doesn't seem to be the same case in three-dimensional space. We can evidently observe more abundant modes and types of pattern in 3D than that of 2D with the same fixed parameters. As for the final pattern state, it yields more sensitive in 3D than in 2D.

It can be speculated that the pattern transformation in 3D will be different at the same point as in 2D with the alteration of parameter e, because the third dimension supplies more freedom for the clustering process. For example, we cannot predict the transition region by bifurcation analysis in the three-dimensional case. The transition occurs at a lower e value, as the nonlinear coupling remains more beneficial for the layered structure than the BCC structure, and increases the connectivity of three-dimensional space. Figure 17.11 presents a bistable pattern state that as $e < 0.299$, we get the stability of the spherical droplet for the BCC structure and the stability for the planar lamellae. When the control parameter e varies, the stable spherical droplet emerges unstable, but the planar lamellae remain stable. At the same time, the state of the system varies from bistable to monostable. Significantly, the planar lamellae pattern remains always stable. Compared with the difference for planar lamellae in Fig. 17.11e, it yields that the distance between the planar lamellae decreases in the same space, and the number of planar lamellae increases when the wavelength of the solution gets smaller shown in Fig. 17.11f–h. Ecologically, prey populations tend to gather tightly to resist predation with the fear effect on reducing sanctuary protection in the BCC structure. In detail, reducing the value of the fear parameter f can re-disperse the prey population into small groups. Figure 17.12 displays

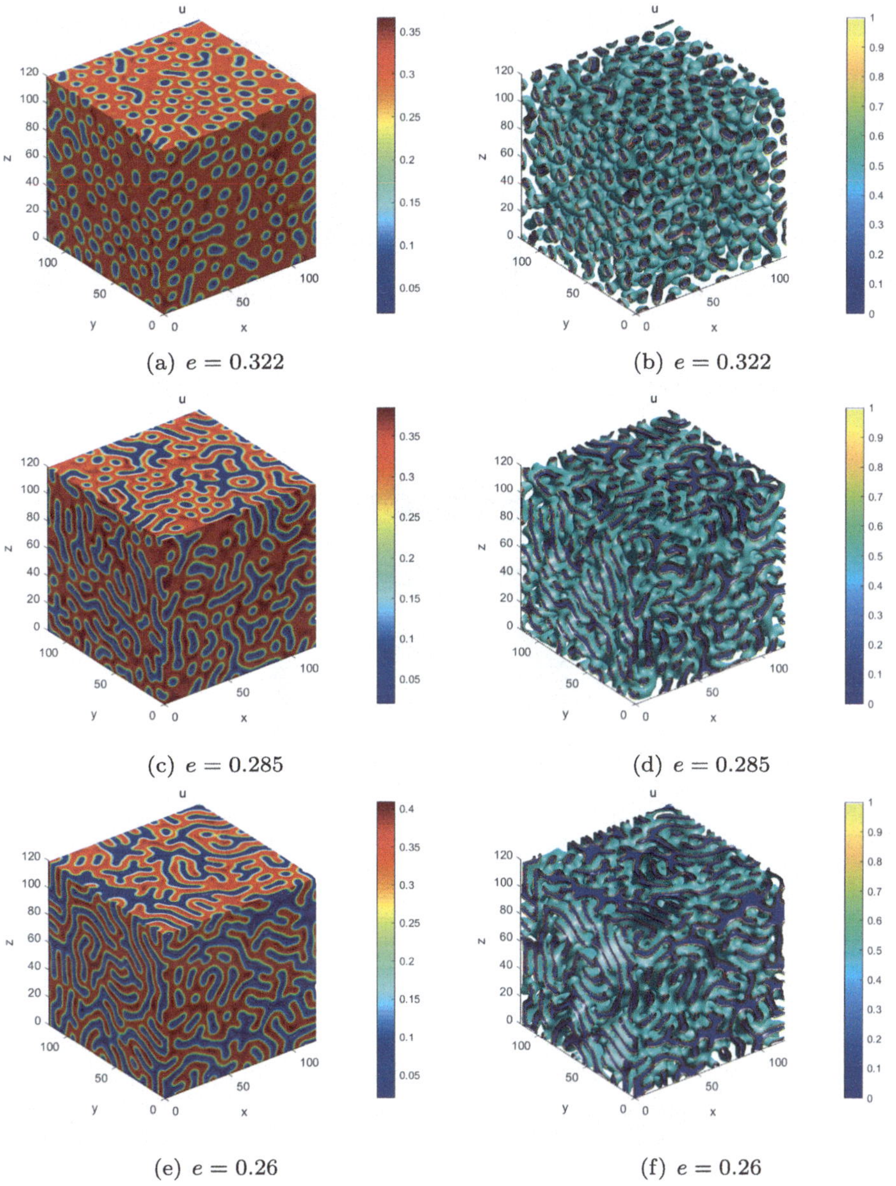

(a) $e = 0.322$ (b) $e = 0.322$

(c) $e = 0.285$ (d) $e = 0.285$

(e) $e = 0.26$ (f) $e = 0.26$

Fig. 17.9 Variation of the Turing patterns of prey population with other parameters of model (17.3) kept fixed at $b = 0.4$, $c = 0.5$, $d = 0.25$, $f = 0.4$, $k_1 = 0.12$, $D_1 = 0.1$, $D_2 = 2$, $m = 0.24$, and $t = 4000$. **a**, **c**, and **e** mean the equivalent surface of prey population; **b**, **d**, and **f** stand for three-dimensional patterns of prey population

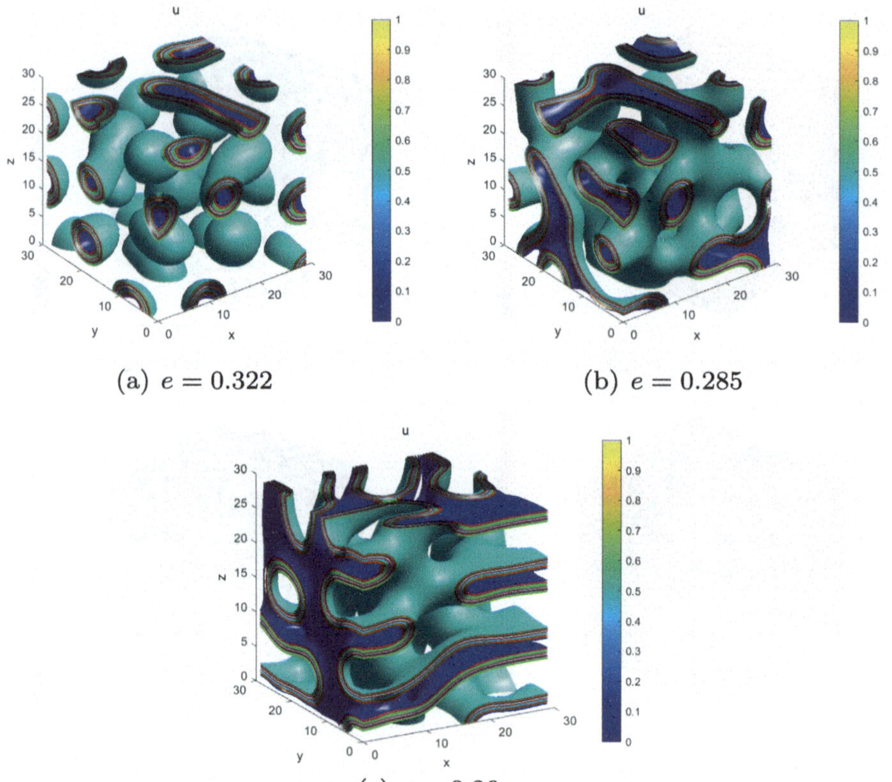

(a) $e = 0.322$

(b) $e = 0.285$

(c) $e = 0.26$

Fig. 17.10 Variation of the Turing patterns of prey population with other parameters of model (17.3) kept fixed at $b = 0.4$, $c = 0.5$, $d = 0.25$, $f = 0.4$, $k_1 = 0.12$, $D_1 = 0.1$, $D_2 = 2$, $m = 0.24$, and $t = 4000$

the phenomenon that the pattern formation is re-dispersed into spherical droplets. Although we can easily obtain the stability of planar lamellae, it becomes impossible for analyzing the stability of twisted layered surfaces with any symmetric conditions.

Moreover, Fig. 17.13 reveals the three-dimensional spatial pattern for the prey population with different fractional diffusion orders. We set the coexistence equilibrium point as $(U_*, V_*) = (0.2630, 0.2964)$, parameters $e = 0.322$, $b = 0.4$, $c = 0.5$, $d = 0.25$, $f = 0.4$, $k_1 = 0.12$, $D_1 = 0.1$, $D_2 = 2$, $m = 0.24$, and get the three-dimensional patterns for four different diffusion indexes α (1.09, 1.42, 1.71, and 2). As can be illustrated from Fig. 17.13, these patterns display spherical droplets. And we assume the critical wavenumber k_c for the small fractional diffusion order small, its wavelength remains large, and the spherical droplets remain sparse and large. Then the wavelength of the non-homogeneous stationary solution for the first bifurcation has negative relationship with α. Therefore, the order of fractional diffusion significantly influences the three-dimensional patterns.

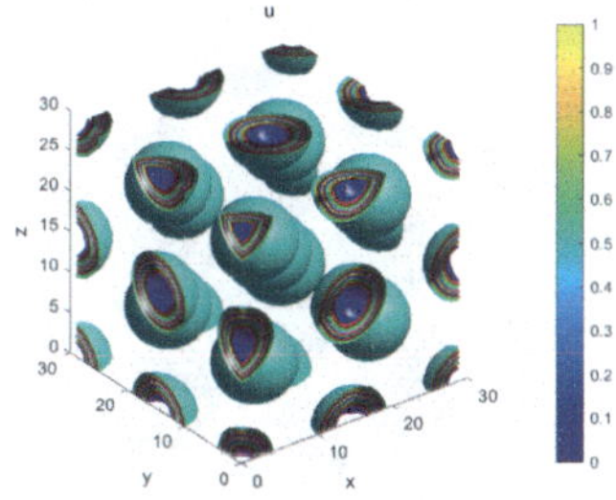

(a) $e = 0.322$

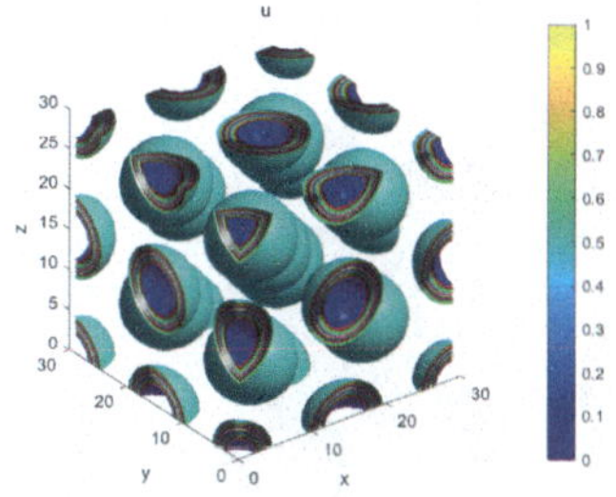

(b) $e = 0.2994$

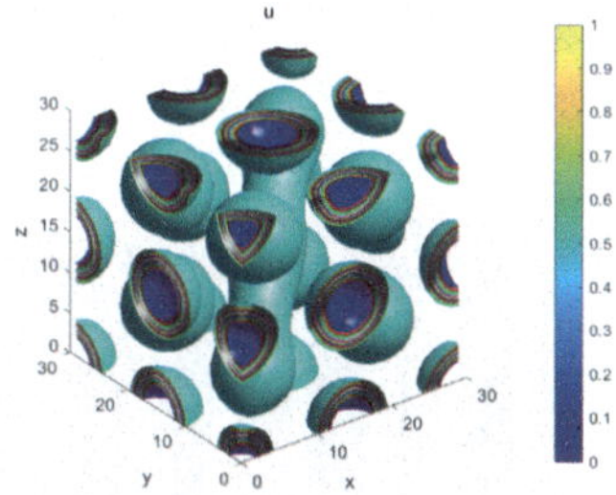

(c) $e = 0.2993$

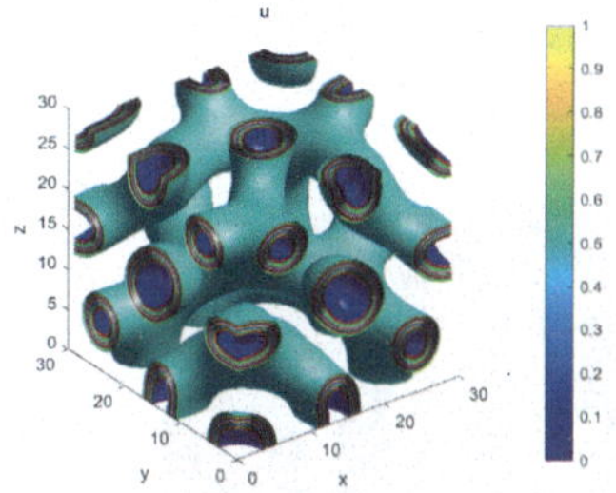

(d) $e = 0.285$

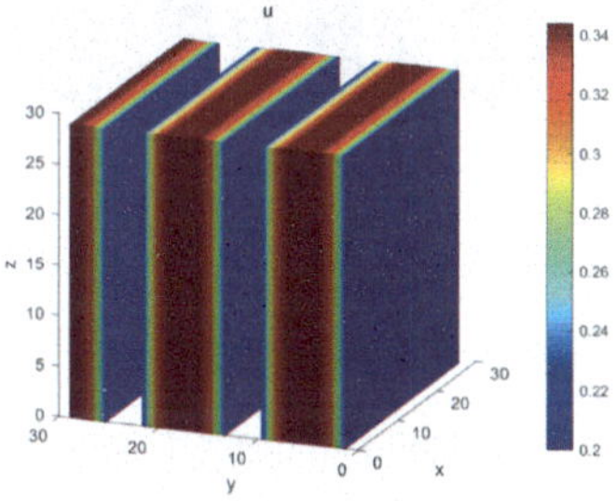

(e) $e = 0.322$

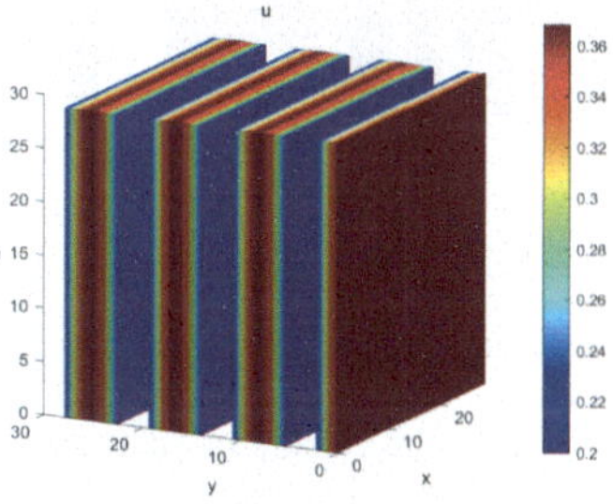

(f) $e = 0.2994$

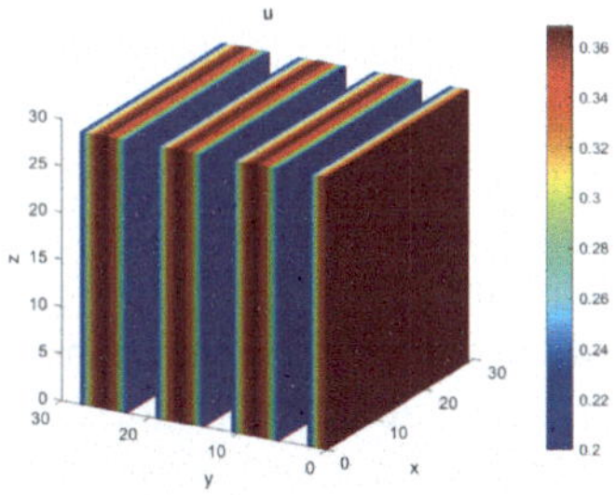

(g) $e = 0.2993$

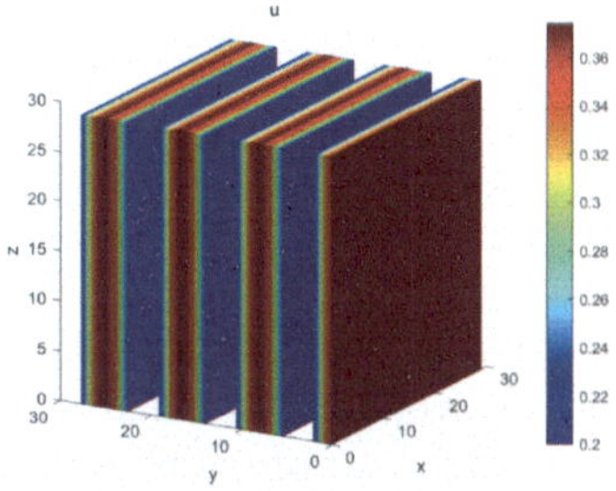

(h) $e = 0.285$

Fig. 17.11 Changes in the three-dimensional Turing model for the prey population. **a**–**d** mean the change of BCC spherical droplet with parameter e; **e**–**f** stand for the change of planar lamellae with parameter e. Other parameters of model (17.3) are kept fixed at $b = 0.4,\ c = 0.5,\ d = 0.25,\ f = 0.4,\ k_1 = 0.12,\ D_1 = 0.1,\ D_2 = 2, m = 0.24$, and $t = 4000$

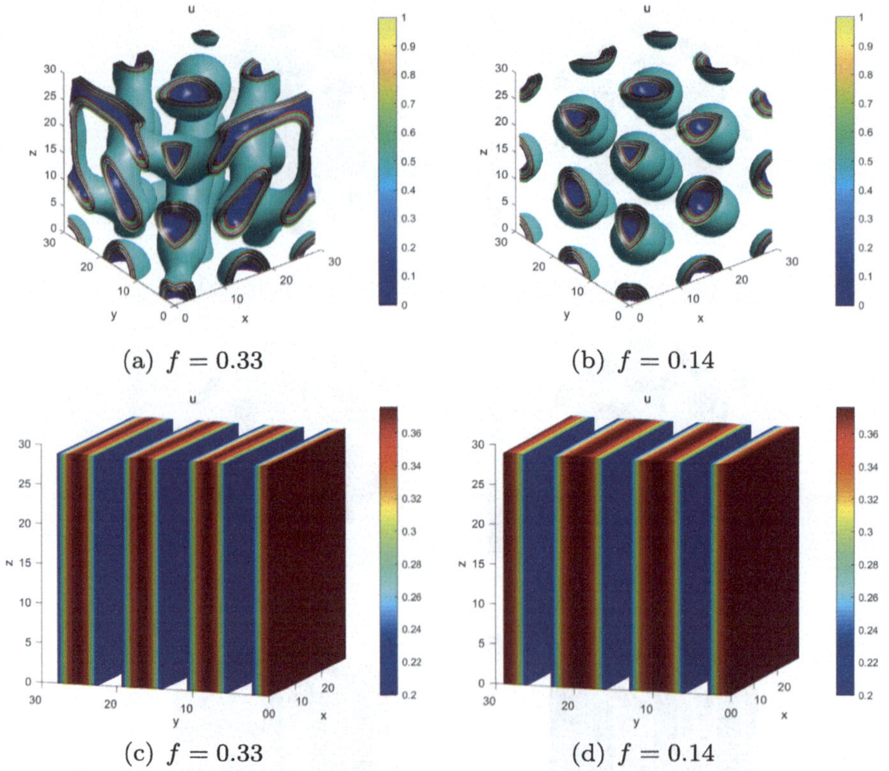

Fig. 17.12 Variation of the Turing patterns of prey population with other parameters of model (17.3) kept fixed at $b = 0.4$, $c = 0.5$, $d = 0.25$, $e = 0.285$, $k_1 = 0.12$, $D_1 = 0.1$, $D_2 = 2$, and $m = 0.24$. **a–b** mean the change of BCC spherical droplet with parameter f; **c–d** stand for the change of planar lamellae with parameter f

Considering the phase space for the system for understanding the pattern selection [450], we observe that the model parameters can control the dynamics and evolution of a particular form. Furthermore, we can find many possible three-dimensional structures to meet the requirements of nonlinear symmetry. The closer of the initial structure to the state satisfying the symmetry requirements gets the smaller of the phase space explored by the system [451]. If we use the initial conditions for spatial heterogeneity in favor of a particular structure, it may yield a limitation of the system dynamics to a small part of the phase space. And we can also obtain the expected optimal species density pattern. Similarly, the droplet structure with BCC symmetry can be lead by the BCC symmetry in the initial condition.

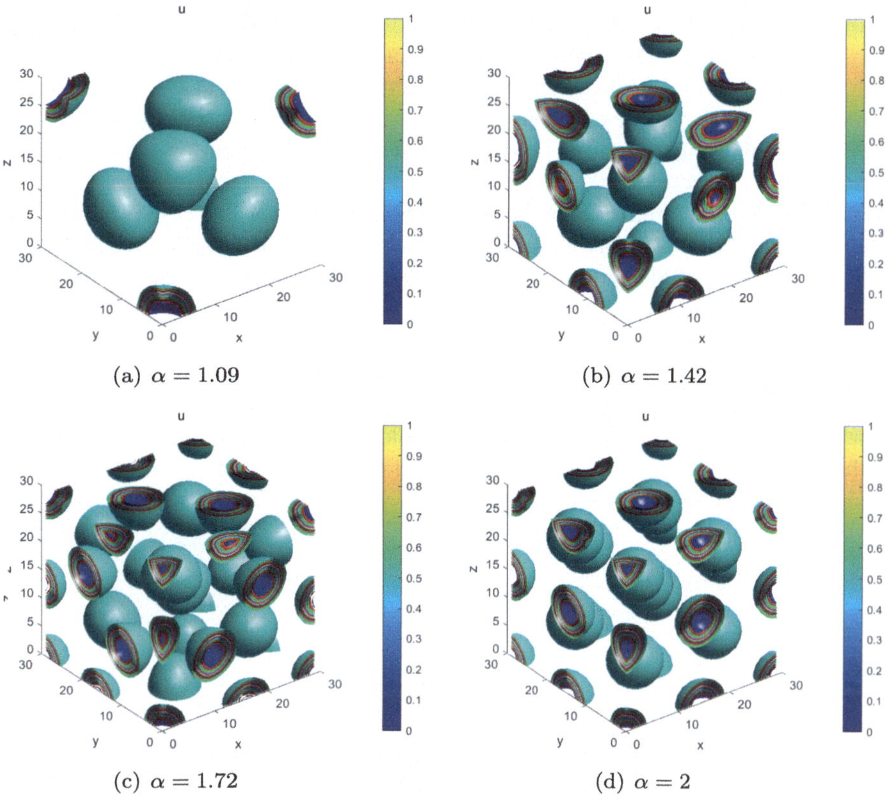

(a) $\alpha = 1.09$ (b) $\alpha = 1.42$

(c) $\alpha = 1.72$ (d) $\alpha = 2$

Fig. 17.13 The variation of spherical droplet ($e = 0.32$) with fractional diffusion order α

17.7 Discussion and Conclusions

In this chapter, the effect of fear, Allee and refuge protection becomes our focus for the pattern dynamics of a fractional predator–prey model. We use the amplitude equation and explore the selection and stability for different modes in 2D and 3D near the beginning of Turing instability. For the sake of understanding the mode selection for the system, we conduct a rigorous nonlinear bifurcation analysis mathematically. And the reduced dynamics of the center manifold is described by the amplitude equation and the 3D Turing bifurcation analysis for studying the whole system dynamics. The results indicate that the bifurcation analysis gives the parameter region according to the specific symmetry, and the numerical simulation results are coincide with the theoretical results in two dimensional and three-dimensional systems.

In summary, the pattern formation has been studied for 2D and 3D Turing models with different control parameters and initial conditions. By changing the nonlinear parameters, we can see the control effect of the general Turing model on the formation of patterns and the exploration of the pattern space generated by the system, so that it yields exciting and rich patterns. For instance, in the density distribution of prey population, as the control parameter refuge e varies, stripes patterns, spots patterns and the mixed transition between them are observed in two-dimensional space. Abundantly, there exist many structures in three dimensions like the regular BCC spherical droplet structure, perfect layered structure, regular tube structure, and twisted tubular structure.

The results exhibit that the consistency in the two-dimensional case is better than that in the three-dimensional case. The reason is that there emerges a simple transition between the monostable and the bistable mode in the two-dimensional case. As for the three-dimensional case, the analytical approximation of connectivity changes remains more difficult as all parameter values can predict a bistable state for the system. Although the BCC structure in the Turing system remains stable, the system seems to be unable as initial configuration is random. Finally, it is clear that we can observe spontaneous formation of various stable patterns and the widespread effect of the Turing system for establishing order. All these modes can be determined and predicted by system parameters, initial Values, and linear analysis, respectively.

Chapter 18
Spatial Dynamics with Time Delay and Allee Effect

18.1 Introduction

As we all know, how to effectively control the long-term coexistence for predator and prey remains essential in protecting biodiversity and economic benefits. Hence, scholars have studied the predator–predator model and achieved fruitful results through ordinary differential equations [465, 486]. But there exist many limitations. For example, spatial diffusion in the real world is ubiquitous and can generate rich temporal and spatial dynamics. In ecology, experimental results indicate that models with diffusion term can achieve rich dynamics such as pattern formations in spatial, temporal, and spatiotemporal spaces, like Turing instability [451], Hopf bifurcation [413], and Turing-Hopf bifurcation [461]. In addition, the effect of the cross-diffusion on the predator–prey system has been studied for pattern formation. For example, the model dynamics of the phytoplankton zooplankton model was produced by introducing diffusion toxin with three-dimensional patches [250]. And the Turing pattern was also generated by cross-diffusion for group predator–prey behavior in the predator–prey model [489].

Within the past decades, researchers pay more attention for the field of fractional calculus in many different fields [483–485]. And there are more interesting studies on the movement of animals with larger spatial scales or longer time scales modeled by a new random walking model called Lévy flight [473, 480], which plays significant role in pattern formation. For example, the Turing instability for pattern formation in the super cross-diffusion predator–prey system can be harvested by Michaelis-Menten predators [380]. The weakly nonlinear dynamics was described in a superdiffusion reaction system with Lévy flight characteristics and the phase diagrams for different states were observed like phase turbulence, chaos and spiral waves generated by super diffusion [470]. Reference [381] considered the role of time delay and cross-fractional diffusion in the dynamic behavior and then studied a predator–prey model with intraguild predation. Finally, Ref. [381] proved that appropriate cross-fractional diffusion term can generate Turing patterns and also indicated that the smaller of the

S. T. Liu and L. Zhang, *Pattern Dynamics of Marine Plankton Behavior*,
https://doi.org/10.1007/978-981-97-5369-7_18

order of fractional diffusion gets the easier of the appearance for Turing patterns. The predator–prey model with superdiffusion was studied and the result like the variation of the superdiffusion index $\alpha(\nabla^{\alpha})$ with the properties of the solution and its controlling and changing effect on the dynamic behavior of the system was also obtained [446]. Therefore, a fractional reaction–diffusion system can be used to best explain the process of pattern formation.

On the other hand, Allee effect, widely observed in nature, may generally increase the extinction risk of low-density populations and also increase the complexity of the model [467]. Its influence has been deeply studied on the spatial pattern for the population [468, 469, 472, 476]. The Hopf bifurcation and Allee effect with time delay were studied for a diffusive predator–prey system with prey chemotaxis and herd behavior [468]. Reference [472] discussed the Allee effect on a nonlinear predator–prey model for two populations, and obtained the spread of various new species for two populations and spatiotemporal chaos led by predator with the Allee effect. Predators cooperative hunting and prey growth were studied for their formation and pattern selection in a predator–prey model by the Allee effect [476]. It is well known that predators' functional response and mortality play a very vital role in the dynamic behavior for the predator–prey system and there have conducted many exciting results [463, 474, 475, 490, 492]. For example, a predator–prey model was studied with Holling-II functional response and hyperbolic mortality [474]. The local stability for the ordinary differential system, the self-diffusion effect on the direction of the Hopf bifurcation, and the stability for the periodic solution were also investigated at the same time.

In recent years, the predator–prey model with time delay has become a hot topic [466, 477, 482, 488]. Importantly, the behavior of a population may be affected not only by its current information but also by the historical information of the past. So introducing the time delay term such as a delay differential equation model is useful to effectively describe the time delay activity in a specific time or space range. Additionally, the time delay term can cause major changes of population dynamics so that it can be considered as a significant factor for better explaining the nonlinear phenomenon of population dynamics, especially the predator-detainee system [464]. First of all, the time delay usually leads the ecological system to be instable and branched like the Hopf bifurcation. Secondly, it can cause multiple attractors in the system. Hence, the delay term has been introduced into the predator–prey system and some interesting results have also obtained [460, 462, 471, 478, 479, 491, 493]. For instance, the nutrient microorganism model with time delay term was presented and then the formation of heterogeneous distribution for bacteria and nutrients in sediments were obtained and the dynamics of Hopf-Hopf bifurcation was had [460]. Reference [461] studied the Turing-Hopf bifurcation characteristics for a time-delay diffusion predator–prey system corresponding to chemotaxis and fear effects. Thus, the time delay term in the related dynamic model plays a necessary role in affecting the dynamic behavior for the population. Based on these literature, time-delay effects have important practical significance for the dynamics for the predator–prey system. Hence, we present a predator–prey model with time delay term and study the role of time delay in pattern formation.

This chapter is organized as below. At first, we present a more realistic mathematical model considering factors like time delay, fractional diffusion, Allee effect, refuge protection, and the relationship between predator–prey population. Then, we study the stability for positive equilibrium and the existence for Hopf bifurcation with or without time delay and fractional diffusion. And the stability and the direction of the bifurcation periodic solution are also determined. Next, we simulate corresponding three-dimensional space pattern by MATLAB which support and prove our theoretical results. Finally, we discussed and gave some conclusions.

18.2 Fractional-Order Mathematical Model with Time Delay and Allee Effect

Pattern formation is significant for better understanding the spatiotemporal population dynamics of the predator–prey model in the real ecological interaction environment. Moreover, the study of Allee effect in the predator–prey model with time delay term has great ecological crucial. So our focus in this chapter is the role of time delay and the Allee effect in the pattern. First of all, based on the predator–prey model with hyperbolic mortality [474], and combined the Allee effect into the model, we get the model as follows

$$\begin{cases} \frac{\partial X}{\partial T} = rX\left(1 - \frac{X}{K}\right) - \frac{a_1 X}{m_1 + X} - \frac{b_1 XY}{X + k_1}, \\ \frac{\partial Y}{\partial T} = \frac{b_2 XY}{X + k_1} - \frac{\mu Y^2}{\xi + \eta Y}, \end{cases} \tag{18.1}$$

where X and Y represent the population density of prey and predator respectively, r means the birth rate, K stands for the carrying capacity, b_1 means the maximum intake rate of the prey, k_1 indicates the degree of protection provided by the predator, and b_2 means the predator's growth rate, μ stands for the death rate of the predator, ξ and η are also constants. Ω is the internal connected region with a smooth boundary $\partial\Omega$, n means the outward unit normal vector on $\partial\Omega$. We take the homogeneous Neumann boundary condition meaning the independence of the above system, and there is no population on the boundary.

Assume

$$X = Ku,\ Y = \frac{K}{b_1}v,\ T = \frac{t}{r},\ b = \frac{a_1}{rK},\ m = \frac{m_1}{K},$$
$$s = \frac{1}{r},\ c = \frac{k_1}{K},\ \mu = b_2\eta,\ \xi = \eta,\ a = \frac{b_2}{r},\ \beta = \frac{K}{b_1}.$$

Then it yields Eq. (18.1) after dimensionless as

$$\begin{cases} \frac{\partial u}{\partial t} = u\left(1 - u - \frac{b}{u+m}\right) - \frac{suv}{u+c}, \\ \frac{\partial v}{\partial t} = \frac{auv}{u+c} - \frac{a\beta v^2}{1+\beta v}. \end{cases} \tag{18.2}$$

Because of the importance of the foraging in the survival of many animal populations, many works study their optimized foraging behavior. For example, data sets comprising more than 10^6 data points displayed convincing evidence of Lévy-like behavior. Random-walk simulations of "model sharks" were used to study the relative success of shark searches comparing with the sharks in a dynamic prey landscape with biomass prey (zooplankton) in the north-east Atlantic Ocean. Biological Lévy flight foraging was applied to concern the flight times for the dinoflagellate Oxyrrhis marina. The results suggested a stronger helical component combination with the Lévy search strategy in a three-dimensional environment which enhanced the relevant encounter rates. This urges us to introduce fractional diffusion term in a reactive diffusion system. According to experimental data and analysis results in Ref. [394], we introduce the three-dimensional fractional diffusion into the reaction–diffusion system, where the fractional diffusion term illustrates the Lévy-walk-like search hunting behavior exhibited by predators in the environment. After the attacks of the predator for the prey, the last diffuses becomes very fast. And then there exists Lévy-like fractal patterns displayed by prey density distributions, which suggests predator response movements according to the prey distributions. Behaviors generated by different marine predators–sharks, turtles, bony fishes, and penguins exhibit Lévy-walk-like search and predation were close to the theoretical optimal. Hence, we rewrite the system (18.2) considering the spatial diffusion of the population as:

$$\begin{cases} \frac{\partial u}{\partial t} = u\left(1 - u - \frac{b}{u+m}\right) - \frac{suv}{u+c} + d_{11}\nabla^{\alpha}u + d_{12}\nabla^{\alpha}v, \\ \frac{\partial v}{\partial t} = \frac{auv}{u+c} - \frac{a\beta v^2}{1+\beta v} + d_{21}\nabla^{\alpha}u + d_{22}\nabla^{\alpha}v, \end{cases} \tag{18.3}$$

where d_{11} and d_{22} mean the self-diffusion coefficients for prey and predator, and d_{12} and d_{21} stand for the cross-diffusion coefficients for predator and prey, respectively. We define the fractional Laplace term $\nabla^{\alpha}u$ as

$$\begin{aligned} \nabla^{\alpha}u &= \frac{\partial^{\alpha}u}{\partial|x|^{\alpha}} + \frac{\partial^{\alpha}u}{\partial|y|^{\alpha}} + \frac{\partial^{\alpha}u}{\partial|z|^{\alpha}} \\ &= -\frac{1}{2\cos(\pi\alpha/2)}\left({}_{RL}D^{\alpha}_{-\infty,x}u + {}_{RL}D^{\alpha}_{x,+\infty}u\right) - \frac{1}{2\cos(\pi\alpha/2)}\left({}_{RL}D^{\alpha}_{-\infty,y}u + {}_{RL}D^{\alpha}_{y,+\infty}u\right) \\ &\quad - \frac{1}{2\cos(\pi\alpha/2)}\left({}_{RL}D^{\alpha}_{-\infty,z}u + {}_{RL}D^{\alpha}_{z,+\infty}u\right), \end{aligned}$$

${}_{RL}D^{\alpha}_{-\infty,x}u$ and ${}_{RL}D^{\alpha}_{x,+\infty}u$ are, respectively, defined as:

$$\begin{aligned} {}_{RL}D^{\alpha}_{-\infty,x}u &= \frac{1}{\Gamma(2-\alpha)}\frac{\partial^2}{\partial x^2}\int_{-\infty}^{x}(x-s)^{1-\alpha}u(s,y,z,t)\,ds, \\ {}_{RL}D^{\alpha}_{x,+\infty}u &= \frac{1}{\Gamma(2-\alpha)}\frac{\partial^2}{\partial x^2}\int_{x}^{+\infty}(s-x)^{1-\alpha}u(s,y,z,t)\,ds. \end{aligned}$$

where $\Gamma(\cdot)$ represents the Gamma function.

Finally, taking into consideration that the predator has a pregnancy period and cannot reproduce immediately after predation, it is more realistic to introduce the Allee effect, hyperbolic mortality and time delay into the predator–prey model (18.3) listed as below:

$$\begin{cases} \frac{\partial u}{\partial t} = u\left(1 - u - \frac{b}{u+m}\right) - \frac{suv}{u+c} + d_{11}\nabla^{\alpha}u + d_{12}\nabla^{\alpha}v, \\ \frac{\partial v}{\partial t} = \frac{au(t-\tau)v}{u(t-\tau)+c} - \frac{a\beta v^2}{1+\beta v} + d_{21}\nabla^{\alpha}u + d_{22}\nabla^{\alpha}v, \end{cases} \tag{18.4}$$

where τ means the time delay.

18.3 Dynamic Analysis with $\tau = 0$

18.3.1 Without Spatial Diffusion Term

First of all, let $\tau = 0$ and we write the system (18.4) as

$$\begin{cases} \frac{\partial u}{\partial t} = u\left(1 - u - \frac{b}{u+m}\right) - \frac{suv}{u+c} \\ \qquad +d_{11}\nabla^{\alpha}u + d_{12}\nabla^{\alpha}v, \\ \frac{\partial v}{\partial t} = \frac{auv}{u+c} - \frac{a\beta v^2}{1+\beta v} + d_{21}\nabla^{\alpha}u + d_{22}\nabla^{\alpha}v. \end{cases} \tag{18.5}$$

Obviously, without no diffusion term in the model (18.5), that is, $d_{11} = d_{12} = d_{21} = d_{22} = 0$, we can obtain the ordinary differential equations as follows:

$$\begin{cases} \frac{du}{dt} = u\left(1 - u - \frac{b}{u+m}\right) - \frac{suv}{u+c}, \\ \frac{dv}{dt} = \frac{auv}{u+c} - \frac{a\beta v^2}{1+\beta v}. \end{cases} \tag{18.6}$$

Let $F(u, v) = u\left(1 - u - \frac{b}{u+m}\right) - \frac{suv}{u+c}$, $G(u, v) = \frac{auv}{u+c} - \frac{a\beta v^2}{1+\beta v}$. And viewing the ecological significance, we mainly investigate the stability of the positive equilibrium point for the model (18.5). After solving it from Descartes' rule, there exists only a positive equilibrium point $E_* = (u_*, v_*)$ for the system (18.5) and the system (18.6) when $b < m < c$ holds, where $F(u_*, v_*) = 0$, $G(u_*, v_*) = 0$, that is

$$\begin{cases} u\left(1 - u - \frac{b}{u+m}\right) - \frac{suv}{u+c} = 0, \\ \frac{auv}{u+c} - \frac{a\beta v^2}{1+\beta v} = 0. \end{cases}$$

Obtained $v_* = \frac{1}{c\beta}u_*$, $1 - u_* - \frac{b}{m+u_*} - \frac{sv_*}{c+u_*} = 0$, and u_* satisfying the following cubic equation:

$$u^3 + B_1u^2 + B_2u + B_3 = 0,$$

where $B_1 = \frac{c^2\beta+c\beta m+s-c\beta}{c\beta}$, $B_2 = \frac{c^2\beta m+bc\beta+ms-c^2\beta-c\beta m}{c\beta}$, $B_3 = c(b-m)$.
The Jacobian matrix at E_* for the model (18.6) is

$$J_* = \begin{pmatrix} a_{11} & a_{12} \\ a_{21} & a_{22} \end{pmatrix}.$$

Here

$$\begin{aligned} a_{11} &= -u_* + \frac{bu_*}{(m+u_*)^2} + \frac{su_*^2}{c\beta(c+u_*)^2}, \\ a_{12} &= -\frac{su_*}{c+u_*}, \ a_{21} = \frac{au_*}{\beta(c+u_*)^2}, \ a_{22} = -\frac{acu_*}{(c+u_*)^2}. \end{aligned}$$

We can calculate the characteristic equation as

$$\lambda^2 - A\lambda + B = 0, \tag{18.7}$$

where

$$A = -u_* + \frac{bu_*}{(m+u_*)^2} + \frac{su_*^2}{c\beta(c+u_*)^2} - \frac{acu_*}{(c+u_*)^2},$$

$$B = -\frac{acu_*}{(c+u_*)^2}\left(\frac{bu_*}{(m+u_*)^2} + \frac{su_*^2}{c\beta(c+u_*)^2} - u_*\right) + \frac{asu_*^2}{\beta(c+u_*)^3}.$$

Let $A < 0$ and $B > 0$, and then we have negative real parts of the roots of the characteristic Eq. (18.7) and the locally asymptotically stability for the equilibrium point $E_* = (u_*, v_*)$ of the system (18.4). For a pair of conjugate complex roots $\lambda = q_1(b) + iq_2(b)$ for the characteristic (18.7), we introduce them into the Eq. (18.7) with the separation for the real part and imaginary part. We obtain

$$\begin{cases} q_1^2 - q_2^2 - Aq_1 + B = 0, \\ 2q_1q_2 - Aq_2 = 0. \end{cases}$$

After deriving the parameter b from both sides for the above equation, we have the following results:

$$\left.\frac{dq_1(b)}{db}\right|_{b=b_H} = \frac{u_*}{2(m+u_*)^2} > 0.$$

Hence, a Hopf bifurcation will be generated from the system (18.6) at the equilibrium point E_* as the parameter $b = b_H$, $A = 0$, $B > 0$ holds. Therefore, we obtain the following conclusions.

Theorem 64 *Assume that $b < m < c$ holds. For model (18.6),*

(i) Its positive equilibrium point E_* becomes unstable if $A > 0$.
(ii) Its positive equilibrium point E_* becomes locally asymptotically stable if $A < 0$ and $B > 0$.
(iii) the system will get a Hopf bifurcation at the positive equilibrium point E_* if $b = b_H$, such that $A < 0$ and $B > 0$.

18.3.2 With the Spatial Diffusion Term

Next, the stability of model (18.5) is considered and the following three-dimensional Fourier form for the solution is chosen :

$$\begin{pmatrix} u \\ v \end{pmatrix} = e^{\lambda_k t + i\mathbf{k}\cdot\mathbf{r}},$$

where, λ_k means the growth rate of the perturbation, $\mathbf{k}$ represents the modulus, and $\mathbf{r}$ stands for the direction vector.

In the following, the spectral definition of the fractional Laplace operator is considered defined by the following Fourier spectral method:

$$\nabla^{\alpha} u(x) = -\sum\nolimits_{j=0}^{\infty} \hat{u}_j \lambda_j^{\frac{\alpha}{2}} \varphi_j,\ u \in I_{\alpha}, \tag{18.8}$$

where, $I_{\alpha} := \left\{ u = \sum\limits_{j=0}^{\infty} \hat{u}_j \varphi_j,\ \hat{u}_j = \langle u, \varphi_j \rangle,\ \sum\limits_{j=0}^{\infty} \left| \hat{u}_j \right|^2 \left| \lambda_j \right|^{\frac{\alpha}{2}} < \infty \right\}$, λ_j and φ_j represent the eigenvalues and orthogonal eigenfunctions for the standard Laplace operator ∇^2, respectively, and the standard boundary condition is on the bounded region $\Omega \subset \mathbb{R}^n$, that is, $\nabla^2 \varphi_j = -\lambda_j \varphi_j$. The conditions are listed as in a one-dimensional domain with length L, with Neumann boundary conditions, and satisfying the fractional Laplace operator with eigenpairs $(\varphi_j, \lambda_j) = (\cos(j\pi x/L), (j\pi/L)^{\alpha})$ in [449].

Then

$$J = \begin{pmatrix} a_{11} - d_{11}k^{\alpha} & a_{12} - d_{12}k^{\alpha} \\ a_{21} - d_{21}k^{\alpha} & a_{22} - d_{22}k^{\alpha} \end{pmatrix}.$$

Hence, we can obtain the dispersion relation:

$$\lambda_k{}^2 - A_k \lambda + B_k = 0, \tag{18.9}$$

where

$$\begin{aligned} A_k &= A - k^{\alpha}\left(d_{11} + d_{22}\right), \\ B_k &= \left(d_{11}d_{22} - d_{12}d_{21}\right)k^{2\alpha} - \left(a_{11}d_{22} + a_{22}d_{11} - a_{12}d_{21} - a_{21}d_{12}\right)k^{\alpha} + B. \end{aligned}$$

When the equilibrium E_* without diffusion term remains stable, $A < 0$, $B > 0$, and $A_k = A - k^{\alpha}(d_{11} + d_{22})$, we get $A_k < 0$ obviously. Therefore, with a fractional diffusion, we can observe the variation from the stable to unstable for the positive equilibrium point E_*, which requires $B_k < 0$. Assume $d_{11}d_{22} - d_{12}d_{21} > 0$, and B_k as a quadratic function about k^{α}:

$$\begin{aligned} B_k &= \left(B - \tfrac{\left(a_{11}d_{22} + a_{22}d_{11} - a_{12}d_{21} - a_{21}d_{12}\right)^2}{4\left(d_{11}d_{22} - d_{12}d_{21}\right)}\right) \\ &+ \left(k^{\alpha}\sqrt{d_{11}d_{22} - d_{12}d_{21}} - \tfrac{a_{11}d_{22} + a_{22}d_{11} - a_{12}d_{21} - a_{21}d_{12}}{2\sqrt{d_{11}d_{22} - d_{12}d_{21}}}\right)^2. \end{aligned}$$

Obviously, when B_k takes the minimum value, we have

$$k^{\alpha}=\frac{a_{11}d_{22}+a_{22}d_{11}-a_{12}d_{21}-a_{21}d_{12}}{2\left(d_{11}d_{22}-d_{12}d_{21}\right)},$$

and the critical wave number of Turing bifurcation as

$$k_c=\left(\frac{a_{11}d_{22}+a_{22}d_{11}-a_{12}d_{21}-a_{21}d_{12}}{2\left(d_{11}d_{22}-d_{12}d_{21}\right)}\right)^{\frac{1}{\alpha}}.$$

Therefore, the necessary conditions for Turing instability can be obtained:

(1) $A<0$;
(2) $B>0$;
(3) $d_{11}d_{22}-d_{12}d_{21}>0$;
(4) $a_{11}d_{22}+a_{22}d_{11}-a_{21}d_{12}-a_{12}d_{21}>2\sqrt{\left(d_{11}d_{22}-d_{12}d_{21}\right)\left(a_{11}a_{22}-a_{12}a_{21}\right)}$,

where the Allee effect parameter b means the bifurcation parameter, and the critical value b_T satisfies the following Turing bifurcation equation:

$$2\sqrt{\left(d_{11}d_{22}-d_{12}d_{21}\right)\left(a_{11}a_{22}-a_{12}a_{21}\right)}-a_{11}d_{22}+a_{22}d_{11}-a_{21}d_{12}-a_{12}d_{21}=0.$$

In the following simulations, we take system parameters as $a=1.5,\ b=0.265,\ c=0.42,\ m=0.28,\ s=0.8,\ \beta=1.2,\ d_{11}=0.1,\ d_{12}=0.05,\ d_{21}=0.3$, and $d_{22}=0.95$. Then Fig. 18.1 reveals that the larger value of the fractional diffusion order deduces the larger value of the critical wave number k_c and the smaller value of the wave number range for Turing instability. Therefore, there emerges a positive relation between the critical wave number k_c of Turing instability and the fractional diffusion order, which indicates that a smaller fractional diffusion order has more possibility to generate Turing bifurcation.

18.4 Dynamic Analysis with $\tau \neq 0$

18.4.1 Hopf Bifurcation

Based on the above results about the stability of the model (18.4) without time delay, we next consider the role of time delay τ and spatial diffusion in the stability of the system. We rewrite the model (18.4) as

$$\begin{pmatrix}\frac{\partial u(x,t)}{\partial t}\\ \frac{\partial v(x,t)}{\partial t}\end{pmatrix}=D\begin{pmatrix}\nabla^{\alpha}u(x,t)\\ \nabla^{\alpha}v(x,t)\end{pmatrix}+A_0\begin{pmatrix}u(x,t)\\ v(x,t)\end{pmatrix}+A_1\begin{pmatrix}u(t-\tau)\\ v(t-\tau)\end{pmatrix},\tag{18.10}$$

where,

$$D = \begin{pmatrix} d_{11} & d_{12} \\ d_{21} & d_{22} \end{pmatrix}, A_0 = \begin{pmatrix} a_{11} & a_{12} \\ 0 & a_{22} \end{pmatrix}, A_1 = \begin{pmatrix} 0 & 0 \\ a_{21} & 0 \end{pmatrix}.$$

Suppose the following form is the solution of system (18.10):

$$u(x,t) = u_* e^{\lambda_\tau t + i\mathbf{k}\cdot\mathbf{r}}, \quad v(x,t) = u_* e^{\lambda_\tau t + i\mathbf{k}\cdot\mathbf{r}}. \tag{18.11}$$

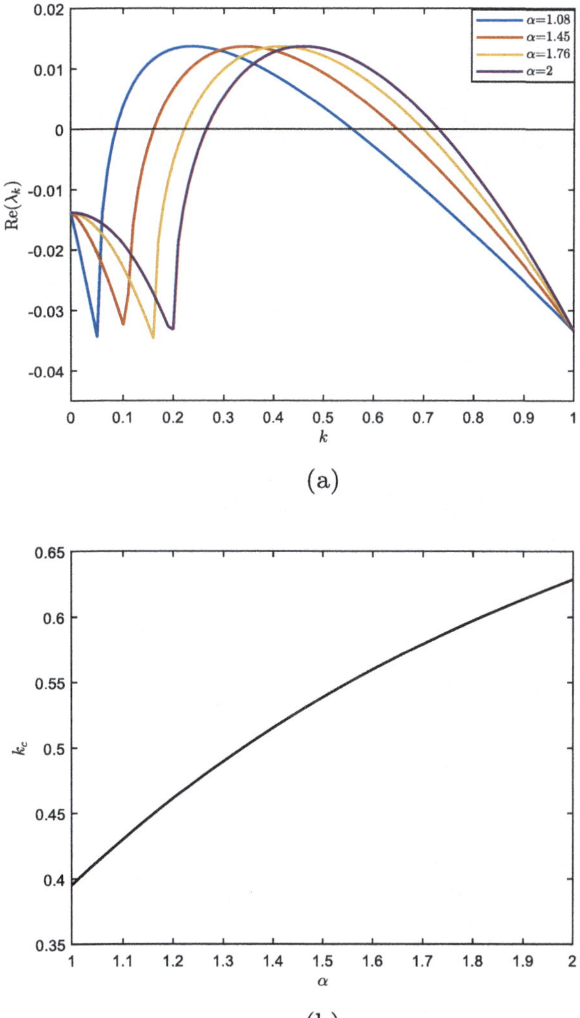

Fig. 18.1 When fractional diffusion order $\alpha = 1.08, 1.45, 1.76$, and 2, **a** the curve of function $Re(\lambda_k)$ with wave number k; **b** the curve of critical wave number k_c with fractional diffusion order α. The parameter are: $a = 1.5$, $b = 0.265$, $c = 0.42$, $m = 0.28$, $s = 0.8$, $\beta = 1.2$, $d_{11} = 0.1$, $d_{12} = 0.05$, $d_{21} = 0.3$, and $d_{22} = 0.95$

We substitute Eq. (18.11) into the system (18.10) and obtain the characteristic matrix at the equilibrium point $E_* = (u_*, v_*)$ for the system (18.10):

$$\begin{pmatrix} \lambda_\tau - a_{11} & -a_{12} \\ -a_{21}e^{-\lambda\tau} & \lambda_\tau - a_{22} \end{pmatrix}\begin{pmatrix} u \\ v \end{pmatrix} = \begin{pmatrix} 0 \\ 0 \end{pmatrix}.$$

Then, we have the characteristic equation of the model (18.10) as follows:

$$\det\begin{pmatrix} \lambda - a_{11} + d_{11}k^\alpha & d_{12}k^\alpha - a_{12} \\ d_{21}k^\alpha - a_{21}e^{-\lambda\tau} & \lambda - a_{22} + d_{22}k^\alpha \end{pmatrix} = 0.$$

Therefore, we get the characteristic equation for the model (18.10) at the equilibrium point $E_* = (u_*, v_*)$ as

$$\lambda_\tau^2 - A_\tau\lambda + B_\tau + C_\tau e^{-\lambda\tau} = 0, \tag{18.12}$$

where

$$A_\tau = a_{11} + a_{22} - k^\alpha\left(d_{11} + d_{22}\right),$$

$$B_\tau = k^{2\alpha}\left(d_{11}d_{22} - d_{12}d_{21}\right) - k^\alpha\left(a_{22}d_{11} - a_{12}d_{21} + a_{11}d_{22}\right) + a_{11}a_{22},$$

$$C_\tau = a_{21}\left(k^\alpha d_{12} - a_{12}\right).$$

Next, the influence of time delay is studied for the system (18.4). Suppose the characteristic Eq. (18.12) has a root as $i\omega > 0$, where ω satisfies the equation as below

$$-\omega^2 - iA_\tau\omega + B_\tau + C_\tau\left(\cos(\omega\tau) - i\sin(\omega\tau)\right) = 0. \tag{18.13}$$

Separating the real part and imaginary part for Eq. (18.13), we can have the relationship as following

$$\begin{cases} -\omega^2 + B_\tau = -C_\tau\cos(\omega\tau), \\ -A_\tau\omega = C_\tau\sin(\omega\tau). \end{cases} \tag{18.14}$$

Through the above formula, we obtain

$$\begin{cases} \cos(\omega\tau) = \frac{\omega^2 - B_\tau}{C_\tau}, \\ \sin(\omega\tau) = -\frac{A_\tau\omega}{C_\tau}. \end{cases}$$

After squaring each equation in Eq. (18.14) and adding it, we obtain the following equation

$$\omega^4 + M\omega^2 + N = 0, \tag{18.15}$$

where $M = A_\tau^2 - 2B_\tau$, $N = B_\tau^2 - C_\tau^2$, and $M = \left(d_{11}^2 + 2d_{12}d_{21} + d_{22}^2\right)k^{2\alpha} - 2\left(a_{11}d_{11} + a_{12}d_{21} + a_{22}d_{22}\right)k^\alpha + a_{11}^2 + a_{22}^2$, $N = \left(B_\tau + C_\tau\right)\left(B_\tau - C_\tau\right)$.
Here,
$B_\tau + C_\tau = \left(d_{11}d_{22} - d_{12}d_{21}\right)k^{2\alpha} - \left(a_{11}d_{22} + a_{22}d_{11} - a_{21}d_{12} - a_{12}d_{21}\right)k^\alpha + a_{11}a_{22} - a_{12}a_{21} = B_k$,
$B_\tau - C_\tau = \left(d_{11}d_{22} - d_{12}d_{21}\right)k^{2\alpha} - \left(a_{11}d_{22} + a_{22}d_{11} + a_{21}d_{12} - a_{12}d_{21}\right)k^\alpha + a_{11}a_{22} + a_{12}a_{21}$.

Regard $B_\tau + C_\tau = B_k$ as a quadratic polynomial with respect to k^α. As $\tau = 0$, $A_k < 0$, there exists $k^0 \in N_0$, so that $B_\tau + C_\tau > 0,\ k \geq k^0 + 1,\ k \in N_0$, and the positive equilibrium E_* becomes locally asymptotically stable, where variable N has the same number as variable $B_\tau - C_\tau$. The followings are assumptions:
(T_1) $a_{11}a_{22} + a_{12}a_{21} < 0$,
(T_2) $a_{11}d_{22} + a_{22}d_{11} + a_{21}d_{12} - a_{12}d_{21} < 0$.
Therefore, when we hold (T_1) and (T_2) conditions, there exists $k^0 \in N_0$, so that

$$B_\tau - C_\tau < 0.$$

At the same time, we have only one positive real root for Eq. (18.15) as follows

$$\omega_k^2 = \frac{-M + \sqrt{M^2 - 4N}}{2},\ k \in \left\{0, 1, ..., k^0\right\}. \tag{18.16}$$

And we obtain the following from Eq. (18.14)

$$\tau_k^j = \frac{1}{\omega_k}\left(\arccos\frac{\omega_k^2 - B_\tau}{C_\tau} + 2j\pi\right),\ j \in \left\{0, 1, ..., k^0\right\}. \tag{18.17}$$

From (18.17), we have $\tau_k^j < \tau_k^{j+1}$. And based on the monotonic property, the complete ordering τ_k^j of bifurcation values are derived as follows:

Lemma 10 *If* $(a_{11} + a_{22} < 0)$ *holds, then*

$$\tau_0^j \leq \tau_1^j \leq ... \leq \tau_k^j \leq \tau_{k+1}^j$$

for $j \in N_0$.

From Lemma 10, there display that $\tau_0^0 = \min\left\{\tau_k^j : 0 \leq k \leq k^0, j \in N_0\right\}$, where τ_0^0 is represented by τ_*, i.e., $\tau_* = \tau_0^0$. Assume $\lambda_\tau = q_3 + iq_4$ being a pair of roots near Eq. (18.12) at $\tau = \tau_k^j$, where $q_3(\tau_k^j) = 0$ and $q_4(\tau_k^j) = \omega_k$. Suppose that $a_{11} + a_{22} < 0$ holds, $A_k < 0$, $B_k > 0$, $k \in 0, 1, ..., k^0$, and we define ω_k and τ_k^j by (18.16) and (18.17) respectively. Then, we have a pair of pure imaginary roots $i\omega_k$ for Eq. (18.12), for each $k \in 0, 1, ..., k^0$, and there exists no pure imaginary root for $k \geq k^0 + 1$. Next, the symbol of $\left.\frac{dRe(\lambda_\tau)}{d\tau}\right|_{\tau=\tau_k^j}$ will be studied.

We derive τ from both sides of Eq. (18.12) and obtain

$$2\lambda_\tau \frac{d\lambda_\tau}{d\tau} - A_\tau \frac{d\lambda_\tau}{d\tau} - C_\tau e^{-\lambda_\tau \tau}\left(\lambda_\tau + \tau \frac{d\lambda_\tau}{d\tau}\right) = 0, \tag{18.18}$$

From (18.18), we have

$$\frac{d\lambda_\tau}{d\tau} = \frac{C_\tau \lambda_\tau e^{-\lambda_\tau \tau}}{2\lambda_\tau - A_\tau - C_\tau e^{-\lambda_\tau \tau}\tau}. \tag{18.19}$$

Substituting $\tau = \tau_k^j$ into the above formula, we obtain its real part

$$Re\left(\left(\frac{d\lambda_\tau}{d\tau}\right)^{-1}\right)\Bigg|_{\tau=\tau_k^j} = Re\left(\frac{2\lambda_\tau - A_\tau - C_\tau e^{-\lambda_\tau \tau}\tau}{C_\tau \lambda_\tau e^{-\lambda_\tau \tau}}\right)\Bigg|_{\tau=\tau_k^j}. \tag{18.20}$$

Considering $\lambda = i\omega$ for Eq. (18.20), we get

$$\begin{aligned} Re\left(\left(\tfrac{d\lambda_\tau}{d\tau}\right)^{-1}\right)\Big|_{\tau=\tau_k^j} &= Re\left(\tfrac{2\omega_k i - A_\tau - C_\tau \tau(\cos(\omega_k \tau) - i\sin(\omega_k \tau))}{i\omega_k C_\tau(\cos(\omega_k \tau) - i\sin(\omega_k \tau))}\right)\Big|_{\tau=\tau_k^j}, \\ &= \tfrac{A_\tau^2 - 2B_\tau + 2\omega_k^2}{C_\tau^2}\Big|_{\tau=\tau_k^j}. \end{aligned}$$

As $A_\tau^2 - 2B_\tau + 2\omega_k^2 = \sqrt{M^2 - 4N} > 0$, $Re\left(\left(\frac{d\lambda_\tau}{d\tau}\right)^{-1}\right)\Big|_{\tau=\tau_k^j} > 0$. Hence, $sign\left\{Re\left(\frac{d\lambda_\tau}{d\tau}\right)\Big|_{\tau=\tau_k^j}\right\}$ $= sign\left\{Re\left(\left(\frac{d\lambda_\tau}{d\tau}\right)^{-1}\right)\Big|_{\tau=\tau_k^j}\right\} > 0.$

Based on the lemma and the qualitative theory for partial functional differential equations, the results of stability and Hopf bifurcation for system (18.4) can be easily given. Therefore, we obtain the conclusions as below.

Theorem 65 *Assume that* (T_1) *and* (T_2) *hold. For model (18.10),*

(1) A Hopf bifurcation limit cycle is generated for the system as the time delay τ *satisfies* $\tau \neq 0$ *for each* $k \in 0, 1, \ldots, k^0$ *and* $j \in N_0$.
(2) The positive equilibrium point $E_* = (u_*, v_*)$ *becomes stable if* $0 \leq \tau < \tau_k^j$.
(3) The positive equilibrium point $E_* = (u_*, v_*)$ *remains unstable if* $\tau > \tau_k^j$.

18.4.2 Direction and Stability of Hopf Bifurcation

In this subsection, we mainly investigate the relevant characteristics for the Hopf branch of the system (18.4) at equilibrium E_*. Utilizing the normal form and the center manifold theory, we obtain the specific formulas controlling the direction for the homogeneous Hopf bifurcation and the stability for the periodic solution at the bifurcation.

Assume the critical value of time delay τ_0^j bing τ_* and introduce new parameters to make $\rho = \tau - \tau_*, \overline{u}(x,t) = u(x,\tau t) - u_*, \overline{v}(x,t) = v(x,\tau t) - v_*$. Define $\overline{U}(t) = (\overline{u}(x,t), \overline{v}(x,t))$ as $u - u_*$ and $v - v_*$, where the horizontal bar symbol is removed to simplify the symbol. In phase space $\mathcal{S} = \mathcal{C}([-1,0], X), X = \left\{(u,v) \in W^{3,3}(0,\pi) \,\middle|\, \frac{\partial u}{\partial x} = \frac{\partial v}{\partial x} = 0 \; at \; x = 0, \pi\right\}$, we can write the system (18.4) as

$$\begin{cases} \frac{\partial u}{\partial t} = \tau((u+u_*)(1-(u+u_*)-\frac{b}{(u+u_*)+m}) \\ -\frac{s(u+u_*)t(v+v_*)}{(u+u_*)+c} + d_{11}\nabla^\alpha u + d_{12}\nabla^\alpha v), \\ \frac{\partial v}{\partial t} = \tau(\frac{au(t-1)(v+v_*)}{u(t-1)+c} - \frac{a\beta(v+v_*)^2}{1+\beta(v+v_*)} + d_{21}\nabla^\alpha u + d_{22}\nabla^\alpha v). \end{cases} \tag{18.21}$$

If $U = (u,v)^T$, we can write the system (18.21) as

$$\dot{U}(t) = \tau_* D\nabla^\alpha U(t) + L(\tau_*)(U_t) + F(U_t, \rho), \tag{18.22}$$

where $D = \begin{pmatrix} d_{11} & d_{12} \\ d_{21} & d_{22} \end{pmatrix}$, $L(\tau)(\cdot) : \mathcal{S} \to X$, and $F : \mathcal{S} \times \mathbb{R} \to X$ are given, respectively, by

$$L(\tau)(\varphi) = \rho \begin{pmatrix} a_{11}\varphi_1(0) + a_{12}\varphi_2(0) \\ a_{21}\varphi_1(-1) + a_{22}\varphi_2(0) \end{pmatrix}, \tag{18.23}$$

$$F(\varphi, \rho) = \rho D\nabla^\alpha \varphi(0) + L(\rho)(\varphi) + f(\varphi, \tau), \tag{18.24}$$

where $\varphi = (\varphi_1, \varphi_2)^T \in \mathcal{S}$.

$$\begin{aligned} f(\varphi,\rho) &= (\tau_* + \rho)\begin{pmatrix} f^{(1)}(\varphi) \\ f^{(2)}(\varphi) \end{pmatrix} \\ &= (\tau_* + \rho)\begin{pmatrix} \sum\limits_{i+j\geq 2} \frac{1}{i!j!} f_{ij}^{(1)} \varphi_1^i(0)\varphi_2^j(0) \\ \sum\limits_{i+j\geq 2} \frac{1}{i!j!} f_{ij}^{(2)} \varphi_1^i(-1)\varphi_2^j(0) \end{pmatrix}, \end{aligned} \tag{18.25}$$

where
$f^{(1)}(u,v) = (u+u_*)\left(1-(u+u_*)-\frac{b}{(u+u_*)+m}\right) - \frac{s(u+u_*)(v+v_*)}{(u+u_*)+c}$,
$f^{(2)}(w,v) = \frac{aw(v+v_*)}{w+c} - \frac{a\beta(v+v_*)^2}{1+\beta(v+v_*)}$,
$f_{ij}^{(1)} = \left.\frac{\partial^{i+j} f^{(1)}(u,v)}{\partial u^i \partial v^j}\right|_{(0,0)}$, $f_{ij}^{(2)} = \left.\frac{\partial^{i+j} f^{(2)}(w,v)}{\partial w^i \partial v^j}\right|_{(0,0)}$.
And

$$f_{20}^{(1)} = 2\left(-1 + \frac{bm}{(m+u_*)^3} + \frac{su_*}{\beta(c+u_*)^3}\right),\ f_{11}^{(1)} = -\frac{cs}{(c+u_*)^2},\ f_{02}^{(1)} = 0,$$
$$f_{30}^{(1)} = -6\left(\frac{bm}{(m+u_*)^4} + \frac{su_*}{\beta(c+u_*)^4}\right),\ f_{21}^{(1)} = \frac{2cs}{(c+u_*)^3},\ f_{12}^{(1)} = 0,\ f_{03}^{(1)} = 0,$$
$$f_{20}^{(2)} = -\frac{2au_*}{\beta(c+u_*)^3},\ f_{11}^{(2)} = \frac{ac}{(c+u_*)^2},$$
$$f_{02}^{(2)} = -\frac{2ac^3\beta}{(c+u_*)^3},\ f_{30}^{(2)} = \frac{6au_*}{\beta(c+u_*)^4},$$
$$f_{21}^{(2)} = -\frac{2ac}{(c+u_*)^3},\ f_{12}^{(2)} = 0,\ f_{03}^{(2)} = \frac{6ac^4\beta^2}{(c+u_*)^4}.$$

After linearizing the system (18.22) at zero point, we get

$$\frac{\partial U(t)}{\partial t} = \tau_* D\nabla^\alpha U(t) + L(\tau_*)(U_t). \tag{18.26}$$

If $\tau = \tau_*$, we have a pair of pure imaginary roots $\lambda_0 = i\omega_0\tau_*, -i\omega_0\tau_*$ for the characteristic Eq. (18.12). Let $\mathcal{S} := \mathcal{C}([-1, 0], \mathbb{R}^3)$ and study the functional differential equation defined in $\mathcal{S}$:

$$\dot{z} = L(\tau_*)(z_t), \tag{18.27}$$

where, $L(\tau_*)$ means a continuous linear functional: $\mathcal{C}([-1, 0], \mathbb{R}^3) \to \mathbb{R}^3$. Based on the Riesz representation theorem, we choose a 2×2 bounded variation function $\eta(\theta, \tau),\ \theta \in [-1, 0]$ satisfying

$$L(\tau_*)(\varphi) = \int_{-1}^{0} [d\eta(\theta, \tau_*)]\varphi(\theta),\ \varphi \in \mathcal{S}. \tag{18.28}$$

In fact, we can express $\eta(\theta, \tau)$ as

$$\eta(\theta, \rho) = (\tau + \rho)\begin{pmatrix} a_{11} & a_{12} \\ 0 & a_{22} \end{pmatrix}\delta(\theta) + (\tau + \rho)\begin{pmatrix} 0 & 0 \\ -a_{21} & 0 \end{pmatrix}\delta(\theta + 1), \tag{18.29}$$

where $\delta(\theta)$ stands for Dirac delta function, $\delta(0) = 1,\ \delta(\theta) = 0$.

For $\varphi \in \mathcal{C}([0, 1], X)$ and $\phi \in \mathcal{S}([0, 1], X)$, assume $R(\tau_*)$ being the infinitesimal generator for the solution semigroup of the system (18.27), and then we prove that $R^*(\tau_*)$ has the formal adjoint of $R(\tau_*)$ with the bilinear pairing. We define

$$R(\rho)\varphi(\theta) = \begin{cases} \frac{d\varphi(\theta)}{d\theta}, & -1 < \theta < 0, \\ \int_{-1}^{0} [d\eta(\theta, \rho)]\varphi(\theta), & \theta = 0, \end{cases}$$

$$R^*(\rho)\phi(\gamma) = \begin{cases} -\frac{d\phi(\gamma)}{d\gamma}, & 0 < \gamma \le 1, \\ \int_{-1}^{0} [d\eta^T(\gamma, \rho)]\phi(-\gamma), & \gamma = 0, \end{cases}$$

and

$$
\begin{aligned}
(\phi(\gamma), \varphi(\theta)) &= \bar{\phi}(0)\varphi(0) - \int_{-1}^{0}\int_{\zeta-0}^{\theta} \bar{\phi}(\zeta-\theta)d\eta(\theta)\varphi(\zeta)d\zeta \\
&= \bar{\phi}(0)\varphi(0) - \tau_0 \int_{-1}^{0} \bar{\phi}(\zeta+1)\begin{pmatrix} 0 & 0 \\ -a_{21} & 0 \end{pmatrix}\varphi(\zeta)d\zeta.
\end{aligned}
\tag{18.30}
$$

According to the previous discussion we have that $i\omega_0\tau_*$ is the eigenvalue of $R(0)$, and $-i\omega_0\tau_*$ is the eigenvalue of $R^*(0)$. Assume

$$
H(\theta) = (h_1, h_2)^T e^{i\omega_0\tau_*\theta} = (1, h_2)^T e^{i\omega_0\tau_*\theta}, \ \theta \in [-1, 0]
$$

be the eigenvector of $R(0)$ for $i\omega_0\tau_*$, and

$$
P(\gamma) = \mu(t_1, t_2)e^{i\omega_0\tau_*\gamma} = \mu(1, t_2)e^{i\omega_0\tau_*\gamma}
$$

be the eigenvector of $R^*(0)$ for $-i\omega_0\tau_*$, where $h_2 = \frac{i\omega_0\tau_* - a_{11}}{a_{12}}$, $t_2 = -\frac{i\omega_0\tau_* + a_{11}}{a_{21}}e^{-i\omega_0\tau_*}$, $\mu = \left(1 + t_2\bar{h}_2 + a_{21}\tau_* t_2 e^{i\omega_0\tau_*}\right)^{-1}$. Then $(P(\gamma), H(\theta)) = 1$, $\left(P(\gamma), \bar{H}(\theta)\right) = 0$.

Assume $\Phi(\theta) = \left(H(\theta), \bar{H}(\theta)\right)$, and $\Psi(\gamma) = \left(P(\gamma), \bar{P}(\gamma)\right)$, we then obtain $(\Phi, \Psi) = I_2$, where I_2 means the identity matrix of 2×2. Moreover, we define $f_0 = (\eta_0^1, \eta_0^2)$ with

$$
\eta_0^1 = \begin{pmatrix} 1 \\ 0 \end{pmatrix}, \eta_0^2 = \begin{pmatrix} 0 \\ 1 \end{pmatrix},
$$

define $s \cdot f_0$ as

$$
s \cdot f_0 = s_1\eta_0^1 + s_2\eta_0^2, \ s = (s_1, s_2)^T, s_j \in \mathbb{R}^3, \ j = 1, 2,
$$

define

$$
(\phi \cdot f_0)(\theta) = \phi(\theta) \cdot f_0, \ \phi(\theta) \in \mathcal{C}([-1, 0], \mathbb{R}^3).
$$

And we give the following inner product for $u = (u_1, u_2)$, $v = (v_1, v_2)$.

$$
\langle u, v \rangle = \frac{1}{\pi}\int_0^{\pi} u_1\bar{v}_1 dx + \frac{1}{\pi}\int_0^{\pi} u_2\bar{v}_2 dx.
$$

Then

$$
\langle \varphi, f_0 \rangle = \left(\langle \varphi, f_0^1 \rangle, \langle \varphi, f_0^2 \rangle\right)^T.
$$

As $\rho = 0$, we have the central subspace of linear equation (18.26) as $P_{cn}\mathcal{S}$, where

$$
P_{cn}\varphi = \Phi(\Psi, \langle \varphi, f_0 \rangle) \cdot f_0, \ \varphi \in \mathcal{S}, P_{cn}\mathcal{S}. \tag{18.31}
$$

And let $\mathcal{S} = P_{cn}\mathcal{S} \oplus P_s\mathcal{S}$, where $P_s\mathcal{S}$ stands for the complement space of $P_{cn}\mathcal{S}$. Define R_U as

$$R_U\varphi(\theta) = \dot{\varphi}(\theta) + X_0(\theta)\left[\tau_* D\nabla^{\alpha}\varphi\left(0\right) + L\left(\tau_*\right)\left(\varphi(\theta)\right) - \dot{\varphi}(0)\right], \quad \varphi \in BC,$$

where

$$X_0\left(\theta\right) = \begin{cases} 0, & -1 \leq \theta < 0, \\ I, & \theta = 0. \end{cases} \tag{18.32}$$

Therefore, as R_U means the infinitesimal generator of the solution for Eq. (18.26), we can write Eq. (18.22) as the operator differential equation as following

$$\dot{U}_t = R_U U_t + X_0 F\left(U_t, \rho\right). \tag{18.33}$$

After utilizing the decomposition $\mathcal{S} = P_{cn}\mathcal{S} \oplus P_s\mathcal{S}$ and Eq. (18.31), we can express the solution of Eq. (18.22) as

$$U_t = \Phi\begin{pmatrix} z(t) \\ \bar{z}(t) \end{pmatrix} \cdot f_0 + W\left(z(t), \bar{z}(t)\right). \tag{18.34}$$

Furthermore, according to Ref. [487], we get z satisfying

$$\dot{z} = i\omega_0\tau_* z + g(z, \bar{z}), \tag{18.35}$$

where

$$g\left(z\left(t\right), \bar{z}\left(t\right)\right) = P\left(0\right)\left\langle F\left(\Phi(z\left(t\right), \bar{z}\left(t\right))^T \cdot f_0 + W\left(z\left(t\right), \bar{z}\left(t\right)\right), 0\right), f_0\right\rangle. \tag{18.36}$$

Let

$$W(z, \bar{z}) = W_{20}\frac{z^2}{2} + W_{11}z\bar{z} + W_{02}\frac{\bar{z}^2}{2} + \cdots, \tag{18.37}$$

$$g(z, \bar{z}) = g_{20}\frac{z^2}{2} + g_{11}z\bar{z} + g_{02}\frac{\bar{z}^2}{2} + \cdots. \tag{18.38}$$

Based on Eqs. (18.35) and (18.36), we obtain

$$\langle F(u_t, 0), f_0\rangle = \frac{\tau_* z^2}{2}\begin{pmatrix} f_{20}^{(1)} + 2f_{11}^{(1)}h_2 \\ f_{20}^{(2)}e^{-2i\omega_0\tau_*} + 2f_{11}^{(2)}h_2e^{-i\omega_0\tau_*} + f_{02}^{(2)}h_2^2 \end{pmatrix}$$
$$+\tau_* z\bar{z}\begin{pmatrix} f_{20}^{(1)} + f_{11}^{(1)}(h_2 + \bar{h_2}) \\ f_{20}^{(2)} + f_{11}^{(2)}\left(h_2e^{i\omega_0\tau_*} + \bar{h_2}e^{-i\omega_0\tau_*}\right) + f_{02}^{(2)}|h_2|^2 \end{pmatrix}$$
$$+\frac{\tau_*\bar{z}^2}{2}\begin{pmatrix} f_{20}^{(1)} + 2f_{11}^{(1)}\bar{h_2} \\ f_{20}^{(2)}e^{2i\omega_0\tau_*} + 2f_{11}^{(2)}\bar{h_2}e^{i\omega_0\tau_*} + f_{02}^{(2)}\bar{h_2}^2 \end{pmatrix}$$
$$+\frac{\tau_* z^2\bar{z}}{2}\begin{pmatrix} P_1 + \left[f_{30}^{(1)} + f_{21}^{(1)}(2h_2 + \bar{h_2})\right] \\ P_2 + P_3 + \left[f_{30}^{(2)}e^{-i\omega_0\tau_*} + f_{21}^{(2)}\left(2h_2 + \bar{h_2}e^{-2i\omega_0\tau_*}\right) + f_{03}^{(2)}h_2|h_2|^2\right] \end{pmatrix} + \cdots,$$

where

$$P_1 = f_{20}^{(1)}\left(2W_{11}^1(0) + W_{20}^1(0)\right)$$
$$+f_{20}^{(1)}\left(2W_{11}^2(0) + 2h_2W_{11}^1(0) + W_{20}^2(0) + \bar{h_2}W_{20}^1(0)\right),$$
$$P_2 = f_{20}^{(2)}\left(2W_{11}^1(-1)e^{-i\omega_0\tau_*} + W_{20}^1(-1)e^{i\omega_0\tau_*}\right)$$
$$+f_{02}^{(2)}\left(2h_2W_{11}^2(0) + \bar{h_2}W_{20}^2(0)\right),$$
$$P_3 = f_{11}^{(2)}\left(2W_{11}^2(0)e^{-i\omega_0\tau_*} + W_{20}^2(0)e^{i\omega_0\tau_*} + 2h_2W_{11}^1(-1) + \bar{h_2}W_{20}^1(-1)\right).$$

According to the above formula, we obtain

$$g_{20} = \tau_*\bar{\mu}(f_{20}^{(1)} + 2f_{11}^{(1)}h_2 + f_{02}^{(1)}h_2^2)$$
$$+\tau_*\bar{\mu}\bar{t_2}(f_{20}^{(2)}e^{-2i\omega_0\tau_*} + 2f_{11}^{(2)}h_2e^{-i\omega_0\tau_*} + f_{02}^{(2)}h_2^2),$$
$$g_{11} = \tau_*\bar{\mu}(f_{20}^{(1)} + 2f_{11}^{(1)}Re(h_2) + f_{02}^{(1)}|h_2|^2)$$
$$+\tau_*\bar{\mu}\bar{t_2}(f_{20}^{(2)} + 2f_{11}^{(2)}Re(h_2e^{i\omega_0\tau_*}) + f_{02}^{(2)}|h_2|^2),$$
$$g_{02} = \tau_*\bar{\mu}(f_{20}^{(1)} + 2f_{11}^{(1)}\bar{h_2} + f_{02}^{(1)}\bar{h_2}^2)$$
$$+\tau_*\bar{\mu}\bar{t_2}(f_{20}^{(2)}e^{2i\omega_0\tau_*} + 2f_{11}^2\bar{h_2}e^{i\omega_0\tau_*} + f_{02}^{(2)}\bar{h_2}^2),$$
$$g_{21} = \tau_*\bar{\mu}(f_{20}^{(1)}(2W_{11}^{(1)}(0) + W_{20}^{(1)}(0))$$
$$+f_{11}^{(1)}(2h_2W_{11}^{(1)}(0) + \bar{h_2}W_{20}^{(1)}(0) + 2W_{11}^{(2)}(0) + W_{20}^{(2)}(0)))$$
$$+\tau_*\bar{\mu}\bar{t_2}\begin{pmatrix} f_{20}^{(2)}(2W_{11}^{(1)}(-1)e^{-i\omega_0\tau_*} + W_{20}^{(1)}(-1)e^{i\omega_0\tau_* t}) \\ +f_{11}^{(2)}(\bar{h_2}W_{20}^{(1)}(-1) + 2h_2W_{11}^{(1)}(-1) + W_{20}^{(2)}(0)e^{i\omega_0\tau_*} + 2W_{11}^{(2)}(0)e^{-i\omega_0\tau_*}) \\ +f_{02}^{(2)}(2h_2W_{11}^{(2)}(0) + \bar{h_2}W_{20}^{(2)}(0)) \end{pmatrix}$$
$$+\tau_*\bar{\mu}(f_{30}^{(1)} + f_{21}^{(1)}(2h_2 + \bar{h_2}))$$
$$+\tau_*\bar{\mu}\bar{t_2}(f_{30}^{(2)}e^{-i\omega_0\tau_*} + f_{21}^{(2)}(\bar{h_2}e^{-2i\omega_0\tau_*} + 2h_2) + f_{03}^{(2)}h_2|h_2|^2).$$

In order to calculate g_{21}, we also need to obtain the expressions of W_{11} and W_{20}. We calculated W_{11} and W_{20} according to the method of Ref. [487], $\theta \in [-1, 0]$. It can be seen from Eq. (18.37).

$$\dot{W} = W_{20}z\dot{z} + W_{11}\left(\dot{z}\bar{z} + z\dot{\bar{z}}\right) + W_{02}\bar{z}\dot{\bar{z}} + \cdots,$$

From the infinitesimal generator R_U of the linearized system (18.26) when $\rho = 0$, we obtain

$$R_U W = R_U W_{20}\frac{z^2}{2} + R_U W_{11} z\bar{z} + R_U W_{02}\frac{\bar{z}^2}{2} + \cdots, \tag{18.39}$$

and W satisfies

$$\dot{W} = R_U W + Q(z, \bar{z}), \tag{18.40}$$

where

$$\begin{aligned} Q(z, \bar{z}) &= X_0 f(u_t, 0) - \Phi(\Psi, \langle X_0 f(u_t, 0), f_0\rangle) \cdot f_0 \\ &= Q_{20}\tfrac{z^2}{2} + Q_{11} z\bar{z} + Q_{02}\tfrac{\bar{z}^2}{2} + \cdots. \end{aligned}$$

According to the above formula, we get

$$\begin{cases} [2i\omega_0\tau_* - R_U] W_{20} = Q_{20}, \\ -R_U W_{11} = Q_{11}, \\ [-2i\omega_0\tau_* - R_U] W_{02} = Q_{02}. \end{cases} \tag{18.41}$$

Therefore, for $-1 \le \theta < 0$, we get

$$Q_{11}(\theta) = -\left[g_{11} H(\theta) + \bar{g}_{11}\bar{H}(\theta)\right], \; Q_{20}(\theta) = -\left[g_{20} H(\theta) + \bar{g}_{02}\bar{H}(\theta)\right],$$

and

$$Q_{20}(0) = \tau_* \begin{pmatrix} f_{20}^{(1)} + 2 f_{11}^{(1)} h_2 \\ f_{20}^{(2)} e^{-2i\omega_0\tau_*} + 2 f_{11}^{(2)} h_2 e^{-i\omega_0\tau_*} + f_{02}^{(2)} h_2^2 \end{pmatrix} - \left[g_{20} H(0) + \bar{g}_{02}\bar{H}(0)\right],$$

$$Q_{11}(0) = \tau_* \begin{pmatrix} f_{20}^{(1)} + 2 f_{11}^{(1)} Re(h_2) \\ f_{20}^{(2)} + 2 f_{11}^{(2)} Re\left(h_2 e^{i\omega_0\tau_*}\right) + f_{02}^{(2)} |h_2|^2 \end{pmatrix} - \left[g_{11} H(0) + \bar{g}_{11}\bar{H}(0)\right].$$

Based on formula (18.39) and formula (18.41), we can calculate as

$$W_{20}(\theta) = \left[\frac{i g_{20}}{\omega_0\tau_*} H(\theta) + \frac{i\bar{g}_{02}}{3\omega_0\tau_*}\bar{H}(\theta)\right] \cdot f_0 + E_1 e^{2i\omega_0\tau_*\theta},$$

$$W_{11}(\theta) = \left[\frac{-i g_{20}}{\omega_0\tau_*} H(\theta) + \frac{i\bar{g}_{11}}{\omega_0\tau_*}\bar{H}(\theta)\right] \cdot f_0 + E_2,$$

where

$$E_1 = \begin{pmatrix} 2i\omega_0 - a_{11} & -a_{12} \\ -a_{21} e^{-2i\omega_0\tau_*} & 2i\omega_0 - a_{22} \end{pmatrix}^{-1} \begin{pmatrix} f_{20}^{(1)} + 2h_2 f_{11}^{(1)} \\ f_{20}^{(2)} e^{-2i\omega_0\tau_*} + 2h_2 f_{11}^{(2)} e^{-i\omega_0\tau_*} + f_{20}^{(2)} h_2^2 \end{pmatrix},$$

$$E_2 = \begin{pmatrix} -a_{11} & -a_{12} \\ -a_{21} & -a_{22} \end{pmatrix}^{-1} \begin{pmatrix} f_{20}^{(1)} + 2\,\mathrm{Re}(h_2)\, f_{11}^{(1)} \\ f_{20}^{(2)} + 2\,\mathrm{Re}\left(h_2 e^{i\omega_0 r_*}\right) f_{11}^{(2)} + f_{20}^{(2)} |h_2|^2 \end{pmatrix}.$$ Thus, we can give the expression of g_{21} explicitly. Based on the Hopf bifurcation theorem, we can obtain the following conclusions.

Theorem 66 *There is the following Poincaré normal form for system (18.4):*

$$\dot{\zeta} = i\omega_0\tau_*\zeta + C(0)\,\zeta|\zeta|^2 + o\left(|\zeta|^5\right).$$

where

$$\begin{cases} C(0) = \frac{i}{2\omega_0\tau_*}\left(g_{11}g_{20} - 2|g_{11}|^2 - \frac{1}{3}|g_{02}|^2\right) + \frac{g_{21}}{2}, \\ \mu_2 = -\frac{Re(C(0))}{Re\left(\lambda'(\tau_*)\right)}, \\ \psi_2 = 2Re\,(C(0)), \\ T_2 = -\frac{Im(C(0)) + \mu_2 Im\left(\lambda'(\tau_*)\right)}{\omega_0\tau_*}. \end{cases} \tag{18.42}$$

The properties of Hopf bifurcation are determined by the above Eq. (18.42), where μ_2 controls the direction of Hopf bifurcation: Hopf bifurcation becomes supercritical (subcritical) if $\mu_2 > 0$ ($\mu_2 < 0$); And the periodic solution at the Hopf bifurcation exists when $\tau > \tau_*$; the stability of Hopf bifurcation periodic solution is determined by β_2: the bifurcation periodic solution remains stable (unstable) if $\psi_2 < 0$ ($\psi_2 > 0$); T_2 controls the period of Hopf bifurcation periodic solution as $T_2 > 0$ ($T_2 < 0$), the period increases (decreases).

18.4.3 Turing-Hopf Bifurcation

When $\tau = \tau_*$ and $b = b_T$, we obtain a pair of purely imaginary roots $i\omega$ and a zero eigenvalue of the characteristic Eq. (18.12) for $k = k_*$. And all other eigenvalues have strictly negative real parts. For the $\tau - b$ plane, when $(\tau, b) = (\tau_*, b_T)$, the system (18.4) generates Turing-Hopf bifurcation at the equilibrium point E_*, where Hopf bifurcation is for $\tau = \tau_*$ and Turing bifurcation is for $b = b_T$ on the boundary of the stable region.

We also get the Hopf bifurcation curve for the system (18.4) as

$$\tau_* = \frac{1}{q_2}\left(\arccos\frac{q_2^2 - B_\tau}{C_\tau}\right), \quad k = 0,$$

and the Turing bifurcation curve for the system (18.4) as

$$a_{11}d_{22} + a_{22}d_{11} - a_{21}d_{12} - a_{12}d_{21} - 2\sqrt{(d_{11}d_{22} - d_{12}d_{21})(a_{11}a_{22} - a_{12}a_{21})} = 0.$$

Figure 18.2 shows the bifurcation relation about additive Allee effect b and delay τ by the linear stability theory and bifurcation theory. The blue Hopf bifurcation line, the red Turing bifurcation line, and $b = m$ as the purple line in Fig. 18.2 depart the whole $\tau - b$ parameter region into five sub-regions. Subsequently, Region I means the range of stability of the equilibrium point for the system in the presence of time delay regardless of whether there exists a fractional diffusion for the system (18.4). Region II indicates the stability of the equilibrium point as the system (18.4) with no fractional diffusion and with a time delay. Contrarily, the equilibrium point of the system remains unstable as the system (18.4) has fractional diffusion. At the same time, region II becomes the Turing space. Region III as the Turing-Hopf space, implies the unstability of the equilibrium point for the system whether there has fractional diffusion or not has time delay for the system (18.4). Region IV as the Hopf space signifies the unstability for the system equilibrium point because of the existence of Hopf bifurcation. And Region V represents the parameter region of population extinction that there exists no positive solution for the system (18.4), where the green dot displays the critical point of Turing-Hopf bifurcation.

18.5 Numerical Simulation

In the following, the numerical simulation about phase diagram three-dimensional (3D) patterns and spatiotemporal patterns is performed to further investigate the dynamic behavior for the reaction–diffusion model (18.4) with time delay at the coexistence equilibrium E_*. Let some parameters of the system (18.4) as

$$m = 0.28, \ s = 0.8, \ \beta = 1.2, \ d_{12} = 0.05. \tag{18.43}$$

18.5.1 With Delay and Without Diffusion

In this part, we carry out the numerical analysis according to $\tau = 0$ and $\tau \neq 0$.
Case 1: No diffusion with the delay $\tau = 0$.

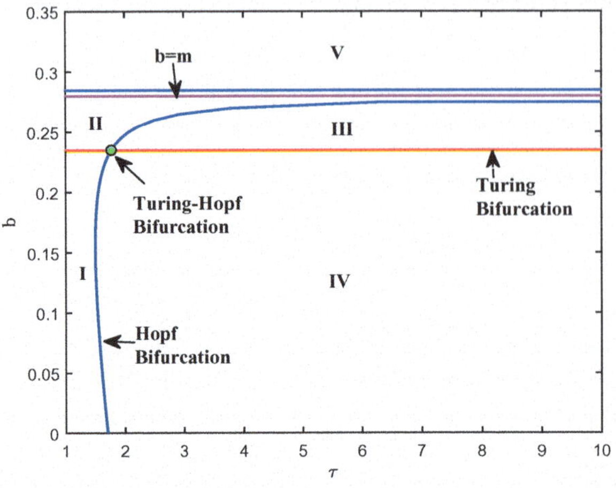

Fig. 18.2 Bifurcation diagram of system (18.4)

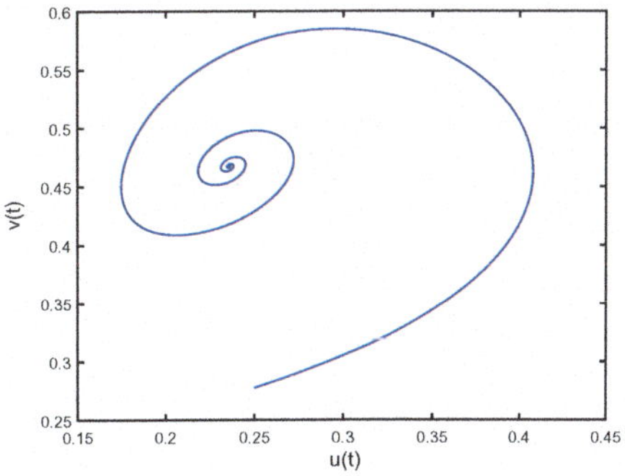

(a) Phase diagram of the system (18.4)

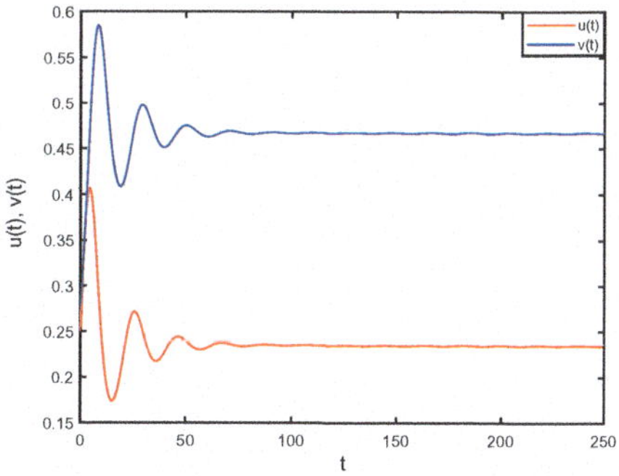

(b) The dynamic behavior of prey and predator changes with time

Fig. 18.3 Phase diagram of the system (18.4) and its dynamic behavior of prey and predator changes with time, where $\tau = 0$, and fixed initial conditions: $(u_0, v_0) = (0.25, 0.278)$

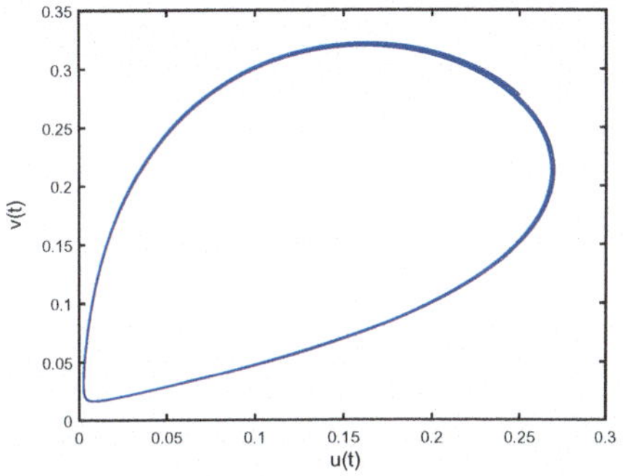

(a) Phase diagram of the system (18.4)

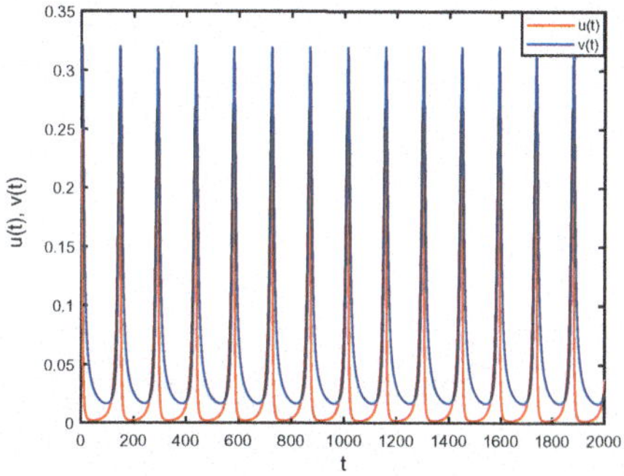

(b) The dynamic behavior of prey and predator changes with time

Fig. 18.4 Phase diagram of the system (18.4) and the dynamic behavior of prey and predator changes with time, where $\tau = 0$, and fixed initial conditions: $(u_0, v_0) = (0.25, 0.278)$

And we take the remaining parameters of the system as $a = 0.8$, $b = 0.1$, $c = 0.42$. Figure 18.3b reveals the locally asymptotically stability of the coexistence equilibrium point (0.2356, 0.4675) where prey and predator populations tend to the coexistence equilibrium values for (18.4) dynamically. It yields a stable spiral converging to equilibrium shown in Fig. 18.3a. Figure 18.4 is simulated with the increasing value of parameter b to 0.265 and keeping other parameter values unchanged. In that case, it can be seen that there emerges oscillation with $E_*(u_*, v_*)$ changing from stable to stable periodic oscillation state, and losing stability with the formation of the limit cycle. Figure 18.4a illustrates the corresponding phase diagram, and Fig. 18.4b shows the periodic oscillation of prey and predator.

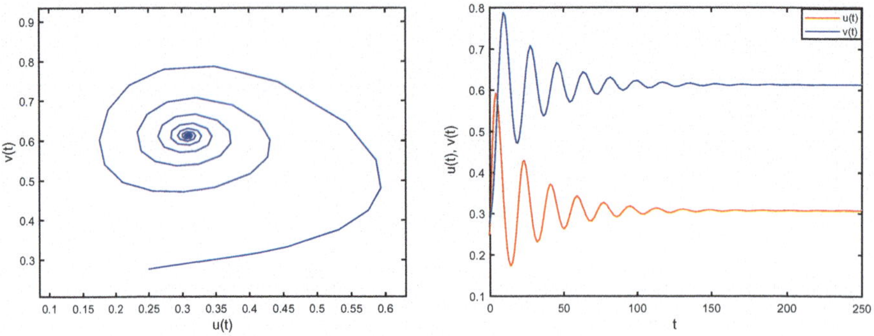

(a) Phase diagram of the system (18.4)

(b) The dynamic behavior of prey and predator changes with time

Fig. 18.5 Phase diagram of the system (18.4) and the dynamic behavior of prey and predator changes with time, where $\tau = 1.12$, and fixed initial conditions: $(u_0, v_0) = (0.25, 0.278)$

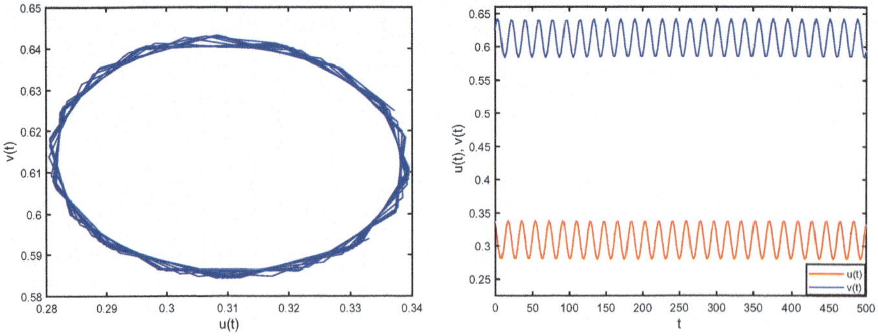

(a) Limit cycle phase diagram of the system (18.4)

(b) The dynamic behavior of prey and predator changes with time

Fig. 18.6 Limit cycle phase diagram of the system (18.4) and the dynamic behavior of prey and predator changes with time, where $\tau_* = 1.6875$, the initial condition of the system (18.4): $(u_0, v_0) = (0.3372, 0.6251)$

Case 2: With delay and no diffusion.

Assume other parameters as $a = 0.8,\ b = 0.01,\ c = 0.42$, and we get a positive equilibrium point $E_*(u_*, v_*) = (0.3095, 0.6142)$ for the system (18.4). As the condition of the dynamic behavior for the system (18.4) changes, the threshold of time delay is calculated as $\tau_* = 1.6875$. When $\tau = 1.12 < \tau_*$. Then based on the above theoretical analysis, Fig. 18.5 displays the locally asymptotically stability of $E_*(u_*, v_*)$. If $\tau = \tau_* = 1.6875$, the system (18.4) gets a stable limit cycle shown in Fig. 18.6a, and a periodic Hopf bifurcation illustrated in Fig. 18.6b. In Fig. 18.7, its initial conditions of the system (18.4) are still the same as in Fig. 18.6. When $\tau = 1.85 > \tau_*$, system (18.4) emerges Hopf bifurcation. Due to large delay, Fig. 18.7a shows the limit cycles

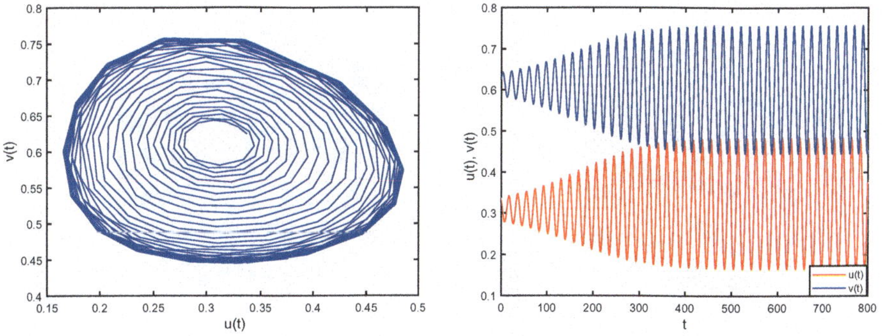

(a) Phase diagram of the system (18.4)

(b) The dynamic behavior of prey and predator changes with time

Fig. 18.7 Phase diagram of the system (18.4) and the dynamic behavior of prey and predator changes with time, where $\tau = 1.85$, the initial condition of the system (18.4): $(u_0, v_0) = (0.3372, 0.6251)$

generated through Hopf bifurcation. This indicates that delay has an important effect on the dynamic behavior and formations of the spatiotemporal patterns.

18.5.2 With Time Delay and with Diffusion

In this part, the three-dimensional numerical simulation is performed to study the dynamic behavior for the predator–prey model (18.4) with time delay term at the coexistence equilibrium E_*. The simulated area is a discrete one of $100 \times 100 \times 100$. We also define the spatial distance between two grids as the step length $\triangle x$, $\triangle y$, and $\triangle z$, and $\triangle x = \triangle y = \triangle z = 1$. And we express the time step as $\triangle t$, and adopt a fixed time step $\triangle t = 0.01$. The boundary condition is the Homogeneous Neumann case for all numerical simulations. A backward Euler discretization for the time derivative is presented [459], where in each time step $[t_n, t_{n+1}]$ the nonlinear factor is considered utilizing the fixed point iteration as below: given u^n, define $u^{n+1,0} := u^n$, and for $m = 1, 2, , L$ find $u^{n+1,m}$ such that

$$\frac{u_i^{n+1,m} - u_i^n}{\triangle t} = -D_i(-\triangle)^{\alpha/2} u_i^{n+1,m} + f_i(u^{n+1,m-1}, t^{n+1}) \tag{18.44}$$

where L can be chosen. Clearly, $L = 1$ means the fully explicit treatment of the nonlinear term. And the method is fully implicit for sufficiently large M. We apply the Fourier transform for both sides of (18.44) and define the fractional Laplacian given by (18.8). Then we get

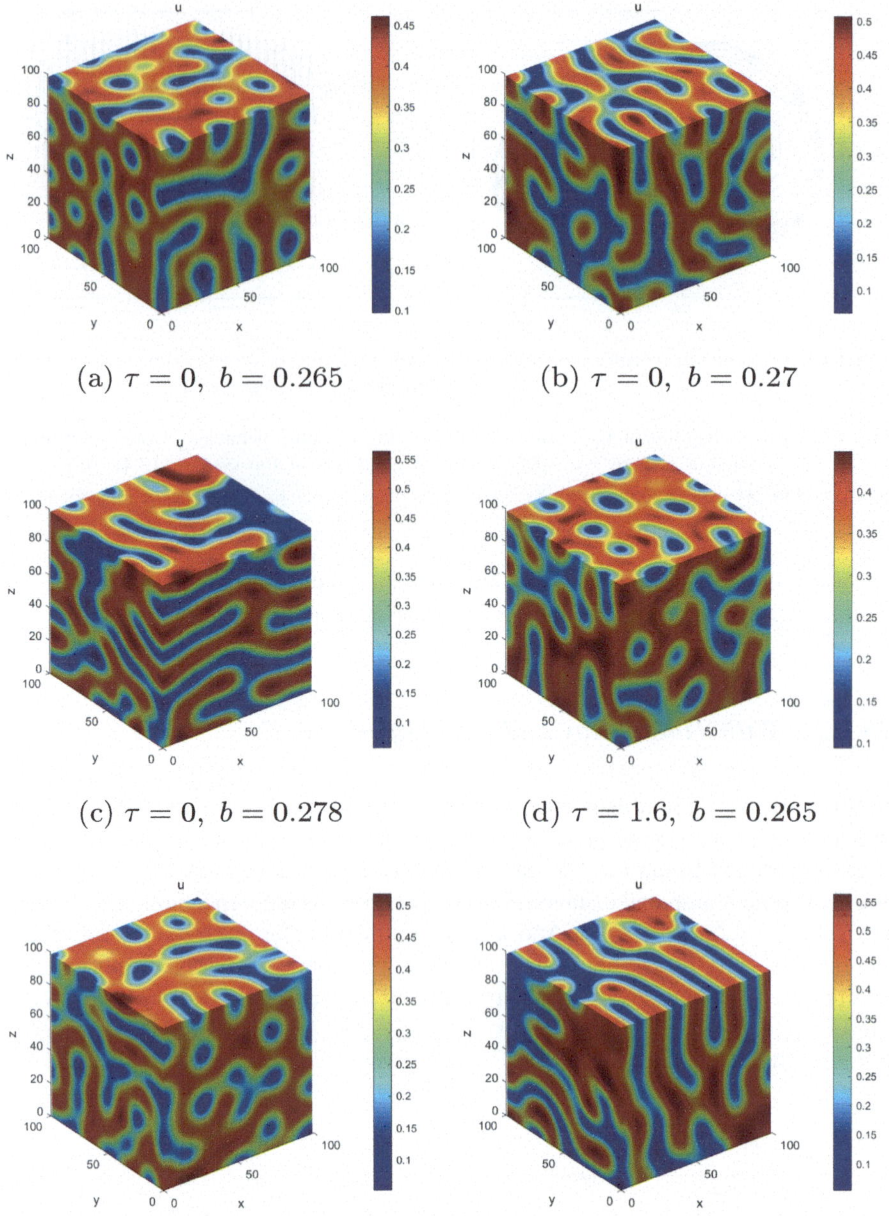

(a) $\tau = 0,\ b = 0.265$ (b) $\tau = 0,\ b = 0.27$

(c) $\tau = 0,\ b = 0.278$ (d) $\tau = 1.6,\ b = 0.265$

(e) $\tau = 1.6,\ b = 0.27$ (f) $\tau = 1.6,\ b = 0.278$

Fig. 18.8 Variation of the population density for the three-dimensional isosurface for the same time, where $\tau = 0$, $\tau = 1.6$, and other parameters remain unchanged

$$\frac{\widehat{u_{ij}}^{n+1,m} - \widehat{u_{ij}}^{n}}{\Delta t} = -D_i \lambda_j^{\alpha/2} \widehat{u_{ij}}^{n+1,m} + \hat{f}_{ij}(u^{n+1,m-1}, t^{n+1})$$

where $\hat{f}_{ij}$ is the j-th Fourier coefficient for the i-th reaction term, and we use the orthogonality of the basis functions to imply their independence for any two of the Fourier coefficients.

Case 3: We consider the effect of Allee effect parameter b and time delay τ on parameter space region II (see Fig. 18.2) for the Turing pattern formation like three-dimensional spherical droplets and tubular patterns. In this simulation, we take the initial condition for the system (18.4) as the random distribution at the equilibrium point E_* for the population. And the other parameters are $a = 1.5,\ c = 0.8,\ d_{11} = 1,\ d_{21} = 11, d_{22} = 11.5, \alpha = 1.8$. Figures 18.8 and 18.9 display the three-dimensional pattern running at the same time $t = 2000$, for $\tau = 0$ and $\tau = 1.6$, and b is 0.265, 0.270, and 0.278, respectively. These patterns are also the patterns generated by parameter region II in bifurcation, shown in Fig. 18.2. First of all, we consider the case without time delay. In other words, when $\tau = 0$, the pattern is generated by using the Turing instability condition. Figures 18.8a–c and 18.9a–c exhibit the three-dimensional isosurface for population density. By combining the patterns displayed in Figs. 18.8a and 18.9a, it yields many spherical (spot) and tube (stripe) patterns and also the patterns combined both of them. With the increase of the Allee parameter b, we can observe its decreasing effect on the number of spherical droplets, and also investigate the deformation of the tube. Figures 18.8c and 18.9c show that as the Allee parameter increases to $b = 0.278$, the Turing pattern varies from the combination of sparse spherical droplets and tube to the distorted tube pattern.

Next, the pattern formation with time delay and the Allee effect will be discussed. Changing the time delay to $\tau = 1.6$ and observing in Figs. 18.8d–f and 18.9d–f, we find that the change of three-dimensional isosurface for prey population density remains similar to that of $\tau = 0$. It also yields that the spot pattern shown in Fig. 18.8d forms faster than that from Fig. 18.8a. In Fig. 18.8e and b, we can observe the coexist case combined both spot pattern and stripe pattern, but the spots in Fig. 18.8e are more than that in Fig. 18.8b. Unchanged patterns are strip patterns that are shown both in Fig. 18.8f and in Fig. 18.8c.

Compared the change of three-dimensional patterns displayed in Fig. 18.9a and d, we can observe the same two structure types: spherical and tube mixed state. But the forming velocity of the spherical pattern in Fig. 18.9d becomes faster and its number remains more than that in Fig. 18.9a. As b increases to 0.27, both spherical and tube patterns are displayed in Fig. 18.9b and e, but the number of tubes in both figures is more than spherical. When b increases to 0.278, there display only unchanged tube patterns in Figs. 18.9c, f dynamically.

Case 4: We consider the role of Allee effect parameter b and time delay τ in parameter space region III named as Turing-Hopf bifurcation region on the Turing pattern formation. Firstly, let other parameters are: $a = 1.5,\ b = 0.1,\ c = 0.42,\ d_{11} = 0.1,\ d_{21} = 0.8,\ d_{22} = 0.95,\ \alpha = 1.8$. At the same time, $\tau_* = 1.4585,\ \tau = 1.5$, the

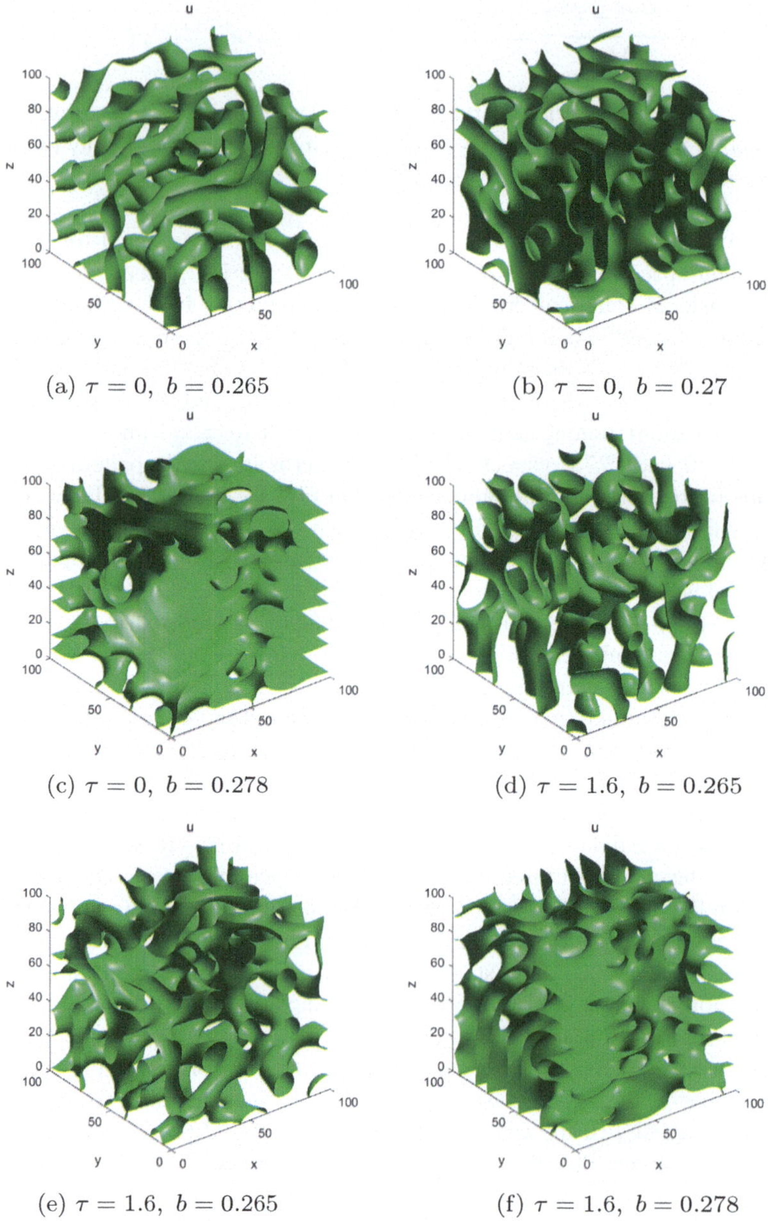

(a) $\tau = 0,\ b = 0.265$

(b) $\tau = 0,\ b = 0.27$

(c) $\tau = 0,\ b = 0.278$

(d) $\tau = 1.6,\ b = 0.265$

(e) $\tau = 1.6,\ b = 0.265$

(f) $\tau = 1.6,\ b = 0.278$

Fig. 18.9 Variation of the three-dimensional isosurface for the prey population systems, where $\tau = 0$, $\tau = 1.6$, other parameters remain unchanged and the value of isosurface parameter is 0.32

initial density of the population is taken as $u_0 = u_* - 0.0002(x - 50) - 0.0002(z - 50)$, $v_0 = v_* - 0.0003(y - 50) - 0.0003(z - 50)$. When $t = 400$, Fig. 18.10a, e shows the pattern tends to spiral. If $t = 500$, Fig. 18.10b, f displays the spherical droplets with equal density value $u = 0.32$) and the spiral pattern for the prey population density. If $t = 2000$, Fig. 18.10c, g show the negative relationship that the increasing of the number of spherical droplets (spot) deduces the decreasing of the spiral pattern significantly. Then as $t = 5000$, Fig. 18.10d, h displays the unchanged phenomenon that the spiral pattern disappears and there only exists a single spherical droplets pattern tending to be stable. Then only change the time delay parameter τ which is still set in the Turing-Hopf bifurcation parameter space (region III shown in Fig. 18.2), and take $\tau = 1.53$ and $\tau = 1.8$. When the time delay τ is 1.53, Fig. 18.11a, b shows some stable spherical droplets. According to the time series of prey population density, Fig. 18.11c displays the weaker of the oscillation with the increasing evolution time. When the time delay τ is set as 1.8, Fig. 18.11d, e still shows a spiral pattern. And as the role of time delay τ in Turing bifurcation is feeble, it also generate a strong effect on Hopf bifurcation. Hence, Fig. 18.11c, f shows the positive relationship that as the greater of time delay τ, we observe the greater oscillation amplitude and the longer oscillation time.

Next, the three-dimensional spiral waves will be considered, which can also be generated from the region IV (Hopf bifurcation) in Fig. 18.2 shown in Figs. 18.12, 18.13, 18.14, 18.15, and 18.16.

Case 5: The parameters are taken as $a = 2.15$, $c = 0.42$, $d_{11} = 0.2$, $d_{21} = 0.4$, $d_{22} = 0.6$, $\alpha = 1.8$, other parameters are determined by (18.43), $\tau_c = 1.5249$, the initial conditions are $u_0 = u_* - 0.0002(x - 50) - 0.0002(z - 50)$, $v_0 = v_* - 0.0002(y - 50) - 0.0002(z - 50)$. When the time delay $\tau = 1.65$, Fig. 18.12 exhibits the time evolution of the three-dimensional spiral pattern for the prey population. Initially, the system did not generate a regular spiral wave pattern shown in Fig. 18.12a, but as time evolves, it gradually evolves to be spirals from the center, and then there exists a regular three-dimensional spiral wave shown in Fig. 18.12b, c. Then, the regular spiral structure is dynamically broken from its periphery and gradually becomes an irregular spiral pattern shown in Fig. 18.12d. Finally, the whole region is occupied by the irregular spiral pattern shown in Fig. 18.12e. Since $\tau = 1.65 > \tau_*$, system (18.4) generates Hopf bifurcation as an oscillation state shown in Fig. 18.12f. Therefore, it yields that there forms the broken spiral pattern gradually from the regular spiral pattern dynamically, and the spatial pattern varies tending to the irregular trend in a long enough time.

Case 6: The effect of time delay τ will be explored for the formation of spiral wave patterns. Figure 18.13 shows the variation of the spiral wave pattern of the system (18.4) with the time delay τ, for $t = 500$. The parameters are taken as $a = 2.15$, $c = 0.05$, $d_{11} = 0.2$, $d_{21} = 0.36$, $d_{22} = 0.6$, $\alpha = 1.8$, and $b = 0.05$, other parameters are deduced from Eq. (18.43). Figure 18.13 displays that for $\tau = 1.6$, the spiral wave pattern remains regular shown in Fig. 18.13a. As the time delay τ increases to 1.7, the system emerges irregular spiral waves shown in

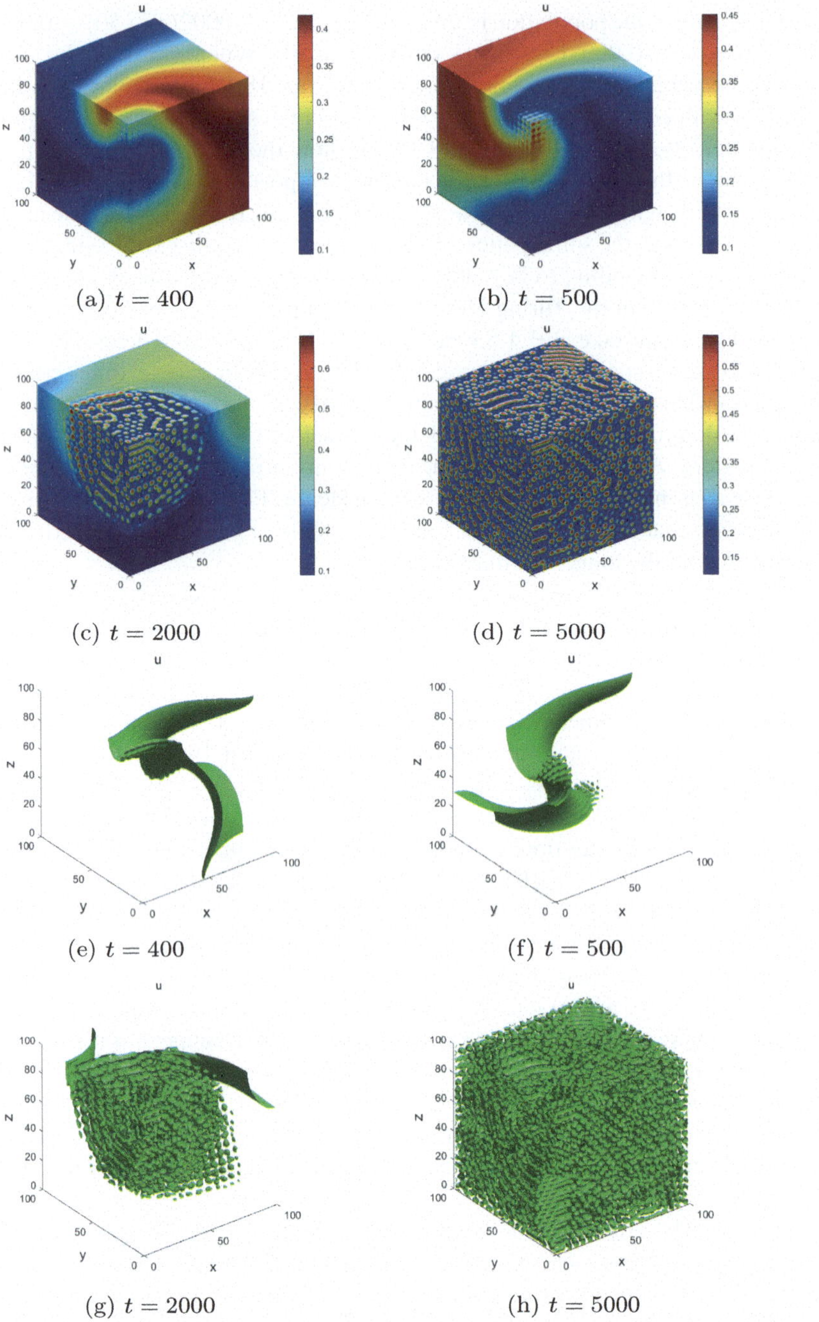

Fig. 18.10 The spatial three-dimensional isosurface and three-dimensional pattern for the prey population, where $\tau = 1.5$, $b = 0.1$, other parameters remain unchanged, and the equal density value of prey population density is 0.32. **e–h** The graphing subroutines were based on the function isosurface with values of the isovalue parameter equal to 0.32

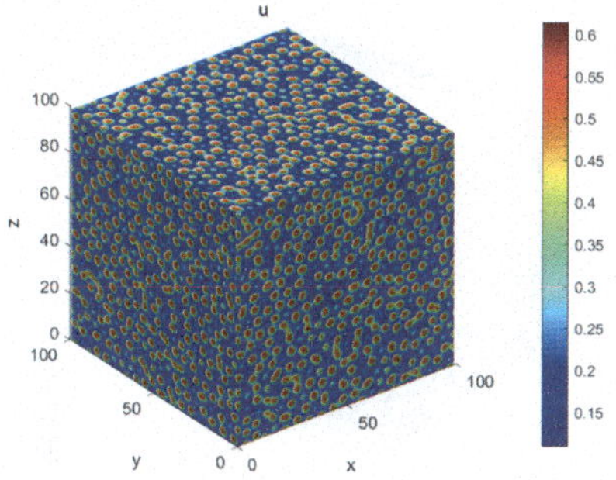

(a) $\tau = 1.57$, isosurface of prey population density

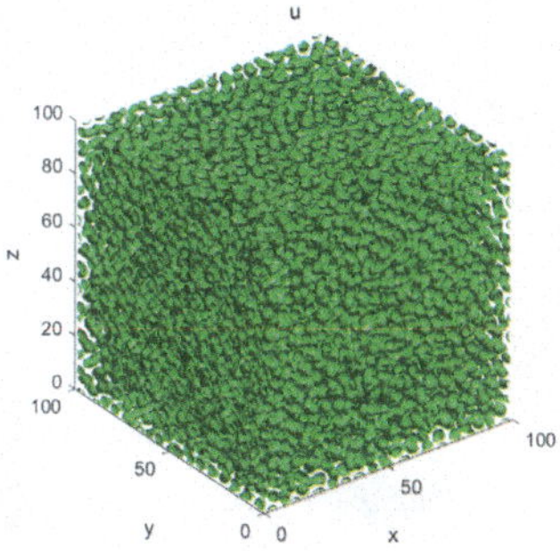

(b) $\tau = 1.57$, three-dimensional pattern of prey population density

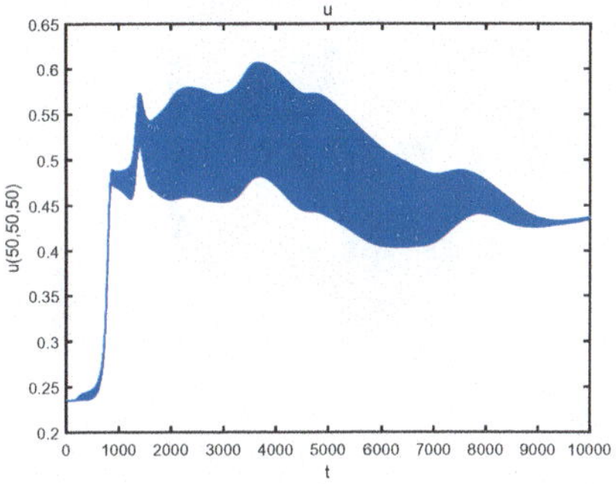

(c) $\tau = 1.57$, time series of predation population density at (50,50,50)

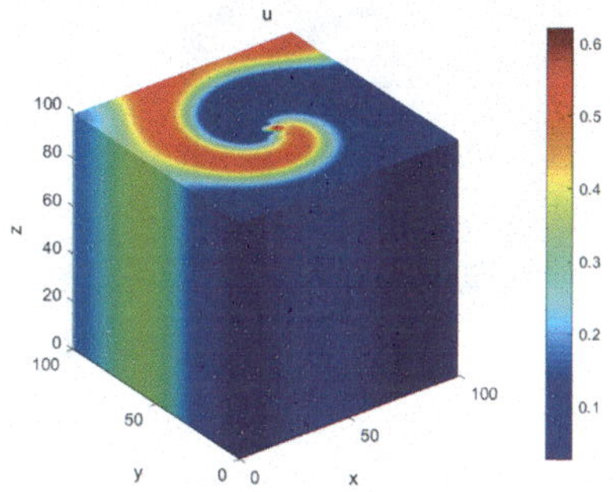

(d) $\tau = 1.8$, isosurface of prey population density

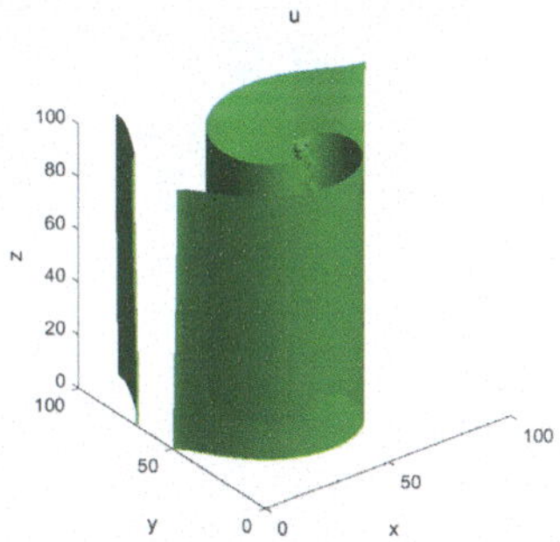

(e) $\tau = 1.8$, three-dimensional pattern of prey population density

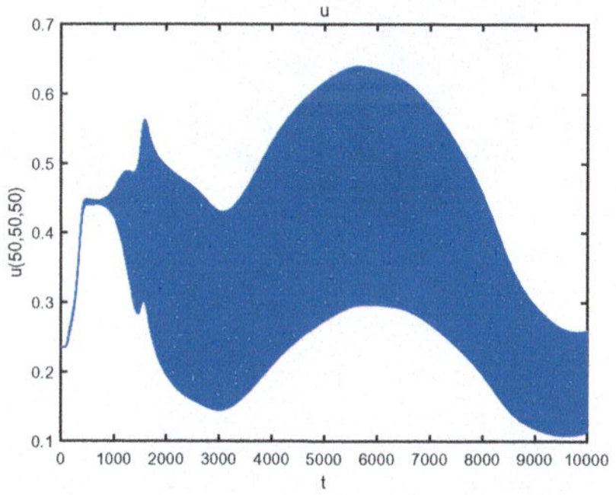

(f) $\tau = 1.8$, time series of prey population density at (50,50,50)

Fig. 18.11 Variation of the prey population density isosurface, three-dimensional spatial pattern, and time series of prey population density with different delay parameters τ. **b**, **e** The value of isosurface parameter is 0.32 and other parameters are $a = 1.5,\ b = 0.1,\ c = 0.42,\ d_{11} = 0.1,\ d_{21} = 0.8,\ d_{22} = 0.95,\ t = 10000$ and formula (4)

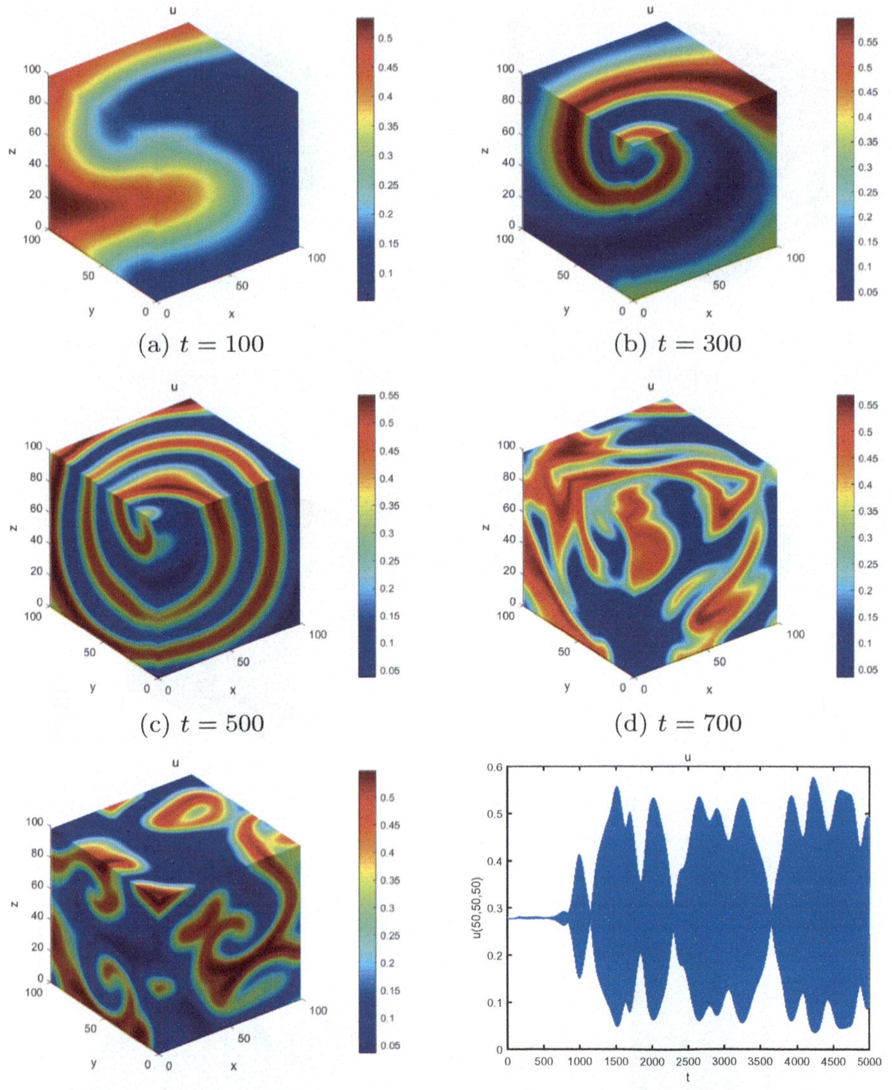

(a) $t = 100$

(b) $t = 300$

(c) $t = 500$

(d) $t = 700$

(e) $t = 4000$

(f) Time series of predation population density at (50,50,50)

Fig. 18.12 Variation of the three-dimensional spiral wave pattern of the prey population at different times

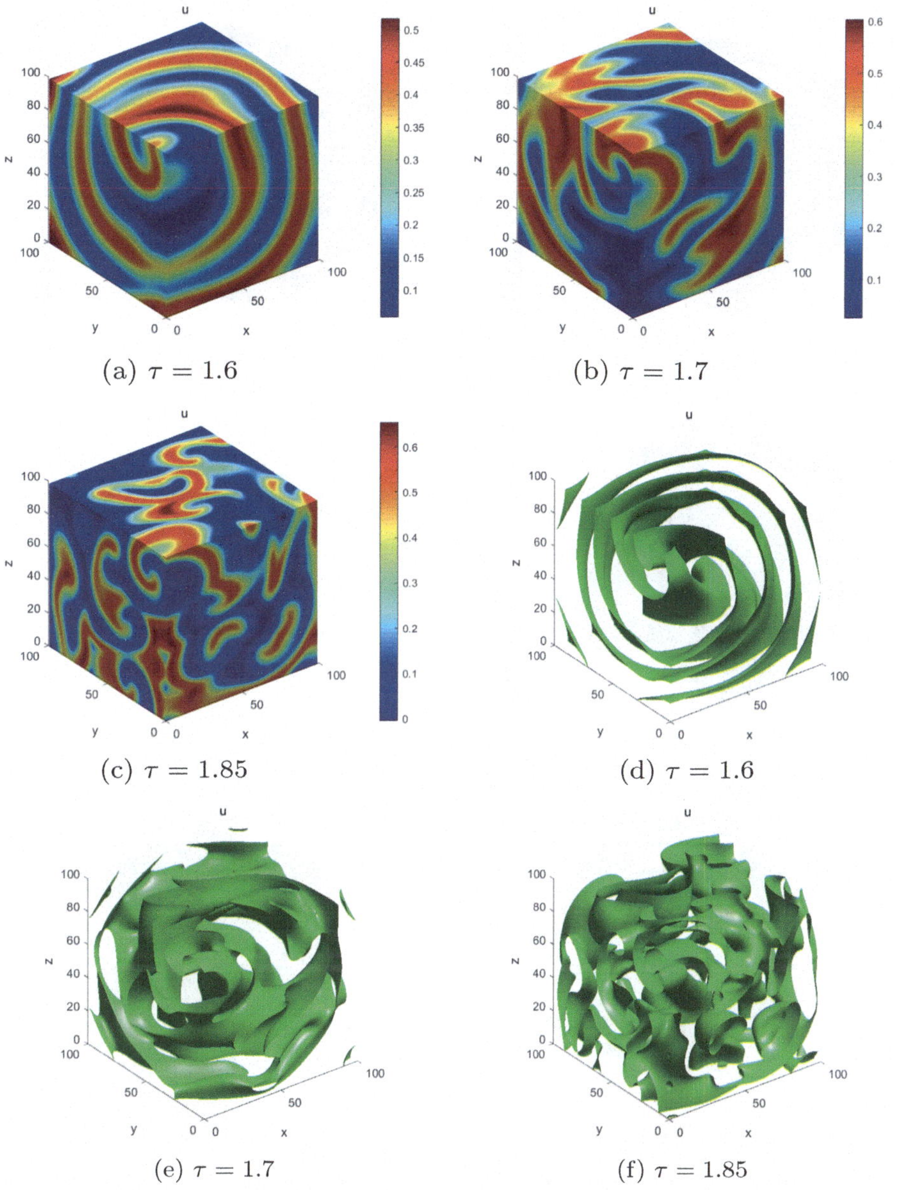

Fig. 18.13 The effect of time delay on the formation of 3D spiral wave patterns, where **a–c** mean the three-dimensional isosurface of the prey population density and **d–f** stand for the three-dimensional pattern of the prey population density; **d–f** The graphing subroutines were based on the function isosurface with values of the isovalue parameter equal to 0.32

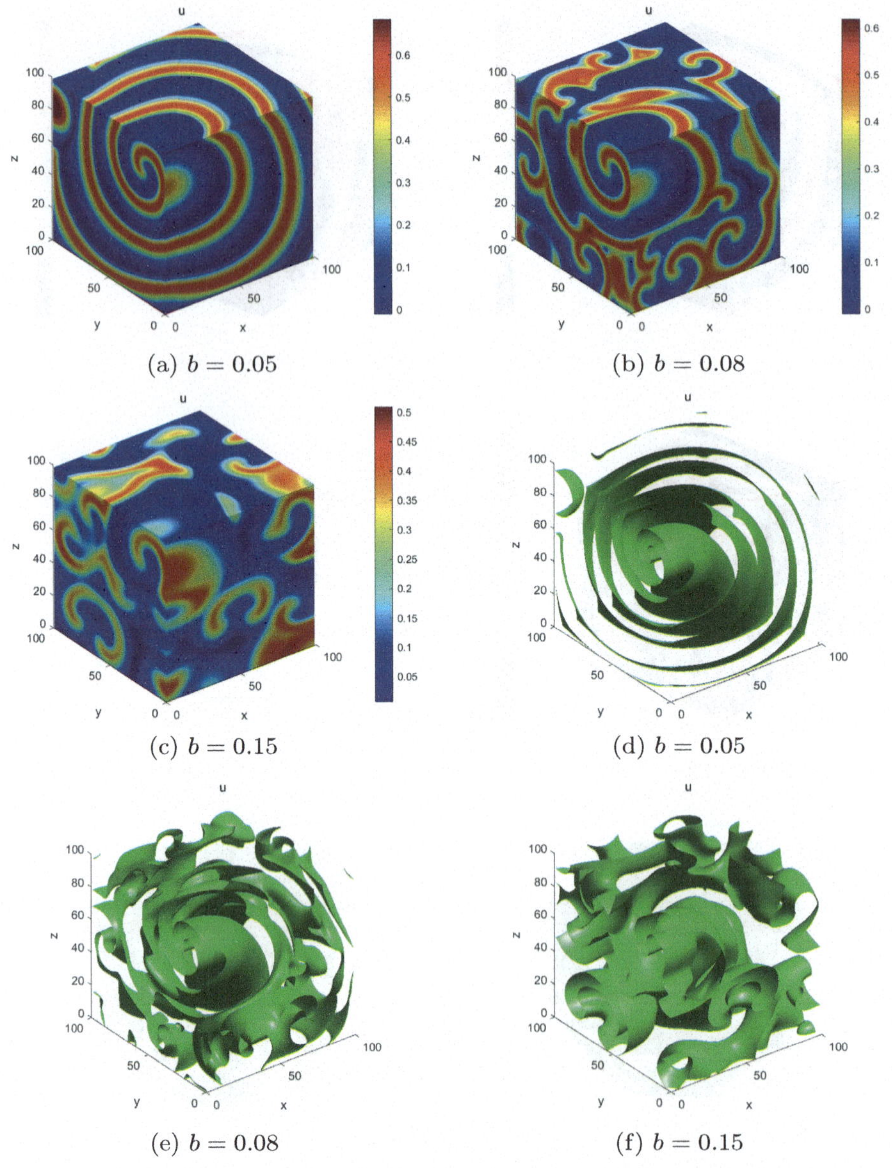

Fig. 18.14 The effect of the Allee effect on the formation of spiral wave patterns, where **a**–**c** mean the three-dimensional isosurface of the prey population density and **d**–**f** stand for the three-dimensional pattern of the prey population density. **d**–**f**—The graphing subroutines were based on the function isosurface with values of the isovalue parameter equal to 0.32

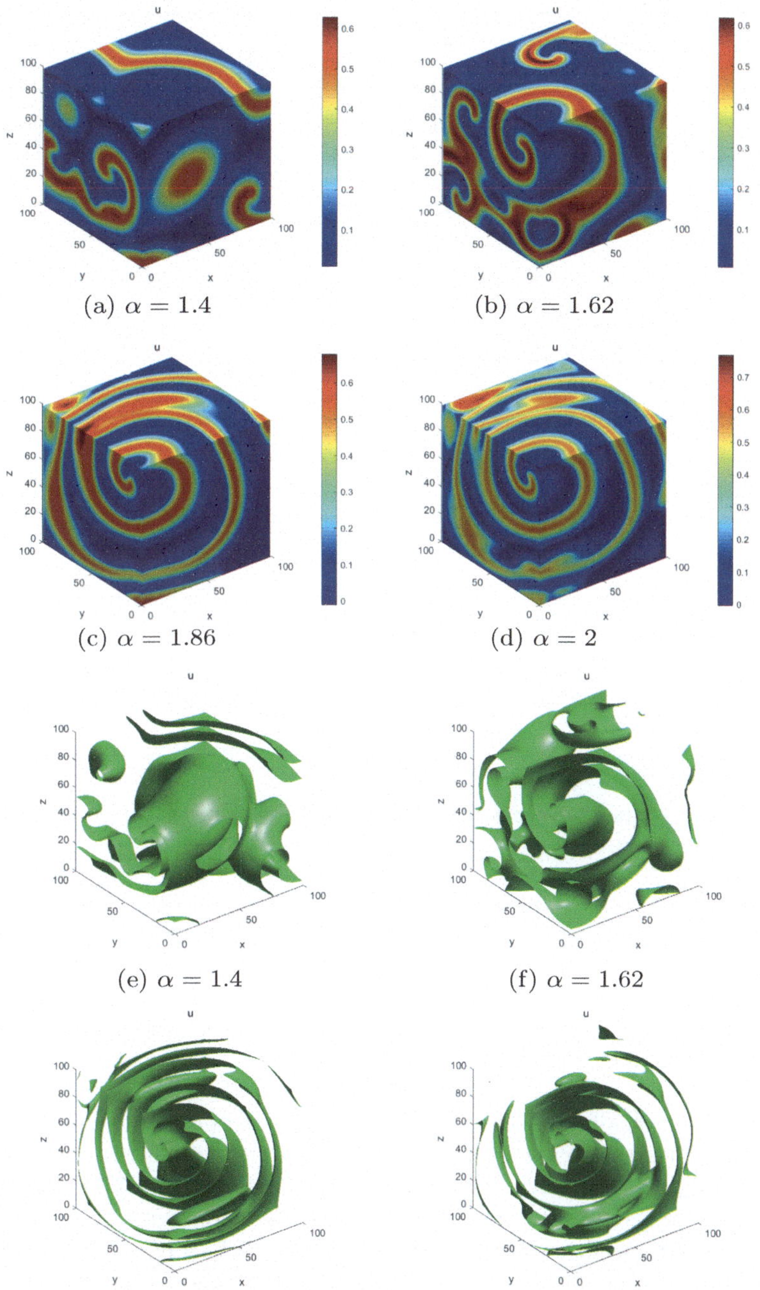

(a) $\alpha = 1.4$ (b) $\alpha = 1.62$

(c) $\alpha = 1.86$ (d) $\alpha = 2$

(e) $\alpha = 1.4$ (f) $\alpha = 1.62$

(g) $\alpha = 1.86$ (h) $\alpha = 2$

Fig. 18.15 The effect of fractional diffusion order α on the formation of the spiral wave pattern. **e–h** The graphing subroutines were based on the function isosurface with values of the isovalue parameter equal to 0.32

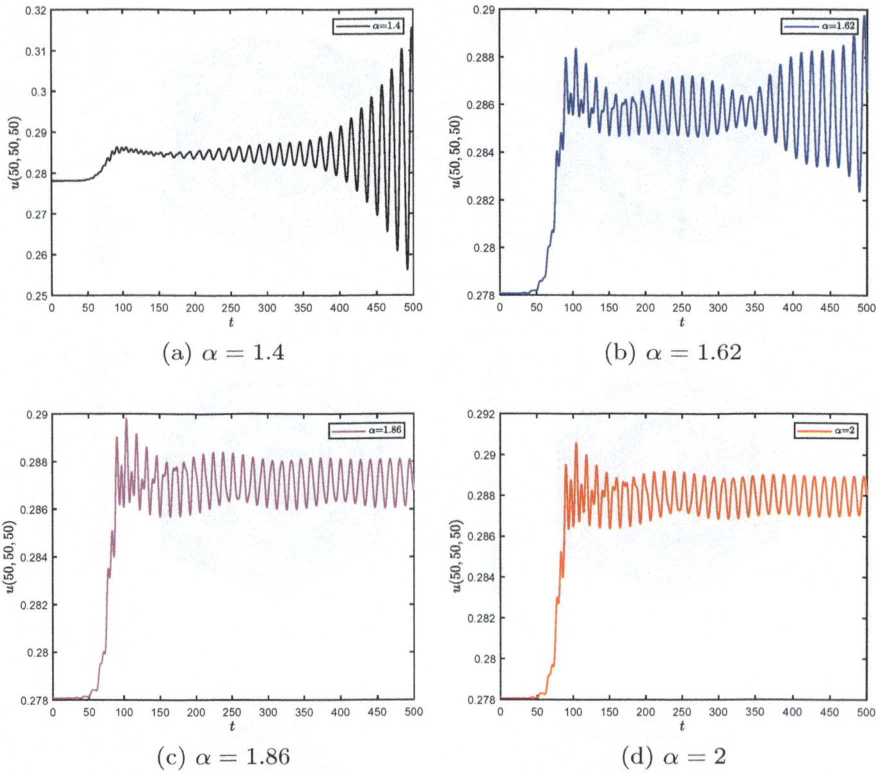

Fig. 18.16 Time series of prey population density at (50, 50, 50), where $a = 2.15$, $b = 0.05$, $c = 0.42$, $d_{11} = 0.2$, $d_{21} = 0.7$, $d_{22} = 0.6$, $t = 500$ and $\tau = 1.9$

Fig. 18.13b extending from the edge to the center in turn. With the time delay τ increasing to 1.98, the whole region is occupied by the completely irregular spiral pattern shown in Fig. 18.13c. Obviously, the increasing of the time delay τ can accelerate the breaking process of the spiral wave. Therefore, the time delay has a protective effect on the population.

Case 7: The role of the Allee effect in the formation of three-dimensional spiral waves is also discussed. Figure 18.14 illustrates the spiral wave pattern of the system (18.4) varying with the Allee effect b at the same time ($t = 500$). Let parameters as $a = 2.15$, $c = 0.42$, $d_{11} = 0.2$, $d_{21} = 0.36$, $d_{22} = 0.6$, $\tau = 2$, $\alpha = 1.8$ and other parameters be determined by Eq. (18.43). Clearly, Fig. 18.14a shows for $b = 0.05$, the running time $t = 500$, a regular three-dimensional spiral wave pattern is generated in system (18.4). With the increasing of the Allee effect b to 0.08, the regular spiral waves break up and then extend from the edge to the center displayed in Fig. 18.14b. It also yields that the whole region is occupied by the completely broken spiral pattern as the parameter b increases to 0.15 illustrated in Fig. 18.14c.

Apparently the increasing of the Allee effect parameter b can accelerate the breaking process of the spiral wave. Therefore, the Allee effect has a protective role in the population, just like the time delay.

Case 8: We also consider the effect of the order for fractional diffusion on pattern formation. Figure 18.15 shows the role of the fractional diffusion order α in the spiral wave, where Fig. 18.15a–d shows the three-dimensional isosurface for prey population density with isosurface parameter is 0.32, and Fig. 18.15e–h displays the three-dimensional patterns for the prey population. The parameters are taken as $a = 2.15,\ b = 0.05,\ c = 0.42,\ d_{11} = 0.2,\ d_{21} = 0.7,\ d_{22} = 0.6, \tau = 1.9$, the initial concentration is assumed as $u_0 = u_* + 0.0002(x - 50) - 0.0002(z - 50)$, $v_0 = v_* - 0.0002(y - 50) - 0.0002(z - 50)$, and the running time is $t = 500$. The fractional diffusion order α in Fig. 18.15a–d or Fig. 18.15e–h is set as 1.4, 1.62, 1.86, and 2 respectively. Shown in Fig. 18.15, there exists the broken of the spiral wave seen in Fig. 18.15a, b (Fig. 18.15e–f) , while there remains unbroken in Fig. 18.15c–d (Fig. 18.15g, h). For the same time t, as the increasing of the index α within a certain range, the system spiral wave becomes less likely to be broken, and also exists the increasing of the number of spiral. Fig. 18.16 also illustrates that as the order of fractional diffusion becomes large, the system will be relatively stable and not prone to oscillation. Thus, at the same operation, the larger of the fractional diffusion order α in a certain range can get more difficult for breaking the spiral wave, and the more turning of the spiral pattern.

18.6 Discussion and Conclusions

In this chapter, delay and Allee effect are introduced into a class of spatial fractional predator-prey systems and also investigated for its spatial dynamics. We establish a model as the super-diffusion extension for certain reaction-diffusion systems in biology as a more general form. The time delay has complicated effect on the system dynamic behavior with usually more realistic and natural.

We discuss the role of the time delay and fractional order diffusion in the stability of positive equilibrium, the direction of Hopf bifurcation, the condition of Hopf bifurcation and the spatial pattern structure. The numerical simulation shows rich dynamic behavior, and different spatial patterns, varying the correctness of the results. The numerical results also display that time delay and fractional diffusion play important role not only in the formation of Turing patterns for three-dimensional spherical droplets and tubes but also in the formation of three-dimensional spiral patterns.

The Allee effect gets plentiful dynamic behavior.

(1) In the Turing bifurcation parameter region, the shape of the pattern varies with the increasing of the Allee effect parameters continuously. A tubular pattern connects with the adjacent spherical droplet patterns signifying more aggregation of the prey

population. Ecologically, the increasing of the extended tube pattern improves the possibility of predation, and the increasing of the Allee effect improves the extinction risk.

(2) In the Hopf bifurcation region, the increase of additive Allee effect can obtain the acceleration of the process for spiral wave generation and extinction.

Taking the role of time delay in the dynamic behavior of the model into account, we obtain the stability varying with the increase of delay parameters.

(1) For the Turing bifurcation region, time delay has no effect on the Turing pattern structure but may have effect on the Turing pattern generation.
(2) In the Turing-Hopf bifurcation region, the larger of the time delay deduces the more stability for the spiral wave relatively.
(3) In the Hopf bifurcation region, with the increase of time delay, the pattern varies from a regular spiral case to an irregular spiral case. Ecologically, the population gradually emerges chaotic from the initial stable coexistence with the increase of delay parameters.

Hence, time delay and the Allee effect play a crucial role in spatial population invasion. Moreover, fractional diffusion has an essential effect on the formation of complex patterns. The smaller of the fractional diffusion order get the easier break up of the spiral pattern, the more chaotic of the state, and the more dispersed for the population distribution. Thus, fractional diffusion may be a significant factor in deeply and further understanding the complexity for biological and ecosystem patterns.

Appendix A

The proof of Theorem 1 is divided into two steps. First, we prove the existence of unique local positive solution of (3.2) on $t \in [0, \tau_e)$. Then, by proving $\tau_e = +\infty$, we can show that the solution is global.

Step 1. Let $u(t) = \ln N(t)$, $v(t) = \ln P(t)$ and $w(t) = \ln Z(t)$, we obtain the following equations

$$\begin{aligned} du(t) &= \left(\tfrac{I}{e^{u(t)}} - a - \tfrac{\mu(x)}{e^{u(t)}+K} e^{v(t)} Q(x) - \tfrac{\sigma_1^2}{2}\right) dt + \sigma_1 dB_1(t), \\ dv(t) &= \left(\mu(x) - \tfrac{e^{u(t)}}{e^{u(t)}+K} - m - s(x) - C(x, y)e^{w(t)} - \tfrac{\sigma_2^2}{2}\right) dt + \sigma_2 dB_2(t), \\ dw(t) &= \left(\phi \tfrac{Q(x)}{q(y)} C(x, y)e^{v(t)} - \delta - \tfrac{\sigma_3^2}{2}\right) dt + \sigma_3 dB_3(t). \end{aligned} \tag{A.1}$$

on $t \geq 0$ with initial values $u(0) = \ln N_0$, $v(0) = \ln P_0$, $w(0) = \ln Z_0$. Obviously, the coefficients of (A.1) satisfy local Lipschitz continuous, hence there is a unique local solution of (A.1) on $t \in [0, \tau_e)$, where τ_e is the explosion time. By Itô's formula, $N(t) = e^{u(t)}$, $P(t) = e^{v(t)}$ and $Z(t) = e^{w(t)}$ is the unique local positive solution of (3.2) with initial value $(N_0, P_0, Z_0) \in \mathbb{R}_+^3$.

Step 2. Now we will prove $\tau_e = +\infty$. Let $k_0 > 0$ be sufficiently large such that N_0, P_0 and Z_0 belong to $\left[\frac{1}{k_0}, k_0\right]$. For each integer $k > k_0$, we define the following stopping time:

$$\tau_k = \inf\left\{t \in [0, \tau_e) : \min\{N(t), P(t), Z(t)\} \leq \frac{1}{k} \; or \; \max\{N(t), P(t), Z(t)\} \geq k\right\}.$$

Obviously, τ_k is increasing as $k \to +\infty$. Set $\tau_\infty = \lim_{k\to+\infty} \tau_k$, where $\tau_\infty \leq \tau_e$ a.s. Now, we prove $\tau_\infty = +\infty$. If this statement is false, there is a pair of constants $T > 0$ and $\varepsilon \in (0, 1)$ such that $\mathbb{P}\{\tau_k \leq T\} > \varepsilon$. Thus there exists an integer $k_1 \geq k_0$ such that $\mathbb{P}\{\tau_k \leq T\} \geq \varepsilon, \; for \; all \; k \geq k_1$. Next, we define a C^2-function $\tilde{V} : \mathbb{R}_+^3 \to \mathbb{R}_+$ by

$$\tilde{V}(N, P, Z) = \frac{1}{Q(x)}(N - 1 - \log N) + (P - 1 - \log P) + \frac{q(y)}{\phi Q(x)}(Z - 1 - \log Z). \tag{A.2}$$

S. T. Liu and L. Zhang, *Pattern Dynamics of Marine Plankton Behavior*,
https://doi.org/10.1007/978-981-97-5369-7

The nonnegativity of (A.2) can be obtained from: $S-1-\log S>0,\ \forall S>0$. By Itô's formula, we get

$$\begin{aligned} &d\tilde{V}(N,P,Z)\\ &=\mathcal{L}\tilde{V}(N,P,Z)dt+\tfrac{1}{Q(x)}\left(1-\tfrac{1}{N}\right)\sigma_1 N dB_1(t)+\left(1-\tfrac{1}{P}\right)\sigma_2 P dB_2(t)\\ &+\tfrac{q(y)}{\phi Q(x)}\left(1-\tfrac{1}{Z}\right)\sigma_3 Z dB_3(t), \end{aligned} \tag{A.3}$$

where

$$\begin{aligned} &\mathcal{L}\tilde{V}(N,P,Z)\\ &=\tfrac{1}{Q(x)}\left(1-\tfrac{1}{N}\right)\left(I-aN-\mu(x)\tfrac{N}{N+K}PQ(x)\right)+\tfrac{1}{2Q(x)}\sigma_1^2\\ &+\left(1-\tfrac{1}{P}\right)P\left(\mu(x)\tfrac{N}{N+K}-m-s(x)-C(x,y)Z\right)+\tfrac{1}{2}\sigma_2^2\\ &+\tfrac{q(y)}{\phi Q(x)}\left(1-\tfrac{1}{Z}\right)Z\left(\phi\tfrac{Q(x)}{q(y)}C(x,y)P-\delta\right)+\tfrac{q(y)}{2\phi Q(x)}\sigma_3^2\\ &\le\tfrac{I+a}{Q(x)}+m+s(x)+\delta\tfrac{q(y)}{\phi Q(x)}+\tfrac{1}{2Q(x)}\sigma_1^2+\tfrac{1}{2}\sigma_2^2+\tfrac{q(y)}{2\phi Q(x)}\sigma_3^2\\ &+\mu(x)\tfrac{P}{N+K}+C(x,y)Z\\ &=K_1+\mu(x)\tfrac{P}{N+K}+C(x,y)Z \end{aligned}$$

with $K_1=\frac{I+a}{Q(x)}+m+s(x)+\frac{\delta q(y)}{\phi Q(x)}+\frac{1}{2Q(x)}\sigma_1^2+\frac{1}{2}\sigma_2^2+\frac{q(y)}{2\phi Q(x)}\sigma_3^2$. In addition, by inequality $S\le 2(S-1-\log S)+2\log 2,\ \forall S>0$, we have

$$\begin{aligned} &\mathcal{L}\tilde{V}(N,P,Z)\le K_1+\mu(x)\tfrac{P}{K}+C(x,y)Z\\ &\le K_1+2\mu(x)\tfrac{1}{K}(P-1-\log P)+2C(x,y)(Z-1-\log Z)\\ &+2\log 2\left(\mu(x)\tfrac{1}{K}+C(x,y)\right)\\ &\le\bar{\lambda}+K_2\tilde{V}(N,P,Z), \end{aligned} \tag{A.4}$$

where $K_2=2\max\left\{\mu(x)\frac{1}{K},C(x,y)\frac{\phi Q(x)}{q(y)}\right\}$ and $\bar{\lambda}=K_1+2\log 2\left(\mu(x)\frac{1}{K}+C(x,y)\right)$.

According to (A.3) and (A.4), we can obtain

$$\begin{aligned} &d\tilde{V}(N,P,Z)\le(\bar{\lambda}+K_2\tilde{V}(N,P,Z))dt+\tfrac{1}{Q(x)}(N-1)\sigma_1 dB_1(t)\\ &+(P-1)\sigma_2 dB_2(t)+\tfrac{q(y)}{\phi Q(x)}(Z-1)\sigma_3 dB_3(t). \end{aligned}$$

Integrating both sides of the above inequality from 0 to $\tau_k\wedge T$ and taking the expectation yield

$$\begin{aligned} &E(\tilde{V}(N(\tau_k\wedge T),P(\tau_k\wedge T),Z(\tau_k\wedge T)))\le\tilde{V}(N_0,P_0,Z_0)+\bar{\lambda}T\\ &+K_2\int_0^{\tau_k\wedge T}E(\tilde{V}(N(t),P(t),Z(t)))dt. \end{aligned}$$

This together with Gronwall's inequality [24] implies

$$E(\tilde{V}(N(\tau_k \wedge T), P(\tau_k \wedge T), Z(\tau_k \wedge T))) \leq (\tilde{V}(N_0, P_0, Z_0) + \bar{\lambda}T)e^{K_2 T}. \quad \text{(A.5)}$$

Set $\Omega_k = \{\omega \in \Omega : \tau_k = \tau_k(\omega) \leq T\}$. Then we obtain $\mathbb{P}(\Omega_k) \geq \varepsilon$. For every $\omega \in \omega_k$, $N(\tau_k, \omega)$, $P(\tau_k, \omega)$ or $Z(\tau_k, \omega)$ equals either k or $1/k$, thus $\tilde{V}(N(\tau_k, \omega), P(\tau_k, \omega), Z(\tau_k, \omega))$ is no less than

$$k - 1 - \log k \ \ or \ \ 1/k - 1 - \log(1/k),$$

which implies

$$\begin{aligned} +\infty > (\tilde{V}(N_0, P_0, Z_0) + \bar{\lambda}T)e^{K_2 T} &\geq E[I_{\Omega_k}(\omega)\tilde{V}(N(\tau_k \wedge T), P(\tau_k \wedge T), Z(\tau_k \wedge T))] \\ &\geq \varepsilon[(k - 1 - \log k) \wedge (1/k - 1 - \log(1/k))], \end{aligned}$$

where I_{Ω_k} is the indicator function of Ω_k. Letting $k \to +\infty$, we get $(\tilde{V}(N_0, P_0, Z_0) + \bar{\lambda}T)e^{K_2 T} > +\infty$, which yields contradiction. Hence, we have $\tau_\infty = +\infty$ $a.s.$ This completes the proof.

Appendix B

In the following, in order to verify condition (B.2) of Lemma 4, we will prove that $\mathcal{L}V \le -1$ on the six domains which mentioned in Sect. 3.2.

Case I: If $(N, P, Z) \in U_1$, then

$$\mathcal{L}V \le M\left(-\lambda + \frac{\mu(x)PQ(x)I}{aK} + \sigma_1 C(x, y)Z\right) - \frac{I}{N} + a + \mu(x)\frac{PQ(x)}{K} + \frac{1}{2}\sigma_1^2 - \frac{1}{4}(2a - \sigma_1^2)N^2$$

$$+\phi\frac{Q(x)}{q(y)}C(x, y)PZ - \frac{1}{4}(2m + 2s(x) - \sigma_2^2)(Q(x)P)^2 - \frac{1}{4}(2\delta - \sigma_3^2)\left(\frac{q(y)}{\phi}Z\right)^2 + E$$

$$\le -\lambda M + \phi\frac{Q(x)}{q(y)}C(x, y)P\varepsilon - \frac{1}{8}(2m + 2s(x) - \sigma_2^2)(Q(x)P)^2 + F$$

$$\le -\lambda M + \phi\frac{Q(x)}{q(y)}C(x, y)\varepsilon\frac{1+P^2}{2} - \frac{1}{8}(2m + 2s(x) - \sigma_2^2)(Q(x)P)^2 + F$$

$$\le -\lambda M + \frac{1}{2}\left(\phi\frac{Q(x)}{q(y)}C(x, y)\varepsilon - \frac{1}{4}(2m + 2s(x) - \sigma_2^2)Q(x)^2\right)P^2 + \phi\frac{Q(x)}{2q(y)}C(x, y)\varepsilon + F$$

$$\le -\lambda M + \phi\frac{Q(x)}{2q(y)}C(x, y)\varepsilon + F.$$

From (3.4), we obtain $\mathcal{L}V \le -1$.

Case II: If $(N, P, Z) \in U_2$, then

$$\mathcal{L}V \le M\left(-\lambda + \frac{I\mu(x)PQ(x)}{aK} + \sigma_1 C(x, y)Z\right) - \frac{I}{N} + a + \mu(x)\frac{PQ(x)}{K} + \frac{1}{2}\sigma_1^2 - \frac{1}{4}(2a - \sigma_1^2)N^2$$
$$+\phi\frac{Q(x)}{q(y)}C(x, y)PZ - \frac{1}{4}(2m + 2s(x) - \sigma_2^2)(Q(x)P)^2 - \frac{1}{4}(2\delta - \sigma_3^2)\left(\frac{q(y)}{\phi}Z\right)^2 + E$$
$$\le -\lambda M + \phi\frac{Q(x)}{q(y)}C(x, y)Z\varepsilon - \frac{1}{8}(2\delta - \sigma_3^2)\left(\frac{q(y)}{\phi}Z\right)^2 + F$$
$$\le -\lambda M + \phi\frac{Q(x)}{q(y)}C(x, y)\varepsilon\frac{1+Z^2}{2} - \frac{1}{8}(2\delta - \sigma_3^2)\left(\frac{q(y)}{\phi}Z\right)^2 + F$$
$$\le -\lambda M + \frac{1}{2}\left(\phi\frac{Q(x)}{q(y)}C(x, y)\varepsilon - \frac{1}{4}(2\delta - \sigma_3^2)\left(\frac{q(y)}{\phi}\right)^2\right)Z^2 + \phi\frac{Q(x)}{2q(y)}C(x, y)\varepsilon + F$$
$$\le -\lambda M + \phi\frac{Q(x)}{2q(y)}C(x, y)\varepsilon + F.$$

S. T. Liu and L. Zhang, *Pattern Dynamics of Marine Plankton Behavior*, https://doi.org/10.1007/978-981-97-5369-7

From (3.4), we obtain $\mathcal{L}V \leq -1$.

Case III: If $(N, P, Z) \in U_3$, then

$$\mathcal{L}V \leq -\frac{I}{N} + M\left(\frac{I\mu(x)PQ(x)}{aK} + \sigma_1 C(x, y)Z\right) + a + \mu(x)\frac{PQ(x)}{K} + \frac{1}{2}\sigma_1^2 - \frac{1}{4}(2a - \sigma_1^2)N^2$$

$$+\phi\frac{Q(x)}{q(y)}C(x, y)PZ - \frac{1}{4}(2m + 2s(x) - \sigma_2^2)(Q(x)P)^2 - \frac{1}{4}(2\delta - \sigma_3^2)\left(\frac{q(y)}{\phi}Z\right)^2 + E$$

$$\leq -\frac{I}{N} + H \leq -\frac{I}{\varepsilon} + H.$$

From (3.4), we obtain $\mathcal{L}V \leq -1$.

Case IV: If $(N, P, Z) \in U_4$, then

$$\mathcal{L}V \leq M\left(\frac{I\mu(x)PQ(x)}{aK} + \sigma_1 C(x, y)Z\right) + a + \mu(x)\frac{PQ(x)}{K} + \frac{1}{2}\sigma_1^2 + \phi\frac{Q(x)}{q(y)}C(x, y)PZ$$

$$-\frac{1}{4}(2a - \sigma_1^2)N^2 - \frac{1}{4}(2m + 2s(x) - \sigma_2^2)(Q(x)P)^2 - \frac{1}{4}(2\delta - \sigma_3^2)\left(\frac{q(y)}{\phi}Z\right)^2 + E$$

$$\leq -\frac{1}{8}(2a - \sigma_1^2)N^2 + H \leq -\frac{1}{8\varepsilon^2}(2a - \sigma_1^2) + H.$$

From (3.4), we obtain $\mathcal{L}V \leq -1$.

Case V: If $(N, P, Z) \in U_5$, then

$$\mathcal{L}V \leq M\left(\frac{I\mu(x)PQ(x)}{aK} + \sigma_1 C(x, y)Z\right) + a + \mu(x)\frac{PQ(x)}{K} + \frac{1}{2}\sigma_1^2 + \phi\frac{Q(x)}{q(y)}C(x, y)PZ$$

$$-\frac{1}{4}(2a - \sigma_1^2)N^2 - \frac{1}{4}(2m + 2s(x) - \sigma_2^2)(Q(x)P)^2 - \frac{1}{4}(2\delta - \sigma_3^2)\left(\frac{q(y)}{\phi}Z\right)^2 + E$$

$$\leq -\frac{1}{8}(2m + 2s(x) - \sigma_2^2)(Q(x)P)^2 + H \leq -\frac{1}{8\varepsilon^2}(2m + 2s(x) - \sigma_2^2)Q(x)^2 + H.$$

From (3.4), we obtain $\mathcal{L}V \leq -1$.

Case VI: If $(N, P, Z) \in U_6$, then

$$\mathcal{L}V \leq M\left(\frac{I\mu(x)PQ(x)}{aK} + \sigma_1 C(x, y)Z\right) + a + \mu(x)\frac{PQ(x)}{K} + \frac{1}{2}\sigma_1^2 + \phi\frac{Q(x)}{q(y)}C(x, y)PZ$$

$$-\frac{1}{4}(2a - \sigma_1^2)N^2 - \frac{1}{4}(2m + 2s(x) - \sigma_2^2)(Q(x)P)^2 - \frac{1}{4}(2\delta - \sigma_3^2)\left(\frac{q(y)}{\phi}Z\right)^2 + E$$

$$\leq -\frac{1}{8}(2\delta - \sigma_3^2)\left(\frac{q(y)}{\phi}Z\right)^2 + H \leq -\frac{1}{8\varepsilon^2}(2\delta - \sigma_3^2)\left(\frac{q(y)}{\phi}\right)^2 + H.$$

From (3.4), we obtain $\mathcal{L}V \leq -1$.

From above cases, we know that

$$\mathcal{L}V \leq -1, \quad \forall (N, P, Z) \in U^c,$$

and condition (B.2) holds.

Appendix C

The expressions of F_1 and F_2 in (7.17) are listed as follows.

$$h_{v_1v_1} = -\alpha + \frac{C(z_1,z_2)Z_*(a-P_*+bZ_*)}{(a+P_*+bZ_*)^3}, \qquad h_{v_1v_2} = \frac{C(z_1,z_2)P_*(a+P_*-bZ_*)}{(a+P_*+bZ_*)^3},$$

$$h_{v_2v_2} = \frac{2bC(z_1,z_2)P_*(a+P_*)}{(a+P_*+bZ_*)^3}, \qquad f_{v_1v_1} = -\frac{2C(z_1,z_2)Z_*(a+bZ_*)}{(a+P_*+bZ_*)^3},$$

$$f_{v_1v_2} = \frac{C(z_1,z_2)(a+2bZ_*-2abZ_*-2b^2Z_*^2)}{(a+P_*+bZ_*)^3}, \qquad f_{v_2v_2} = -\frac{2bC(z_1,z_2)P_*(a+P_*)}{(a+P_*+bZ_*)^3},$$

$$h_{v_1v_1v_1} = -\frac{2C(z_1,z_2)Z_*(2a-P_*+2bZ_*)}{(a+P_*+bZ_*)^4}, \qquad h_{v_1v_1v_2} = \frac{C(z_1,z_2)(a^2-P_*^2-b^2Z_*^2+4bP_*Z_*)}{(a+P_*+bZ_*)^4}, \; h_{v_1v_2v_2}$$

$$= \frac{2bC(z_1,z_2)(a+P_*-2bZ_*)}{(a+P_*+bZ_*)^4}, \qquad h_{v_2v_2v_2} = -\frac{6b^2C(z_1,z_2)P_*(a+P_*)}{(a+P_*+bZ_*)^4}, \; f_{v_1v_1v_1}$$

$$= \frac{6Z_*C(z_1,z_2)(a+bZ_*)}{(a+P_*+bZ_*)^4}, \qquad f_{v_2v_2v_2} = -\frac{6b^2C(z_1,z_2)P_*(a+P_*)}{(a+P_*+bZ_*)^4}, \; f_{v_1v_2v_2}$$

$$= \frac{C(z_1,z_2)b(2P_*-2a^2-2aP_*-4bZ_*P_*-a-4Z_*+2b^2Z_*^2)}{(a+P_*+bZ_*)^4},$$

$$f_{v_1v_1v_2} = -\frac{2C(z_1,z_2)(a^2+aP_*+2bZ_*P_*-b^2Z_*^2)}{(a+P_*+bZ_*)^4}.$$

S. T. Liu and L. Zhang, *Pattern Dynamics of Marine Plankton Behavior*,
https://doi.org/10.1007/978-981-97-5369-7

Appendix D

The expressions of (7.32) and (7.33) are listed as follows.

$$
\begin{aligned}
&\kappa_1 = \tfrac{\chi_1}{l\pi}\int_0^{l\pi}\cos^2\left(\tfrac{nx}{l}\right)dx + \tfrac{\chi_2}{l\pi}\int_0^{l\pi}\cos^4\left(\tfrac{nx}{l}\right)dx,\\
&\kappa_2\\
&= \tfrac{\varsigma_1}{l\pi}\int_0^{l\pi}\cos^2\left(\tfrac{nx}{l}\right)dx + \tfrac{\varsigma_2}{l\pi}\int_0^{l\pi}\cos^4\left(\tfrac{nx}{l}\right)dx,\\
&\chi_{20} = \tfrac{1}{2}(h_{v_1v_1} + g(2h_{v_1v_2} + gh_{v_2v_2})),\\
&\chi_{11} = \tfrac{1}{4}(2h_{v_1v_1} + (g+\bar{g})h_{v_1v_2} + g\bar{g}h_{v_2v_2}),\\
&\chi_1 = W_{11}^{(1)}(0)(2h_{v_1v_1} + gh_{v_1v_2}) + W_{11}^{(2)}(0)(h_{v_1v_2} + 2gh_{v_2v_2})\\
&+W_{20}^{(1)}(0)\tfrac{2h_{v_1v_1}+\bar{g}h_{v_1v_2}}{2} + W_{20}^{(2)}(0)\tfrac{h_{v_1v_2}+2\bar{g}h_{v_2v_2}}{2},\\
&\chi_2 = \tfrac{1}{4}(3h_{v_1v_1v_1} + (2g+\bar{g})h_{v_1v_1v_2} + g(2\bar{g}h_{v_1v_2v_2} + gh_{v_1v_2v_2} + 3h_{v_2v_2v_2}\bar{g}g)),\\
&\varsigma_{20} = \tfrac{1}{2}e^{-2i\omega_n\tau_0}(f_{v_1v_1} + g(f_{v_1v_2} + f_{v_2v_2}g)),\\
&\varsigma_{11} = \tfrac{1}{4}(2f_{v_1v_1} + (g+\bar{g})f_{v_1v_2} + 2f_{v_2v_2}g\bar{g}),\\
&\varsigma_1 = e^{-i\omega_n\tau_0}W_{11}^{(1)}(-1)(2h_{v_1v_1} + gh_{v_1v_2}) + \tfrac{1}{2}W_{20}^{(1)}(-1)e^{i\omega_n\tau_0}(2h_{v_1v_1} + \bar{g}h_{v_1v_2})\\
&+\tfrac{1}{2}W_{20}^{(2)}(-1)e^{i\omega_n\tau_0}(h_{v_1v_2} + 2\bar{g}h_{v_2v_2}) + e^{-i\omega_n\tau_0}W_{11}^{(2)}(-1)(h_{v_1v_2} + 2gh_{v_2v_2}),\\
&\varsigma_2 = \tfrac{1}{4}e^{-i\omega_n\tau_0}(3h_{v_1v_1v_1} + (2g+\bar{g})h_{v_1v_1v_2} + g(2\bar{g}h_{v_1v_2v_2} + gh_{v_1v_2v_2} + 3h_{v_2v_2v_2}\bar{g}g)).
\end{aligned}
$$

S. T. Liu and L. Zhang, *Pattern Dynamics of Marine Plankton Behavior*,
https://doi.org/10.1007/978-981-97-5369-7

Appendix E

In this section, we give some detailed presentations produced in Sect. 15.3.

We set $\mathrm{P_n} = \begin{bmatrix} p_n \\ z_n \end{bmatrix}$ and $A = \begin{bmatrix} a_{11} & a_{12} \\ a_{21} & a_{22} \end{bmatrix}$, then

$$\mathrm{P}_{21} = -2\begin{bmatrix} J_{11} & J_{12} \\ J_{21} & J_{22} \end{bmatrix}^{-1} A \begin{bmatrix} (J_{12} - d_{12}q_c^2)^2 \\ d_{11}q_c^2(J_{12} - d_{12}q_c^2) \end{bmatrix}, \mathrm{P}_{22} = -2\begin{bmatrix} J_{11} & J_{12} \\ J_{21} & J_{22} \end{bmatrix}^{-1} A \begin{bmatrix} J_{12}^2 \\ 0 \end{bmatrix},$$

$$\mathrm{P}_{23} = 2\begin{bmatrix} i\omega_c + d_{11}q_c^2 & -J_{12} + d_{12}q_c^2 \\ -J_{21} + d_c q_c^2 & i\omega_c + d_{22}q_c^2 \end{bmatrix}^{-1} A \begin{bmatrix} J_{12}^2 - d_{12}J_{12}q_c^2 \\ \frac{1}{2}(J_{12} - d_{12}q_c^2)i\omega_c + \frac{1}{2}d_{11}J_{12}q_c^2 \end{bmatrix},$$

$$\mathrm{P}_{25} = -\begin{bmatrix} -4d_{11}q_c^2 & J_{12} - 4d_{12}q_c^2 \\ J_{21} - 4d_c q_c^2 & -4d_{22}q_c^2 \end{bmatrix}^{-1} A \begin{bmatrix} (J_{12} - d_{12}q_c^2)^2 \\ d_{11}q_c^2(J_{12} - d_{12}q_c^2) \end{bmatrix},$$

$$\mathrm{P}_{26} = \begin{bmatrix} 2i\omega_c & -J_{12} \\ -J_{21} & 2i\omega_c \end{bmatrix}^{-1} A \begin{bmatrix} J_{12}^2 \\ iJ_{12}\omega_c \end{bmatrix}, \; \mathrm{P}_{28} = -2\begin{bmatrix} -3d_{11}q_c^2 & J_{12} \\ J_{21} & -3d_{22}q_c^2 \end{bmatrix}^{-1} A \begin{bmatrix} (J_{12} - d_{12}q_c^2)^2 \\ d_{11}q_c^2(J_{12} - d_{12}q_c^2) \end{bmatrix}.$$

Subsequently, we introduce following notations

$$L_2(q) = \begin{bmatrix} (\frac{K+h}{(\alpha-\delta_c)^2} - \frac{2h}{(\alpha-\delta_c-m)^2})\frac{mr\delta_2}{K} & -\frac{m\beta\delta_2}{(\alpha-\delta_c)^2} \\ -d_2q^2 - \frac{r\delta_2}{\beta} & 0 \end{bmatrix},$$

$$I_{1H} = \begin{bmatrix} 2a_{11}(p_{22} + p_{26})\psi + a_{12}(z_{22} + z_{26})\psi + ia_{12}(p_{22} - p_{26})\omega_c \\ 2a_{21}(p_{22} + p_{26})\psi + a_{22}(z_{22} + z_{26})\psi + ia_{22}(p_{22} - p_{26})\omega_c \end{bmatrix}, S_{1H} = \begin{bmatrix} 3b_{11}\psi^3 + b_{12}\psi^2 \\ 3b_{21}\psi^3 + b_{22}\psi^2 \end{bmatrix},$$

$$I_{2H} = \begin{bmatrix} (4a_{11}p_{23} + 2a_{12}z_{23})\varphi + 2(a_{11}p_{21} + a_{12}z_{21})\psi + a_{12}(ip_{21}\omega_c + 2d_{11}p_{23}q_c^2) \\ (4a_{21}p_{23} + 2a_{22}z_{23})\varphi + 2(a_{21}p_{21} + a_{22}z_{21})\psi + a_{22}(ip_{21}\omega_c + 2d_{11}p_{23}q_c^2) \end{bmatrix},$$

S. T. Liu and L. Zhang, *Pattern Dynamics of Marine Plankton Behavior*,
https://doi.org/10.1007/978-981-97-5369-7

$$S_{2H} = \begin{bmatrix} (6b_{11}\psi + 2ib_{12}\omega_c)\varphi^2 + 4d_{11}b_{12}q_c^2\varphi \\ (6b_{21}\psi + 2ib_{22}\omega_c)\varphi^2 + 4d_{11}b_{22}q_c^2\varphi \end{bmatrix}, \ I_{1T} = \begin{bmatrix} (2a_{11}\varphi + d_{11}a_{12}q_c^2)(p_{21} + p_{25}) + a_{12}(z_{21} + z_{25})\varphi \\ (2a_{21}\varphi + d_{11}a_{22}q_c^2)(p_{21} + p_{25}) + a_{22}(z_{21} + z_{25})\varphi \end{bmatrix},$$

$$S_{1T} = \begin{bmatrix} 3(b_{11}\varphi + d_{11}b_{12}q_c^2)\varphi^2 \\ 3(b_{21}\varphi + d_{11}b_{22}q_c^2)\varphi^2 \end{bmatrix}, \ I_{2T} = \begin{bmatrix} (2a_{11}\varphi + d_{11}a_{12}q_c^2)(p_{21} + p_{28}) + a_{12}(z_{21} + z_{28})\varphi \\ (2a_{21}\varphi + d_{11}a_{22}q_c^2)(p_{21} + p_{28}) + a_{22}(z_{21} + z_{28})\varphi \end{bmatrix},$$

$$S_{2T} = \begin{bmatrix} 6(b_{11}\varphi + d_{11}b_{12}q_c^2)\varphi^2 \\ 6(b_{21}\varphi + d_{11}b_{22}q_c^2)\varphi^2 \end{bmatrix}, \ I_{3T} = \begin{bmatrix} 2a_{11}\varphi^2 + 2d_{11}a_{12}q_c^2\varphi \\ 2a_{21}\varphi^2 + 2d_{11}a_{22}q_c^2\varphi \end{bmatrix}, \ S_{4T} = \begin{bmatrix} 6b_{11}\psi^2\varphi + 2b_{12}d_{11}q_c^2\psi^2 \\ 6b_{21}\psi^2\varphi + 2b_{22}d_{11}q_c^2\psi^2 \end{bmatrix},$$

$$I_{4T} = \begin{bmatrix} 4a_{11}\psi \mathrm{Re}\, p_{23} + 2a_{12}(\mathrm{Re}\,(ip_{23}\omega_c) + \psi \mathrm{Re} z_{23}) + \varphi(2a_{11}p_{22} + a_{12}z_{22}) + a_{12}d_{11}q_c^2 p_{22} \\ 4a_{21}\psi \mathrm{Re}\, p_{23} + 2a_{22}(\mathrm{Re}\,(ip_{23}\omega_c) + \psi \mathrm{Re} z_{23}) + \varphi(2a_{21}p_{22} + a_{22}z_{22}) + a_{22}d_{11}q_c^2 p_{22} \end{bmatrix},$$

then

$$\eta = \frac{L_H L_2(0) R_H}{L_H R_H}, \ \xi = \frac{L_T L_2(q_c) R_T}{L_T R_T}, \ \tilde{\gamma} = \frac{L_H(I_{1H} + S_{1H})}{L_H R_H}, \ \tilde{\sigma} = \frac{L_H(I_{2H} + S_{2H})}{L_H R_H},$$
$$\tilde{\mu} = \frac{L_T(I_{1T} + S_{1T})}{L_T R_T}, \ \tilde{g} = \frac{L_T(I_{2T} + S_{2T})}{L_T R_T}, \ \tilde{b} = \frac{L_T(I_{3T} + S_{3T})}{L_T R_T}, \ \tilde{s} = \frac{L_T(I_{4T} + S_{4T})}{L_T R_T},$$

and

$$\gamma = \frac{\tilde{\gamma}}{\psi^2}, \ \sigma = \frac{\tilde{\sigma}}{\varphi^2}, \ \mu = \frac{\tilde{\mu}}{\varphi^2}, \ g = \frac{\tilde{g}}{\varphi^2}, \ b = \frac{\tilde{b}}{\varphi}, \ s = \frac{\tilde{s}}{\psi^2}.$$

References

1. H Malchow, S Petrovskii and A Medvinsky. 2001. Pattern formation in models of plankton dynamics. A synthesis. *Oceanologica Acta*, 24(5): 479–487.
2. P D'Odorico, F Laio and L Ridolfi. 2015. Patterns as indicators of productivity enhancement by facilitation and competition in dryland vegetation. *Journal of Geophysical Research Biogeosciences*, 111.
3. J K Hale. 2009. Ordinary differential equations. *Dover Publications*.
4. Q Ouyang. 2010. Pattern formation in reaction-diffusion systems. *Peking University Press*.
5. L P Shil'nikov. 2001. Methods of qualitative theory in nonlinear dynamics. *World Scientific*.
6. B Brogliato, R Lozano and B Maschke, et al. 2007. Dissipative systems analysis and control-theory and applications. *Springer-Verlag London*.
7. M H Holmes. 2013. Introduction to perturbation methods. *Springer-Verlag New York*.
8. S Chow, C Li, D Wang. 2009. Normal forms and bifurcation of planar vector fields. *Cambridge University Press*.
9. J Jost. 2013. Partial differential equations. *Springer, New York*.
10. H K. Khalil. 2007. Nonlinear systems. *Publishing House of Electronics Industry*.
11. M Han and P Yu. 2012. Normal forms, Melnikov functions and bifurcations of limit cycles. *Springer-Verlag London*.
12. Y Ye. 1984. The theory of limit cycles. *Shanghai Science and Technology Press*.
13. Y Kuang. 1993. Delay differential equations: with applications in population dynamics. *Academic Press*.
14. L Arnold. 1974. Stochastic differential equations: theory and applications. *Wiley Interscience*.
15. R Shi and J Yu. 2017. Hopf bifurcation analysis of two zooplankton-phytoplankton model with two delays.*Chaos, Solitons & Fractals*. 100:62–73.
16. Zhang T and Wang W. 2012. Hopf bifurcation and biostability of a nutrient-phytoplankton-zooplankton model. *Applied Mathematical Modelling*. 36(12):6225–6235.
17. R Yang, M Liu and C Zhang. 2017. A diffusive toxin producing phytoplankton model with maturation delay and three-dimensional patch. *Computers & Mathematics with Applications*. 73(5):824–837.
18. Q Zhao,S Liu and X Niu. 2019. Dynamic behavior analysis of a diffusive plankton model with defensive and offensive effects. *Chaos, Solitons & Fractals*. 129:94–102.
19. M He, Z Li and F Chen. 2017. Dynamics of an impulsive model of plankton allelopathy with delays. *Journal of Applied Mathematics and Computing*. 55(1–2):749–762.
20. S Chen, J Wei and J Yu. 2018. Stationary patterns of a diffusive predator-preymodel with Crowley-Martin functional response. *Nonlinear Analysis-Real World Applications*. 39:33–57.

S. T. Liu and L. Zhang, *Pattern Dynamics of Marine Plankton Behavior*,
https://doi.org/10.1007/978-981-97-5369-7

21. R Han and B Dai. 2019.Spatiotemporal pattern formation and selection induced by nonlinear cross-diffusion in a toxic-phytoplankton-zooplankton model with Allee effect. *Nonlinear Analysis- Real World Applications*. 45:822–853.
22. C Xu, M S Boyce and D J Daley. 2005. Harvesting in seasonal environments. *Journal of Mathematical Biology*. 50(6):663–682.
23. W Uszko, S Diehl and G Englund, et al. 2017. Effects of warming on predator-prey interactions-a resource-based approach and a theoretical synthesis. *Ecology Letters*. 20(4):513–523.
24. X R Mao, G Marion and E Renshaw. 2002. Environmental Brownian noise suppresses explosions in population dynamics. *Stochastic Processes and Their Applications*. 97(1):95–110.
25. Y Zhao, S L Yuan and T H Zhang. 2016. The stationary distribution and ergodicity of a stochastic phytoplankton allelopathy model under regime switching. *Communications in Nonlinear Science and Numerical Simulation*. 37:131–142.
26. B Mukhopadhyay and R Bhattacharyya. 2008. Role of gestation delay in a plankton-fish model under stochastic fluctuations. *Mathematical Biosciences*. 215(1):26–34.
27. C Liu, L F Yu and Q L Zhang QL, et al. Dynamic analysis of a hybrid bioeconomic plankton system with double time delays and stochastic fluctuations. *Applied Mathematics and Computation*. 316:115–137.
28. Z W Chen, S W Zhang and C J Wei. 2019. Dynamics of a stochastic phytoplankton-toxin phytoplankton-zooplankton model. *Advances in Difference Equations*. 2019(1):347.
29. Q Liu, D Q Jiang and T Hayat, et al. 2018. Stationary distribution and extinction of a stochastic predator-prey model with additional food and nonlinear perturbation. *Applied Mathematics and Computation*. 320(3):226–239.
30. A P Maiti, B Dubey and J Tushar. 2017. A delayed prey-predator model with Crowley-Martin-type functional response including prey refuge. *Mathematical Methods in the Applied Sciences*. 40(16):5792–5809.
31. Y L Song, T Yin and H Y Shu. 2017. Dynamics of a ratio-dependent stage-structured predator-prey model with delay. *Mathematical Methods in the Applied Sciences*. 2017;40(18):6451–6467.
32. R C Hart and E A Bychek. 2011. Body size in freshwater planktonic crustaceans: an overview of extrinsic determinants and modifying influences of biotic interactions. *Hydrobiologia*. 668(1):61–108.
33. E Marañón. 2015. Cell size as a key determinant of phytoplankton metabolism and community structure. *Annual Review of Marine Science*. 7(1):241–264.
34. Q Y Zhao, S T Liu and D D Tian. 2018. Dynamic behavior analysis of phytoplankton-zooplankton system with cell size and time delay. *Chaos, Solitons & Fractals*. 113:160–168.
35. Z C Pu, M H Cortez and L Jiang. 2017. Predator-prey coevolution drives productivity-richness relationships in planktonic systems. *The American Naturalist*. 2017;189(1):28–42.
36. L Jiang, O M E Schofield and P G Falkowski. 2005. Adaptive evolution of phytoplankton cell size. *The American Naturalist*. 166(4):496–505.
37. R Khasminskii. 2011. Stochastic stability of differential equations. *Springer Science & Business Media*.
38. L Zu, D Q Jiang and D O'Regan. 2015. Conditions for persistence and ergodicity of a stochastic Lotka-Volterra predator-prey model with regime switching. *Communications in Nonlinear Science and Numerical Simulation*. 2015;29(1–3):1–11.
39. S R J Jang and E J Allen. 2015. Deterministic and stochastic nutrient-phytoplankton-zooplankton models with periodic toxin producing phytoplankton. *Applied Mathematics and Computation*. 271:52–67.
40. X H Zhang, D Q JianG and T Hayat, et al. 2017. Dynamical behavior of a stochastic SVIR epidemic model with vaccination. *Physica A: Statistical Mechanics and Its Applications*. 483:94–108.
41. V Lakshmikantham, D D Bainov and P S Simeonov. 1989. Theory of impulsive differential equations.*World scientific*.

42. A M Samoilenko, N A Perestyuk and Y Chapovsky. 1995. Impulsive differential equations. *Singapore: World Scientific.*
43. Z Liu,J Wu and Y Chen, et al. 2010. Impulsive perturbations in a periodic delay differential equation model of plankton allelopathy. *Nonlinear Analysis: Real World Applications*. 11(1):432–445.
44. S Li, Z Xiong and X Wang. 2010. The study of a predator-prey system with group defense and impulsive control strategy. *Applied Mathematical Modelling*. 34(9):2546–2561.
45. H Guo, L Chen and X Song. 2015. Qualitative analysis of impulsive state feedback control to an algae-fish system with bistable property. *Applied Mathematics and Computation*. 271:905–922.
46. L Chen, J Sun and F Chen, et al. 2016. Extinction in a Lotka-Volterra competitive system with impulse and the effect of toxic substances. *Applied Mathematical Modelling*. 40(3):2015–2024.
47. J P Tripathi, S Abbas and M Thakur. 2014. Local and global stability analysis of a two prey one predator model with help. *Communications in Nonlinear Science and Numerical Simulation*. 19(9):3284–3297.
48. M He, F Chen and Z Li. 2010. Almost periodic solution of an impulsive differential equation model of plankton allelopathy. *Nonlinear Analysis: Real World Applications*. 11(4):2296–2301.
49. J Hou, Z Teng and S Gao. 2010. Permanence and global stability for nonautonomous N-species Lotka-Valterra competitive system with impulses. *Nonlinear Analysis: Real World Applications*. 11(3):1882–1896.
50. A Morozov, E Arashkevich and A Nikishina, et al. 2011. Nutrient-rich plankton communities stabilized via predator-prey interactions: revisiting the role of vertical heterogeneity. *Mathematical Medicine and Biology: A Journal of the IMA*. 28(2):185–215.
51. S Dutkiewicz, M J Follows and J G Bragg. 2009. Modeling the coupling of ocean ecology and biogeochemistry.*Global Biogeochemical Cycles*. 23(4).
52. D Valenti, G Denaro and F Giarratana, et al. 2016. Modeling of sensory characteristics based on the growth of food spoilage bacteria. *Mathematical Modelling of Natural Phenomena*. 11(5):119–136.
53. C Dai, M Zhao and H Yu. 2016. Dynamics induced by delay in a nutrient-phytoplankton model with diffusion. *Ecological Complexity*. 26:29–36.
54. E González-Olivares, J Mena-Lorca and A Rojas-Palma A, et al. 2011. Dynamical complexities in the Leslie-Gower predator-prey model as consequences of the Allee effect on prey. *Applied Mathematical Modelling*. 35(1):366–381.
55. Z Zhao, C Luo and L Pang, et al. 2016. Nonlinear modelling of the interaction between phytoplankton and zooplankton with the impulsive feedback control. *Chaos, Solitons & Fractals*. 87:255–261.
56. M Gao, H Shi and Z Li. 2009. Chaos in a seasonally and periodically forced phytoplankton-zooplankton system. *Nonlinear Analysis: Real World Applications*. 10(3):1643–1650.
57. L Chen, F Chen and L Chen. 2010. Qualitative analysis of a predator-prey model with Holling type II functional response incorporating a constant prey refuge. *Nonlinear Analysis: Real World Applications*. 11(1):246–252.
58. Y Tian and P Weng. 2011. Stability analysis of diffusive predator-prey model with modified Leslie-Gower and Holling-type III schemes. *Applied Mathematics and Computation*. 218(7):3733–3745.
59. A Sharma, A K Sharma and K Agnihotri. 2015. Analysis of a toxin producing phytoplankton-zooplankton interaction with Holling IV type scheme and time delay. *Nonlinear Dynamics*. 81(1–2):13–25.
60. W Wang, L Zhang and H Wang, et al. 2010. Pattern formation of a predator-prey system with Ivlev-type functional response. *Ecological Modelling*. 221(2):131–140.
61. M Banerjee and S Petrovskii. 2011. Self-organised spatial patterns and chaos in a ratio-dependent predator-prey system. *Theoretical Ecology*. 4(1):37–53.

62. J P Tripathi, S Tyagi and S Abbas. 2016. Global analysis of a delayed density dependent predator-prey model with Crowley-Martin functional response. *Communications in Nonlinear Science and Numerical Simulation*. 30(1):45–69.
63. S Sarwardi, M Haque and P K Mandal. 2014. Persistence and global stability of Bazykin predator-prey model with Beddington-DeAngelis response function. *Communications in Nonlinear Science and Numerical Simulation*. 19(1):189–209.
64. J P Tripathi, S Abbas and M Thakur. 2015. A density dependent delayed predator-prey model with Beddington-DeAngelis type function response incorporating a prey refuge. *Communications in Nonlinear Science and Numerical Simulation*. 22(1):427–450.
65. G Zhang and Y Shen. 2015. Periodic solutions for a neutral delay Hassell-Varley type predator-prey system. *Applied Mathematics and Computation*. 264:443–452.
66. G T Skalski and J F Gilliam. 2001. Functional responses with predator interference: viable alternatives to the Holling type II model. *Ecology*. 82(11):3083–3092.
67. J P Delong and D A Vasseur. 2013. Linked exploitation and interference competition drives the variable behavior of a classic predator-prey system. *Oikos*. 122(10):1393–1400.
68. J S Weitz and S A Levin. 2006. Size and scaling of predator-prey dynamics. *Ecology Letters*. 9(5):548–557.
69. D L DeAngelis, R A Goldstein and R V O'neill. 1975. A model for tropic interaction. *Ecology*. 56(4):881–892.
70. B Hansen, P K Bjornsen and P J Hansen. 1994. The size ratio between planktonic predators and their prey. *Limnology and Oceanography*. 9(2):395–403.
71. S L Nielsen. 2006. Size-dependent growth rates in eukaryotic and prokaryotic algae exemplified by green algae and cyanobacteria: comparisons between unicells and colonial growth forms. *Journal of Plankton Research*. 28(5):489–498.
72. B BÉatrice, C Yves and V AndrÉ, et al. 2008. Growth rate peaks at intermediate cell size in marine photosynthetic picoeukaryotes. *Limnology and Oceanography*. 53(2):863–867.
73. M Emilio, C Pedro and C L S Daffne, et al. 2013. Unimodal size scaling of phytoplankton growth and the size dependence of nutrient uptake and use. *Ecology Letters*. 16(3):371–379.
74. A Verdy, M Follows and G Flierl. 2009. Optimal phytoplankton cell size in an allometric model. *Marine Ecology Progress Series*. 379:1–12.
75. L A Segel and J L Jackson. 1972. Dissipative structure: an explanation and an ecological example. *Journal of Theoretical Biology*. 37:545–559.
76. D Manna, A Maiti and G P Samanta. 2018. Analysis of a predator-prey model for exploited fish populations with schooling behavior. *Applied Mathematics and Computation*. 317:35–48.
77. Y Wang, H Wang and W Jiang. 2014. Hopf-transcritical bifurcation in toxic phytoplankton-zooplankton model with delay. *Journal of Mathematical Analysis and Applications* . 415(2):574–594.
78. D Hu and H Cao. 2016. Stability and Hopf bifurcation analysis in Hindmarsh-Rose neuron model with multiple time delays. *International Journal of Bifurcation and Chaos*. 26(11):165–187.
79. B D Hassard, N D Kazarinoff and Y H Wan. 1981. Theory and applications of Hopf bifurcation. *Cambridge University*.
80. G Denaro, D Valenti and B Spagnolo, et al. 2013. Dynamics of two picophytoplankton groups in Mediterranean Sea: Analysis of the deep chlorophyll maximum by a stochastic advection-reaction-diffusion model. *PloS one*. 8(6):e66765.
81. D Valenti, G Denaro and R Ferreri, et al. 2017. Spatio-temporal dynamics of a planktonic system and chlorophyll distribution in a 2D spatial domain: matching model and data. *Scientific Reports*. 7(1):220.
82. B C Rall, Q Vucic-pestic and R B Ehnes, et al. 2010. Temperature, predator-prey interaction strength and population stability. *Global Change Biology*. 16(8):2145–2157.
83. J L Lessard and D B Hayes. 2003. Effects of elevated water temperature on fish and macroinvertebrate communities below small dams. *River Research and Applications*. 19(7):721–732.
84. J C Morrill, R C Bales and M H Conklin. 2005. Estimating stream temperature from air temperature: implications for future water quality. *Journal of Environmental Engineering*. 131(1):139–146.

85. L S Peck, K E Webb and D M Bailey. 2004. Extreme sensitivity of biological function to temperature in Antarctic marine species. *Functional Ecology*. 18(5):625–630.
86. A Clarke. 2006. Temperature and the metabolic theory of ecology. *Functional Ecology*. 20(2):405–412.
87. J M Donelson, P L Munday and M I McCormick, et al. 2010. Effects of elevated water temperature and food availability on the reproductive performance of a coral reef fish. *Marine Ecology Progress Series*. 401:233–243.
88. J L Johansen and G P Jones. 2011. Increasing ocean temperature reduces the metabolic performance and swimming ability of coral reef damselfishes. *Global Change Biology*. 17(9):2971–2979.
89. O T Hogg, D K A Barnes and H J Griffiths. 2011. Highly diverse, poorly studied and uniquely threatened by climate change: an assessment of marine biodiversity on South Georgia's continental shelf. *PLoS One*. 6(5):e19795.
90. B E Beisner, E McCauley and F J Wrona. 1997. The influence of temperature and food chain length on plankton predator prey dynamics. *Canadian Journal of Fisheries and Aquatic Sciences*. 54(3):586–595.
91. P Amarasekare. 2015. Effects of temperature on consumer-resource interactions. *Journal of Animal Ecology*. 84(3):665–679.
92. P A Staehr, K A J Sand-Jensen. 2006. Seasonal changes in temperature and nutrient control of photosynthesis, respiration and growth of natural phytoplankton communities. *Freshwater Biology*. 51(2):249–262.
93. E Bestion, B García-Carreras and C E Schaum, et al.2018. Metabolic traits predict the effects of warming on phytoplankton competition. *Ecology Letters*. 21(5):655–664.
94. A M Lewandowska, P Breithaupt and H Hillebrand, et al. 2012. Responses of primary productivity to increased temperature and phytoplankton diversity. *Journal of Sea Research*. 72:87–93.
95. A Toseland, S J Daines and J R Clark, et al. 2013. The impact of temperature on marine phytoplankton resource allocation and metabolis. *Nature Climate Change*. 3(11):979.
96. C E Hare, K Leblanc and G R DiTullio, et al. 2007. Consequences of increased temperature and CO_2 for phytoplankton community structure in the Bering Sea. *Marine Ecology Progress Series*. 352:9–16.
97. K E Havens, R M Pinto-Coelho and M Beklioğlu, et al. 2015. Temperature effects on body size of freshwater crustacean zooplankton from Greenland to the tropics. *Hydrobiologia*. 743(1):27–35.
98. A M Kramer, O Sarnelle and J Yen. 2011. The effect of mating behavior and temperature variation on the critical population density of a freshwater copepod. *Limnology and Oceanography*. 56(2):707–715.
99. G Kalinkat, F D Schneider and C Digel, et al. 2013. Body masses, functional responses and predator-prey stability. *Ecology Letters*. 16(9):1126–1134.
100. O Sarnelle and A E Wilson. 2008. Type III functional response in Daphnia. *Ecology*. 89(6):1723–1732.
101. M Lindmark, M Huss and J Ohlberger, et al. 2018. Temperature-dependent body size effects determine population responses to climate warming. *Ecology Letters*. 21(2):181–189.
102. M K Thomas, C T Kremer and C A Klausmeier, et al. 2012. A global pattern of thermal adaptation in marine phytoplankton. *Science*. 338:1085–1088.
103. G Englund, G Öhlund and C L Hein, et al. 2011. Temperature dependence of the functional response. *Ecology Letters*. 14(9):914–921.
104. N Chitnis, J M Hyman and J M Cushing. 2008. Determining important parameters in the spread of malaria through the sensitivity analysis of a mathematical model. *Bulletin of Mathematical Biology*. 70(5):1272.
105. Z C Jiang, X H Bi and T Q Zhang et al. 2019. Sampath Aruna Pradeep B.G. Global Hopf bifurcation of a delayed phytoplankton-zooplankton system considering toxin producing effect and delay dependent coefficient. *Mathematical Biosciences and Engineering*. 16(5):3807–3829.

106. J C Jiang, W Z Zhang and J Zhang, et al. 2018. Dynamical analysis of a phytoplankton-zooplankton system with harvesting term and Holling III functional response. *International Journal of Bifurcation and Chaos*. 28(13):1850162.
107. A J Richardson and D S Schoeman. 2004. Climate impact on plankton ecosystems in the Northeast Atlantic. *Science*. 305(5690):1609–1612.
108. M A MacNeil, N A J Graham and J E Cinner, et al. 2010. Transitional states in marine fisheries: adapting to predicted global change. *Philosophical Transactions of the Royal Society B: Biological Sciences*. 365(1558):3753–3763.
109. T Tanioka and K Matsumoto. 2018. Effects of incorporating age-specific traits of zooplankton into a marine ecosystem model. *Ecological Modelling*. 368:257–264.
110. A K Sharma, A Sharma and K Agnihotri. 2016. Bifurcation behaviors analysis of a plankton model with multiple delays. *International Journal of Biomathematics*. 9(06):1650086.
111. T Nguyen-Quang and F Guichard. 2010. The role of bioconvection in plankton population with thermal stratification. *International Journal of Bifurcation and Chaos*. 20(06):1761–1778.
112. X Irigoien, J Huisman and R P Harris. 2004. Global biodiversity patterns of marine phytoplankton and zooplankton. *Nature*. 429(6994):863.
113. D Gerten and R Adrian. 2000. Climate-driven changes in spring plankton dynamics and the sensitivity of shallow polymictic lakes to the North Atlantic Oscillation. *Limnology and Oceanography*. 45(5):1058–1066.
114. E Giricheva. 2018. Spatiotemporal dynamics of an NPZ model with prey-taxis and intratrophic predation. *Nonlinear Dynamics*. 2018:1–18.
115. F D Chen, X D Xie and H N Wang. 2015. Global stability in a competition model of plankton allelopathy with infinite delay. *Journal of Systems Science and Complexity*. 28(5):1070–1079.
116. M Kloosterman, S A Campbell and F J Poulin. 2016. An NPZ model with state-dependent delay due to size-structure in juvenile zooplankton. *SIAM Journal on Applied Mathematics*. 76(2):551–577.
117. Z P Ma, J Liu and J L Yue. 2018. Spatiotemporal dynamics induced by delay and diffusion in a predator-prey model with mutual interference among the predator. *Computers and Mathematics with Applications*. 75(10):3488–3507.
118. J Li, Y Z Song and H Wan, et al. 2017. Dynamical analysis of a toxin-producing phytoplankton-zooplankton model with refuge. *Mathematical Biosciences and Engineering*. 14(2):529–557.
119. R Z Yang and C R Zhang. 2016. Dynamics in a diffusive predator-prey system with a constant prey refuge and delay. *Nonlinear Analysis: Real World Applications*. 31:1–22.
120. T G Nielsen and B Hansen. 1995. Plankton community structure and carbon cycling on the western coast of Greenland during and after the sedimentation of a diatom bloom. *Marine Ecology Progress Series*. 125:239–257.
121. E Wahlströ and E Westman. 1999. Planktivory by the predacious cladoceran Bythotrephes longimanus: effects on zooplankton size structure and abundance. *Canadian Journal of Fisheries and Aquatic Sciences*. 56(10):1865–1872.
122. M T Bremigan and R A Stein. 1994. Gape-dependent larval foraging and zooplankton size: implications for fish recruitment across systems. *Canadian Journal of Fisheries and Aquatic Sciences*. 51(4):913–922.
123. R K Upadhyay, W Wang and N K Thakur. 2010. Spatiotemporal dynamics in a spatial plankton system. *Mathematical Modelling of Natural Phenomena*. 5(5):102–122.
124. E Giricheva. 2019. Spatiotemporal dynamics of an NPZ model with prey-taxis and intratrophic predation. *Nonlinear Dynamics*. 95(2):875–892.
125. Y F Li, H C Liu and R Z Yang. 2018. A delayed diffusive predator-prey system with predator cannibalism. *Computers and Mathematics with Applications*. 75(4):1355–1367.
126. J T Turner and P A Tester. 1997. Toxic marine phytoplankton, zooplankton grazers, and pelagic food webs. *Limnology and Oceanography*. 42(5):1203–1213.
127. A M Edwards. 2001. Adding detritus to a nutrient-phytoplankton-zooplankton model: a dynamical-systems approach. *Journal of Plankton Research*. 23(4):389–413.
128. D Chaturvedi and O P Misra. 2019. Modeling impact of varying pH due to carbondioxide on the dynamics of prey-predator species system. *Nonlinear Analysis: Real World Applications*. 46:374–402.

129. M P Delaney. 2003. Effects of temperature and turbulence on the predator-prey interactions between a heterotrophic flagellate and a marine bacterium. *Microbial Ecology*. 45(3):218–225.
130. X Y Meng and Y Q Wu. 2018. Bifurcation and control in a singular phytoplankton-zooplankton-fish model with nonlinear fish harvesting and taxation. *International Journal of Bifurcation and Chaos*. 28(03):1850042.
131. D W Huang, H L Wang and J F Feng, et al. 2006. Hopf bifurcation of the stochastic model on HAB nonlinear stochastic dynamics. *Chaos, Solitons & Fractals*. 27(4):1072–1079.
132. A Giuffrida, D Valenti and G Ziino, et al. 2009. A stochastic interspecific competition model to predict the behaviour of Listeria monocytogenes in the fermentation process of a traditional Sicilian salami. *European Food Research and Technology*. 228(5):767–775.
133. D Valenti, G Denaro and A La Cognata, et al. 2012. Picophytoplankton dynamics in noisy marine environment. *Acta Physica Polonica B*. 43(5):1227–1240.
134. A Mougi and Y Iwasa. 2011. Unique coevolutionary dynamics in a predator-prey system. *Journal of Theoretical Biology*. 277(1):83–89.
135. T Kiørboe, E Saiz and P Tiselius, et al. 2018. Adaptive feeding behavior and functional responses in zooplankton. *Limnology and Oceanography*. 63(1):308–321.
136. M Pančić and T Kiørboe. 2018. Phytoplankton defence mechanisms: traits and trade-offs.*Biological Reviews*. 93(2):1269–1303.
137. J M Jeschke. 2006. Density-dependent effects of prey defenses and predator offenses. *Journal of Theoretical Biology*. 242(4):900–907.
138. E Velzen and U Gaedke. 2018. Reversed predator-prey cycles are driven by the amplitude of prey oscillations. *Ecology and Evolution*. 2018.
139. E Velzen and U Gaedke. 2017. Disentangling eco-evolutionary dynamics of predator-prey coevolution: the case of antiphase cycles. *Scientific Reports*. 7(1):17125.
140. A Mougi. 2012. Predator-prey coevolution driven by size selective predation can cause anti-synchronized and cryptic population dynamics. *Theoretical Population Biology*. 81(2):113–118.
141. R C Wu, Y Zhou and Y Shao, et al. 2017. Bifurcation and Turing patterns of reaction-diffusion activator-inhibitor model. *Physica A: Statistical Mechanics and its Applications*. 482:597–610.
142. R C Wu, M X Chen and B Liu, et al. 2018. Hopf bifurcation and Turing instability in a predator-prey model with Michaelis-Menten functional response. *Nonlinear Dynamics*. 91(3):2033–2047.
143. F Q Yi, J J Wei and J P Shi. 2009. Bifurcation and spatiotemporal patterns in a homogeneous diffusive predator-prey system. *Journal of Differential Equations*. 246(5):1944–1977.
144. W Tan, W W Yu and T Hayat, et al. 2018. Turing Instability and Bifurcation in a Diffusion Predator-Prey Model with Beddington-DeAngelis Functional Response. *International Journal of Bifurcation and Chaos*. 28(09):1830029.
145. X R Mao. 1994. Stochastic stabilization and destabilization. *Systems & Control Letters*. 23(4):279–290.
146. A M Turing. 1952. The chemical basis of morphogenesis. *Philosophical Transactions of the Royal Society of London. Series B, Biological Sciences*. 237:37–72.
147. C A lausmeier. 1999. Regular and irregular patterns in semiarid vegetation. *Science*. 284:1826–1828.
148. J A Sherratt. 2013. Pattern solutions of the Klausmeier model for banded vegetation in semiarid environments V: the transition from patterns to desert. *Siam Journal on Applied Mathematics*. 73:1347–1367.
149. G Sun, C H Wang and L L Chang, et al. 2018. Effects of feedback regulation on vegetation patterns in semi-arid environments. *Applied Mathematical Modelling*. 61:200–215.
150. G Sun, M Jusup and Z Jin, et al. 2016. Pattern transitions in spatial epidemics: Mechanisms and emergent properties. *Physics of Life Reviews*. 19:43–73.
151. T Zhang, Y Xing and H Zang. 2014. Spatio-temporal dynamics of a reaction-diffusion system for a predator-prey model with hyperbolic mortality. *Nonlinear Dynamics*. 78:265–277.

152. T Zhang and H Zang. 2014. Delay-induced Turing instability in reaction-diffusion equations. *Physical Review E*. 90:052908.
153. G Sun, L Li and Z K Zhang. 2013. Spatial dynamics of a vegetation model in an arid flat environment. *Nonlinear Dynamics*. 73:2207–2219.
154. G Sun. 2012. Pattern formation of an epidemic model with diffusion. *Nonlinear Dynamics*. 69:1097–1104.
155. G Sun, Z Jin and L Li, et al. 2010. Self-organized wave pattern in a predator-prey model. *Nonlinear Dynamics*. 60:265–275.
156. M Sambath, K Balachandran and L N Guin. 2018. Spatiotemporal Patterns in a Predator-Prey Model with Cross-Diffusion Effect. *International Journal of Bifurcation and Chaos*. 28:1830004.
157. A J Lotka. 1925. Elements of Physical Biology. *William and Wilkins, Baltimore*.
158. V Volterra. 1926. Variazione e fluttuazini del numero dindividui in specie animali conviventi. *Memoria Della Reale Accademia Nazionale dei Lincei*. 2:31–113.
159. F Yi, E A Gaffney and S S Lee. 2017. The bifurcation analysis of Turing pattern formation induced by delay and diffusion in the Schnakenberg system. *Discrete and Continuous Dynamical Systems-SerIES B*. 22:647–668.
160. S Chen,J Shi and J Wei. 2013. The effect of delay on a diffusive predator-prey system with Holling type-II predator functional response. *Communications on Pure and Applied Analysis*. 12:481–501.
161. S Chen, J Shi and J Wei. 2012. Global stability and Hopf bifurcation in a delayed diffusive Leslie-Gower predator-prey system. *International Journal of Bifurcation and Chaos in Applied Sciences & Engineering*. 22:1250061.
162. S Chen, and J Yu. 2016. Dynamics of a diffusive predator-prey system with a nonlinear growth rate for the predator. *Journal of Differential Equations*. 260:7923–7939.
163. J Wang, J Wei and J Shi. 2016. Global bifurcation analysis and pattern formation in homogeneous diffusive predator-prey systems. *Journal of Differential Equations*. 260:3495–3523.
164. C S Holling. 1959. The components of predation as revealed by a study of small-mammal predation of the European pine sawfly. *Canadian Entomologist*. 91:293–320.
165. C S Holling. 1966. The functional response of predator to prey density and its role in mimicry and population regulation. *Memoirs of the Entomolotical Society of Canada*. 97:5–60.
166. A D Bazykin. 1998. Nonlinear Dynamics of Interacting Populations. *World Scientific, Singapore*.
167. J B Collings. 1997. The effects of the functional response on the bifurcation behavior of a mite predator-prey interaction model. *Journal of Mathematical Biology*. 36:149–168.
168. J M Smith. 1974. Models in Ecology. *Cambridge University Press, Cambridge*.
169. R Arditi and L R Ginzburg. 1989. Coupling in predator-prey dynamics: ratio-dependence. *Journal of Theoretical Biology*. 139:311–326.
170. P H Leslie, and J C Gower. 1960. The properties of a stochastic model for the predator-prey type of interaction between two species. *Biometrika*. 47:219–234.
171. J R Beddington. 1975. Mutual interference between parasites or predators and its effect on searching efficiency. *Journal of Animal Ecology*. 331–340.
172. P H Crowley, and E K Martin. 1989. Functional responses and interference within and between year classes of a dragonfly population. *Journal of the North American Benthological Society*. 8:211–221.
173. M P Hassell, and G C Varley. 1969. New inductive population model for insect parasites and its bearing on biological control. *Nature*. 223:1133.
174. P A Braza. 2012. Predator-prey dynamics with square root functional responses. *Nonlinear Analysis-Real World Applications*. 13:1837–1843.
175. X Tang, and Y Song. 2015. Cross-diffusion induced spatiotemporal patterns in a predator-prey model with herd behavior. *Nonlinear Analysis-Real World Applications*. 24:36–49.
176. S Yuan, C Xu, and T Zhang. 2013. Spatial dynamics in a predator-prey model with herd behavior. *Chaos*. 23:033102.

177. M Haque. 2010. A predator-prey model with disease in the predator species only. *Nonlinear Analysis-Real World Applications*. 11:2224–2236.
178. B Sahoo, and S Poria. 2014. Effects of supplying alternative food in a predator-prey model with harvesting. *Applied Mathematics and Computation*. 234:150–166.
179. A Basheer, E Quansah and S Bhowmick, et al. 2016. Prey cannibalism alters the dynamics of Holling-Tanner-type predator-prey models. *Nonlinear Dynamics*. 85:2549–2567.
180. C Tian, Z Ling, and Z Lin. 2011. Turing pattern formation in a predator-prey-mutualist system. *Nonlinear Analysis.-Real World Applications*. 12:3224–3237.
181. Y Chen, and F Zhang. 2013. Dynamics of a delayed predator-prey model with predator migration. *Applied Mathematical Modelling*. 37:1400–1412.
182. P Aguirre, E Gonzalez-Olivares, and E Saez. 2009. Three limit cycles in a Leslie-Gower predator-prey model with additive Allee effect. *Siam Journal on Applied Mathematics*. 69:1244–1262.
183. Y Song, and X Tang. 2017. Stability, Steady-State Bifurcations, and Turing Patterns in a Predator-Prey Model with Herd Behavior and Prey-taxis. *Studies in Applied Mathematics*. 139:371–404.
184. X Tang, Y Song and T Zhang. 2016. Turing-Hopf bifurcation analysis of a predator-prey model with herd behavior and cross-diffusion. *Nonlinear Dynamics*. 86:73–89.
185. J Brattey. 1997. Biological characteristics of Atlantic cod (Gadus morhua) from three inshore areas of northeastern Newfoundland. *NAFO Scientific Council Studies*. 29:31–42.
186. E D Brown, J H Churnside, and R L Collins, et al. 2002. Remote sensing of capelin and other biological features in the North Pacific using lidar and video technology. *Ices Journal of Marine Science*. 59:1120–1130.
187. M D'Elia, B Patti, and A Sulli, et al. 2009. Distribution and spatial structure of pelagic fish schools in relation to the nature of the seabed in the Sicily Straits (Central Mediterranean). *Marine Ecology*. 30:151–160.
188. C Tian, L Lin, and L Zhang. 2017. Additive noise driven phase transitions in a predator-prey system. *Applied Mathematical Modelling*. 46:423–432.
189. D Manna, A Maiti, and G P Samanta. 2018. Analysis of a predator-prey model for exploited fish populations with schooling behavior.*Applied Mathematics and Computation*. 317:35–48.
190. Y Almirantis and S Papageorgiou. 1991. Cross-diffusion effects on chemical and biological pattern formation. *Journal of Theoretical Biology*. 151.3:289–311.
191. V K Vanag, and I R Epstein. 2009. Cross-diffusion and pattern formation in reaction-diffusion systems. *Physical Chemistry Chemical Physics*. 11:897–912.
192. G Gambino, M C Lombardo and M Sammartino. 2013. Pattern formation driven by cross-diffusion in a 2D domain. *Nonlinear Analysis-Real World Applications*. 14.3:1755–1779.
193. M C Cross, and P C Hohenberg. 1993. Pattern formation outside of equilibrium. *Review of Modern Physics*. 65:851–1112.
194. Q Ouyang. 2010. Nonlinear Science and the Pattern Dynamics Introduction. *Peking University Press, Beijing, China*.
195. Q Ouyang, G H Gunaratne and H L Swinney. 1993. Rhombic patterns: Broken hexagonal symmetry. *Chaos*. 3:707–711.
196. S Chakraborty, P K Tiwari and S K Sasmal, et al. 2017. Interactive effects of prey refuge and additional food for predator in a diffusive predator-prey system. *Applied Mathematical Modelling*. 47:128–140.
197. S V Petrovskii, and H Malchow. 1999. A minimal model for pattern formation in a prey-predator system. *Mathematical and Computer Modelling*. 29:49–63.
198. J C Tzou, A Bayliss and B J Matkowsky, et al. 2011. Interaction of Turing and Hopf modes in the superdiffusive Brusselator model near a codimension two bifurcation point. *Mathematical Modelling of Natural Phenomena*. 6:87–118.
199. F Wu, J Ma and G Ren. 2018. Synchronization stability between initial-dependent oscillators with periodical and chaotic oscillation. *Journal of Zhejiang University-Science A*. 19:889–903.

200. J Ma, Y Xu and C Wang, et al. 2016. Pattern selection and self-organization induced by random boundary initial values in a neuronal network. *Physica A: Statistical Mechanics and Its Applications*. 461:586–594.
201. J Chen, J Xiao and L Qiao, et al. 2018. Dynamics of scroll waves with time-delay propagation in excitable media. *Communications in Nonlinear Science and Numerical Simulation*. 59:331–337.
202. J Shi, R Cui and J Xiao, et al. 2018. Pair interaction of catalytical sphere dimers in chemically active media. *Micromachines*. 9:35.
203. R H Clayton, O Bernus and E M Cherry, et al. 2011. Models of cardiac tissue electrophysiology: progress, challenges and open questions. *Progress in Biophysics & Molecular Biology*. 104:22–48.
204. J Ma and J Tang. 2015. A review for dynamics of collective behaviors of network of neurons. *Science China Technological Sciences*. 58:2038–2045.
205. J Ma and J Tang. 2017. A review for dynamics in neuron and neuronal network. *Nonlinear Dynamics*. 89:1569–1578.
206. J Ma, H Qin and X Song, et al. 2015. Pattern selection in neuronal network driven by electric autapses with diversity in time delays. *International Journal of Modern Physics B*. 29:1450239.
207. Jr R Erichsen and L G Brunnet. 2008. Multistability in networks of Hindmarsh-Rose neurons. *Physical Review E*. 78:061917.
208. A De Wit, D Lima and G Dewel, et al. 1996. Spatiotemporal dynamics near a codimension-two point. *Physical Review E*. 54:261.
209. M L Steyn-Ross, D A Steyn-Ross and J W Sleigh. 2013. Interacting Turing-Hopf instabilities drive symmetry-breaking transitions in a mean-field model of the cortex: a mechanism for the slow oscillation. *Physical Review X*. 3:021005.
210. M Meixner, S Bose and E Scholl. 1997. Analysis of complex and chaotic patterns near a codimension-2 Turing-Hopf point in a reaction-diffusion model. *Physica D-Nonlinear Phenomena*. 109:128–138.
211. C M Topaz and A J Catll. 2010. Forced patterns near a Turing-Hopf bifurcation. *Physical Review E*. 81:026213.
212. W Just, M Bose and S Bose, et al. 2001. Spatiotemporal dynamics near a supercritical Turing-Hopf bifurcation in a two-dimensional reaction-diffusion system. *Physical Review E*. 64:026219.
213. F D S Silva, R L Viana and T D L Prado, et al. 2014. Characterization of spatial patterns produced by a Turing instability in coupled dynamical systems. *Communications in Nonlinear Science and Numerical Simulation*. 19:1055–1071.
214. K J Lee, W D McCormick and Q Ouyang, et al. 1993. Pattern formation by interacting chemical fronts. *Science*. 261(5118):192–194.
215. V Castets, E Dulos and J Boissonade, et al. 1990. Experimental evidence of a sustained standing Turing-type nonequilibrium chemical pattern. *Physical Review Letters*. 64(24):2953–2956.
216. T Thorsen, R W Roberts and F H Arnold, et al. 2001. Dynamic pattern formation in a vesicle-generating microfluidic device. *Physical Review Letters*. 86(18):4163–4166.
217. A Gierer and H Meinhardt. 1972. A theory of biological pattern formation. *Biological Cybernetics*. 12(1):30–39.
218. S Kondo and T Miura. 2010. Reaction-diffusion model as a framework for understanding biological pattern formation. *Science*. 329(5999):1616–1620.
219. L A Segel and J L Jackson. 1972. Dissipative structure: an explanation and an ecological example. *Journal of Theoretical Biology*. 37(3):545–559.
220. S A Levin and L A Segel. 1976. Hypothesis for origin of planktonic patchiness. *Nature*. 259(5545):659–659.
221. T Huang, H Zhang and H Yang, et al. 2017. Complex patterns in a space-and time-discrete predator-prey model with Beddington-DeAngelis functional response. *Communications in Nonlinear Science and Numerical Simulation*. 43:182–199.

222. D Alonso, F Bartumeus and J Catalan. 2002. Mutual interference between predators can give rise to Turing spatial patterns. *Ecology*. 83(1):28–34.
223. X Zhang, G Sun and Z Jin. 2012. Spatial dynamics in a predator-prey model with Beddington-DeAngelis functional response. *Physical Review E*. 85:021924.
224. S Chakraborty, A Ramesh and P S Dutta. 2016. Toxic phytoplankton as a keystone species in aquatic ecosystems: stable coexistence to biodiversity. *Oikos*. 125(5):735–746.
225. J D Harwood and J J Obrycki. 2005. The role of alternative prey in sustaining predator populations.*Second International Symposium on Biological Control of Arthropods*. 2:453–462.
226. P D N Srinivasu, B Prasad and M Venkatesulu. 2007. Biological control through provision of additional food to predators: a theoretical study. *Theoretical Population Biology*. 72(1):111–120.
227. J Chattopadhayay, R R Sarkar, and S Mandal. 2002. Toxin-producing plankton may act as a biological control for planktonic blooms-field study and mathematical modelling. *Journal of Theoretical Biology*. 215(3):333–344.
228. P D N Srinivasu and B Prasad. 2010. Time optimal control of an additional food provided predator-prey system with applications to pest management and biological conservation. *Journal of Mathematical Biology*. 60(4):591–613.
229. E H Kerner. 1957. A statistical mechanics of interacting biological species. *Bulletin of Mathematical Biology*. 19(2):121–146.
230. S V Petrovskii and H Malchow. 1999. A minimal model of pattern formation in a prey-predator system. *Mathematical and Computer Modelling*. 29(8):49–63.
231. S Alvain, C Moulin and Y Dandonneau, et al. 2008. Seasonal distribution and succession of dominant phytoplankton groups in the global ocean: A satellite view. *Global Biogeochemical Cycles*. 22(3).
232. R Arditi, L Ginzburg and H Akcakaya. 1991. Variation in plankton densities among lakes: a case for ratio-dependent predation models. *The American Naturalist*. 138(5):1287–1296.
233. M J Behrenfeld and E S Boss. 2013. Resurrecting the ecological underpinnings of ocean plankton blooms. *Annual Review of Marine Science*. 6(1):167.
234. M Bengfort, U Feudel and F M Hilker, et al. 2014. Plankton blooms and patchiness generated by heterogeneous physical environments. *Ecological complexity*. 20:185–194.
235. D G Boyce, K T Frank and W C Leggett. 2015. From mice to elephants: overturning the 'one size fits all' paradigm in marine plankton food chains. *Ecology Letters*. 18(6):504–515.
236. R Campbell, F Diaz and Z Hu, et al. 2013. Nutrients and plankton spatial distributions induced by a coastal eddy in the gulf of lion. Insights from a numerical model. *Progress in Oceanography*. 109:47–69.
237. S Chakraborty, M Pančić and K H Andersen, et al. 2019. The cost of toxin production in phytoplankton: the case of PST producing dinoflagellates. *The ISME Journal*. 13(1):64–75.
238. J Derot, H Yajima and F G Schmitt. 2020. Benefits of machine learning and sampling frequency on phytoplankton bloom forecasts in coastal areas. *Ecological Informatics*. 60:101174.
239. E M Dickman, M J Vanni and M J Horgan. 2006. Interactive effects of light and nutrients on phytoplankton stoichiometry. *Oecologia*. 149(4):676–689.
240. S Diehl, S Berger and R Ptacnik, et al. 2002. Phytoplankton, light, and nutrients in a gradient of mixing depths: field experiments. *Ecology*. 83(2):399–411.
241. P Falkowski. 2012. Ocean science: the power of plankton. *Nature*. 483(7387):S17–S20.
242. K J Flynn and A Mitra. 2016. Why plankton modelers should reconsider using rectangular hyperbolic (michaelis-menten, monod) descriptions of predator-prey interactions. *Frontiers in Marine Science*. 3:165.
243. K J Flynn, D K Stoecker and A Mitra, et al. 2013. Misuse of the phytoplankton–zooplankton dichotomy: the need to assign organisms as mixotrophs within plankton functional types. *Journal of Plankton Research*. 35(1):3–11.
244. H L Fuchs and P JS Franks. 2010. Plankton community properties determined by nutrients and size-selective feeding. *Marine Ecology Progress Series*. 413:1–15.

245. K A Ger, P U Cordero and P C Frost, et al. 2016. The interaction between cyanobacteria and zooplankton in a more eutrophic world. *Harmful Algae*. 54:128–144.
246. R Han and B Dai. 2017. Cross-diffusion induced turing instability and amplitude equation for a toxic-phytoplankton–zooplankton model with nonmonotonic functional response. *International Journal of Bifurcation and Chaos*. 27(06):1750088.
247. I H Carrasco, V Rossi and E H García, et al. 2014. The reduction of plankton biomass induced by mesoscale stirring: A modeling study in the Benguela upwelling. *Deep Sea Research Part I: Oceanographic Research Papers*. 83:65–80.
248. J Huisman, N N P Thi and D M Karl, et al. 2006. Reduced mixing generates oscillations and chaos in the oceanic deep chlorophyll maximum. *Nature*. 439(7074):322–325.
249. X Irigoien, K J Flynn and R P Harris. 2005. Phytoplankton blooms: a 'loophole' in microzooplankton grazing impact? *Journal of Plankton Research*. 27(4):313–321.
250. D Jia, T Zhang and S Yuan. 2019. Pattern dynamics of a diffusive toxin producing phytoplankton–zooplankton model with three-dimensional patch. *International Journal of Bifurcation and Chaos*. 29(04):1930011.
251. Y Kuang and E Beretta. 1998. Global qualitative analysis of a ratio-dependent predator–prey system. *Journal of Mathematical Biology*. 36(4):389–406.
252. M Lévy, D Iovino and L Resplandy, et al. 2012. Large-scale impacts of submesoscale dynamics on phytoplankton: Local and remote effects. *Ocean Modelling*. 43:77–93.
253. N D Lewis, M N Breckels and M Steinke, et al. 2013. Role of infochemical mediated zooplankton grazing in a phytoplankton competition model. *Ecological complexity*. 16:41–50.
254. X-You Meng, Y Q Wu and J Li. 2020. Bifurcation analysis of a singular nutrient-plankton-fish model with taxation, protected zone and multiple delays. *Numerical Algebra, Control & Optimization*. 10(3):391.
255. X Pan, A Mannino and M E Russ, et al. 2010. Remote sensing of phytoplankton pigment distribution in the united states northeast coast. *Remote Sensing of Environment*. 114(11):2403–2416.
256. J C Prairie, K R Sutherland and K J Nickols, et al. 2012. Biophysical interactions in the plankton: A cross-scale review. *Limnology and Oceanography: Fluids and Environments*. 2(1):121–145.
257. C M Prokopenko, K Turgeon and J M Fryxell. 2017. Evaluation of alternative prey-, predator-, and ratio-dependent functional response models in a zooplankton microcosm. *Canadian Journal of Zoology*. 95(3):177–182.
258. F Rao. 2013. Spatiotemporal dynamics in a reaction–diffusion toxic-phytoplankton–zooplankton model. *Journal of Statistical Mechanics: Theory and Experiment*. 2013(08):P08014.
259. J J Ruzicka, T C Wainwright and W T Peterson. 2011. A simple plankton model for the Oregon upwelling ecosystem: Sensitivity and validation against time-series ocean data. *Ecological Modelling*. 222(6):1222–1235.
260. A B Ryabov and B Blasius. 2008. Population growth and persistence in a heterogeneous environment: the role of diffusion and advection. *Mathematical Modelling of Natural Phenomena*. 3(3):42–86.
261. Y Sekerci and S Petrovskii. 2015. Mathematical modelling of spatiotemporal dynamics of oxygen in a plankton system. *Mathematical Modelling of Natural Phenomena*. 10(2):96–114.
262. H Serizawa, T Amemiya and K Itoh. 2009. Patchiness and bistability in the comprehensive cyanobacterial model (CCM). *Ecological Modelling*. 220(6):764–773.
263. T J Smayda. 1997. What is a bloom? A commentary. *Limnology and Oceanography*. 42(5 part 2):1132–1136.
264. J H Steele and E W Henderson. The role of predation in plankton models. *Journal of Plankton Research*. 14(1):157–172.
265. Y Tanaka and H Mano. 2012. Functional traits of herbivores and food chain efficiency in a simple aquatic community model. *Ecological Modelling*. 237:88–100.
266. L Truong, I M Suthers and D O Cruz, et al. 2017. Plankton supports the majority of fish biomass on temperate rocky reefs. *Marine Biology*. 164(4):73.

267. E L Turner, D A Bruesewitz and R F Mooney, et al. 2014. Comparing performance of five nutrient phytoplankton zooplankton (NPZ) models in coastal lagoons. *Ecological Modelling*. 277:13–26.
268. Hn A Verhoef and H Olff. 2010. Trophic dynamics of communities. *Community Ecology. Processes, Models and Applications*. 25–36.
269. G Volf, N Atanasova and B Kompare, et al. 2011. Descriptive and prediction models of phytoplankton in the northern Adriatic. *Ecological Modelling*. 222(14):2502–2511.
270. W Wang, S Liu and Z Liu, et al. 2019. Pattern dynamics in a predator–prey model with schooling behavior and cross-diffusion. *International Journal of Bifurcation and Chaos*. 29(11):1950146.
271. W Wang, S Liu and D Tian, et al. 2017. Permanence and extinction of a nonautonomous impulsive plankton model with help. *Mathematical Methods in the Applied Sciences*. 40(18):7175–7184.
272. Q Zhao, S Liu and X Niu. 2020. Effect of water temperature on the dynamic behavior of phytoplankton–zooplankton model. *Applied Mathematics and Computation*. 378:125211.
273. P A Abrams. 2015. Why ratio dependence is (still) a bad model of predation. *Biological Reviews*. 90(3):794–814.
274. J M Beman, K R Arrigo and P A Matson. 2005. Agricultural runoff fuels large phytoplankton blooms in vulnerable areas of the ocean. *Nature*. 434(7030):211–214.
275. B Chen, M R Landry and B Huang, et al. 2012. Does warming enhance the effect of microzooplankton grazing on marine phytoplankton in the ocean? *Limnology and Oceanography*. 57(2):519–526.
276. P C Fraser and W G Sprules. 1992. Type-3 functional response in limnetic suspension-feeders, as demonstrated by in situ grazing rates. *Hydrobiologia*. 232(3):175–191.
277. C Cosner, D L DeAngelis and J S Ault, et al. 1999. Effects of spatial grouping on the functional response of predators. *Theoretical Population Biology*. 56(1):65–75.
278. V David, P Chardy and B Sautour. 2006. Fitting a predator–prey model to zooplankton time-series data in the Gironde estuary (France): ecological significance of the parameters. *Estuarine, Coastal and Shelf Science*. 67(4):605–617.
279. C L Follett, S Dutkiewicz and D M Karl, et al. 2018. Seasonal resource conditions favor a summertime increase in north pacific diatom–diazotroph associations. *The ISME Journal*. 12(6):1543–1557.
280. I A Garibotti, M Vernet and M E Ferrario, et al. 2003. Phytoplankton spatial distribution patterns along the western Antarctic peninsula (southern ocean). *Marine Ecology Progress Series*. 261:21–39.
281. K E Helliwell. 2017. The roles of b vitamins in phytoplankton nutrition: new perspectives and prospects. *New Phytologist*. 216(1):62–68.
282. A Kalmbach, Y Girdhar and H M Sosik, et al. 2017. Phytoplankton hotspot prediction with an unsupervised spatial community model. In *2017 IEEE International Conference on Robotics and Automation (ICRA)*. 4906–4913.
283. S Lee and D Lee. 2018. Improved prediction of harmful algal blooms in four major south Korea's rivers using deep learning models. *International Journal of Environmental Research and Public Health*. 15(7):1322.
284. Y Lehahn, F d'Ovidio and I Koren. 2018. A satellite-based Lagrangian view on phytoplankton dynamics. *Annual Review of Marine Science*. 10:99–119.
285. J P Mellard, K Yoshiyama and E Litchman, et al. 2011. The vertical distribution of phytoplankton in stratified water columns. *Journal of Theoretical Biology*. 269(1):16–30.
286. Y Peng, Y Li and T Zhang. 2021. Global bifurcation in a toxin producing phytoplankton–zooplankton system with prey-taxis. *Nonlinear Analysis: Real World Applications*. 61:103326.
287. I V Telesh. 2004. Plankton of the baltic estuarine ecosystems with emphasis on neva estuary: a review of present knowledge and research perspectives. *Marine Pollution Bulletin*. 49(3):206–219.

288. S M Vallina, B A Ward and S Dutkiewicz, et al. 2014. Maximal feeding with active prey-switching: A kill-the-winner functional response and its effect on global diversity and biogeography. *Progress in Oceanography*. 120:93–109.
289. W Wang, S Liu and Z Liu. 2019. Spatiotemporal dynamics near the Turing–Hopf bifurcation in a toxic-phytoplankton–zooplankton model with cross-diffusion. *Nonlinear Dynamics*. 98(1):27–37.
290. X Yu, S Yuan and T Zhang. 2018. The effects of toxin-producing phytoplankton and environmental fluctuations on the planktonic blooms. *Nonlinear Dynamics*. 91(3):1653–1668.
291. X Yu, S Yuan and T Zhang. 2019. Asymptotic properties of stochastic nutrient-plankton food chain models with nutrient recycling. *Nonlinear Analysis: Hybrid Systems*. 34:209–225.
292. X Yu, S Yuan and T Zhang. 2019. Survival and ergodicity of a stochastic phytoplankton–zooplankton model with toxin-producing phytoplankton in an impulsive polluted environment. *Applied Mathematics and Computation*. 347:249–264.
293. F Zhang, Y Wang and M Cao, et al. 2016. Deep-learning-based approach for prediction of algal blooms. *Sustainability*. 8(10):1060.
294. T Zhang, X Liu and X Meng, et al. 2018. Spatio-temporal dynamics near the steady state of a planktonic system. *Computers & Mathematics with Applications*. 75(12):4490–4504.
295. T Zohary, J Padisák and L N Flores. 2010. Phytoplankton in the physical environment: beyond nutrients, at the end, there is some light. *Hydrobiologia*. 639(1):261–269.
296. M Banerjee and E Venturino. 2011. A phytoplankton–toxic phytoplankton–zooplankton model. *Ecological Complexity*. 8(3):239–248.
297. A S Brierley. 2017. Plankton. *Current Biology*. 27(11):R478–R483.
298. J X Chen and B Hu. 2008. Spiral breakup and consequent patterns induced by strong polarized advective field. *EPL (Europhysics Letters)*. 84(3):34002.
299. J X Chen, L Peng and J Ma, et al. 2014. Liberation of a pinned spiral wave by a rotating electric pulse. *EPL (Europhysics Letters)*. 107(3):38001.
300. F Cordoleani, D Nerini and A Morozov, et al. 2013. Scaling up the predator functional response in heterogeneous environment: When holling type iii can emerge? *Journal of Theoretical Biology*. 336:200–208.
301. F Diotallevi. 2007. *The physics of cellulose biosynthesis: polymerization and self-organization, from plants to bacteria. Wur Wagenigen Ur*. 2007.
302. W M Durham, E Climent and M Barry, et al. 2013. Turbulence drives microscale patches of motile phytoplankton. *Nature Communications*. 4(1):1–7.
303. H I Egilmez and A Y Morozov. 2016. Tri-trophic plankton models revised: importance of space, food web structure and functional response parametrisation. *Mathematical Modelling of Natural Phenomena*. 11(4):16–33.
304. W Gentleman, A Leising and B Frost, et al. 2003. Functional responses for zooplankton feeding on multiple resources: a review of assumptions and biological dynamics. *Deep Sea Research Part II: Topical Studies in Oceanography*. 50(22–26):2847–2875.
305. R E Goldstein. 2015. Green algae as model organisms for biological fluid dynamics. *Annual Review of Fluid Mechanics*. 47:343–375.
306. J Halatek and E Frey. 2018. Rethinking pattern formation in reaction–diffusion systems. *Nature Physics*. 14(5):507–514.
307. A Hastings, K C Abbott and K Cuddington, et al. 2018. Transient phenomena in ecology. *Science*. 361(6406).
308. T Kiørboe. 2011. How zooplankton feed: mechanisms, traits and trade-offs. *Biological Reviews*. 86(2):311–339.
309. Y Lv, Y Pei and S Gao, et al. 2010. Harvesting of a phytoplankton–zooplankton model. *Nonlinear Analysis: Real World Applications*. 11(5):3608–3619.
310. A Morozov and E Arashkevich. 2008. Patterns of zooplankton functional response in communities with vertical heterogeneity: a model study. *Mathematical Modelling of Natural Phenomena*. 3(3):131–148.
311. A Morozov, K Abbott and K Cuddington, et al. 2020. Long transients in ecology: theory and applications. *Physics of Life Reviews*. 32:1–40.

312. A Morozov, E Arashkevich and M Reigstad, et al. 2008. Influence of spatial heterogeneity on the type of zooplankton functional response: a study based on field observations. *Deep Sea Research Part II: Topical Studies in Oceanography*. 55(20–21):2285–2291.
313. A Morozov, G Denaro and B Spagnolo, et al. 2019. Revisiting the role of top-down and bottom-up controls in stabilisation of nutrient-rich plankton communities. *Communications in Nonlinear Science and Numerical Simulation*. 79:104885.
314. S Pal, M Banerjee and S Ghorai. 2020. Effects of boundary conditions on pattern formation in a nonlocal prey–predator model. *Applied Mathematical Modelling*. 79:809–823.
315. J C Poggiale, M Gauduchon and P Auger. 2008. Enrichment paradox induced by spatial heterogeneity in a phytoplankton-zooplankton system. *Mathematical Modelling of Natural Phenomena*. 3(3):87–102.
316. S Roy, S Sathyendranath and H Bouman, et al. 2013. The global distribution of phytoplankton size spectrum and size classes from their light-absorption spectra derived from satellite data. *Remote Sensing of Environment*. 139:185–197.
317. A B Ryabov, L Rudolf and B Blasius. 2010. Vertical distribution and composition of phytoplankton under the influence of an upper mixed layer. *Journal of Theoretical Biology*. 263(1):120–133.
318. S J Taipale, K K Kahilainen and G W Holtgrieve, et al. 2018. Simulated eutrophication and browning alters zooplankton nutritional quality and determines juvenile fish growth and survival. *Ecology and Evolution*. 8(5):2671–2687.
319. C Tian and L Zhang. 2013. Delay-driven irregular spatiotemporal patterns in a plankton system. *Physical Review E*. 88(1):012713.
320. W Uszko, S Diehl and N Pitsch, et al. 2015. When is a type iii functional response stabilizing? Theory and practice of predicting plankton dynamics under enrichment. *Ecology*. 96(12):3243–3256.
321. X Wang and F Lutscher. 2019. Turing patterns in a predator–prey model with seasonality. *Journal of Mathematical Biology*. 78(3):711–737.
322. Y Wang, H Wang and W Jiang. 2014. Stability switches and global Hopf bifurcation in a nutrient-plankton model. *Nonlinear Dynamics*. 78(2):981–994.
323. G Zhang, F Wu and T Hayat, et al. 2018. Selection of spatial pattern on resonant network of coupled memristor and Josephson junction. *Communications in Nonlinear Science and Numerical Simulation*. 65:79–90.
324. W Zhu, L Wan and L Zhao. 2010. Effect of nutrient level on phytoplankton community structure in different water bodies. *Journal of Environmental Sciences*. 22(1):32–39.
325. A Zincenko, S Petrovskii and V Volpert. 2020. Turing instability in an economic-demographic dynamical system can lead to pattern formation on geographical scale. *arXiv preprint* arXiv:2006.01664.
326. P Znachor and J Nedoma. 2010. Importance of dissolved organic carbon for phytoplankton nutrition in a eutrophic reservoir. *Journal of Plankton Research*. 32(3):367–376.
327. T Caraballo, R Colucci and X Han. 2016. Non-autonomous dynamics of a semi-Kolmogorov population model with periodic forcing. *Nonlinear Analysis-Real World Applications*. 31:661–680.
328. R H Fleming. 1939. The control of diatom populations by grazing. *ICES Journal of Marince Science*. 14:210–227.
329. C S Holling. 1959. Some characteristics of simple types of predation and parasitism. *Canadian Entomologist*. 91:385–398.
330. C S Holling. 1973. Resilience and stability of ecological systems. *Annunal Review of Ecology and Systematics*. 4:1–23.
331. H F Huo and W T Li. 2004. Permanence and global stability for nonautonomous discrete model of plankton allelopathy. *Applied Mathematical Modelling*. 17:1007–1013.
332. A Hutt, A Longtin and L Schimansky-Geier. 2007. Additive global noise delays Turing bifurcations. *Physical Review Letters*. 98:230601.
333. V Ivlev. 1945. Biologicheskaya produktivnost' vodoemov. *Uspekhi Sovremennoi Biologii*. 19:98–120.

334. J Ma, C N Wang and W Y Jin, et al. 2010. Transition from spiral wave to target wave and other coherent structures in the networks of Hodgkin-Huxley neurons. *Applied Mathematics and Computation*. 217:3844–3852.
335. P Mayzaud and S A Poulet. 1978. The importance of the time factor in the response of zooplankton to varying concentrations of naturally occurring particulate matter. *Limnology and Oceanography*. 23:1144–1154.
336. A B Medvinsky, S V Petrovskii and I A Tikhonova, et al. 2002. Spatiotemporal complexity of plankton and fish dynamics. *Siam Review*. 44:311–370.
337. S Mishra and R K Upadhyay. 2020. Strategies for the existence of spatial patterns in predator-prey communities generated by cross-diffusion. *Nonlinear Analysis-Real World Applications*. 51:103018.
338. I M Moroz, R Cropp and J Norbury. 2016. Chaos in plankton models: foraging strategy and seasonal forcing. *Ecological Modelling*. 332:103–111.
339. E E Popova, M J R Fasham, and A V Osipov, et al. 1997. Chaotic behaviour of an ocean ecosystem model under seasonal external forcing. *Journal of Plankton Research*. 19:1495–1515.
340. A Priyadarshi, S Mandal, and S Smith, et al. 2017. Micro-scale variability enhances trophic transfer and potentially sustains biodiversity in plankton ecosystems. *Journal of Theoretical Biology*. 412:86–93.
341. F Rao, W Wang and Z Li. 2009. Spatiotemporal complexity of a predator-prey system with the effect of noise and external forcing. *Chaos Solitons & Fractals*. 41:1634–1644.
342. V A Ryabchenko, M J R Fasham and B A Kagan, et al. 1997. What causes short-term oscillations in ecosystem models of the ocean mixed layer? *Journal of Marine Systems*. 13:33–50.
343. N Shigesada, K Kawasaki and E Teramoto. 1979. Spatial segregation of interacting species. *Journal of Theoretical Biology*. 79:83–99.
344. G Q Sun, Z Jin, and Q X Liu, et al. 2010. Rich dynamics in a predator-prey model with both noise and periodic force. *Biosystems*. 100:14–22.
345. G Q Sun, Z Jin and L Li, et al. 2012. Spatial patterns of a predator-prey model with cross diffusion. *Nonlinear Dynamics*. 69:1631–1638.
346. N K Thakur, S K Tiwari and R K Upadhyay. 2016. Harmful algal blooms in fresh and marine water systems: The role of toxin producing phytoplankton. *International Journal of Biomathmatics*. 9:1650043.
347. R K Upadhyay, J Datta and Y Dong, et al. 2018. Emergence of spatial patterns in a damaged diffusive eco-epidemiological system. *International Journal of Bifurcation and Chaos*. 28:1830028.
348. R K Upadhyay, S Kumari and P Kumar, et al. 2019. Spatial distribution of microalgae in marine systems: A reaction-diffusion model. *Ecological Complexity*. 39:100771.
349. R K Upadhyay, N Kumari and V Rai. 2009. Wave of chaos in a diffusive system; Generating realistic patterns of patchiness in plankton-fish dynamics. *Chaos Solitons & Fractals*. 40:262–276.
350. L Wang, Y Tang and R W Wang, et al. 2019. Re-evaluating the 'plankton paradox' using an interlinked empirical data and a food web model. *Ecological Modelling*. 407:108721.
351. F Wu, C Wang, and W Jin, et al. 2017. Dynamical responses in a new neuron model subjected to electromagnetic induction and phase noise. *Physica A: Statistical Mechanics and Its Applications*. 469:81–88.
352. X Tang and Y Song. 2015. Stability, Hopf bifurcations and spatial patterns in a delayed diffusive predator-prey model with herd behavior. *Applied Mathematics and Computation*. 254:375–391.
353. S Kinast, Y Zelnik and G Bel, et al. 2014. Interplay between Turing mechanisms can increase pattern diversity. *Physical Review Letters*. 112:078701.
354. D Schüler, S Alonso and A Torcini, et al. 2014. Spatio-temporal dynamics induced by competing instabilities in two asymmetrically coupled nonlinear evolution equations. *Chaos*. 24:043142.

355. J Perraud, A Wit and E Dulos, et al. 1993. One-dimensional "spirals": Novel asynchronous chemical wave sources. *Physical Review Letters*. 71:1272.
356. H Malchow. 1993. Spatio-temporal pattern formation in nonlinear non-equilibrium plankton dynamics. *Philosophical Transactions of the Royal Society of London Series B-Biological Sciences*. 251:103–109.
357. Y Song, H Jiang and Q Liu, et al. 2017. Spatiotemporal dynamics of the diffusive mussel-algae model near Turing-Hopf bifurcation. *Siam Journal on Applied Dynamical Systems*. 16:2030–2062.
358. Y Lv, T Li and Y Pei, et al. 2016. A complete analysis of the global dynamics of a diffusive predator and toxic prey model. *Applied Mathematics and Computation*. 291:182–196.
359. E Kerner. 1957. A statistical mechanics of interacting biological species. *Bulletin of Mathematical Biology*. 19:121–146.
360. R Baier and C Tian. 2013. Mathematical analysis and numerical simulation of pattern formation under cross-diffusion. *Nonlinear Analysis-Real World Applications*. 14:601–612.
361. C Tian, L Zhang and Z Lin. 2011. Pattern formation for a model of plankton allelopathy with cross-diffusion. *Journal of the Franklin Institute*. 348:1947–1964.
362. M Banerjee, S Ghorai, N Mukherjee. 2018. Study of cross-diffusion induced Turing patterns in a ratio-dependent prey-predator model via amplitude equations. *Applied Mathematical Modelling*. 55:383–399.
363. E Zemskov, V Vanag and I Epstein. 2011. Amplitude equations for reaction-diffusion systems with cross diffusion. *Physical Review E*. 84:036216.
364. L Bai and G Zhang. 2009. Nontrivial solutions for a nonlinear discrete elliptic equation with periodic boundary conditions. *Applied Mathematics and Computation*. 210:321–333.
365. T Huang, H Zhang and H Yang. 2017. Spatiotemporal complexity of a discrete space-time predator-prey system with self- and cross-diffusion. *Applied Mathematical Modelling*. 47:637–655.
366. E J Crampin, E A Gaffney and P K Maini. 1999. Reaction and diffusion on growing domains: scenarios for robust pattern formation. *Bulletin of Mathematical Biology*. 61:1093–1120.
367. Wit A D. 1999. Spatial patterns and spatiotemporal dynamics in chemical systems. *Advances in Chemical Physics*. 109:435–514.
368. A S Alnahdi, J Niesen and A M Rucklidge. 2014. Localized patterns in periodically forced systems. *SIAM Journal on Applied Dynamical Systems*. 13:1311–1327.
369. C Wang, M Lv and A Alsaedi, et al. 2017. Synchronization stability and pattern selection in a memristive neuronal network. *Chaos*. 27:113108.
370. J Ma, Y Xu and G Ren, et al. 2016. Prediction for breakup of spiral wave in a regular neuronal network. *Nonlinear Dynamics*. 84:497–509.
371. F H Fenton, E M Cherry and H M Hastings, et al. 2002. Multiple mechanisms of spiral wave breakup in a model of cardiac electrical activity. *Chaos* 12:852–892.
372. D Barman, J Roy and H Alrabaiah, et al. 2021. Impact of predator incited fear and prey refuge in a fractional order prey predator model. *Chaos, Solitons & Fractals*. 142:110420.
373. A M Boustany, R Matteson and M Castleton, et al. 2010. Movements of pacific bluefin tuna (thunnus orientalis) in the eastern north pacific revealed with archival tags. *Progress in Oceanography*. 86(1-2):94–104.
374. T K Callahan and E Knobloch. 1999. Pattern formation in three-dimensional reaction-diffusion systems. *Physica D: Nonlinear Phenomena*. 132(3):339–362.
375. M Das and G P Samanta. 2020. A delayed fractional order food chain model with fear effect and prey refuge. *Mathematics and Computers in Simulation*. 178:218–245.
376. O Flaaten. 1998. On the bioeconomics of predator and prey fishing. *Fisheries research*. 37(1-3):179–191.
377. G C Hays, T Bastian and T K Doyle, et al. 2012. High activity and Lévy searches: jellyfish can search the water column like fish. *Proceedings of the Royal Society B: Biological Sciences*. 279(1728):465–473.
378. J E Herbert-Read, J Buhl and F Hu, et al. 2015. Initiation and spread of escape waves within animal groups. *Royal Society open science*. 2(4):140355.

379. N E Humphries, N Queiroz and J R M Dyer, et al. 2010. Environmental context explains Lévy and Brownian movement patterns of marine predators. *Nature*. 465(7301):1066–1069.
380. B Liu, R Wu and L Chen. 2018. Patterns induced by super cross-diffusion in a predator-prey system with Michaelis-Menten type harvesting. *Mathematical Biosciences*. 298:71–79.
381. Z P Ma, H F Huo and H Xiang. 2020. Spatiotemporal patterns induced by delay and cross-fractional diffusion in a predator-prey model describing intraguild predation. *Mathematical Methods in the Applied Sciences*. 43(8):5179–5196.
382. A Marzano. 2013. Harvesting the sea: the exploitation of marine resources in the Roman Mediterranean. *Oxford University Press*.
383. Y Mitsunaga, C Endo and R P Babaran. 2013. Schooling behavior of juvenile yellowfin tuna Thunnus albacares around a fish aggregating device (FAD) in the Philippines. *Aquatic Living Resources*. 26(1):79–84.
384. D Mukherjee. 2014. The effect of prey refuges on a three species food chain model. *Differential Equations and Dynamical Systems*. 22(4):413–426.
385. D Mukherjee. 2016. The effect of refuge and immigration in a predator-prey system in the presence of a competitor for the prey. *Nonlinear Analysis: Real World Applications*. 31:277–287.
386. D Mukherjee. 2017. Study of refuge use on a predator-prey system with a competitor for the prey. *International Journal of Biomathematics*. 10(02):1750023.
387. H Murakami, T Niizato and T Tomaru, et al. 2015. Inherent noise appears as a Lévy walk in fish schools. *Scientific Reports*. 5(1):1–11.
388. I Podlubny. 1998. Fractional differential equations: an introduction to fractional derivatives, fractional differential equations, to methods of their solution and some of their applications. *Elsevier*.
389. H Qi and X Meng. 2021. Threshold behavior of a stochastic predator-prey system with prey refuge and fear effect. *Applied Mathematics Letters*. 113:106846.
390. I I Rodriguez-Pinto, G Rieucau and N O Handegard, et al. 2020. Environmental context elicits behavioural modification of collective state in schooling fish. *Animal Behaviour*. 165:107–116.
391. G D Ruxton. 1995. Short term refuge use and stability of predator-prey models. *Theoretical Population Biology*. 47(1):1–17.
392. A Sih. 1987. Prey refuges and predator-prey stability. *Theoretical Population Biology*. 31(1):1–12.
393. D W Sims and V A Quayle. 1998. Selective foraging behaviour of basking sharks on zooplankton in a small-scale front. *Nature*. 393(6684):460–464.
394. D W Sims, E J Southall and N E Humphries, et al. 2008. Scaling laws of marine predator search behaviour. *Nature*. 451(7182):1098–1102.
395. D P Tittensor, C Mora and W Jetz, et al. 2010. Global patterns and predictors of marine biodiversity across taxa. *Nature*. 466(7310):1098–1101.
396. Y Wang, S Liu and H Li. 2020. On fractional difference logistic maps: Dynamic analysis and synchronous control. *Nonlinear Dynamics*. 102(1):579–588.
397. F Wei and Q Fu. 2016. Hopf bifurcation and stability for predator-prey systems with Beddington-DeAngelis type functional response and stage structure for prey incorporating refuge. *Applied Mathematical Modelling*. 40(1):126–134.
398. Y Xie, J Lu and Z Wang. 2019. Stability analysis of a fractional-order diffused prey-predator model with prey refuges. *Physica A: Statistical Mechanics and its Applications*. 526:120773.
399. S W Yao, Z P Ma and J L Yue. 2018. Biostability and Turing pattern induced by cross fraction diffusion in a predator-prey model. *Physica A: Statistical Mechanics and its Applications*. 509:982–988.
400. P Zhou, J Ma and J Tang. 2020. Clarify the physical process for fractional dynamical systems. *Nonlinear Dynamics*. 100(3):2353–2364.
401. W Wang, S Liu and D Tian, et al. 2018. Pattern dynamics in a toxin-producing phytoplankton-zooplankton model with additional food. *Nonlinear Dynamics*. 94(1):211–228.

402. J Bhattacharyya and J Chattopadhyay. 2021. Non-smooth dynamics emerging from predator-driven discontinuous prey dispersal. *Nonlinear Dynamics*. 106(4):3647–3668.
403. X Zhang, Q An, and L Wang. 2021. Spatiotemporal dynamics of a delayed diffusive ratio-dependent predator-prey model with fear effect. *Nonlinear Dynamics*. 105(4):3775–3790.
404. M Belabbas, A Ouahab and F Souna. 2021. Rich dynamics in a stochastic predator-prey model with protection zone for the prey and multiplicative noise applied on both species. *Nonlinear Dynamics*. 106(3):2761–2780.
405. Y Zhou, X P Yan and C H Zhang. 2021. Turing patterns induced by self-diffusion in a predator-prey model with schooling behavior in predator and prey. *Nonlinear Dynamics*. 105(4):3731–3747.
406. Z Sabir, M A Z Raja and D Baleanu, et al. 2021. Design of Gudermannian Neuroswarming to solve the singular Emden-Fowler nonlinear model numerically. *Nonlinear Dynamics*. 106(4):3199–3214.
407. J H Choi, H Kim and R Sakthivel. 2016. On certain exact solutions of diffusive predator-prey system of fractional order. *Chinese Journal of Physics*. 54(1):135–146.
408. A E Allahverdyan, S G Babajanyan, C K Hu. 2019. Polymorphism in rapidly changing cyclic environment. *Physical Review E*. 100(3):032401.
409. L Chen and F Chen. 2010. Global analysis of a harvested predator-prey model incorporating a constant prey refuge. *International Journal of Biomathematics*. 3(02):205–223.
410. F Chen, L Chen and X Xie. 2009. On a Leslie-Gower predator-prey model incorporating a prey refuge. *Nonlinear Analysis: Real World Applications*. 10(5):2905–2908.
411. B Tiwari, S N Raw and P Mishra. 2020. Qualitative analysis of a spatiotemporal prey-predator model with multiple Allee effect and schooling behaviour. *Nonlinear Dynamics*. 102(4):3013–3038.
412. J Jiao and C Chen. 2021. Bogdanov-Takens bifurcation analysis of a delayed predator-prey system with double Allee effect. *Nonlinear Dynamics*. 104(2):1697–1707.
413. F B Yousef, A Yousef and C Maji. 2021. Effects of fear in a fractional-order predator-prey system with predator density-dependent prey mortality. *Chaos, Solitons & Fractals*. 145:110711.
414. P K Tiwari, K A N A Amri and S Samanta, et al. 2021. A systematic study of autonomous and nonautonomous predator-prey models with combined effects of fear, migration and switching. *Nonlinear Dynamics*. 103(2):2125–2162.
415. S Mondal and G P Samanta. 2021. Impact of fear on a predator-prey system with prey-dependent search rate in deterministic and stochastic environment. *Nonlinear Dynamics*. 104(3):2931–2959.
416. X Wang, L Zanette, and X Zou. 2016. Modelling the fear effect in predator-prey interactions. *Journal of Mathematical Biology*. 73(5):1179–1204.
417. J Wang, Y Cai and S Fu, et al. 2019. The effect of the fear factor on the dynamics of a predator-prey model incorporating the prey refuge. *Chaos: An Interdisciplinary Journal of Nonlinear Science*. 29(8):083109.
418. Z Ma, W Li and Y Zhao, et al. 2009. Effects of prey refuges on a predator-prey model with a class of functional responses: the role of refuges. *Mathematical Biosciences*. 218(2):73–79.
419. J N McNair. 1986. The effects of refuges on predator-prey interactions: a reconsideration. *Theoretical Population Biology*. 29(1):38–63.
420. J N McNair. 1987. Stability effects of prey refuges with entry-exit dynamics. *Journal of Theoretical Biology*. 125(4):449–464.
421. G Tang, S Tang and R A Cheke. 2014. Global analysis of a Holling type II predator-prey model with a constant prey refuge. *Nonlinear Dynamics*. 76(1):635–647.
422. H Wang, W Morrison and A Singh, et al. 2009. Modeling inverted biomass pyramids and refuges in ecosystems. *Ecological Modelling*. 220(11):1376–1382.
423. S Creel and D Christianson. 2008. Relationships between direct predation and risk effects. *Trends in Ecology & Evolution*. 23(4):194–201.
424. E L Preisser, D I Bolnick and M F Benard. 2005. Scared to death? The effects of intimidation and consumption in predator-prey interactions. *Ecology*. 86(2):501–509.

425. J Martín, P López and W E Cooper Jr. 2003. When to come out from a refuge: balancing predation risk and foraging opportunities in an alpine lizard. *Ethology*. 109(1):77–87.
426. L Persson and P Eklov. 1995. Prey refuges affecting interactions between piscivorous perch and juvenile perch and roach. *Ecology*. 76(1):70–81.
427. A Sih. 1992. Prey uncertainty and the balancing of antipredator and feeding needs. *The American Naturalist*. 139(5):1052–1069.
428. B I Henry and S L Wearne. 2002. Existence of Turing instabilities in a two-species fractional reaction-diffusion system. *SIAM Journal on Applied Mathematics*. 62(3):870–887.
429. A D Wit, P Borckmans and G Dewel. 1997. Twist grain boundaries in three-dimensional lamellar Turing structures. *Proceedings of the National Academy of Sciences*. 94(24):12765–12768.
430. H Shoji, K Yamada and D Ueyama, et al. 2007. Turing patterns in three dimensions. *Physical Review E*. 75(4):046212.
431. T Bánsági, V K Vanag and I R Epstein. 2011. Tomography of reaction-diffusion microemulsions reveals three-dimensional Turing patterns. *Science*. 331(6022):1309–1312.
432. M Alber, T Glimm and H G E Hentschel, et al. 2004. Stability of n-dimensional patterns in a generalized Turing system: implications for biological pattern formation. *Nonlinearity*. 18(1):125.
433. T Glimm and H G E Hentschel. 2008. On isoconcentration surfaces of three-dimensional Turing patterns. *International Journal of Bifurcation and Chaos*. 18(02):391–406.
434. D Wetzel. 2014. Pattern analysis in a benthic bacteria-nutrient system. *arXiv preprint* arXiv:1403.7146.
435. H Uecker. 2019. Pattern formation with pde2path-a tutorial. *arXiv preprint* arXiv:1908.05211.
436. P Borckmans, A D Wit and G Dewel. 1992. Competition in ramped Turing structures. *Physica A: Statistical Mechanics and Its Applications*. 188(1–3):137–157.
437. T Leppänen, M Karttunen and K Kaski, et al. 2002. A new dimension to Turing patterns. *Physica D: Nonlinear Phenomena*. 168:35–44.
438. H Shoji, K Yamada and T Ohta. 2005. Interconnected Turing patterns in three dimensions. *Physical Review E*. 72(6):065202.
439. H Shoji and T Ohta. 2015. Computer simulations of three-dimensional Turing patterns in the Lengyel-Epstein model. *Physical Review E*. 91(3):032913.
440. S Puri. 2004. Kinetics of phase transitions. *Phase Transitions*. 77(5–7):407–431.
441. T Huang and H Zhang. 2016. Bifurcation, chaos and pattern formation in a space-and time-discrete predator-prey system. *Chaos, Solitons & Fractals*. 91:92–107.
442. R Metzler and J Klafter. 2004. The restaurant at the end of the random walk: recent developments in the description of anomalous transport by fractional dynamics. *Journal of Physics A: Mathematical and General*. 37(31):R161.
443. M Buchanan. 2008. Ecological modelling: the mathematical mirror to animal nature. *Nature News*. 453(7196):714–716.
444. T K Callahan and E Knobloch. 1997. Symmetry-breaking bifurcations on cubic lattices. *Nonlinearity*. 10(5):1179.
445. T K Callahan and E Knobloch. 1999. Pattern formation in three-dimensional reaction-diffusion systems. *Physica D: Nonlinear Phenomena*. 132(3):339–362.
446. B Liu, R Wu and L Chen. 2018. Turing-Hopf bifurcation analysis in a superdiffusive predator-prey model. *Chaos: An Interdisciplinary Journal of Nonlinear Science*. 28(11):113118.
447. Y Huang, Z Zhu and L Zhong. 2020. Modeling the Allee effect and fear effect in predator-prey system incorporating a prey refuge. *Advances in Difference Equations*. 2020(1):1–13.
448. H Uecker. 2021. Numerical continuation and bifurcation in Nonlinear PDEs. *SIAM*.
449. N Ehstand, C Kuehn and C Soresina. 2021. Numerical continuation for fractional PDEs: sharp teeth and bloated snakes. *Communications in Nonlinear Science and Numerical Simulation*. 98:105762.
450. M Golubitsky and I Stewart. 2002. The symmetry perspective. *Avances En Psicologia Latinoamericana*. 27(2):252–262.

451. Z Bi, S Liu, and M Ouyang. 2022. Three-dimensional pattern dynamics of a fractional predator-prey model with cross-diffusion and herd behavior. *Applied Mathematics and Computation*. 421:126955.
452. R Hoyle and R B Hoyle. 2006. Pattern formation: an introduction to methods. *Cambridge University Press*.
453. A Faramarzi, M Heidarinejad and S Mirjalili, et al. 2020. Marine Predators Algorithm: A nature-inspired metaheuristic. *Expert Systems with Applications*. 152:113377.
454. G M Viswanathan, E P Raposo, and M G E Da Luz. et al. Lévy flights and superdiffusion in the context of biological encounters and random searches. *Physics of Life Reviews*. 5(3):133–150.
455. W Campeau, A M Simons and B Stevens. 2022. The evolutionary maintenance of Lévy flight foraging. *PLOS Computational Biology*. 18(1):e1009490.
456. F Bartumeus, F Peters and S Pueyo, et al. 2003. Helical Lévy walks: adjusting searching statistics to resource availability in microzooplankton. *Proceedings of the National Academy of Sciences*. 100(22):12771–12775.
457. T Dannemann, D Boyer and O Miramontes. 2018. Lévy flight movements prevent extinctions and maximize population abundances in fragile Lotka-Volterra systems. *Proceedings of the National Academy of Sciences*. 115(15):3794–3799.
458. D W Sims, M J Witt and A J Richardson, et al. 2006. Encounter success of free-ranging marine predator movements across a dynamic prey landscape. *Proceedings of the Royal Society B: Biological Sciences*. 273(1591):1195–1201.
459. A Bueno-Orovio, D Kay and K Burrage. 2014. Fourier spectral methods for fractional-in-space reaction-diffusion equations. *BIT Numerical Mathematics*. 54(4):937–954.
460. M Chen, R Wu and B Liu, et al. 2020. Hopf-hopf bifurcation in the delayed nutrient-microorganism model. *Applied Mathematical Modelling*. 86:460–483.
461. B Dai and G Sun. 2021. Turing-hopf bifurcation of a delayed diffusive predator-prey system with chemotaxis and fear effect. *Applied Mathematics Letters*. 111:106644.
462. B K Das, D Sahoo and G P Samanta. 2022. Impact of fear in a delay-induced predator-prey system with intraspecific competition within predator species. *Mathematics and Computers in Simulation*. 191:134–156.
463. M H Cruz, I Kriest and Y S José, et al. 2021. Zooplankton mortality effects on the plankton community of the northern humboldt current system: sensitivity of a regional biogeochemical model. *Biogeosciences*. 18(9):2891–2916.
464. Q Hu and J Shen. 2022. Delay-induced self-organization dynamics in a prey-predator network with diffusion. *Nonlinear Dynamics*. 1–12.
465. W Li, T Zhang and Y Wang, et al. 2022. Dynamic analysis of a plankton-herbivore state-dependent impulsive model with action threshold depending on the density and its changing rate. *Nonlinear Dynamics*. 107(3):2951–2963.
466. C X Liu, Y Yan and W Q Wang. 2019. Primary and secondary resonance analyses of a cantilever beam carrying an intermediate lumped mass with time-delay feedback. *Nonlinear Dynamics*. 97(2):1175–1195.
467. M Liu and M Deng. 2020. Analysis of a stochastic hybrid population model with Allee effect. *Applied Mathematics and Computation*. 364:124582.
468. T Ma, X Meng and T Hayat, et al. 2022. Hopf bifurcation induced by time delay and influence of Allee effect in a diffusive predator-prey system with herd behavior and prey chemotaxis. *Nonlinear Dynamics*. 1–18.
469. K Manna and M Banerjee. 2018. Stationary, non-stationary and invasive patterns for a prey-predator system with additive Allee effect in prey growth. *Ecological Complexity*. 36:206–217.
470. Y Nec, A A Nepomnyashchy and A A Golovin. 2008. Oscillatory instability in super-diffusive reaction-diffusion systems: Fractional amplitude and phase diffusion equations. *Europhysics Letters*. 82(5):58003.
471. R R Patra, S Kundu and S Maitra. 2022. Effect of delay and control on a predator-prey ecosystem with generalist predator and group defence in the prey species. *The European Physical Journal Plus*. 137(1):28.

472. S Rana, A R Bhowmick and T Sardar. 2021. Invasive dynamics for a predator-prey system with Allee effect in both populations and a special emphasis on predator mortality. *Chaos: An Interdisciplinary Journal of Nonlinear Science*. 31(3):033150.
473. C Ross, L Pacheco-Cobos and B Winterhalder. 2018. A general model of forager search: Adaptive encounter-conditional heuristics outperform Lévy flights in the search for patchily distributed prey. *Journal of Theoretical Biology*. 455:357–369.
474. M Sambath, K Balachandran and M Suvinthra. 2016. Stability and hopf bifurcation of a diffusive predator-prey model with hyperbolic mortality. *Complexity*. 21(S1):34–43.
475. T Singh and S Banerjee. 2019. Spatiotemporal model of a predator-prey system with herd behavior and quadratic mortality. *International Journal of Bifurcation and Chaos*. 29(04):1950049.
476. D Song, Y Song and C Li. 2020. Stability and Turing patterns in a predator-prey model with hunting cooperation and Allee effect in prey population. *International Journal of Bifurcation and Chaos*. 30(09):2050137.
477. Y Song, Y Peng and T Zhang. 2021. The spatially inhomogeneous hopf bifurcation induced by memory delay in a memory-based diffusion system. *Journal of Differential Equations*. 300:597–624.
478. Y Su, J Wei and J Shi. 2010. Bifurcation analysis in a delayed diffusive Nicholson's blowflies equation. *Nonlinear Analysis: Real World Applications*. 11(3):1692–1703.
479. X Tao and L Zhu. 2021. Study of periodic diffusion and time delay induced spatiotemporal patterns in a predator-prey system. *Chaos, Solitons & Fractals*. 150:111101.
480. G M Viswanathan, V Afanasyev and S V Buldyrev, et al. 1996. Lévy flight search patterns of wandering albatrosses. *Nature*. 381(6581):413–415.
481. G M Viswanathan, E P Raposo and M G E D Luz. 2008. Lévy flights and superdiffusion in the context of biological encounters and random searches. *Physics of Life Reviews*. 5(3):133–150.
482. J Wang, H Zheng and Y Jia. 2021. Dynamical analysis on a bacteria-phages model with delay and diffusion. *Chaos, Solitons & Fractals*. 143:110597.
483. Y Wang, S Liu and H Li. 2019. Fractional diffusion-limited aggregation: Anisotropy originating from memory. *Fractals*. 27(08):1950137.
484. Y Wang, S Liu and H Li, et al. 2021. New fractal sets coined from fractional maps. *Fractals*. 29(08):1–12.
485. Y Wang, S Liu and H Li, et al. 2019. On the spatial Julia set generated by fractional Lotka-Volterra system with noise. *Chaos, Solitons & Fractals*. 128:129–138.
486. Z Wei, Y Xia and T Zhang. 2021. Stability and bifurcation analysis of a commensal model with additive Allee effect and nonlinear growth rate. *International Journal of Bifurcation and Chaos*. 31(13):2150204.
487. J Wu. 1996. Theory and applications of partial functional differential equations. 119.
488. J Yang, S Yuan and T Zhang. 2021. Complex dynamics of a predator-prey system with herd and schooling behavior: with or without delay and diffusion. *Nonlinear Dynamics*. 104(2):1709–1735.
489. J Yang, T Zhang and S Yuan. 2020. Turing pattern induced by cross-diffusion in a predator-prey model with pack predation-herd behavior. *International Journal of Bifurcation and Chaos*. 30(07):2050103.
490. R Yang and J Wei. 2015. Bifurcation analysis of a diffusive predator-prey system with non-constant death rate and Holling III functional response. *Chaos, Solitons & Fractals*. 70:1–13.
491. T Yi and X Zou. 2013. On Dirichlet problem for a class of delayed reaction-diffusion equations with spatial non-locality. *Journal of Dynamics and Differential Equations*. 25(4):959–979.
492. J Zhao and J Wei. 2015. Dynamics in a diffusive plankton system with delay and toxic substances effect. *Nonlinear Analysis: Real World Applications*. 22:66–83.
493. W Zuo and J Wei. 2011. Stability and hopf bifurcation in a diffusive predator-prey system with delay effect. *Nonlinear Analysis: Real World Applications*, 12(4):1998–2011.

The manufacturer's authorised representative in the EU is Springer Nature Customer Service Centre GmbH, Europaplatz 3, 69115 Heidelberg, Germany. If you have any concerns regarding our products, please contact ProductSafety@springernature.com

Printed and bound by CPI Group (UK) Ltd, Croydon, CR0 4YY

01/10/2026

02225941-0012